MAKING USE OF MATHEMATICS for GCSE

Other titles in this series

Making Use of Biology for GCSE, P. Alderson and M. Rowland
Making Use of Physics for GCSE, R. Kibble

MAKING USE OF MATHEMATICS
for GCSE

G. D. Buckwell and A. D. Ball

MACMILLAN

First published 1990

Published by
MACMILLAN EDUCATION LTD
Houndmills, Basingstoke, Hampshire RG21 2XS
and London
Companies and representatives
throughout the world

Typeset by P&R Typesetters Ltd, Salisbury, Wiltshire

Printed in Hong Kong

British Library Cataloguing in Publication Data
Buckwell, G. D. (Geoffrey D.)
Making use of mathematics for GCSE
1. Mathematics
I. Title II. Ball, A. D. (Alan D.)
510
ISBN 0–333–47436–8

CONTENTS

PREFACE

Many students find at the end of five years of secondary education that they still have not reached grade C standard in GCSE Mathematics. It is for these people that this book has been written. It is not aimed at any particular syllabus, but covers all of the topics needed at intermediate level for each examination group both at 16+ and 17+ levels, and also enough extra material to cover the 17+ modular courses such as those offered by LEAG and SEG. Harder questions have also been included in the later chapters for students who are hoping to tackle the higher level of the GCSE examination.

The book does *not* have to be worked through from cover to cover, but it can certainly be used as a course text in this way if required. The approach adopted will help to rekindle the enthusiasm of the reader by putting the mathematics into context, and demonstrating its uses.

As coursework assignments are an essential part of the GCSE examination, a selection of ideas has been included at the end of each unit, though it is expected that many students may be continuing or improving investigational work which they have already presented in a previous examination.

The book contains many worked examples and exercises. A large number of actual examination questions are provided at the end of each unit, so that students may become experienced in approaching and tackling such questions.

The authors would like to record their thanks to Ann Ball and Josie Buckwell for their work in the preparation of the manuscript and also to Colin Prior for his super artwork.

Alan Ball
Geoff Buckwell

ACKNOWLEDGEMENTS

The authors and publishers wish to thank the following who have kindly given permission for the use of copyright material: London East Anglian Group; Midland Examining Group; Northern Examining Association comprised of Associated Lancashire Schools Examining Board, Joint Matriculation Board, North Regional Examinations Board, North West Regional Examining Board and Yorkshire and Humberside Regional Examinations Board; Oxford and Cambridge Schools Examination Board; and Southern Examining Group for questions from specimen and past examination papers.

Every effort has been made to trace all the copyright holders, but if any have been inadvertently overlooked the publishers will be pleased to make the necessary arrangement at the first opportunity.

Please note that the answers given to the examination questions are the responsibility of the authors. They have not been provided or approved by the examination boards.

BASIC SKILLS WITH NUMBERS

1

The purpose of this unit is to cover the basic skills of working with numbers and using a calculator. It is important that you have a good working knowledge of the topics in this unit before progressing further.

1.1 The Four Rules and the calculator

(i) What does $4 + 3 \times 5$ equal? The answer is *not* 35. You have to find $3 \times 5 = 15$ first, and then add 4.

Hence: $4 + 3 \times 5 = 19$

Most calculators would carry out the order of operations correctly as follows (check with yours):

4 [+] 3 [×] 5 [=] 19

If, however, it was required to find $(4 + 3) \times 5$, then the answer is $7 \times 5 = 35$. In order to find this on the calculator, it would be necessary to press the [=] button after $4 + 3$.

Hence: 4 [+] 3 [=] [×] 5 [=] Display [35]

If your calculator has brackets on it, then proceed as follows:

[(] 4 [+] 3 [)] [×] 5 [=] Display [35]

(ii) $8 + 16 \div 4 - 2$
This time, the division must be done first.

Hence: $8 + 16 \div 4 - 2 = 8 + 4 - 2 = 12 - 2 = 10$

Using the calculator

8 [+] 16 [÷] 4 [−] 2 [=] Display [10]

If it was required to find $(8+16)\div(4-2)$, then the brackets must be worked out first. The answer to the second bracket should be found first, and stored in the memory. Hence:

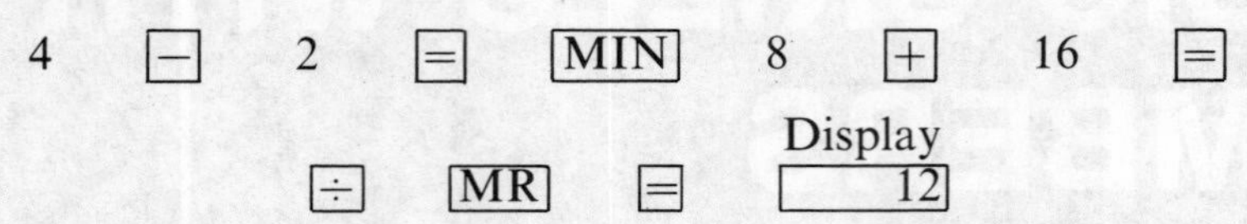

If your calculator has brackets, it is much easier to feed in directly, although it still requires 12 key pushes.

(8 + 16) ÷ (4 −

Display

2) = 12

The order of priority in which operations are carried out is summarised by the BODMAS rule. This gives the order as

Brackets, Of, Division, Multiplication, Addition, Subtraction

The following example should show how this rule is applied.

Worked example 1.1

Find $18-2$ of $3+16\div4+12\times3+5\times(23-8)$. Try and check your answer by using a calculator.

This is simplified as follows:

Brackets: $18-2$ of $3+16\div4+12\times3+5\times15$

Of: $18-6+16\div4+12\times3+5\times15$

Division: $18-6+4+12\times3+5\times15$

Multiplication: $18-6+4+36+75$

Addition: $133-6$

Subtraction: $=127$

On a calculator, the word 'of' is replaced by the × key. Proceeding on a calculator, the above example becomes

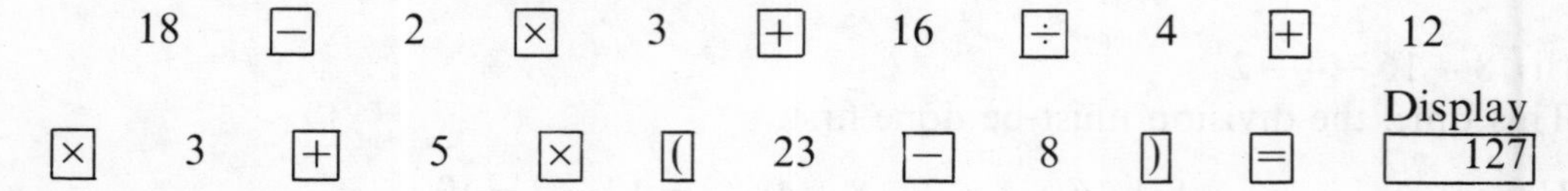

Exercise 1a

(1) The following exercise is practice in using the BODMAS RULE. Without using a calculator, find what you think is the correct value of the following expressions.

(i) $8 \times 3 + 6$
(ii) $8 + 3 \times 6$
(iii) $4 \text{ of } 3 - 2$
(iv) $7 \text{ of } 8 + 4 \times 3$
(v) $6 \times (3 \text{ of } 8 \div 2)$
(vi) $4 \times (8 \div (2 \div 4))$
(vii) $4 - 3 \div 7 + 2$
(viii) $6 \times (8 \times (3 + 2))$
(ix) $6 + 8 \times (3 \times (16 \div 2))$
(x) $5 + (8 \times (16 \div 4 \times 3 - 5) - 6 \times 8 \div 5 \text{ of } 3 \times 6)$

(2) The following examples are correct, but the brackets have been omitted. Insert the necessary brackets to make the sum correct.

(i) $7 + 4 \times 3 = 33$
(ii) $6 + 3 \times 5 = 21$
(iii) $3 + 5 \times 5 - 3 = 16$
(iv) $6 + 8 \div 2 = 10$
(v) $12 - 2 \div 5 = 2$
(vi) $4 \times 12 \times 3 + 2 = 240$
(vii) $12 - 3 \times 2 + 4 = 10$
(viii) $16 + 7 \times 3 + 2 = 51$
(ix) $6 \div 8 \times 12 = 9$
(x) $7 + 8 \div 5 \times 5 \times 10 = 6$

(3) Use your calculator to evaluate the following:

(i) $(6 + 3) \times (8 - 5)$
(ii) $6 \times 5 \times (4 + 7)$
(iii) $5 \text{ of } 3 - 2 \div 4 \times 8$
(iv) $6 \times 3 \times (4 \div 3)$
(v) $12 \div (5 \div 6) \times 10$
(vi) $7 \text{ of } 8 + 5 \text{ of } 4 - 3 \text{ of } 8$
(vii) $28 \div (8 + 3 \times 4) \times 20$
(viii) $((8 \div 3) \div (6 \times 2 \div 4)) \times 9$
(ix) $(12 \times (4 - 3)) \times (8 \div 6 \times 12)$
(x) $(16 \div 3 \times (5 + 4)) \times (8 \div 3 \div 4)$

1.2 Types of whole numbers

There are various categories of whole numbers we use, that are given special names. You will have met some of them before, but a list is given here partly as a reminder.

Natural numbers: 1, 2, 3, 4, 5, 6, ...

Integers: ... −5, −4, −3, −2, −1, 0, 1, 2, 3, 4, 5, ...

negative integers (... −5, −4, −3, −2, −1) positive integers (1, 2, 3, 4, 5, ...)

Prime numbers: 2, 3, 5, 7, 11, 13, ...

A prime number can be divided exactly only by 1 and itself. Remember that 1 is *not a* prime number.

Square numbers: 1, 4, 9, 16, 25, ...

These take their name from the associated dot patterns:

triangular numbers: 1, 3, 6, 10, 15, ...

Once again, the dot pattern gives us the description:

1.3 Factors

The numbers that divide exactly into 40 are given in the set {1, 2, 4, 5, 8, 10, 20, 40}. We call this the **set of factors** of 40. The set of factors of 64 would be {1, 2, 4, 8, 16, 32, 64}.

The largest number that appears in both sets is the number 8. We say that 8 is the highest common factor (HCF) of 64 and 40. Look at the multiples of 40:

40, 80, 120, 160, 200, 240, 280, 320, 360, ...

Look also at the multiples of 64:

64, 128, 192, 256, 320, 384, ...

The smallest number in both sets is 320. We say that 320 is the lowest common multiple (LCM) of 40 and 64.

Exercise 1b

Find the HCF and the LCM of the following pairs of numbers.

(1) 20, 30	(2) 16, 30	(3) 25, 40
(4) 24, 36	(5) 50, 60	(6) 12, 28
(7) 60, 80	(8) 32, 50	(9) 100, 120
(10) 56, 14	(11) 108, 45	(12) 60, 24
(13) 12, 9	(14) 24, 18	(15) 38, 50

1.4 Prime factors

If a prime number divides exactly into a given number, then that prime number is called a **prime factor** of the given number. For example, since 5 divides exactly into 105, then 5 is a prime factor of 105.

Prime factors are used to split up numbers into **products** of prime factors as shown in the following example:

Worked example 1.2

Express 105 and 60 as a product of their prime factors and hence find the HCF and LCM of 105 and 60.

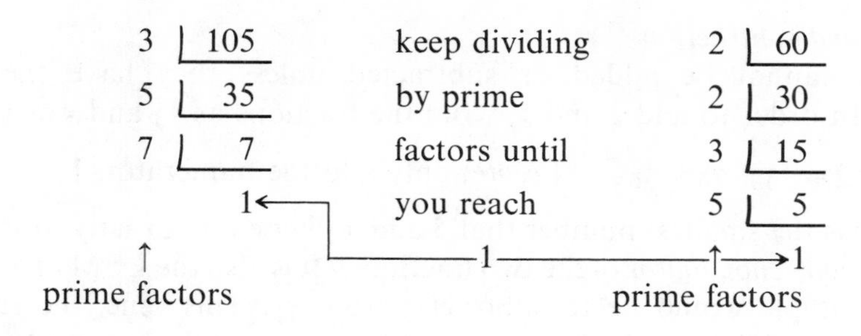

Hence $105 = 3 \times 5 \times 7$ $\qquad 60 = 2 \times 2 \times 3 \times 5$
The HCF $= 3 \times 5 = 15$
The LCM $= 2 \times 2 \times 3 \times 5 \times 7 = 420$

Exercise 1c
Express all numbers in the following questions as a product of their prime factors. Use your answers to find the HCF and LCM of the numbers.

(1) 20, 30	(2) 48, 30	(3) 50, 64	(4) 66, 44
(5) 125, 40	(6) 63, 36	(7) 75, 40	(8) 24, 30
(9) 120, 80	(10) 28, 40	(11) 36, 48	(12) 84, 60
(13) 150, 70	(14) 136, 100	(15) 65, 120	

1.5 Working with fractions

(i) Equivalent fractions
Figure 1.1 shows diagrammatic representations of the fractions $\frac{2}{3}$ and $\frac{4}{6}$. Remember that $\frac{2}{3}$ means 2 parts out of a total of 3 equal parts, $\frac{4}{6}$ means 4 parts out of a total of 6 equal parts. The diagram shows clearly that $\frac{2}{3} = \frac{4}{6}$. They are said to be *equivalent* fractions. The set of fractions $\{\frac{2}{3}, \frac{4}{6}, \frac{6}{9}, \frac{8}{12}, \frac{10}{15}, \dots\}$ are all equivalent to $\frac{2}{3}$. We say that $\frac{10}{15}$ in its *lowest terms* is $\frac{2}{3}$. It can be seen that in this set, the top line (the *numerator*) goes up in *multiples* of 2, the bottom line (the *denominator*) goes up in *multiples* of 3.

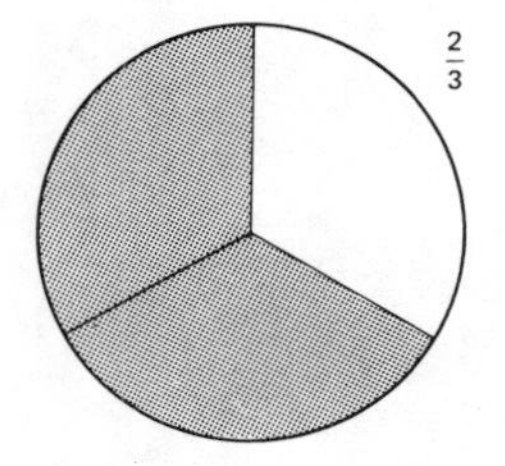

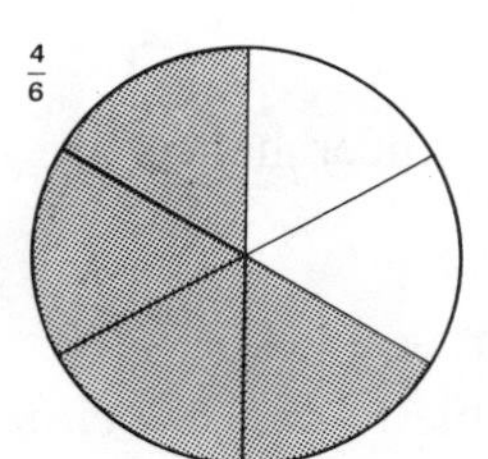

Figure 1.1

Exercise 1d
Write down the first 5 fractions in the set which are equivalent to the following fractions:
(1) $\frac{2}{5}$ $\quad$ (2) $\frac{5}{6}$ $\quad$ (3) $\frac{2}{7}$ $\quad$ (4) $\frac{5}{9}$ $\quad$ (5) $\frac{3}{8}$

(ii) Addition and subtraction

Two fractions cannot be added or subtracted unless they have the same denominator. In order to add $\frac{3}{5}$ and $\frac{1}{4}$, write the fractions as $\frac{12}{20}$ and $\frac{5}{20}$.

Hence: $\frac{3}{5}+\frac{1}{4}=\frac{12}{20}+\frac{5}{20}=\frac{17}{20}$ [*Note*: only add the numerators]

In this case, 20 is the smallest number that 5 and 4 divide into exactly. It is called the *lowest common denominator* of the two fractions. (It is also the LCM of 5 and 4.)

A second example would be to subtract $\frac{1}{3}$ from $\frac{5}{12}$. This time, the smallest number that 3 and 12 divide into exactly is 12, hence you do not need to alter $\frac{5}{12}$, but $\frac{1}{3}$ becomes $\frac{4}{12}$.

Hence: $\frac{5}{12}-\frac{1}{3}=\frac{5}{12}-\frac{4}{12}=\frac{1}{12}$

In each of the above examples, the answer was in its lowest terms. If this is not the case, make sure you simplify your answer.

(iii) Cancelling

If we are given a fraction, and we want to express it in its lowest terms, we can proceed as follows:

$$\frac{112}{252}=\frac{\overset{56}{\cancel{112}}}{\underset{126}{\cancel{252}}}=\frac{\overset{28}{\cancel{56}}}{\underset{63}{\cancel{126}}}=\frac{\overset{4}{\cancel{28}}}{\underset{9}{\cancel{63}}}=\frac{4}{9}$$

(÷ numerator and denominator by 2) (÷ top and bottom by 2) (÷ top and bottom by 7)

This method is often called cancelling.

The order in which each step is carried out does not matter. If you are good at arithmetic, you could have divided top and bottom by 28 to get the answer $\frac{4}{9}$ straight away.

(iv) Multiplication

In order to multiply two fractions, you just multiply the two numerators together, and the two denominators,

Hence: $\frac{3}{8}\times\frac{4}{9}=\frac{12}{72}$

You can often cancel, hence:

$$\frac{\overset{2}{\cancel{12}}}{\underset{12}{\cancel{72}}}=\frac{\overset{1}{\cancel{2}}}{\underset{6}{\cancel{12}}}=\frac{1}{6}$$

It is also possible to cancel before muliplying, hence:

$$\frac{\overset{1}{\cancel{3}}}{\underset{2}{\cancel{8}}}\times\frac{\overset{1}{\cancel{4}}}{\underset{3}{\cancel{9}}}=\frac{1}{6}$$

You should use whichever method you find easier.

(v) Division

In order to divide $\frac{4}{9}$ by $\frac{2}{3}$, the second fraction is 'turned upside down' and the fractions are multiplied. Hence:

$$\frac{4}{9} \div \frac{2}{3} = \frac{4}{9} \times \frac{3}{2} = \frac{\overset{2}{\cancel{12}}}{\underset{3}{\cancel{18}}} = \frac{2}{3}$$

↑

You must remember to invert the second fraction

(vi) Mixed numbers

When dealing with a number such as $4\frac{1}{4}$ which is called a mixed number, it is advisable if you find working with fractions difficult to change them into **top heavy** fractions as follows:

$$4\tfrac{1}{4} = 4 + \tfrac{1}{4} = \tfrac{4}{1} + \tfrac{1}{4} = \tfrac{16}{4} + \tfrac{1}{4} = \tfrac{17}{4}$$

Another example:

$$3\tfrac{2}{5} = 3 + \tfrac{2}{5} = \tfrac{3}{1} + \tfrac{2}{5} = \tfrac{15}{5} + \tfrac{2}{5} = \tfrac{17}{5}$$

It is then possible to carry out calculations. Look at the following example:

Worked example 1.3

Evaluate $(2\frac{1}{2} + 1\frac{1}{3}) \div (1\frac{1}{4} \times \frac{2}{5})$.

Since $2\frac{1}{2} = \frac{5}{2}$, $1\frac{1}{3} = \frac{4}{3}$ and $1\frac{1}{4} = \frac{5}{4}$, the sum can be written:

$$\begin{aligned}(\tfrac{5}{2} + \tfrac{4}{3}) \div (\tfrac{5}{4} \times \tfrac{2}{5}) &= (\tfrac{15}{6} + \tfrac{8}{6}) \div (\tfrac{10}{20}) \\ &= \tfrac{23}{6} \div \tfrac{1}{2} \\ &= \tfrac{23}{6} \times \tfrac{2}{1} = \tfrac{46}{6} = \tfrac{23}{3}\end{aligned}$$

The top heavy fraction should be changed into a mixed number to give $7\frac{2}{3}$.

Exercise 1e

(1) In each of the following diagrams, state in its simplest form the fraction of the total area that has been shaded.

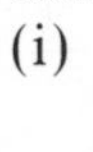
(i)

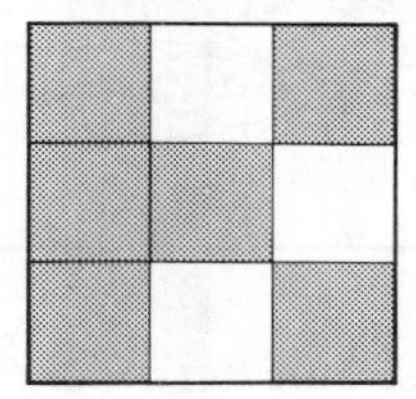

(ii)

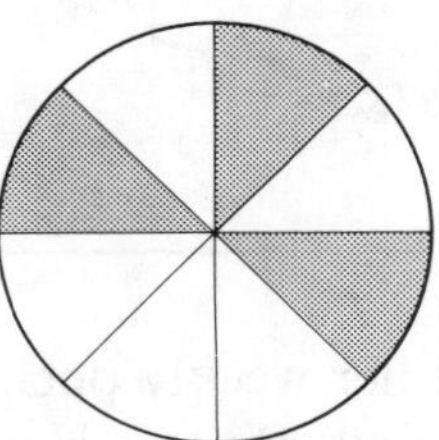

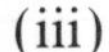
(iii)

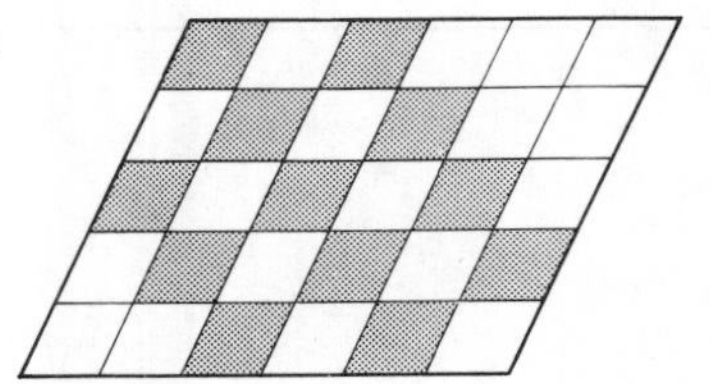

(iv)

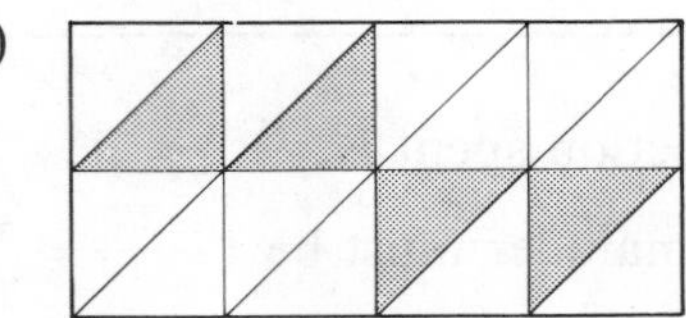

(v)

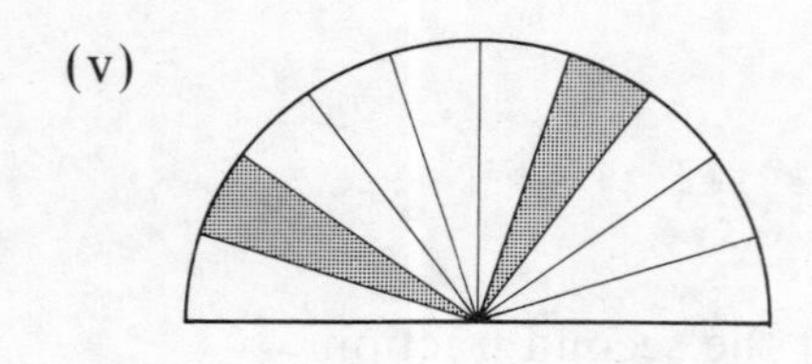

(vi)

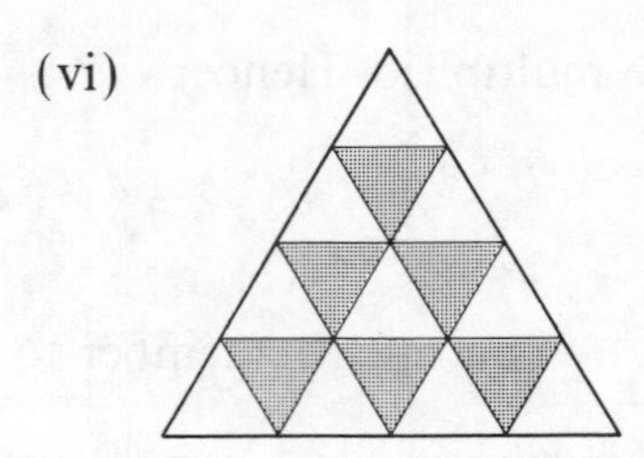

(2) Reduce the following fractions to their lowest terms:

(i) $\frac{6}{9}$ (ii) $\frac{12}{20}$ (iii) $\frac{20}{25}$ (iv) $\frac{12}{54}$
(v) $\frac{21}{56}$ (vi) $\frac{3}{12}$ (vii) $\frac{10}{70}$ (viii) $\frac{9}{45}$
(ix) $\frac{16}{24}$ (x) $\frac{48}{108}$ (xi) $\frac{24}{28}$ (xii) $\frac{21}{24}$
(xiii) $\frac{40}{48}$ (xiv) $\frac{17}{51}$ (xv) $\frac{40}{140}$ (xvi) $\frac{16}{80}$
(xvii) $\frac{24}{216}$ (xviii) $\frac{9}{72}$ (xix) $\frac{30}{66}$ (xx) $\frac{54}{180}$

(3) Work out the following, checking if possible your answers on a calculator. Simplify your answers where possible.

(i) $\frac{1}{2}+\frac{1}{3}$ (ii) $\frac{2}{5}+\frac{1}{3}$ (iii) $\frac{1}{4}+\frac{1}{2}+\frac{1}{3}$
(iv) $\frac{2}{5}-\frac{1}{3}$ (v) $1\frac{2}{3}-\frac{1}{2}$ (vi) $2\frac{1}{4}\times 1\frac{1}{2}$
(vii) $\frac{2}{3}\div\frac{1}{2}$ (viii) $2\frac{1}{2}\div 1\frac{1}{4}$ (ix) $\frac{2}{3}$ of 18
(x) $\frac{1}{4}\times(\frac{1}{2}+\frac{1}{3})$ (xi) $2\frac{3}{5}-1\frac{1}{2}$ (xii) $\frac{2}{7}\div 1\frac{1}{4}$
(xiii) $\frac{2}{3}\div(\frac{1}{2}+\frac{1}{4})$ (xiv) $\frac{1}{2}-\frac{1}{4}\times\frac{1}{3}$ (xv) $\frac{2}{3}\div(1-\frac{1}{5})$
(xvi) $\frac{1}{9}\times\frac{1}{2}\times 2\frac{1}{4}$ (xvii) $\frac{1}{4}\times(\frac{2}{3}\div 1\frac{1}{2})$ (xviii) $\frac{2}{5}+1\frac{3}{7}$
(xix) $12\frac{1}{2}-7\frac{3}{5}$ (xx) $2\frac{3}{5}+1\frac{1}{2}+6\frac{1}{4}$

1.6 Using fractions

At this stage, only certain types of problem can be considered. There are others, where units of measurement or percentages are needed, which appear throughout the book. The following examples indicate what you should be able to do at this stage.

Worked example 1.4

Joan spent $\frac{1}{2}$ of her weekly pocket money on magazines, $\frac{1}{5}$ of her pocket money on sweets, and had 60p left. How much pocket money does she get?

The fraction spent is $\frac{1}{2}+\frac{1}{5}=\frac{5}{10}+\frac{2}{10}=\frac{7}{10}$

The remainder must be $1-\frac{7}{10}=\frac{3}{10}$

Hence, $\frac{3}{10}$ of her money is 60p

Hence, $\frac{1}{10}$ of her money is $60\text{p} \div 3 = 20\text{p}$

The pocket money must be $10 \times 20\text{p} = 200\text{p}$

$= £2$

Worked example 1.5

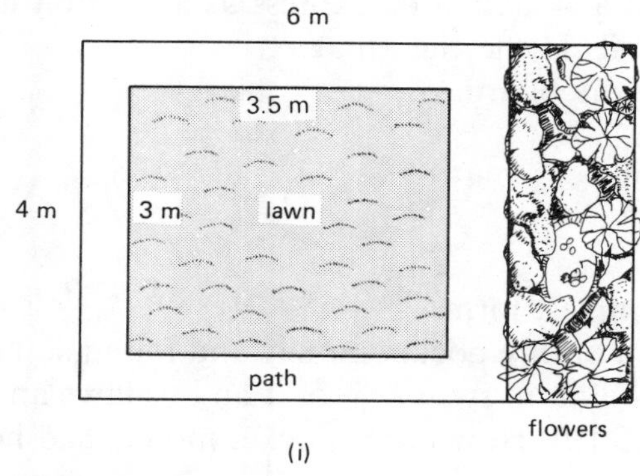

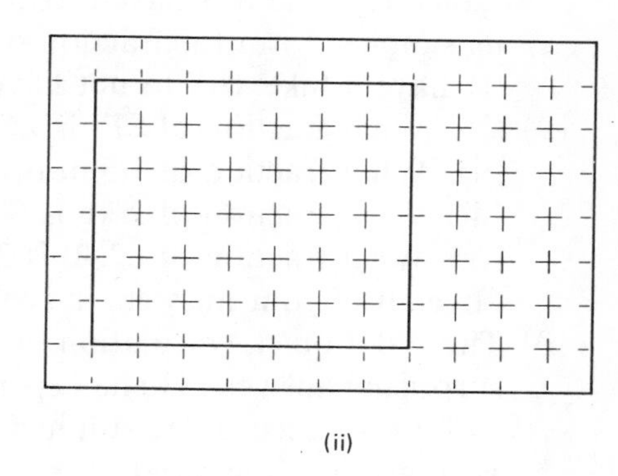

Figure 1.2

David's garden is a rectangle measuring 6 m by 4 m. He has a small lawn which measures 3 m by 3.5 m. What fraction of the garden is not covered by lawn. (Simplify your answer.)

If you are not good at finding areas, this problem can be solved by drawing the garden on a $\frac{1}{2}$ m grid as shown in Figure 1.2(ii).

The number of squares covered by the garden = 96

The number of squares covered by the lawn = 42

Hence the number of squares not covered by the lawn = 96 − 42 = 54

The fraction not covered by the lawn $= \frac{54}{96} = \frac{9}{16}$

Exercise 1f

(1) David has £8.50 in his pocket. He gives $\frac{2}{5}$ of it to his friend John and spends $\frac{2}{3}$ of the remainder on a magazine. How much does he have left?

(2) A jug holds 6.5 litres when full. If $\frac{4}{13}$ of the jug contains water, what volume can still be added to it before it overflows?

(3) When mixing up some mortar, Jim uses 60 kg of sand and 720 kg of cement. What fraction of the mortar is cement? (Give your answer in its lowest terms.)

(4) The petrol tank of a car holds 36 litres. If it is $\frac{2}{9}$ full, how many more litres can be put into the tank.

(5) The distance between London and Brighton is approximately 52 miles. If I am travelling from London to Brighton and stop when I am 13 miles from Brighton, what fraction of the journey have I completed?

(6) John has a piece of card in the shape of a rectangle measuring 30 cm by 16 cm. He cuts a piece of card in the shape of a square of sides 12 cm to make a small model. What fraction of the card remains?

(7) A bag contains 50 coloured balls. 18 are red, 20 are blue, and the rest are green. Find in its simplest form, the fraction of balls that are (i) red, (ii) blue, (iii) not green.

(8) In a class of 32 children, $\frac{1}{4}$ do not have a bike or a skateboard, 18 do have a bike and the rest have a bike and a skateboard. Express as a fraction in its simplest form the fraction of pupils in the class that:
(i) have a bike, (ii) do not have a skateboard.

(9) (i) What fraction of £18 is £6?
(ii) What fraction of 1.5 m is 60 cm?
(iii) What fraction of 2 kg is 800 g?
(iv) What fraction of £2.40 is 36p.
Give all of your answers in their simplest form.

(10) Tim did a quick calculation on his monthly pocket money and found some surprising answers. He had spent $\frac{2}{15}$ of it on sweets, $\frac{5}{17}$ of it on bus fares and $\frac{1}{4}$ of it on a record. He still had £3.29 left. How much pocket money had he been given in the month?

1.7 Percentages

Percentages are used in many areas in our everyday lives. The percentage symbol % means 'out of 100'. Its use need not be as frightening as many people find it, if you approach its use in a systematic way. You should work through all the following examples if you are unsure about this topic.

(a) Changing a fraction to a percentage

(i) $\frac{4}{5}$ means 4 out of 5.

But $\frac{4}{5} = \frac{4 \times 20}{5 \times 20} = \frac{80}{100}$ which means 80 out of 100.

Hence, $\frac{4}{5}$ is the same as 80%.

This method works well when it is easy to turn the bottom line into 100. Alternatively, to change a fraction into a percentage, it should be multiplied by 100/1.

(ii) $\frac{2}{7}$ as a fraction.

$\frac{2}{7} \times \frac{100}{1} = \frac{200}{7}$

$= 24\frac{4}{7}\%$.

(iii) Since 0.35 is really $\frac{35}{100}$, then a decimal simply becomes a % if multiplied by 100.

Hence $0.35 = 0.35 \times 100 = 35\%$.

(b) Changing a percentage into a fraction

(iv) 35% means 35 out of 100.

Hence, $35\% = \frac{\overset{7}{\cancel{35}}}{\underset{20}{\cancel{100}}} = \frac{7}{20}$ (after cancelling).

(v) $67\frac{1}{2}\%$ means $67\frac{1}{2}$ out of 100.

Hence, $67\frac{1}{2}\% = \frac{67\frac{1}{2}}{100} = \frac{135}{200} = \frac{27}{40}$ (after cancelling).

Exercise 1g

(1) Change the following fractions into percentages:

(i) $\frac{3}{5}$	(ii) $\frac{1}{4}$	(iii) $\frac{3}{4}$	(iv) $\frac{1}{8}$
(v) 0.12	(vi) $\frac{3}{20}$	(vii) $\frac{5}{8}$	(viii) $\frac{3}{10}$
(ix) $2\frac{1}{2}$	(x) 0.465	(xi) $\frac{2}{3}$	(xii) $\frac{5}{9}$
(xiii) $\frac{1}{15}$	(xiv) $\frac{1}{11}$	(xv) $3\frac{1}{3}$	

(2) Change the following percentages into fractions, simplifying your answer if possible.

(i) 24%	(ii) 15%	(iii) $33\frac{1}{3}\%$	(iv) 12%
(v) 60%	(vi) 150%	(vii) 45%	(viii) $67\frac{1}{2}\%$
(ix) 28%	(x) 33%	(xi) 120%	(xii) 8%
(xiii) 0.5%	(xiv) $2\frac{1}{4}\%$		

1.8 Place value

(i) Place value in whole numbers

When we write a number such as 3850, we really mean 3000 + 800 + 50. The 8 actually stands for 800. If you now consider the number 6485, this means 6000 + 400 + 80 + 5. This time, the 8 stands for 80. The value of the 8 clearly depends on which place it occupies. This is what we mean by *place value.*

(ii) Place value in decimals

The number 6.23 really means $6 + \frac{2}{10} + \frac{3}{100}$. The number 2 actually stands for $\frac{2}{10}$ (two tenths). The number 604.028 really stands for $600 + 4 + \frac{2}{100} + \frac{8}{1000}$. The number 2 this time stands for $\frac{2}{100}$ (two hundredths).

Exercise 1h

For the following numbers, write them out in full as shown in the previous section.

(1) 684	(2) 906	(3) 40.2	(4) 3.68
(5) 10 109	(6) 28.094	(7) 63.705	(8) 1.001
(9) 6 000 402	(10) 6.0101		

Write out in words what the 8 stands for in the following examples:

(11) 836	(12) 481	(13) 48.6	(14) 5.081
(15) 50.83	(16) 8496	(17) 60 857	(18) 30.089
(19) 2.008	(20) 1.0804		

1.9 Decimals

Winston has been asked to split £100 equally between 6 people. He is not very good at dividing and has decided to use a calculator. He presses the correct buttons to obtain:

Display

$100 \div 6 =$ [16.66666667]

His answer is clearly not very helpful. He decided to round his answer correct to 2 decimal places to get £16.67. Unfortunately, when he counted out the money, he found that six lots of £16.67 came to £100.02, and he was 2p short. In this particular example, he would have been better to choose £16.66 and he would have had 4p left. The process of ignoring decimal places is called *truncation*, and it is clearly sometimes more appropriate than rounding.

The following examples show the points to watch out for when *rounding*.

(i) 14.648 rounded to 2 decimal places (2 d.p.) = 14.65.
↑
greater than 5 therefore increase 4 by 1

(ii) 2.985 rounded to 2 d.p. = 2.98 or 2.99.
↑
5 means that digit in front can be increased or left as it is

(iii) 1.998 rounded to 2 d.p. = 2.00.
↑
greater than 5 hence increase 9 in front by 1 which causes a 'knock on' effect

(iv) 4876 rounded to 3 significant figures (3 s.f.) = 4880.
↑ greater than 5 hence increase 7 by 1
↑ count significant figures from here
↑ do not forget this 0

(v) 37.89 rounded to 3 s.f. = 37.9.

(vi) 0.04853 rounded to 3 s.f. = 0.0485.
↑ count 3 significant figures from here
↑ ignore this 0 in counting significant figures

(vii) 80099 rounded to 3 s.f. = 80 100.

↑ ↑ greater than 5 so increase zero in front to 1

this 0 counts as
a significant digit
when it is between non-zero digits

You should now try exercise 1i, checking your answers carefully.

Exercise 1i

Express the following numbers rounded correct to 2 decimal places.

(1) 6.874 (2) 3.096 (3) 28.178
(4) 63.099 (5) 20.085 (6) 31.977
(7) 5.0908 (8) 0.0102 (9) 0.0034
(10) 6.899

Express the following numbers rounded correct to 3 significant figures.

(11) 81.63 (12) 2497 (13) 8999
(14) 10.07 (15) 0.0483 (16) 63.785
(17) 184 930 (18) 88.88 (19) 90 909
(20) 12.043

Truncate the following numbers to leave 3 digits in the significant places.

(21) 6897 (22) 64.35 (23) 0.004 89
(24) 0.1249 (25) 668.7

1.10 Fractions to decimals

(i) Since $\frac{4}{5}$ also means $4 \div 5$, then a fraction can be changed to a decimal by either straightforward division without a calculator or with a calculator.

Hence:

$$5\overline{)4.{}^{4}0} = 0.8 \qquad \text{i.e. } \frac{4}{5} = 0.8$$

Or: 4 [÷] 5 [=] Display [0.8]

Exercise 1j

Change the following fractions into decimals using any method you like.

(1) $\frac{2}{5}$ (2) $\frac{5}{8}$ (3) $\frac{1}{4}$ (4) $\frac{3}{8}$ (5) $\frac{7}{10}$ (6) $\frac{5}{16}$
(7) $\frac{12}{100}$ (8) $\frac{9}{20}$ (9) $\frac{7}{16}$ (10) $\frac{11}{20}$ (11) $\frac{27}{1000}$ (12) $\frac{16}{125}$

(ii) What happens if you try to change $\frac{1}{9}$ into a decimal?

$$9\overline{)1.0{}^{1}0{}^{1}0{}^{1}0{}^{1}0} = 0.1\,1\,1\,1\,1$$

The process does not stop and we have a repeating decimal. This is indicated by a · above the number 1. Hence:

$\frac{1}{9} = 0.\dot{1}$ (The word recurring is used often instead of repeating)

Another example might be to express $\frac{3}{11}$ as a decimal. If we try this on the calculator, we get

				Display
3	÷	11	=	0.27272727

Although we are not absolutely certain, it looks as though it does repeat.

This time we write $\frac{3}{11} = 0.\dot{2}\dot{7}$

Similarly $\frac{4}{37} = 0.108108\ldots$

is written $\frac{4}{37} = 0.\dot{1}0\dot{8}$ (only 2 dots are required).

Any fraction can either be written as an exact decimal, or a repeating decimal. All fractions are called *rational numbers.* A number which gives a non-repeating decimal is called an *irrational number.*

Worked example 1.6

State whether you think the following numbers are rational or irrational:

(i) 6.40 (ii) $4\frac{7}{8}$ (iii) $3.1 \div 1.6$ (iv) π

(v) $\sqrt{8}$ (vi) $3\pi \div 2\pi$ (vii) 0.012 (viii) $(0.1)^2$

(i) 6.40 is $6\frac{2}{5}$ hence it is rational.

(ii) $4\frac{7}{8}$ is a fraction and is rational.

(iii) $\frac{3.1}{1.6} = \frac{31}{16}$ which is a fraction and is rational.

(iv) π cannot be written as an exact fraction, and is irrational.

(v) $\sqrt{8}$ is not an exact decimal; it is irrational.

(vi) $3\pi \div 2\pi$ is exactly 1.5 and is rational.

(vii) 0.012 is $\frac{12}{1000}$ which is rational.

(viii) $(0.1)^2 = \frac{1}{10} \times \frac{1}{10} = \frac{1}{100}$ which is rational.

Exercise 1k

State whether you think the following numbers are rational or irrational, giving a clear reason in each case.

(1) $4\frac{1}{3}$ (2) 0.1 (3) $\sqrt{5}$ (4) $4.2 \div 2.3$

(5) $(0.11)^2$ (6) 6π (7) $\sqrt{64}$ (8) $\frac{5}{9}$

(9) $\sqrt{\frac{1}{4}}$ (10) 0.095

1.11 Directed numbers

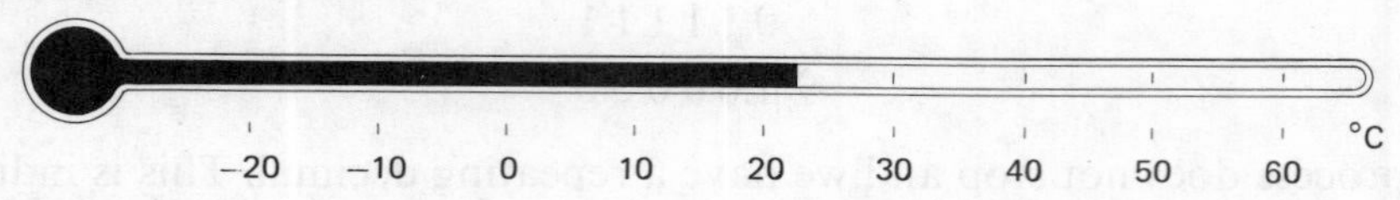

 Figure 1.3

Figure 1.3 shows a thermometer which measures temperature in °C. The freezing point of water is 0°C. When the temperature falls below freezing point, we say the temperature has become *negative*. To distinguish these from numbers greater than zero, the numbers greater than zero are referred to as *positive*. Positive and negative numbers are collectively known as *directed numbers*.

Another situation in which directed numbers occur is that of a bank statement, see Table 1.1.

Table 1.1

			BALANCE £
18 OCT			84.20
20 OCT	WITHDRAWAL	60.00	24.20
22 OCT	WITHDRAWAL	28.00	3.80 DR
25 OCT	WITHDRAWAL	16.00	19.80 DR
26 OCT	DEPOSIT	40.00	20.20

The DR indicates that the account has become overdrawn (it has become negative).

1.12 Working with directed numbers and the calculator

There are a number of rules for working with directed numbers. It is beyond the scope of this book to justify these rules, and the situations are illustrated by the following examples:

(i) $-4 + -3 = -7$ (negative + negative becomes more negative)
(ii) $-8 + 11 = +3$ (the + sign is not necessary)
(iii) $8 + -6 = +2$
(iv) $6 - -2 = 6 + 2 = 8$ (to subtract a negative number, change the sign and add)
(v) $-8 - -4 = -8 + 4 = -4$
(vi) $2 \times -3 = -6$
(vii) $-4 \times 3 = -12$ } negative × positive = negative
(viii) $-2 \times -8 = +16$ (negative × negative = positive)
(ix) $-12 \div 3 = -4$ (negative ÷ positive = negative)
(x) $24 \div -8 = -3$ (positive ÷ negative = negative)
(xi) $-27 \div -3 = +9$ (negative ÷ negative = positive)

To enter a negative number into the calculator, the [+/−] button is used. Hence -8×-12 on the calculator becomes:

8 [+/−] [×] 12 [+/−] [=] Display [96]

The student should have come across these ideas before, and it is envisaged that the following exercise will be revision.

Exercise 1l

(1) Arrange in order of size, smallest first, the numbers $-8, 10, -4, -2.5, 6, 3$.

(2) What is the difference in temperature between $-6°C$ and $8°C$?

(3) The temperature at 2000 hr is 12°C. John observes that the temperature falls 2°C every hour. What will be the temperature at 0700 hr the following morning?

(4) Tim is overdrawn in his bank account by £25.80. If he pays £38 into his account, what is his new balance?

(5) The temperature overnight dropped from 12°C to $-7°C$. What was the temperature drop?

(6) Ann's bank account is £15.45 overdrawn. How much does she need to pay in to cover this and a bill of £16.38 that she has to pay?

(7) Joanna is trying to climb up an icy slope. Every 5 seconds, she manages to go forward 5 feet and then slides back 2 feet. How far does she climb in 1 minute?

(8) Work out the following, trying not to use a calculator:

(i) $6 + -5$ (ii) $-3 + -8$ (iii) $8 - 11$

(iv) $16 + -3$ (v) $5 - -3$ (vi) $8 + -7$

(vii) $-8 - -3$ (viii) $4\frac{1}{2} + -6$ (ix) $12 - -5\frac{1}{2}$

(x) $-6\frac{1}{4} + 8\frac{1}{2}$ (xi) $6\frac{3}{4} + -4\frac{1}{2}$ (xii) $3\frac{1}{2} - -2\frac{1}{2}$

(xiii) $-5 - 3$ (xiv) -11×3 (xv) -8×-2

(xvi) $-26 \div 2$ (xvii) $-3 \times 4\frac{1}{2}$ (xviii) $6.2 - 14.8$

(xix) 1.2×-4 (xx) $-8 - -3.6$ (xxi) $-4 + -3 + -5$

(xxii) $-6 + 8 + -3$ (xxiii) $15 \div (-3 + -2)$ (xxiv) $-2\frac{1}{2} \times -3\frac{1}{2}$

1.13 Indices

Indices are a form of mathematical shorthand which enable us to simplify calculations.

For example,

(i) $6 \times 6 \times 6 \times 6 \times 6 = 6^5$ (6 raised to the power 5)

(ii) $2 \times 3 \times 2 \times 2 \times 3 \times 3 \times 2$
$= 2 \times 2 \times 2 \times 2 \times 3 \times 3 \times 3$
$= (2 \times 2 \times 2 \times 2) \times (3 \times 3 \times 3)$
$= 2^4 \times 3^3$ (This does not simplify any more.)

(iii) $4^5 \times 4^3 = (4 \times 4 \times 4 \times 4 \times 4) \times (4 \times 4 \times 4)$
$= 4 \times 4 \times 4 \times 4 \times 4 \times 4 \times 4 \times 4 = 4^8$

It can be seen that the powers have been **added**.

(iv) $3^4 \div 3^3 = \dfrac{\overset{1}{\cancel{3}} \times \overset{1}{\cancel{3}} \times 3 \times 3}{\underset{1}{\cancel{3}} \times \underset{1}{\cancel{3}}} = 3 \times 3 = 3^2$

It can be seen that the powers have been **subtracted**. Your calculator may have a button for working out powers. It usually looks like this $\boxed{x^y}$.

Worked example 1.7

Use a calculator to work out:

(i) 4^6 (ii) $3^2 \times 2^4$ (iii) $4^3 + 7^2$.

(i) 4^6: 4 $\boxed{x^y}$ 6 $\boxed{=}$ Display $\boxed{4096}$

(ii) $3^2 \times 2^4$: 3 $\boxed{x^y}$ 2 $\boxed{=}$ $\boxed{\times}$ 2 $\boxed{x^y}$ 4 = Display $\boxed{144}$

(iii) $4^3 + 7^2$: 4 $\boxed{x^y}$ 3 $\boxed{=}$ $\boxed{+}$ 7 $\boxed{x^y}$ 2 $\boxed{=}$ Display $\boxed{113}$

Exercise 1m

Evaluate

(1) 4^3 (2) 8^3 (3) $3^2 \times 2^2$ (4) $4^3 \times 2^2$
(5) $8^6 \div 2^5$ (6) $3^2 + 4^2$ (7) $2^3 \times 4^2$ (8) $3^3 + 2^3$
(9) $5^2 - 3^2$ (10) $5^4 \times 2^3$ (11) $6^3 \div 2^2$ (12) $5^3 \div 5^3$
(13) $6^4 - 5^4$ (14) $1^3 + 2^3 + 3^3$ (15) $4^3 \div 2^4$

1.14 Negative indices

Using the rules from the previous section, what is $10^3 \div 10^5$?

$$10^3 \div 10^5 = 10^{-2} \text{ (subtract the indices)}$$

What does 10^{-2} mean?

$$10^3 = 1000 \text{ and } 10^5 = 100\,000$$

So $10^3 \div 10^5 = \frac{1000}{100000} = \frac{1}{100}$.

Hence $10^{-2} = \frac{1}{100}$, also $10^{-1} = \frac{1}{10}$, $10^{-3} = \frac{1}{1000}$ etc.

1.15 Large and small numbers

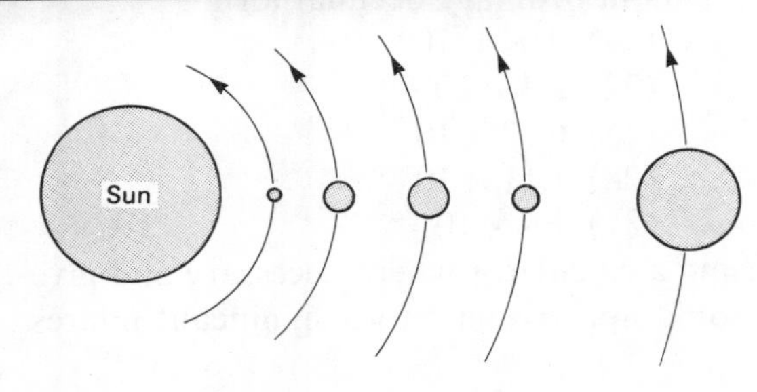

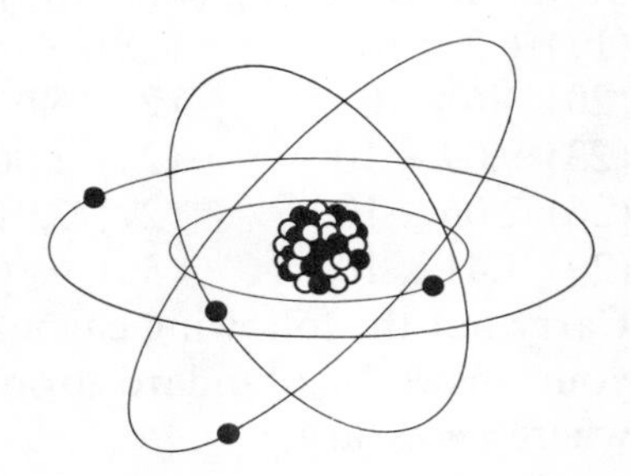

The distance between Earth and the Sun is approximately 93 million miles, i.e. 93 000 000 miles. This number can be written 9.3×10^7. This convenient way of writing a large number is called *standard form.* The number at the front must be greater than or equal to 1, and less than 10. The same number could be written as 93×10^6. This is not in standard form.

The mass of an atomic particle is roughly

0.000 000 000 000 000 000 000 001 7 g!

Clearly we need a better way of writing this.

The number is $1.7 \div 1\,000\,000\,000\,000\,000\,000\,000\,000$ g

$= 1.7 \times 10^{-24}$ g (There are 23 zeros, not 24, between the decimal point and the first digit.)

1.16 Calculating in standard form

To enter 2.95×10^6 into the calculator, proceed as follows:

2.95 [EXP] 6 — Display [2.95 06]

To enter 3.2×10^{-2} into the calculator, proceed as follows:

3.2 [EXP] 2 [+/−] — Display [3.2 −02]

Now check $(2.95 \times 10^6) \times (3.2 \times 10^{-2})$ gives

Display [94400]

In standard form, this is 9.44×10^4.

Exercise 1n

Write the following numbers in standard form:

(1) 630	(2) 2800	(3) 4875	(4) 68 000
(5) 126 000	(6) 9 600 000	(7) 125.9	(8) 0.12
(9) 0.003	(10) 0.000 465	(11) 0.111	(12) 85 000 000
(13) 0.000 006 4	(14) 840	(15) 0.000 000 078 4	(16) 998 000 000

Write the following standard form numbers in ordinary decimal form:

(17) 6.2×10^3	(18) 3.8×10^2	(19) 2.04×10^4
(20) 1.6×10^{-2}	(21) 3.6×10^{-3}	(22) 6.4×10^{-4}
(23) 9.09×10^3	(24) 1.06×10^{-1}	(25) 6.33×10^2
(26) 2.06×10^{-2}	(27) 3.01×10^4	(28) 1.11×10^{-3}
(29) 1.41×10^{-5}	(30) 6.86×10^6	(31) 1.9×10^{-4}

Carry out the following calculations using a calculator where necessary and give your answer in standard form. You should approximate to 3 significant figures where necessary.

(32) $6.3 \times 10^4 + 2.1 \times 10^4$

(33) $8.3 \times 10^5 + 4.1 \times 10^4$

(34) $2.3 \times 10^5 + 8.9 \times 10^3$

(35) $2 \times (3.6 \times 10^5)$

(36) $(2.8 \times 10^7) \times (1.2 \times 10^4)$

(37) $(8 \times 10^6) \times (4 \times 10^6)$

(38) $5 \times 10^{-2} + 2 \times 10^{-3}$

(39) $3 \times 10^{-3} + 2 \times 10^{-5}$

(40) $6.5 \times 10^{-2} + 3.8 \times 10^{-4}$

(41) $(3.8 \times 10^{-5}) \times (2 \times 10^3)$

(42) $(4.8 \times 10^6) \div (1.4 \times 10^3)$

(43) $(5.7 \times 10^5) \div (3.6 \times 10^2)$

(44) $(6.5 \times 10^2) \div (2.3 \times 10^{-2})$

(45) $3.2 \times 10^2 + 1.2 \times 10^2 + 3.6 \times 10^2$

(46) $6.8 \times 10^{-1} + 8.5 \times 10^{-1} + 3.4 \times 10^{-1}$

(47) $(3 \times 10^2)^2$

(48) $(2.4 \times 10^3)^2$

(49) $(1.2 \times 10^{-2})^2$

(50) $(2.85 \times 10^6) \div (1.27 \times 10^2)$

(51) $(3.52 \times 10^{-3}) \div (2.87 \times 10^4)$

(52) Light travels at approximately 300 000 km per second. Write this in standard form, and hence find the distance that light travels in 1 year. This distance is known as a light year.

(53) The diameter of Earth is approximately 12800 km. Using the fact that the circumference of a circle is roughly 3 × the diameter, estimate the length of the equator, giving your answer in standard form correct to 1 decimal place.

Suggestions for coursework 1

1. Find out a range of quantities in life that are measured in large or small numbers. Try and compare them, and set up a scale of sizes which would help you to understand measurements in the world around you.
2. If you are able to write a simple computer program, try and devise a way of finding prime numbers.
3. Investigate the number series $\frac{1}{2} + \frac{1}{4} + \frac{1}{8} + \frac{1}{16} + \ldots$. Look at similar sequences of numbers.
4. Find out as many places as you can where fractions are used in everyday life. Justify to your teacher what sort of fraction calculations they need to teach you.
5. The following problem is fairly well known. Can you explain it and extend the ideas involved.

Take any 3 digit number	966
Reverse the digits	669
and subtract	297
Now reverse the digits again	792
and add	1089

Do you always get 1089?

6. A security code consists of 3 digits. For example, 231, 442, 888, etc. How many different codes can you make? What about car number plates?

Miscellaneous exercise 1

1. From the set of whole numbers 63, 71, 84, 95, 121, 130, 169, write down:
   (i) any multiples of 3; (ii) any prime numbers;
   (iii) any perfect squares.

2. Jim spent $\frac{1}{3}$ of his pocket money on a record, and $\frac{1}{4}$ of his pocket money on a present for his mum. If he had £3.50 left, how much pocket money did he start with?
3. The diagram in Figure 1.4 shows the crops grown by a farmer in his field.

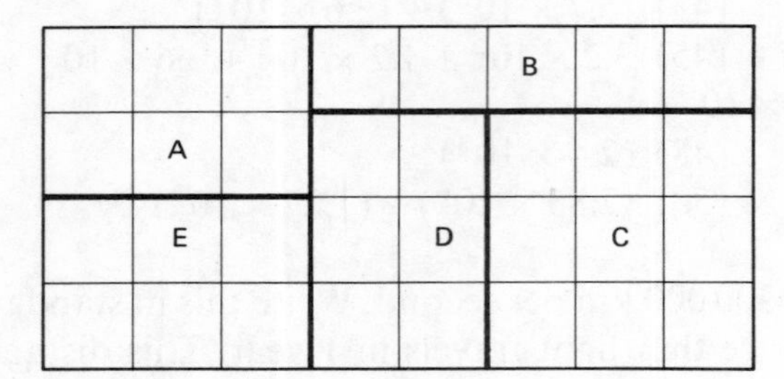

A Potatoes
B Peas
C Runner beans
D Cabbages
E Soft fruits

Figure 1.4

(i) State in its simplest form the fraction of the field used for runner beans.

(ii) Give in its simplest form, $\frac{\text{fraction of field used for cabbages}}{\text{fraction of field used for runner beans}}$.

(iii) What percentage of the field is used for peas?

4. A train stopped in Sheffield where 86 passengers got off and 145 passengers boarded the train. When the train left Sheffield, it had 329 passengers. How many passengers were on the train when it arrived in Sheffield? (NEA)
5. The distance between Earth and the Sun is 150 million km. The distance between Venus and the Sun is 108 million km. Express the distance of Earth from the Sun as a percentage of the distance of Venus from the Sun.
6. Copy the shapes shown in Figure 1.5 and shade an appropriate amount to give the fraction shown.

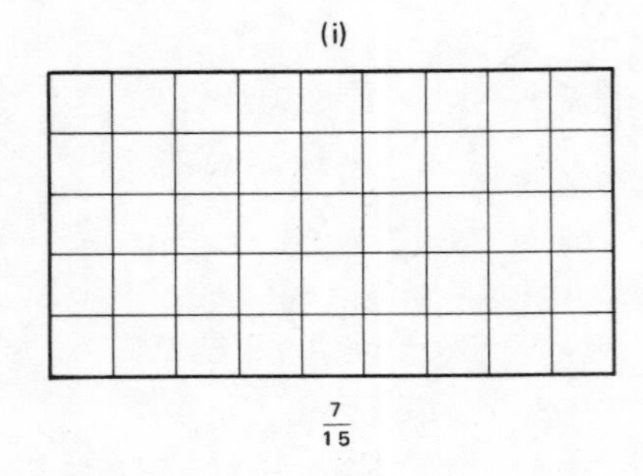

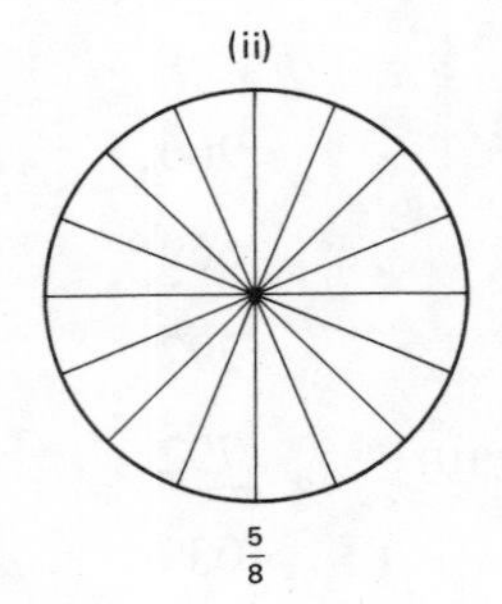

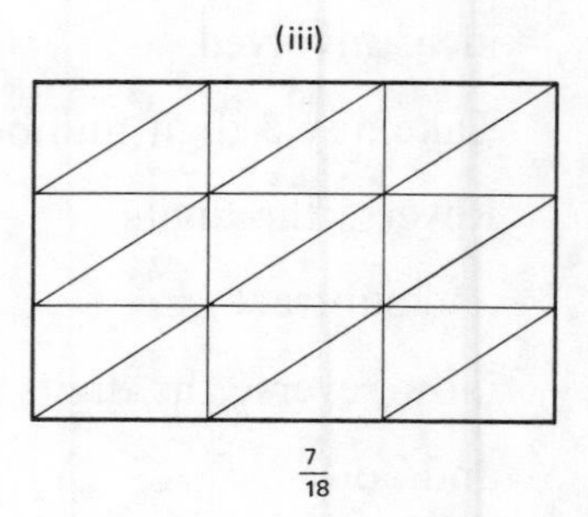

Figure 1.5

7. 1980 people work in a local factory. A survey was taken by Tony at the factory gates to see how people travelled to work. Unfortunately, it was raining at the time and Tony's results became a bit blurred. However, he remembered that the number who travelled by car was three times the number that travelled by train.

Car	[illegible]
Train	[illegible]
Bus	1471
Walk	17

(i) Find the missing numbers.
(ii) Find the percentage of people who walked to work.
(Round your answer to 1 decimal place.)

8. A packet of typing paper contains 484 sheets of paper. If the thickness of the packet is 5.1 cm, find the thickness of one sheet of paper, giving your answer in mm using standard form notation.
9. At the start of an experiment in biology, there is a single cell in a dish. A cell can divide into 2 every 15 minutes.
(i) Calculate, as a power of 2, the number of cells in the dish after 6 hours.
(ii) At a certain time, the number of cells in the dish is thought to be 2^{20}. These cells are divided equally into 4 smaller dishes. How many cells will there be in each of the smaller dishes?
10. Write the following numbers in figures:
(i) A quarter of a million;
(ii) fifty three thousand and ten;
(iii) two hundred and five thousand;
(iv) twenty five million and twenty thousand.
Then express each of your answers in standard form.
11. A cinema seats 550 people when full. There are 25 rows of seats, each with the same number of seats.
(a) Find how many seats there are in a row.
(b) Find the percentage of seats filled when 330 people are seated in the cinema.
(MEG)
12. In an experiment, the temperature of a liquid was steadily reduced by 3°C per minute for 4 minutes. At the beginning of the 4 minutes its temperature was 5°C. What was its temperature at the end of the 4 minutes?
13. A cinema has 300 seats costing £4 each and 100 seats costing £3 each. For the performance on June 1, $\frac{3}{5}$ of the £4 seats were sold, and $\frac{4}{5}$ of the £3 seats were sold.
(a) How many £4 seats were sold?
(b) How many £3 seats were sold?
(c) What percentage of all the seats in the cinema was sold?
(d) How much money was taken for this performance?
(e) How much money would be taken if all the seats were sold?
(f) Express the takings for the performance held on June 1 as a percentage of the takings on a night when all the seats were sold.
(g) Comment on your answers to parts (c) and (f). .(SEG)

14. Describe three number patterns in this table.

18	9	9	81
27	24	3	72
49	47	2	94
499	497	2	994

(NEA)

ALGEBRA 1

2

> *This unit takes you through all the basic processes of algebra, without going beyond the demands of intermediate level GCSE. The ideas developed are fundamental to those used in the rest of the book.*

2.1 The purpose of algebra

Many people find the ideas of algebra difficult to manage, but it should be realised that it is really only mathematical shorthand. For example, the area of a rectangle is found by multiplying its length by its breadth, i.e.

$$\text{Area} = \text{length} \times \text{breadth}$$

If we abbreviate this further, we get

$$A = l \times b$$

or $\quad A = lb$ (if we omit the $\times$ sign)

The final line we have obtained is called a *formula* for A in terms of l and b.

If the length of the rectangle is 6 cm, and the breadth is 2.4 cm, then we can substitute $l = 6$ and $b = 2.4$ into the formula. Hence:

$$A = 6 \times 2.4 = 14.4$$

The area of the rectangle $= 14.4\ \text{cm}^2$.

In using any formula like this, you must have consistent units. For example, if the length of the rectangle is 1.2 m, and the breadth is 85 cm, then $l = 120$ and $b = 85$ (using cm as the unit). Hence:

$$A = 120 \times 85 = 10\,200$$

The area of the rectangle is $10\,200\ \text{cm}^2$.

If, however, we use metres as the unit, then, $l = 1.2$ and $b = 0.85$. Hence:

$$A = 1.2 \times 0.85 = 1.02$$

The area of the rectangle is $1.02\ \text{m}^2$.

Worked example 2.1

In the formula $A = 2a + 3b$, find A if

(i) $a = 6, b = 3$ (ii) $a = 4, b = 0$

(iii) $a = \frac{1}{2}, b = 4$ (iv) $a = \frac{1}{4}, b = \frac{1}{4}$

(i) Remember that $2a$ means $2 \times a$ and $3b$ means $3 \times b$.

So, $A = 2 \times a + 3 \times b$

$= 2 \times 6 + 3 \times 3 = 12 + 9 = 21.$

(ii) $A = 2 \times 4 + 3 \times 0 = 8 + 0 = 8.$
(iii) $A = 2 \times \frac{1}{2} + 3 \times 4 = 1 + 12 = 13.$
(iv) $A = 2 \times \frac{1}{4} + 3 \times \frac{1}{4} = \frac{1}{2} + \frac{3}{4} = 1\frac{1}{4}.$

Worked example 2.2

H is given by the formula $H = 4xt$. Find H when
(i) $x = 3, t = 2$ (ii) $x = 5, t = 0$
(iii) $x = 6, t = \frac{1}{2}$ (iv) $x = \frac{1}{3}, t = \frac{1}{5}$

(i) $4xt$ means $4 \times x \times t$, so
$H = 4 \times x \times t$

$= 4 \times 3 \times 2 = 12 \times 2 = 24.$

(ii) $H = 4 \times 5 \times 0 = 20 \times 0 = 0.$
(iii) $H = 4 \times 6 \times \frac{1}{2} = 24 \times \frac{1}{2} = 12.$
(iv) $H = 4 \times \frac{1}{3} \times \frac{1}{5} = \frac{4}{3} \times \frac{1}{5} = \frac{4}{15}.$

The following example is a little more complicated, and includes negative numbers.

Worked example 2.3

If $v = u + at$, find
(i) v when $u = 8$, $a = -2$, $t = 6$
(ii) v when $u = 10$, $a = -3$, $t = -4$;
(iii) v when $u = -4$, $a = -3$, $t = 6$.

(i) $v = 8 + -2 \times 6 = 8 + -12 = -4$
(ii) $v = 10 + -3 \times -4 = 10 + 12 = 22$
(iii) $v = -4 + -3 \times 6 = -4 + -18 = -22$

Exercise 2a

(1) T is given by the formula $T = 2x + 5y$. Find T when
(i) $x = 3, y = 2$ (ii) $x = 5, y = 0$
(iii) $x = \frac{1}{2}, y = 4$ (iv) $x = \frac{1}{4}, y = \frac{1}{4}$

(2) Z is given by the formula $Z = 3pq$. Use this formula to find Z if
(i) $p = 2, q = 3$ (ii) $p = 7, q = \frac{1}{2}$
(iii) $p = 800, q = 0$ (iv) $p = \frac{1}{6}, q = 4$

(3) The cost C in pence of producing a magazine is given by the formula $C = 18 + 3N$ where N is the number of colour pictures.

(i) Find the cost of producing a magazine with 10 colour pictures.
(ii) Find the cost of producing a magazine with 18 colour pictures.
(iii) A magazine sells for £1.50. How much profit is made on each copy if it contains 25 colour pictures?
(iv) It is planned to sell a magazine for £1.10. What is the greatest number of pictures that you can have in the magazine if the publisher wants to make at least 40p profit on each copy?

(4) If T is related to x and y by the formula $T = \frac{4x}{y}$, find T in the following cases:

(i) $x = 3, y = 2$ (ii) $x = 12, y = 6$
(iii) $x = 4, y = 8$ (iv) $x = 5, y = 8$
(v) $x = 7, y = 16$ (vi) $x = 0, y = 100$

(5) If $s = 3u + 9a$, find s in the following cases:
(i) $u = -6, a = 2$ (ii) $u = 3, a = -4$
(iii) $u = 5, a = 9$ (iv) $u = -3, a = -2$
(v) $u = 1.6, a = 2.4$

(6) If $y = 4pq$, find y when
(i) $p = 3, q = -2$ (ii) $p = 6, q = 8$ (iii) $p = -2, q = -3$

2.2 Like and unlike terms

Expressions such as $4x$, x^2, $3xy$, $4t$ are called **terms** in algebra.

$3t$ and $7t$ are **like terms** and $3t + 7t = 10t$

$4x^2$ and $2x^2$ are **like terms** and $4x^2 - 2x^2 = 2x^2$

$6t$ and $4t^2$ are **unlike terms** and $6t + 4t^2$ cannot be simplified

$6pq$ and $2qp$ are **like terms** and $6pq + 2qp = 8pq$

When we simplify expressions like this, we are collecting like terms.

Exercise 2b
Simplify, if possible, the following:
(1) $4x + 6x$ (2) $3x + x^2$ (3) $2t + 5t - 3t$
(4) $2t^2 + 3t^2$ (5) $3pq + 2qp$ (6) $5m + 8m + 2m$
(7) $3n + 2n - n$ (8) $6t^2 + 5t - 3t^2$ (9) $5xy + 7yx$
(10) $3q + 2q^2 - 3q$

2.3 Coping with brackets

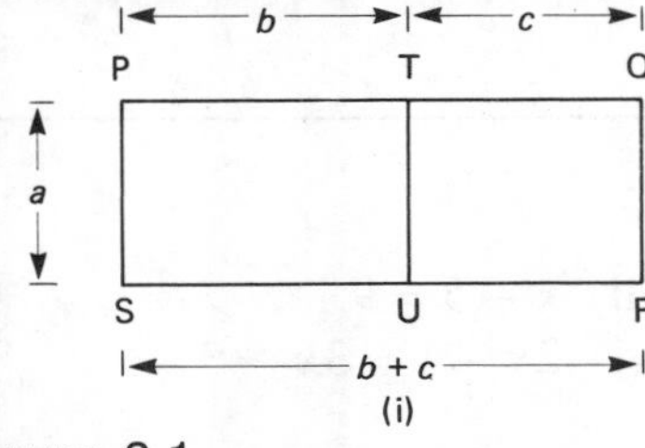

(i)

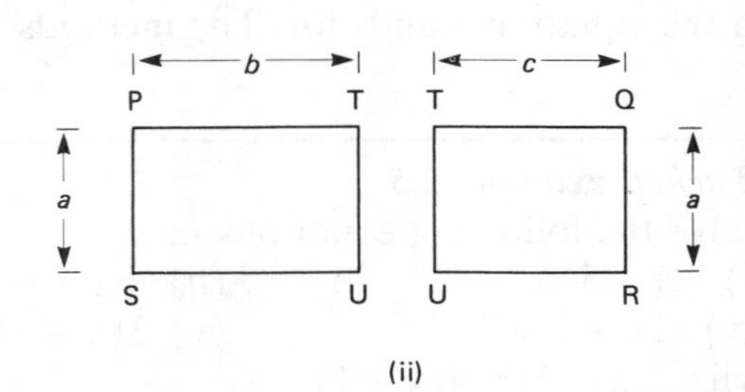

(ii)

Figure 2.1

The expression $a(b+c)$ really means $a \times (b+c)$. If you look at Figure 2.1, the rectangle PQRS has been divided into 2 rectangles by the line TU.

The area of PQRS $= a \times (b+c)$.

In Figure 2.1(ii) the rectangles have been separated. Hence:

The area of PTUS $= a \times b$

The area of TQRU $= a \times c$

Hence $a \times (b+c) = a \times b + a \times c$

This simple process enables us to remove brackets in simple expressions.

Worked example 2.4

Write the following expressions without brackets, simplifying where possible:

(i) $4(x+y)$ (ii) $3(x-2y)$

(iii) $2(x+y)+3(x+2y)$

(iv) $4(t+2p)+3(t-2p)$

(i) $4 \times (x+y) = 4 \times x + 4 \times y = 4x + 4y$

(ii) $3 \times (x-2y) = 3 \times x - 3 \times 2y = 3x - 6y$

(iii) $2 \times (x+y) + 3 \times (x+2y) = 2x + 2y + 3x + 6y = 5x + 8y$

(iv) $4(t+2p) + 3(t-2p) = 4t + 8p + 3t - 6p = 7t + 2p$

Exercise 2c

Rewrite the following expressions without brackets, simplifying your answers where possible.

(1) $4(x+y)$ (2) $2(p-2q)$ (3) $3(t+2x)$

(4) $4q(x+y)$ (5) $3(5t-2q)$ (6) $3(a+b)+2(a+2b)$

(7) $6(x-5y)$ (8) $3(z+2t)$ (9) $4(x+2y)+2(x-y)$

(10) $5(2a+b)$ (11) $3(2-2b)$ (12) $4(x+1)+5(x+2)$

(13) $5(2t-1)$ (14) $6t(x+y)$ (15) $3(q+p)+2(q+2p)$

(16) $6(3x+y)$ (17) $x(a+b)$ (18) $6(x-2)+4(x+10)$

(19) $3(2x+7)$ (20) $4(x+3)+2(x+3)$

2.4 Simple equations

When solving an equation, we are trying to find out what value a particular letter in the equation stands for. The methods are illustrated by Worked example 2.5.

Worked example 2.5

Solve the following equations in x.

(i) $\frac{x}{4} = 5$ (ii) $x + 7 = 11$ (iii) $x - 5 = 9$

(iv) $2x + 1 = 7$ (v) $3(x+1) = 4$ (vi) $2x - 1 = 3x - 5$

(vii) $2(x-5) = 4(x+1)$

(i) $\frac{x}{4}$ means $x \div 4$, hence the equation says what number $x \div 4$ gives an answer 5.

Clearly, $x = 20$. The correct way to lay this out is:

$$\frac{x}{4} = 5$$

Multiply each side by 4 $\quad \overset{1}{\not 4} \times \frac{x}{\underset{1}{\not 4}} = 4 \times 5$

$$x = 20$$

(ii) $x + 7 = 11$

Subtract 7 from each side $\quad x + 7 - 7 = 11 - 7$

Collect terms $\quad x = 4$

(iii) $x - 5 = 9$

Add 5 to each side $\quad x - 5 + 5 = 9 + 5$

Collect terms $\quad x = 14$

(iv) $2x + 1 = 7$

Subtract 1 from each side $\quad 2x + 1 - 1 = 7 - 1$

Collect terms $\quad 2x = 6$

Divide each side by 2 $\quad x = 3$

(v) $3(x + 1) = 4$

Remove the bracket $\quad 3x + 3 = 4$

Subtract 3 from each side $\quad 3x + 3 - 3 = 4 - 3$

Collect terms $\quad 3x = 1$

Divide each side by 3 $\quad \frac{\overset{1}{\not 3}x}{\underset{1}{\not 3}} = \frac{1}{3}$

Hence $\quad x = \frac{1}{3}$ [x does not have to be a whole number]

(vi) $2x - 1 = 3x - 5$

Subtract $2x$ from each side $2x - 1 - 2x = 3x - 5 - 2x$

Collect terms $\quad -1 = x - 5$

Add 5 to each side $\quad -1 + 5 = x - 5 + 5$

Collect terms $\quad 4 = x$ [notice that x can end up on the right-hand side]

(vii) $2(x-5)=4(x+1)$

Remove the brackets	$2x-10=4x+4$
Subtract $2x$ from each side	$2x-10-2x=4x+4-2x$
Collect terms	$-10=2x+4$
Subtract 4 from each side	$-14=2x$
Divide each side by 2	$\frac{-\overset{7}{\cancel{14}}}{\underset{1}{\cancel{2}}}=\frac{\overset{1}{\cancel{2}}x}{\underset{1}{\cancel{2}}}$
Hence:	$x=-7$ [x can be negative]

Exercise 2d

Solve the following equations, some of the answers are negative numbers.

(1) $\frac{x}{4}=2$ (2) $\frac{x}{5}=7$ (3) $5x=10$ (4) $x+5=16$
(5) $x+15=6$ (6) $x-7=8$ (7) $2x+1=11$
(8) $5x-3=12$ (9) $2x+1=x+5$
(10) $3x-1=2x+6$ (11) $4x-1=3x-2$
(12) $2(x+3)=12$ (13) $4(x+1)=3(x-2)$
(14) $2(x+5)=3(x+1)$ (15) $4-2x=1+x$
(16) $7-3x=12x-23$ (17) $5x+8=4+x$
(18) $8-3x=8+x$ (19) $6x+3=4x+4$
(20) $\frac{1}{4}x+\frac{1}{3}x=14$

2.5 Forming expressions and formulae

(i) If I buy 8 m of material costing £2 per metre, the total cost is $8 \times £2 = £16$.
If I buy L m of material costing £C per metre, the total cost is $L \times £C = £LC$.
LC is called an *expression.*
If we let T stand for the total cost, then we can write $T = LC$.
This gives us a *formula* for T.

(ii) If I share £24 between 8 people, how much does each person receive? I have to divide £24 by 8. So each person receives $£24 \div 8 = £3$.
If I share £A between n people, how much does each person receive? I have to divide A by n, written

$$A \div n = \frac{A}{n}$$

If T stands for the amount each person receives, we have

$$T = \frac{A}{n}.$$

(iii) If I buy 3 pens at x pence each, and 8 books at £y each, what is the total cost?
The cost of the pens is $3 \times x = 3x$ pence.

The cost of the books is $8 \times £y = £8y$.
I can change the cost of the pens to £ by dividing by 100.

Hence the cost of the pens is $\frac{£3x}{100}$.

The total cost is $£\left(8y + \frac{3x}{100}\right)$.

The following exercise gives you a number of situations similar to these. Some readers may find these difficult at first.

Exercise 2e

(1) What is the cost of 4 tyres at £y each?
(2) Tom earns £E each week. How much does he earn in a year?
(3) What are the next two whole numbers after N?
(4) How many days are there in S weeks?
(5) What is the cost of 2 pairs of trousers at £t a pair and 3 shirts at £s each?
(6) Zak earns £e a week, and his wife Paula earns £E a year. How much do they earn between them (i) in a year, (ii) each week?
(7) A prize of £P is shared between Afsal and Petra, so that Petra receives twice as much as Afsal. How much does Afsal receive?
(8) A farmer buys cows at £80 each, and sheep at £23 each. Write down a formula for the total cost of c cows and s sheep. Use your formula to find the cost of buying 120 cows and 240 sheep.
(9) Tim buys prints at £x each, and sells them for £$(x + 5)$. What profit does he make if he sells 45 prints?
(10) Grange Hill is holding a disco. There are two types of ticket available. One costs x pence each and admits one, the other, which admits 2 people, costs y pence each. Write down an expression in £ for how much they take if they sell 100 of the single tickets and 250 of the double tickets. Work out this total if $x = 40$, and $y = 75$.
(11) Find an expression for the interior angle of a regular polygon of n sides.
(12) What is the perimeter of a rectangle of sides $(x - 1)$ cm and $(2x + 3)$ cm. Simplify your answer.
(13) How many 5p pieces can I get with £x and y 50p pieces?
(14) How many seconds are there in x hours, y minutes and t seconds?
(15) A batsman scored x 4s, y 6s and 15 singles in an innings. What is his total score?

2.6 Problem solving

We have so far learnt how to form expressions in algebra, and how to solve equations. These two skills can now be combined to solve problems. The following example should illustrate the ideas, but harder examples will be found in unit 11.

Worked example 2.6

In a week, Peter earns twice as much as John, and George earns £3 less than

Peter. Altogether, the 3 boys earn £17 a week. If John earns £x a week:

(a) write down, in terms of x

 (i) how much Peter earns a week

 (ii) how much George earns a week.

(b) Write down an equation in x, and solve it.

Hence work out how much each boy earns in a week.

(a) (i) Twice as much, means $\times 2$, hence Peter earns £$2x$.

 (ii) George earns £3 less than Peter, so we have to subtract 3 from Peter's earnings to get that George earns £$(2x - 3)$. [*Note*: it is safer to enclose it with brackets.]

(b) The total earned by all three is

$$£(x + 2x + (2x - 3))$$

This must equal £17, and so we can write down the equation:

$$x + 2x + (2x - 3) = 17$$

Collect terms

$$5x - 3 = 17$$

Hence: $5x = 20$ and $x = 4$.

We have John earns £4 a week; Peter earns £$2 \times 4 = $ £8 a week; and George earns £$2 \times 4 - 3 = $ £5 a week.

Worked example 2.7

If three consecutive odd numbers add up to 51, what are the numbers?

The interesting thing about this question is that we do not need to worry about the fact that the numbers are odd, rather than even.

If the smallest number is x, since odd numbers go up by 2 (so of course do even numbers), the other numbers must be $x + 2$ and $x + 4$. So

$$x + x + 2 + x + 4 = 51$$

$$3x + 6 = 51$$

$$3x = 45,\ x = 15$$

The numbers are 15, 17, 19.

Exercise 2f

(1) Rebecca's age is x years. Tim is 4 years older than Rebecca, and the total of their ages is 38 years.

 (i) Write down Tim's age in terms of x.

 (ii) Write down an equation satisfied by x.

 (iii) Find Rebecca's age and Tim's age.

(2) Toya's age is x years. Her mother is twice as old as she is. Their combined age is 54 years.
   (i) Write down the mother's age in terms of x.
   (ii) Write down an equation in x.
   (iii) Find their ages.

(3) (i) What is the next whole number after the whole number x?
   (ii) If two consecutive whole numbers add up to 47, what are the numbers?

(4) The sides of a rectangle are of length x cm and $(2x + 1)$ cm. If the perimeter of the rectangle is 20 cm, write down an equation in x and solve it.

(5) Matt is 5 years older than Peter, and Peter is twice as old as Jane. If their combined ages total 60 years, find their ages.

2.7 Changing the subject of a formula

(i) Consider the formula $y = 2x + 1$.
If $x = 3$, $y = 2 \times 3 + 1 = 7$.
What do you need to put x equal to, so that $y = 15$?
A little mental juggling will probably give you the correct answer of $x = 7$.
It is better, however, to make x the subject of a formula as follows:

$$y = 2x + 1$$

Subtract 1 from each side $y - 1 = 2x + 1 - 1$

So $\quad y - 1 = 2x$

$\div 2 \quad \dfrac{y-1}{2} = x \quad$ or $\quad x = \dfrac{y-1}{2}$

If you put $y = 15$, $x = \dfrac{15-1}{2} = \dfrac{14}{2} = 7$.

(ii) If H is related to p and q by the formula:

$$H = \frac{p}{2q}$$

Find q when $p = 12$, and $H = 4$.
Again, it is easier to make q the subject as follows:

$$H = \frac{p}{2q}$$

$\times$ each side by $2q \quad 2qH = \dfrac{p}{\cancel{2q}_1} \times \overset{1}{\cancel{2q}} = p$

$\div$ each side by $2H \quad q = \dfrac{p}{2H}$

So $q = \dfrac{12}{2 \times 4} = \dfrac{12}{8} = 1\frac{1}{2}$.

In the following exercise, examples with negative numbers are not included; these can be found in unit 11.

Exercise 2g

(1) If $y = 4x + 3$, make x the subject and find x when $y = 23$.
(2) If $t = 3p - 2$, make p the subject and find p when $t = 22$.
(3) If $q = \dfrac{2}{x}$, make x the subject and find x when (i) $q = 6$; (ii) $q = \frac{1}{2}$.
(4) If $t = \dfrac{3}{x+1}$, find x when $t = 1$.
(5) If $v = 2t + 7$, find t when $v = 29$.
(6) If $y = 13 - 2x$, make x the subject and find x when $y = 4$.
(7) If $N = \dfrac{300}{q}$, find q when $N = 20$.
(8) If $x = 4t + 3q$, find (i) t, when $x = 20$ and $q = 2$; (ii) q, when $x = 40$ and $t = 1$.
(9) If $x = 4pq$, find (i) p when $x = 40$, $q = 2$; (ii) q when $x = 15$, $p = 3$.
(10) If $y = \dfrac{400}{3x+1}$, find (i) y when $x = 5$; (ii) x when $y = 200$.

2.8 Venn diagrams

A Venn diagram is a device for solving problems which at first sight may appear quite difficult. They are included in this unit on algebra because sometimes they involve solving simple equations. The techniques involved are illustrated in the following examples.

Worked example 2.8

The houses in Coronation Street were surveyed to see if they had: a television set; a video recorder; or an automatic washing machine. The results have been summarised by David who put them into a Venn diagram, see Figure 2.2.

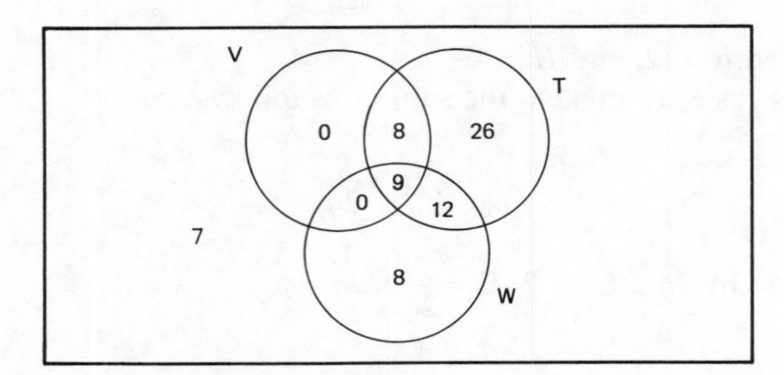

Figure 2.2

Assuming everyone answered the survey:

(i) How many houses are there in Coronation Street?
(ii) How many houses had a television?

(iii) How many people did not have a washing machine?
(iv) Why do you think that two of the regions have zero in them?
(v) How many had a television and a washing machine but not a video recorder?

(i) You need to add all the numbers including the outside region:

$$8 + 9 + 12 + 8 + 26 + 7 = 70 \text{ houses}$$

(ii) You need to add all the numbers inside the T curve:

$$8 + 9 + 12 + 26 = 55 \text{ with a television}$$

(iii) Everybody outside W:

$$8 + 26 + 7 = 41$$

did not have an automatic washing machine.
(iv) Presumably because if you do not have a television, there is no point in having a video recorder.
(v) This must be inside T and W, but outside V. The answer is 12.

Worked example 2.9
A group of 70 girls were asked about whether or not they liked playing hockey, tennis or badminton. 10 said they did not like playing any of the sports, 28 said they liked hockey, 23 liked tennis, and 36 liked badminton.
Also, 7 said they liked hockey and tennis only, 8 liked tennis and badminton only, and 2 liked hockey and badminton only. Let the number that liked all three equal x. Draw a Venn diagram to represent this problem, and hence find x.

The sets for hockey (H), tennis (T) and badminton (B) are represented by ovals or circles.
The numbers that can be entered straight away in the diagram are shown in Figure 2.3. A little thought is required to complete this diagram. The empty space remaining in T is not those that like tennis (23), but those that like tennis only.

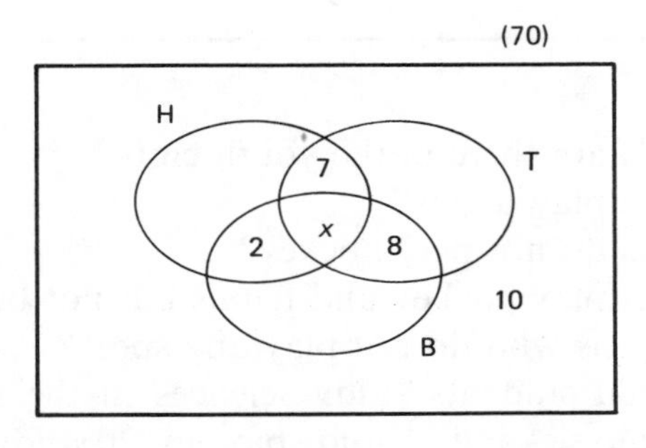

Figure 2.3

The number that like tennis only is given by

$$23 - (7 + x + 8) = 23 - 7 - x - 8 = 8 - x$$

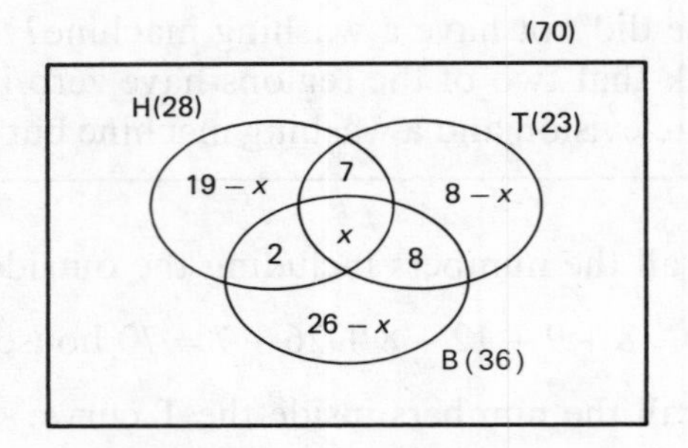

Figure 2.4

Similar arguments enable you to fill in the diagram as shown in Figure 2.4.

You know that the total of all the regions is 70, so we can write down an equation:

$$(19 - x) + 7 + x + 2 + (8 - x) + 8 + (26 - x) + 10 = 70$$

Simplify the left-hand side:

$$80 - 2x = 70$$

i.e.

$$10 = 2x$$

$$x = 5$$

Exercise 2h

(1) The Venn diagram in Figure 2.5 is about the girls in the local youth club. H is the set who play hockey, T is the set who play tennis, and B is the set who play badminton.

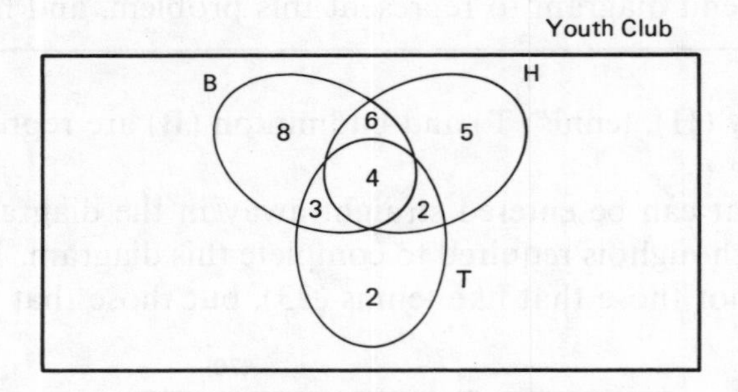

Figure 2.5

(i) How many girls are there in the youth club?
(ii) How many girls play tennis?
(iii) How many girls do not play hockey?
(iv) How many girls play hockey and tennis but not badminton?
(v) Are there any girls who do not play any sport?

(2) In a school where 80 students study sciences in the sixth form, 40 study chemistry, 46 study physics and 31 study biology. 20 study physics and biology, and the number who study physics and chemistry is three times the number who study chemistry and biology. If only 3 students take all three sciences, how many study only physics?

2.9 Number patterns

Look at the following *sequence* of numbers:

$$2, 5, 8, 11, 14, 17, \ldots$$

What can we say about these numbers? One obvious statement is that they go up by 3 at a time. We say that 3 is the *common difference* of this sequence.

A more difficult question would be, what is the twentieth number in this sequence? We could of course list all of the numbers until we came to the twentieth, the answer would in fact be 59. Look at the sequence 3, 6, 9, 12, 15, ... These numbers are all multiples of 3. The twentieth number in this sequence is $3 \times 20 = 60$. Each number in this sequence is 1 greater than those in the first sequence, and so we get $60 - 1 = 59$.

Each number in the sequence is called a *term* of the sequence. We can write the sequence as follows:

$T_1, T_2, T_3, T_4, \ldots$ where T_1 stands for the first term, T_2 the second term, and so on.

Hence, in the first sequence

$$T_{20} = 3 \times 20 - 1$$

It is now possible to write down a formula to find any term in the sequence. If we want to find the nth term T_n, then it follows that

$$T_n = 3 \times n - 1 = 3n - 1$$

We would now easily find the one hundredth number in this sequence, because

$$T_{100} = 3 \times 100 - 1$$
$$= 300 - 1 = 299$$

Exercise 2i

For the following number sequences, find the twentieth number and, where possible a rule for the nth term.

(1) 4, 7, 10, 13, 16, ...
(2) 1, 6, 11, 16, 21, ...
(3) 5, 9, 13, 17, 21, 25, ...
(4) 2, 9, 16, 23, 30, ...
(5) $1, \frac{1}{2}, \frac{1}{4}, \frac{1}{8}, \ldots$
(6) 1, 3, 6, 10, 15, ...
(7) 1, 1, 2, 3, 5, 8, ...
(8) 1, 3, 2, 4, 3, 5, 4, ...
(9) 1, 4, 9, 16, ...
(10) 2, 3, 5, 7, 11, ...
(11)

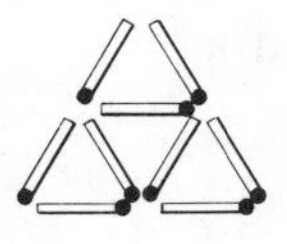
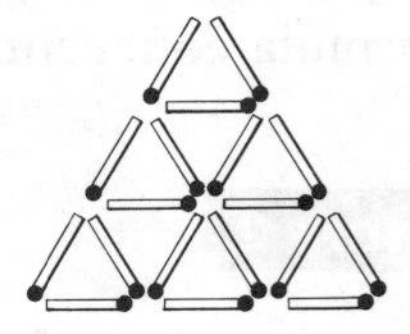
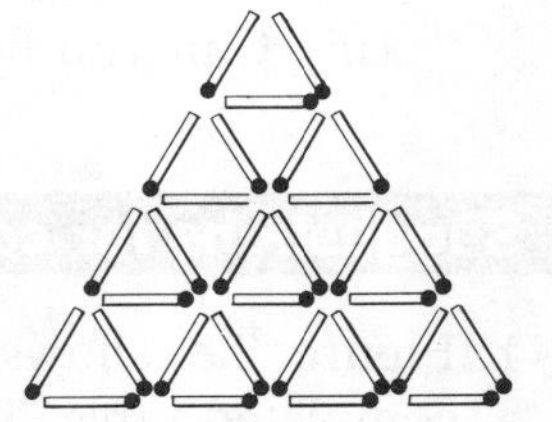

Figure 2.6

Figure 2.6 shows a series of patterns made out of matchsticks. Each matchstick is 3 cm long. Mary carried out an investigation for each pattern by counting the number of matches (n), the perimeter of each pattern (P cm), and the number of equilateral triangles of side 3 cm (N). Her answers for the first three patterns are given in Table 2.1.

Table 2.1

n	3	9	18
N	1	4	16
P cm	9	18	27

(i) One of her answers is wrong. Which one is it, and what is the correct answer?
(ii) What answers should she get for the next 2 columns?
(iii) Describe in words the patterns that you can see in the numbers in each row of the table if it was continued.

(12)

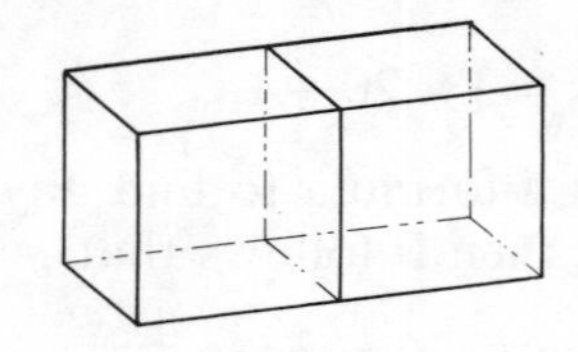

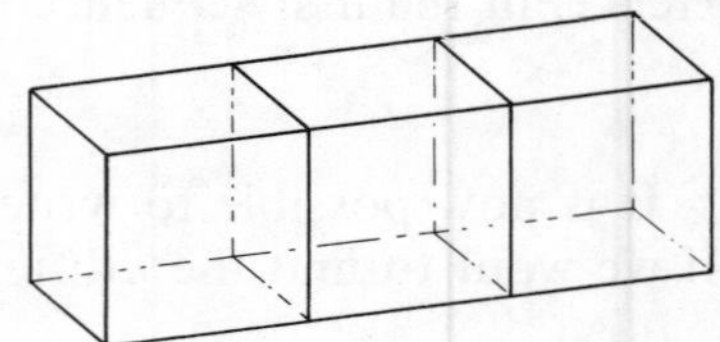

Figure 2.7

Small cubes are joined together to form a cuboid increasing in length as shown in Figure 2.7. The length of the side of each cube is 2 cm.

Table 2.2

n	1	2	3	4	5	6
S cm^2	24	40	56			

Table 2.2 shows the values of the number of cubes n and the total surface area of the solid S cm^2. The first few values have been entered already.
(i) Complete the table.
(ii) How many cubes are needed to give a total area of 168 cm^2?
(iii) Can you find a formula connecting S and n?

Suggestions for coursework 2

1. Figure 2.8 shows somes examples of matchstick patterns, with a partly completed table. Try to investigate any pattern you can find; try a pattern of your own. Can you predict how the pattern continues?

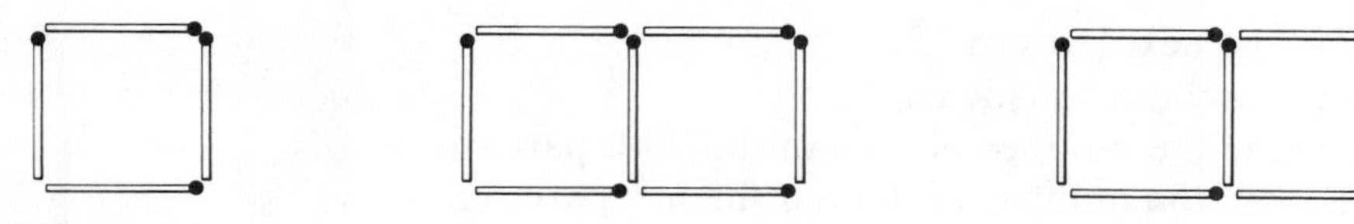

Figure 2.8

Table 2.3

Number of squares	1	2	3			
Number of matches	4	7	10			

2. Investigate some of the following number sequences:
   (i) 1, 2, 3, 5, 8, 13, 21, ...
   (ii) 1, 3, 6, 10, 15, 21, ...
   (iii) 1, 4, 9, 16, 25, ...
   (iv) 1, 5, 12, 22, 35, ...
   (v) 1, 2, 4, 8, 16, 32, ...

Miscellaneous exercise 2

1. A temperature can be measured as $F°$ Fahrenheit or $C°$ Celsius. The exact relationship between F and C is given by

$$F = \frac{9}{5}C + 32$$

   (a) Find the value of F when $C = 20$.
   (b) Find the value of C when $F = 14$.
   An approximate relationship between F and C is given by the following rule. 'To find F, add 15 to C and double your answer.'
   (c) Write this relationship as a formula.
   (d) Use this relationship to find an approximate value of F when $C = 20$.
   (MEG)

2. Lilian asked her uncle how old he was. 'In 13 years, I'll be twice as old as I was 7 years ago', he replied.
   (a) Taking his age now to be x years, write down:
      (i) his age in 13 years, in terms of x;
      (ii) an equation in x.
   (b) Solve your equation and find Lilian's uncle's age. (MEG)

3. Here are some dot patterns

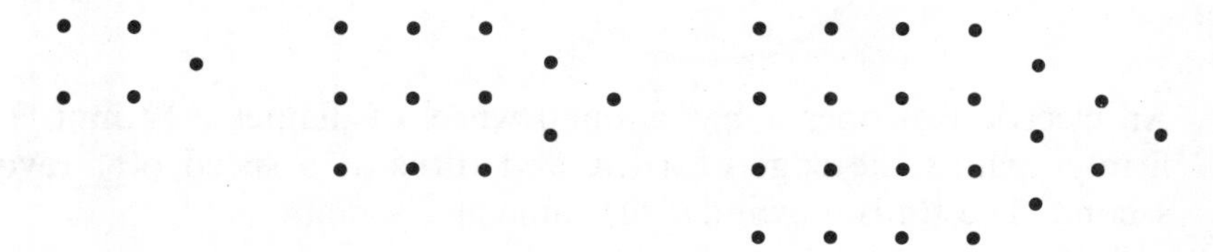

(i) Draw the next pattern.
(ii) How many dots has it got?
(iii) Calculate the number of dots in the 20th pattern.
(iv) Calculate the number of dots in the nth pattern.
(v) Write down the first four terms in the series of differences between the number of dots in succeeding patterns. (NEA)

4. 100 adults were asked whether they had visited a church, a library, or a public house in the previous week. 39 had not been to any of the three, 12 had been to church, 25 to a library, and 37 to a public house. 6 had been to church and a library only, 2 had been to a library and a public house only, and 1 to a church and public house only.

   x is the number that had visited all three.

   Draw a Venn diagram to illustrate this information. Mark on it the numbers in each of the subsets, using x where necessary. Hence write down an equation, and solve it to find x. (SEG)

5. If $x = 2.4 \times 10^4$, calculate:
   (i) $14x$ (ii) $\sqrt{2x+1}$
   In each case, give your answer in standard form.

6. The surface area A of the zone of a sphere of radius R between two parallel planes distance h apart (see Figure 2.9), is given by the formula $A = 6.2Rh$.

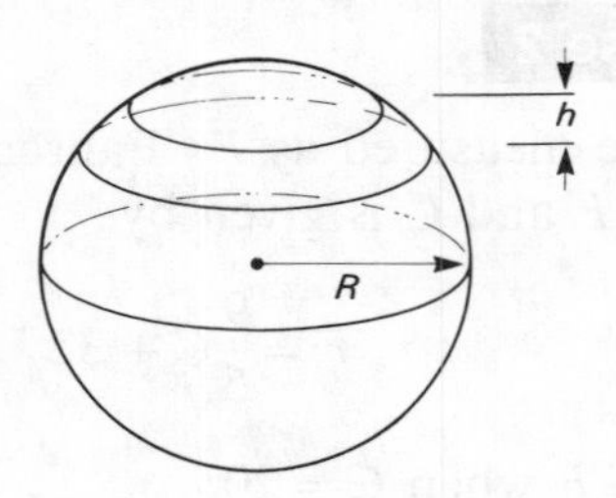

Figure 2.9

(i) Find A if $R = 10$ cm, $h = 5$ cm.
(ii) Find A if $R = 6370$ km, $h = 100$ km.
(iii) Find the surface area of a hemisphere of radius 10 cm.
(iv) Find h if $A = 100$ cm^2 and the diameter of the sphere is 20 cm.

7.

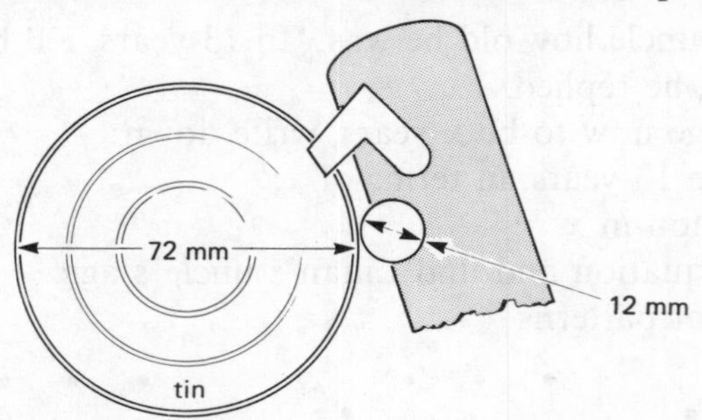

An electric can opener has a small wheel of diameter 12 mm. This is held firmly against the edge of a tin, and turns at a speed of 1 revolution per second. The tin is a cylinder of diameter 72 mm.

(a) The circumference of a circle is calculated using the formula:

CIRUMFERENCE = 3.14 × DIAMETER

Calculate the circumference of
(i) the small wheel,
(ii) the tin.

(b) How many seconds does it take to open the tin completely? (NEA)

8. A car salesman is paid a basic monthly salary of £450. He is also paid a bonus for £x for each car sold.
   (a) If he sells 10 cars in one month write down an expression for his total pay for the month, in terms of x.
   (b) If he sells N cars in one month write down an expression for the total amount he is paid for the month, in terms of N and x.
   (c) If $N = 12$ and $x = 35$, find the salesman's total pay for the month. (LEAG)

9. From a mixed pack of seeds, 22 plants grew. 11 of these plants were tall. 25% of those that had red flowers were also tall. 5 plants were neither tall nor had red flowers.
   Let x denote the number of plants which were both tall and had red flowers. Mark on a Venn diagram the numbers, in terms of x where necessary, of plants in the subsets represented by each region. Label your diagram clearly.
   Hence find x. (SEG)

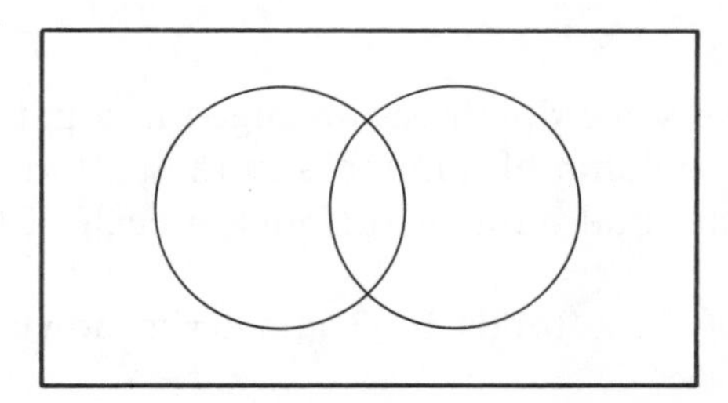

10. Figure 2.10 shows the first three of a sequence of shapes made up of rectangles of height 1 cm and base 2 cm.

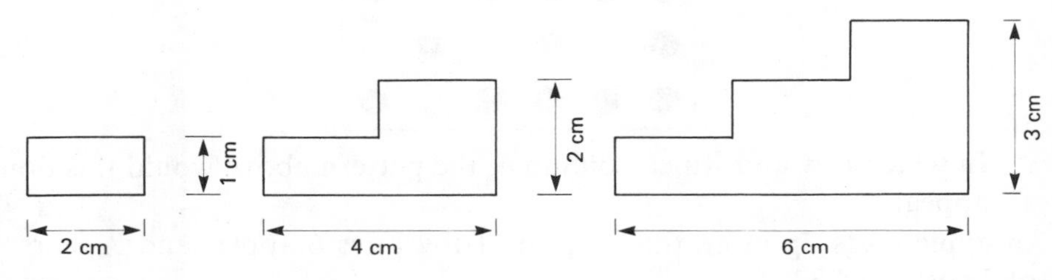

Figure 2.10

(a) Copy and complete the following table to show the perimeters and areas of the first six shapes.

Height (cm)	1	2	3	4	5	6
Perimeter (cm)	6	12				
Area (cm^2)	2	6				

(b) The shape which has height H cm has a perimeter of P cm and an area of A cm^2.
 (i) Write down a formula for P in terms of H.
 (ii) Write down a formula for A in terms of H.

(c) How many of the shapes in the sequence have a perimeter of less than 82 cm?

(d) What is the height of the shape in the sequence which has an area 380 m^2?

(LEAG)

11.

The diagram shows some dominoes arranged in a pattern.

(a) Draw the next column of dominoes in the pattern.

(b) Write down the total number of spots in each of these four columns of dominoes.

(c) Divide each of these totals by 3 and write down the new sequence of numbers.

(d) Write down the next two numbers in the sequence you obtained in part (c).

(e) Using your previous answers, find the total number of spots in the column in which the double 6 domino appears.

(f)

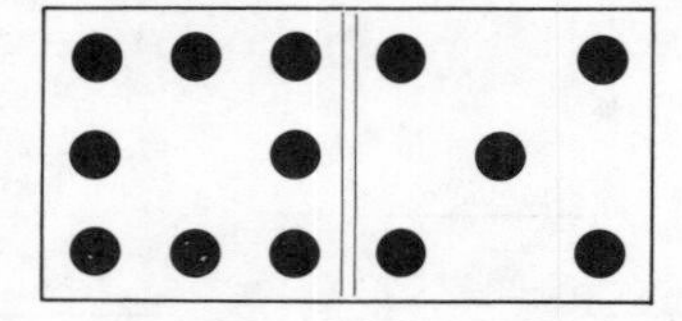

In which row and which column of the pattern above would this domino appear?

(NEA)

12. An apple costs 7p more than a pear. Toby buys 6 apples and 7 pears. The total cost is £2.37.

Let x pence be the cost of a pear.

(a) Write down, in terms of x,
 (i) the cost of **all** the pears,
 (ii) the cost of **one** apple,

(iii) the cost of **all** the apples.

(b) Write down an equation for the total cost of the fruit. Solve the equation to find x.

What is the cost of an apple? (SEG)

13. In a week, Joan earns three times as much as Mavis, and Elsie earns £5 less than Joan. Altogether the 3 girls earn £54.50. If Mavis earns £E a week:

(a) Write down, in terms of E,

(i) how much Joan earns in a week,

(ii) how much Elsie earns in a week.

(b) Write down an equation in E, and solve it.

Hence find how much each girl earns in a week.

14. A sample of 60 viewers were asked what they had watched on television. 34 had seen 'Eastenders', 33 'Coronation Street', and 16 'Top of the Pops'. All 60 had seen at least one of these three programmes.
12 had watched 'Eastenders' and 'Coronation Street' only,
2 had watched 'Eastenders' and 'Top of the Pops' only,
and 3 had watched 'Coronation Street' and 'Top of the Pops' only.
x is the number that watched all three programmes.
Draw a Venn diagram to illustrate this information, marking on it the numbers in each of the subsets, using x where necessary. Hence find x. (SEG)

15. (a) Using a 1 cm grid of dots, a square of side 3 cm is drawn as shown in the figure.

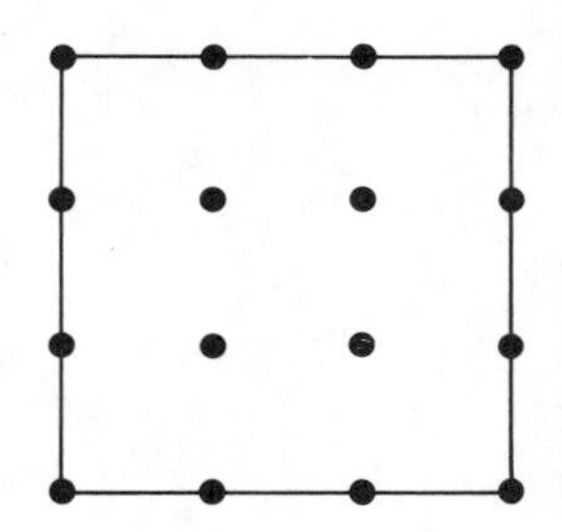

(i) How many dots are inside the square?

(ii) How many dots are on the border of the square?

(b) In a similar way, draw squares of different sizes on a grid and complete the following table.

Length of side (cm)	1	2	3	4	5
Number of dots inside (I)	0				
Number of dots on border (B)	4				
Total number of dots ($I + B$)	4				

(c) (i) Write down the special name which is given to the numbers that appear in the 'I' row of the table.

(ii) Describe the pattern of the numbers that appear in the 'B' row of the table.

(iii) What do you notice about the numbers that appear in the '$I + B$' row of the table?

(d) A square of side 12 cm is drawn on a 1 cm grid.

Find (i) the number of dots inside the square,

(ii) the number of dots on the border of the square. (MEG)

16. The formula

$$R = 231 - 0.45t$$

is remarkably accurate in estimating the world record in the men's 1500 metres race. R seconds is the record time and t is the number of years after 1930.

(a) (i) Use the formula to find the value of R when $t = 30$.

(ii) The world record in 1960 was 3 minutes 35.6 seconds. Calculate the difference, in seconds, between this time and the time given by the formula.

(b) (i) Use the formula to find the value of t when $R = 195$.

(ii) Write down the year in which, according to the formula, the world record will be 3 minutes 15 seconds.

(c) Explain why the formula must eventually go wrong. (MEG)

UNITS OF MEASUREMENT AND THE METRIC SYSTEM

3

This unit covers all of the problems that we encounter in everyday situations concerned with using units. Although in this country we are now using a metric system, there are some non-metric units still in use, and we sometimes need to be able to convert from one to another.

3.1 Definitions

The units most commonly used in the metric system are listed below, together with their abbreviations.

length: kilometre (km), metre (m), centimetre (cm), millimetre (mm)
1 km = 1000 m, 1 m = 100 cm, 1 m = 1000 mm

mass: kilogram (kg), gram (g), milligram (mg)
1 kg = 1000 g, 1 g = 1000 mg

time: seconds (s), hours (h), minutes (min)
1 h = 60 min, 1 min = 60 s

area: square kilometres (km^2), square metres (m^2), square millimetres (mm^2), hectare (ha)
$1\ km^2 = 1000\ m \times 1000\ m = 1000000\ m^2$
$1\ m^2 = 100\ cm \times 100\ cm = 10000\ cm^2$
$1\ ha = 10000\ m^2$ [This unit is commonly used in measuring land areas.]

Volume: cubic metres (m^3), millilitres (ml), litres (l), cubic centimetres (cm^3)
$1 l = 1000\ cm^3$ or 1000 ml, since $1\ cm^3 = 1$ ml
$1\ m^3 = 100\ cm \times 100\ cm \times 100\ cm = 1000\,000\ cm^3$ or $10^6\ cm^3 = 1000$ L

Speed: metres per second (m/s), kilometres per hour (km/h)

There are also a number of units used in more specialised situations. A series of prefixes is available to describe these units. Table 3.1 gives you a full list.

Table 3.1

mega means 1 000 000 used in megawatt or megatonne	micro means $\frac{1}{1\,000\,000}$ used in microsecond
kilo means 1000	milli means $\frac{1}{1000}$
hecto means 100	centi means $\frac{1}{100}$
deca means 10	deci means $\frac{1}{10}$

3.2 Changing units

When changing from one unit into another, you have to find the conversion factor for that change. The following examples should show you how this works.

Worked example 3.1

(i) Change 6.8 m into mm

Since 1 m = 1000 mm, the conversion factor = $\times 1000$.
$6.8 \times 1000 = 6800$ mm.

(ii) Change 0.8 kg into mg

Since 1 kg = 1000 g and 1 g = 1000 mg,
1 kg = $1000 \times 1000 = 1000\,000$ mg, and the conversion factor is $\times 1000\,000$.
$0.8 \times 1000\,000 = 800\,000$ mg or 8×10^5 mg.

(iii) Change 985 mm into m

Since 1 m = 1000 mm, 1 mm = 0.001 m ($\frac{1}{1000}$), and the conversion factor is $\times 0.001$ or $\div 1000$.
Hence $985 \div 1000 = 0.985$ m.

(iv) Change 80 m/s into km/h

This type of question is not quite so easy.
80 m/s means 80 m are covered in 1 second.
Hence in 1 minute, distance = $80 \times 60 = 4800$ m.
Hence in 1 hour, distance = $4800 \times 60 = 288000$ m.
Change this to km by dividing by 1000, giving the distance travelled in 1 hour as 288 km.
Hence the speed is 288 km/h.

(v) Change 4 km^2 into hectares

$1\ km^2 = 1000\ m \times 1000\ m = 1000000\ m^2$.
But $1\ ha = 10000\ m^2$.
So $1\ km^2 = 1000000 \div 10000 = 100$ ha.
Hence $4\ km^2 = 4 \times 100 = 400$ ha.

Exercise 3a

Carry out the following conversions:

(1) Into cm
(i) 8.4 m (ii) 6 km (iii) 8.4 mm (iv) 280 m
(v) 640 mm

(2) Into km
(i) 8684 m (ii) 10^6 cm (iii) 85000 mm (iv) 6×10^5 m

(3) Into g
(i) 4.6 kg (ii) 840 mg (iii) 5×10^8 mg (iv) 620 kg

(4) Into mm
(i) 60 cm (ii) 5.8 m (iii) 0.8 km (iv) 6×10^2 cm

(5) Into cm^2
(i) 2 m^2 (ii) 84 mm^2 (iii) 0.4 km^2 (iv) $8 \times 10^3\ m^2$

(6) Into ha
(i) 60000 m^2 (ii) 3.6 km^2 (iii) $3.8 \times 10^6\ m^2$
(iv) $2.8 \times 10^{-2}\ km^2$

(7) Into m/s
(i) 60 km/h (ii) 35 km/h (iii) 12 cm/s (iv) 2000 cm/h

(8) Into litres
(i) 600 cm^3 (ii) 2 m^3 (iii) 8.4×10^4 ml (iv) $4.3 \times 10^6\ mm^3$

3.3 Other units

It is not possible to give a complete list of other units used, but we give those that are most commonly found in practice in Table 3.2.

Table 3.2

		Approximate conversion factor	
1 yard	= 0.91 m	1 metre	= 1 yard
1 mile	= 1.61 km	1 km	= $\frac{5}{8}$ mile
1 gallon	= 4.55 litres	1 gallon	= 5 litres
1 mile an hour (abbreviation m.p.h.)	= 1.61 km/h	70 m.p.h.	= 110 km/h
1 nautical mile	= 1.85 km	1 n mile	= 2 km
1 pound (lb)	= 0.45 kg	2 lb	= 1 kg
1 ounce (oz)	= 28.35 g	1 oz	= 30 g
1 pound	= 16 oz		
1 gallon	= 8 pints		

Worked example 3.2

Using the information given in Table 3.2, carry out the following conversions, first of all giving an approximate answer, and then finding the exact answer.

(i) 100 l into gallons; (ii) 40 m.p.h. into km/h;
(iii) 1000 g into lbs and ounces.

(i) Using 1 gallon = 5 litres

$$100 \text{ l} = 100 \div 5 = 20 \text{ gallons}$$

Exactly: $100 \div 4.55 = 22$ gallons

(ii) Using 70 m.p.h. = 110 km/h

$$10 \text{ m.p.h.} = 110 \div 7 = 16 \text{ roughly}$$

So $40 \text{ m.p.h.} = 4 \times 16 = 64 \text{ km/h}$

Exactly: $40 \times 1.61 = 64.4 \text{ km/h}$

(iii) Using 30 g = 1 oz

$$1000 \text{ g} = 1000 \div 30 = 33 \text{ oz roughly}$$

Since 1 lb = 16 oz, we have 2 lb 1 oz

Exactly: $1000 \div 28.35 = 35.3 \text{ oz} = 2 \text{ lb } 3 \text{ oz}$

Exercise 3b

Carry out the following conversions, firstly giving an approximate value, and then giving the correct value, as accurately as possible using the information in Table 3.2.

(1) 15 pints into litres
(2) 8 yards into metres
(3) 6 square yards into m^2
(4) 10 gallons into litres
(5) 60 m.p.h. into km/h
(6) $4\frac{1}{2}$ lb into g
(7) 12 litres into pints
(8) 6.5 m into yards
(9) 50 km/h into m.p.h.
(10) 200 g into lb
(11) 100 km into n miles
(12) 8 oz into kg

3.4 Reading scales

There are many occasions when using our systems of units, that measurements have to be taken from some form of calibrated scale. Most of you will have had plenty of practice at this, but a few examples are given here as a reminder. The crucial point is that you realise what each division on the scale represents.

(i) Electricity meter

It is important to remember that the direction of reading the dial is alternately clockwise and anticlockwise. In Figure 3.1 the reading is 54666 units.

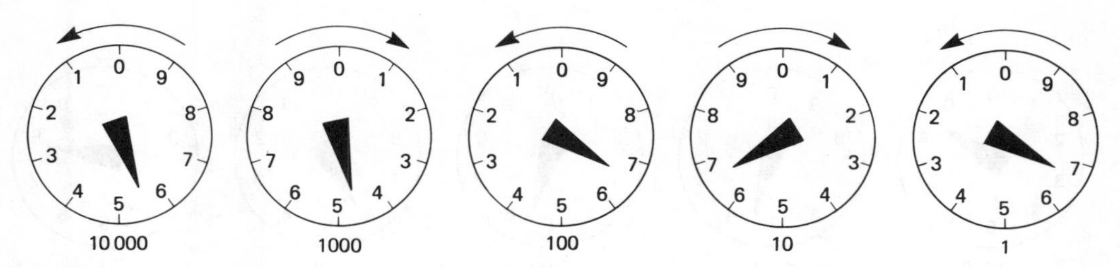

Figure 3.1

(ii) Thermometer

This is an example of a scale which may be double sided. In Figure 3.2, on the °C scale the reading is between 20 and 21, and you would probably say 20.5°C.

On the °F scale, it is about half-way between 68 and 70, so we call it 69°F.

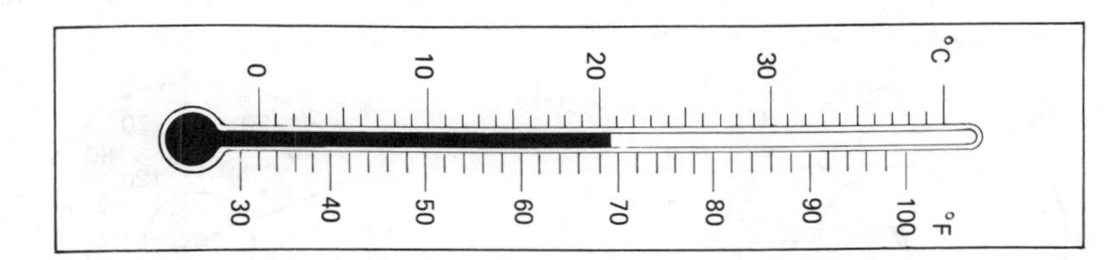

Figure 3.2

(iii) Awkward scales

Sometimes, the scale is not well chosen, and a little work needs to be done to take the reading. In Figure 3.3, you can see that there are 10 divisions between 0 and 80 kg. So one division is 80 kg ÷ 10 = 8 kg.

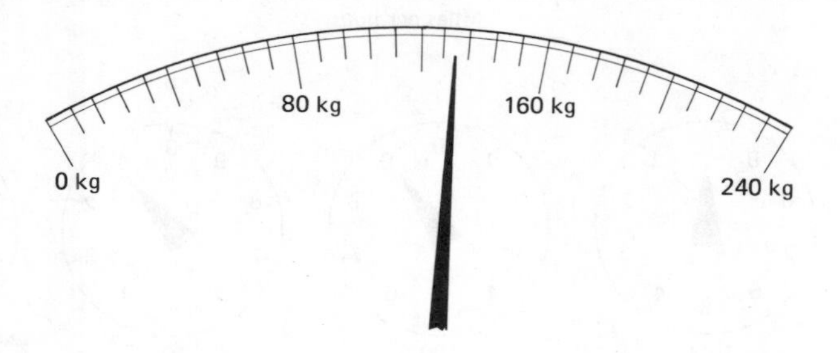

Figure 3.3

The reading shown is $6\frac{1}{2}$ divisions beyond 80 kg:

$$6\tfrac{1}{2} \times 8 = 52 \text{ kg}$$

So the reading is 80 kg + 52 kg = 132 kg.

Exercise 3c

Write down, stating clearly the units used, the measurements shown on the following scales.

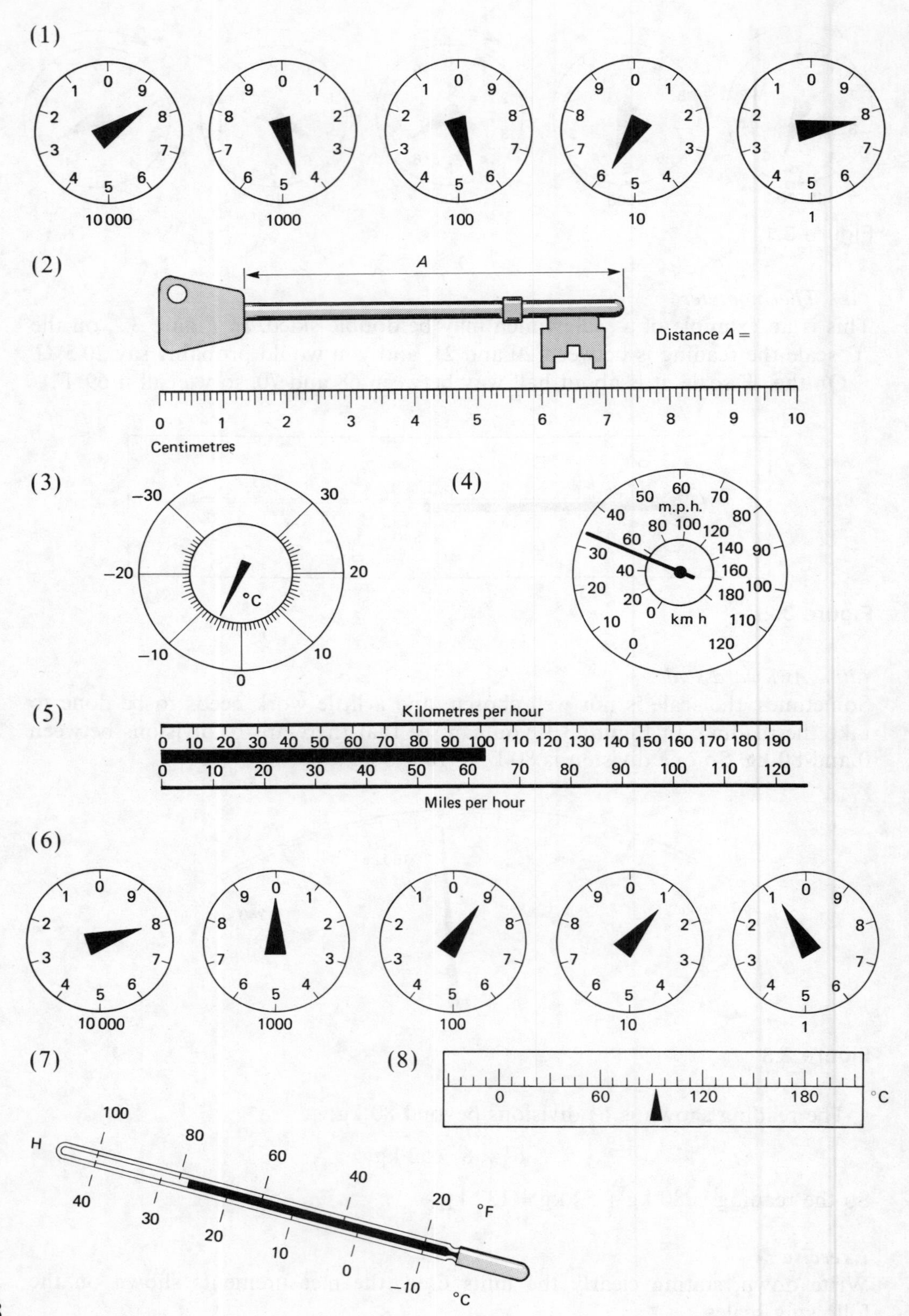
(1)
10 000
1000
100
10
1
(2)
A
Distance A =
Centimetres
(3)
°C
(4)
m.p.h.
km h
(5)
Kilometres per hour
Miles per hour
(6)
10 000
1000
100
10
1
(7)
H
°F
°C
(8)
°C

3.5 Time

Working with time is often not as easy as people think. There can be confusion between using the 24 hour clock and using a.m. or p.m. for morning and afternoon. The following examples indicate ways of coping with problems about time.

Worked example 3.3
A bus service runs every 8 minutes. A bus leaves the depot at 7.40 p.m. How many buses leave the depot between 6.35 p.m. and 8.15 p.m. inclusive?

The time difference between 6.35 p.m. and 8.15 p.m. is 8 h 15 min − 6 h 35 min = 1 h 40 min = 100 min.

If we divide 100 by 8, we get $12\frac{1}{2}$. Can we now say that 12 buses leave the depot? The danger of this can easily be shown. The actual departure times are as follows:

6.36, 6.44, 6.52, 7.00, 7.08, 7.16, 7.24, 7.32, 7.40, 7.48, 7.56, 8.04, 8.12

It can be seen that in fact the correct answer is 13.

Worked example 3.4
A train leaves London at 2015 h and arrives in Edinburgh at 0125 h the next day. How long did the journey take?

When calculating a time difference across midnight, it is usually safer to work with before and after midnight separately.

Hence, from 2015 h to 2400 h is 3 h 45 min. From 2400 h to 0125 h is 1 h 25 min.

$$\text{The total time} = \begin{array}{rr} \text{h} & \text{min} \\ 3 & 45 \\ +1 & 25 \\ \hline 5 & 10 \\ \hline \end{array} = 5 \text{ h } 10 \text{ min}$$

Remember that there are only 60 min in 1 hour!

Exercise 3d

(1) How long is it between 7.43 a.m. and 5.21 p.m.?
(2) A film which lasts 1 h 25 min starts at 5.40 p.m. What time does it finish?
(3) A bus service runs every 20 min. There is one bus at 7.20 p.m. How many buses run between 5.00 p.m. and 8.00 p.m. inclusive?
(4) The cooking instructions for a joint of meat are 20 min plus 35 min for each pound (or part). If Jean put a joint weighing 3 lb 6 oz into the oven at 10.40 a.m., what time should it be taken out of the oven?
(5) How many minutes are there between 1845 on Tuesday, and 0420 on Wednesday.
(6) How many days are there between 15 April and 24 May inclusive?
(7) How many hours are there in a leap year?
(8) An amateur meteorologist had worked out that the temperature fell 1°C during the night every 20 min after 2200 h. If the temperature at 2300 h was 6°C, what was the temperature at 0400 h the following morning?

3.6 Tables and timetables

There are many situations in our lives where we have to read information that is presented to us in the form of a table. This usually means lining up information in a certain row and column. It is very important that you read carefully any information given around the table before deciding which row and column to choose.

Worked example 3.5

A family of 2 adults and 2 children, aged 15 and 11, wish to go on holiday to France and take their car, length 4.35 m, with them. The cost of the ferry crossing is given below:

Table 3.3

Fare for Single Journey			Tariff A £	Tariff B £	Tariff C £	Tariff D £
Passengers	Adults		12	12	12	12
	Children, 4 and under 14 years (under 4 free)		6	6	6	6
Cars	Overall length not exceeding	4.00 m	56	47	32	20
		4.50 m	65	54	38	20
		5.50 m	67	57	41	20
		over 5.50 m	77	69	53	20

The tariff code depends on the time, and day, of their crossing. For the week in which they intend to travel the code is as follows:

Table 3.4

	M	T	W	Th	F	Sat	Sun
July	18	19	20	21	22	23	24
0030	D	D	D	D	D	C	B
0300	C	C	C	C	C	B	B
0530	C	C	C	C	C	B	B
0800	C	C	C	C	C	B	B
1030	B	B	B	B	B	A	A
1300	B	B	B	B	A	A	A
1500	B	B	B	B	A	A	B
1730	C	C	C	B	A	B	C
2000	C	C	C	C	B	B	C
2230	D	D	D	D	B	C	C

What would be the cost of the crossing if they left England at 10.30 a.m. on Saturday, 23 July? How much would they save if they decided to leave at 8 a.m. on Friday, 22 July? (SEG)

Referring to Table 3.4, look along the row labelled 1030 until you come to the column headed Sat 23. You find a letter A, which means the family will have to pay tariff A. In Table 3.3, 4.35 m for the car means not exceeding 4.50 m.

Under tariff A, this is £65.

$$2 \text{ adults under A is } 2 \times £12 = £24.$$

Also, the child aged 15 counts as an *adult* hence must pay the full rate which is £12.
The child pays £6.

$$\text{The total} = £65 + 36 + 6 = £107.$$

Using Table 3.3 again, the tariff for 8 a.m. on 22 July is tariff C.
The only difference is the cost of the car, which is £38 instead of £65.
The saving is £65 − 38 = £27.

Worked example 3.6

Table 3.5 gives the times of the train services between Euston (London) and Piccadilly (Manchester).

Table 3.5

London–Manchester

MONDAYS to FRIDAYS			SATURDAYS			SUNDAYS		
Euston depart		Piccdly arrive	Euston depart		Piccdly arrive	Euston depart		Piccdly arrive
06.50	wj	09.42	06.50	wj	09.42	09.50	fj	14.02
08.00	wj	10.33	08.15	mj	10.57	11.50	fj	16.04
08.15	mj	10.57	09.50	mj	12.35	14.10	m	17.37
09.50	mj	12.53	11.15	w	13.53	16.15	wj	19.10
11.15	w	13.35	12.30	m	15.20	17.00	m	19.41
12.30	m	15.20	13.55	w	16.34	17.50	mj	20.29
13.55	w	16.34				18.30	w	21.09
15.05	m	18.00	16.15	wj	18.49	19.30	m	22.10
16.15	wj	18.49				20.20	w	23.09
17.00	m	19.38				23.50		03.21
17.50	mj	20.28	18.30	w	21.09			
18.30	w	21.09						
18.10	m	21.38	20.20	w	23.09			
19.30	m	22.10	23.50		04.38			
20.20	w	23.09						
19.40	m	23.14						
23.50		03.21						

Essential engineering work on Sundays may result in these trains being retimed. Please check before travelling

All trains are due STOCKPORT 9 minutes before Piccadilly time

f calls at Macclesfield 39 mins before Piccadilly time shown

j calls at Watford Junction (to pick up only) 16 mins after Euston time

m calls at Macclesfield (to set down only) 27 mins before Piccadilly time

w calls at Wilmslow (to set down only) 20 mins before Piccadilly time

(i) What times does the 12.30 from Euston on a Tuesday arrive at Macclesfield?
(ii) David wants to arrive at Piccadilly on Saturday no later than 4 p.m. What is the latest time he can leave Euston?
(iii) How long is the journey between London and Manchester on the Sunday service leaving Euston at 17.50?
(iv) What times does the 16.15 from Euston on a Wednesday leave Watford, and how long does it then take to travel to Wilmslow?

This table has a number of code letters that must be checked carefully.

(i) The 12.30 from Euston on a Tuesday has code m alongside. At the bottom of the table, this says it arrives at Macclesfield 27 min before the Piccadilly time, which is 27 min before 15.20, i.e. 14.53.
(ii) Since 4 p.m. is 16.00 h, he must catch the *12.30* from Euston which arrives at 15.20.
(iii) Leaves 17.50, arrives 20.29. The time difference is 2 h 39 min (check it).
(iv) The 16.15 from Euston has code wj. The w means it arrives at Wilmslow 20 min before 18.49, i.e. 18.29. The j means it calls at Watford 16 min after the Euston time, i.e. 16 min after 16.15, which is 16.31.
The length of the journey is 18.29 – 16.31, which is 2 min less than 2 hours, i.e. 1 h 58 min.

Exercise 3e

(1) The following questions refer to the information given in Tables 3.3 and 3.4.
(i) How much would it cost to take a family of 5 consisting of 3 adults and 2 children, aged 8 and 14, together with a car of length 5.20 m, on holiday to France, if they travelled on 22 July leaving England at 1300 h.
(ii) How much would it cost David to take his two children aged 3 and 7, together with their car of length 3.60 m, across to France on 21 July at the cheapest possible rate.

(2) The following questions refer to the information given in Table 3.5.
(i) How long does the 16.15 on a Sunday take to travel from London to Manchester?
(ii) What times does the 18.30 from Euston on a Friday arrive at Wilmslow?
(iii) How long does the 09.50 from Euston on a Monday take to travel between Watford and Stockport?
(iv) What would be the latest train that Jane could catch from Watford to arrive in Stockport by 3.00 p.m. on a Wednesday?
(v) How many of the Euston to Manchester trains do not stop at Watford on a Saturday?

3.7 Estimation

It is probably true that many people are far more likely in everyday life to need a quick approximate answer to a problem than to calculate the exact answer, which may need a calculator that they have not got with them. Similarly, if they are using a calculator, they need to have some idea of the correct answer to check they have not pressed a wrong button.

The following examples should indicate how you go about it and then there is an exercise for you to try.

Worked example 3.7

(i) What is the approximate cost of 14 stamps costing 19p each?

(ii) What is the approximate time it takes to travel 185 km at an average speed of 38 km/h?

(iii) What should the calculator show if you try to muliply 6.9×17.6?

(i) 14 @ 19p is nearly the same as 14 @ 20p = £2.80.

(ii) 185 km is nearly 200 km
38 km/h is nearly 40 km/h
Hence the time it takes is $200 \div 40 = 5$ hours.

(iii) Taking 6.9 as 7 and 17.6 as 20:

$$7 \times 20 = 140$$

Seeing what we have increased both numbers, your calculator should show a value less than 140.

Exercise 3f

Using whatever approximations you consider necessary, find approximate values to the following:

(1) The cost of 17 articles at 93p each.
(2) $(999 + 888) \times 4.8$.
(3) The area of a rectangle which measures 6.2 cm by 5.8 cm.
(4) The time it takes to travel 88 km at an average speed of 36 km/h.
(5) The number of pieces of wood of length 5.8 cm that can be cut from a piece of wood $2\frac{1}{2}$ m long.
(6) The circumference of a circle of radius 6 cm.
(7) $3.9 \times 4.1 \times 2.8$.
(8) The cost of eighty-one 19p stamps.
(9) The number of wine glasses each holding 45 cm^3 of liquid that you can fill from a 1 litre bottle.
(10) $(99^2 \div 58^2) \times 0.973$.
(11) The total cost of 20 shirts at £16.95 each and 20 pairs of shorts at £2.68 each.
(12) $48 \times 9.65 + 48 \times 0.35$.
(13) The speed in m/s of an ordinary cyclist travelling along a level road.
(14) The volume of a mars bar in cm^3.

Try and discuss your answers to some of these questions with other members of your group, and justify which is the best approximation.

3.8 Rate

If Asif earns £440 in a period of 4 weeks, then we could also say that he earns £110 per week, or that he is paid at a *rate* of £110 per week. If we then want to know how much Asif earns in 8 weeks, we multiply the rate by 8, hence Asif earns $8 \times £110 = £880$. The idea of rate is a common one in everyday life and is illustrated by the following examples.

Worked example 3.8

Jim can run round 3 laps of his local running track in 3 min 42 s. If he continues to run at the same speed, (i) how long would he take to run 10 laps; (ii) how many complete laps could he run in 10 min?

The rate is 3 min 42 s $\div$ 3 = 1 min 14 s/lap

= 74 s/lap.

(i) $10 \times \text{rate} = 10 \times 74 = 740$ s

= 12 min 20 s.

(ii) 10 min = 600 s.

Number of laps = $600 \div$ rate

$= 600 \div 74 = 8.1$ laps or 8 complete laps.

Worked example 3.9

The West Indies scored 437 in a total of $6\frac{1}{2}$ hours. Assuming they scored at the same rate throughout the innings, calculate:

(i) How long it took to score 100 runs.

(ii) How many runs they scored between 2.15 p.m. and 3.30 p.m.

The scoring rate $= 437 \div 6.5 = 67.2$ runs per hour.

(i) The number of hours to score 100 runs will be

$$100 \div \text{rate} = 100 \div 67.2 = 1.49 \text{ hours}$$

Now 0.49 h $= 0.49 \times 60 = 29.4$ min.
The time required is 1 h 29 min.

(ii) 2.15 – 3.30 is $1\frac{1}{4}$ h.
The number of runs scored $= 1\frac{1}{4} \times \text{rate} = 1\frac{1}{4} \times 67.2$

= 84 runs.

Exercise 3g

(1) Alex earns £30.80 for working 8 hours. Find:

(i) How much he earns for working 35 hours.
(ii) How many hours he has to work to earn £165.55?

(2) Fifteen gallons of petrol cost £26.70.
(i) How many gallons can be bought for £21.36?
(ii) How much would 22 gallons cost?

(3) A firm has calculated that a large number of toys can be packaged in 10 hours by 8 packers. Assuming that everyone is working at roughly the same rate, how long would it take for 10 packers to package the same number of toys?

(4) Joseph can run 4 laps of his local running tack on average in about $5\frac{1}{2}$ minutes. How long would it take him to run 10 laps if he can run at the same average speed?

(5) June had bought 5 m of material which had cost £14.80. If she had bought $6\frac{1}{2}$ m of the same material, how much extra would that cost?

(6) A taxi meter shows 90p when you first get in. After a journey of $3\frac{1}{2}$ miles, Paul was charged £3. How much do you think he would have been charged for a journey of $4\frac{1}{4}$ miles? Explain carefully how you obtain your answer.

(7) It takes 5 minutes for one tap to fill a bath. If two taps are running at the same rate, how long would it take to half-fill the bath?

(8) Jennie can type 8 pages in $1\frac{1}{2}$ hours. If she types always at roughly the same rate, how long would she expect to take to type a book of 150 pages? If she charges 75p to type a page, how much will she earn?

3.9 Best buys

When shopping in a supermarket, it is often difficult to decide what to buy to get the best value for money. The simplest approach is to try and find a *unit* cost. This is illustrated by the following example.

Worked example 3.10

'Sudso' biological soap powder can be bought in three different sized packets in the local supermarket. The 1 kg size costs £1.64, the $1\frac{1}{2}$ kg size costs £2.40, and the $2\frac{1}{2}$ kg size costs £3.80.

Which packet is the best value for money?
The 'unit cost' here means how much is it per kg.

For the small size the cost = £1.64/kg.
For the second size £2.40 ÷ 1.5 = £1.60/kg.
For the large size £3.80 ÷ 2.5 = £1.52/kg.
The best buy is clearly the large size which has the lowest unit cost.

Exercise 3h

In the following questions, you must explain carefully the reason for your answers.

(1) Baked beans can be bought in the following sizes: 225 g for 21p, 450 g for 38p, and 1000 g for 86p. Which size is the best buy?

(2) Dried apricots can be bought in the following sized packets: 250 g for 70p, 500 g for £1.30, and 2000 g for £5. Which appears to be the best buy? What else might influence your answer?

(3) Corn flakes are available in the following sized packets: 300 g at 70p, 500 g at £1.10, 650 g at £1.40. Which is the best buy?

(4) Orange juice can be bought in small 200 ml cartons for 25p, large 500 ml cartons for 60p, or a 1 litre bottle for £1.08. Which is the best value for money?

(5) A PVA glue can be bought in 5 fl oz bottles for 86p, an 8 fl oz dispenser for £1.20 or a large tin containing 25 fl oz for £3.60. Which would you buy and why?

(6) Apples are available in packs of 4 for 58p, packs of 6 for 80p and bags of 3 kg containing about 14 apples for £1.70. Which would you buy?

3.10 Mixtures

Worked example 3.11

Mr. Patel mixes together two types of herb, X and Y. If X costs £3.20 per kg, and Y costs £1.80 per kg, how much per kg would a mixture of 3 parts of X to 2 parts of Y cost per kg?

The important thing to realise with this type of problem is that initially you do not have to worry about the cost of 1 kg.

Assume that you use 3 kg of X; total cost = 3 × £3.20
= £9.60
then use 2 kg of Y; total cost = 2 × £1.80
= £3.60

The total cost of 5 kg is therefore £9.60 + £3.60 = £13.20.
Then you can find the cost of 1 kg.
Hence the cost per kg = £13.20 ÷ 5 = £2.64.

Exercise 3i

(1) A brand of tea is mixed from two types A and B. A costs £4 per kg and B costs £5 per kg. How much per kg would a mix of tea cost made from 4 parts of A to 1 part of B?

(2) A paint is made by mixing white and blue. If the white costs £4 per litre, and

the blue costs £7 per litre, how much per litre would the paint cost if 2 parts of white are mixed with 5 parts of blue?

(3) Oil is mixed equally from an expensive brand costing £6 for 5 litres, and a cheap brand costing £3 for 4 litres. How much will the oil cost to make per litre?

3.11 Ratio

If you have been to a race course, you will probably have seen a board rather like the illustration above. You can see that Lucky Jim is quoted as 8/1. We normally say 8 to 1. What does this mean? It means that if you stake a £1 bet on Lucky Jim winning, then if the horse wins, you win £8. We say that the ratio of the winnings to the stake money is £8 to £1 or just 8 to 1.

In mathematics, this is written 8:1. If you had put £3 on Lucky Jim winning, then if the horse had won, you would have won £24. The ratio is then 24:3. Each number has been multiplied by 3, and we say that the ratios are equal. Hence:

$$24:3 = 8:1$$

We say that the ratio 24:3 in its lowest terms is 8:1. This is rather like simplifying a fraction.

Worked example 3.12

Reduce the following ratios to their simplest terms where possible:

(i) 30:5 (ii) 26:14 (iii) 9:4 (iv) $7\frac{1}{2}:2\frac{1}{2}$ (v) $2\frac{1}{3}:1\frac{1}{2}$

(i) 30:5 Dividing each number by 5 it becomes 6:1.

(ii) 26:14 Dividing each number by 2 it becomes 13:7.

(iii) 9:4 There is no number that divides exactly into 9 and 4, hence it cannot be simplified.

(iv) $7\frac{1}{2}:2\frac{1}{2}$ Dividing each number by $2\frac{1}{2}$ it becomes 3:1.

There are many people who would not see this, and it can be tackled in the following way:

Multiply each number by 2 gives 15:5

Now divide each number by 5 to give 3:1

(v) $2\frac{1}{3}:1\frac{1}{2}$ Multiply each number by 6 (why?).

$$6 \times 2\tfrac{1}{3} = 14 \qquad 6 \times 1\tfrac{1}{2} = 9$$

Hence: $2\frac{1}{3}:1\frac{1}{2} = 14:9$.

Exercise 3j

Reduce the following ratios to their simplest form:

(1) 28:4 (2) 18:3 (3) 40:60 (4) 16:40 (5) $12\frac{1}{2}:5$
(6) $\frac{1}{3}:\frac{1}{2}$ (7) $2\frac{1}{3}:3\frac{1}{2}$ (8) 20:14:12 (9) $7\frac{1}{2}:10:12\frac{1}{2}$ (10) $1\frac{1}{4}:3\frac{3}{4}$

3.12 Proportional parts

There are many situations where a given quantity has to be shared out in a certain way such that the parts are in a given ratio. This idea is known as dividing an amount into proportional parts.

Worked example 3.13

Jacob, David and Sulim have just finished tidying up Mr Jones' garden. Mr Jones had agreed that the price for the job should be £45. The number of hours worked by Jacob, David and Sulim was 8, 12 and 10 respectively. How should the payment be divided fairly between them?

The fairest way is if the money is in the same ratio as the hours worked, which is

$$8:12:10$$

This can be simplified by dividing by 2, i.e.

$$4:6:5$$

The total number of *parts* required is $4 + 6 + 5 = 15$, and each part is $£45 \div 15 = £3$.

Hence

Jacob receives $4 \times £3 = £12$

David receives $6 \times £3 = £18$

and Sulim receives $5 \times £3 = £15$

Total $= £45$

Exercise 3k

(1) Abdul, 10 years old, Mohammed, 12 years old, and Dave, 8 years old, are given £45 to share in the ratio of their ages. How much does each receive?

(2) A farmer has three fields, of areas 15 acres, 42 acres and 13 acres. If he has

14 tons of fertiliser to spread evenly on the fields, how much fertilizer is spread on each field?

(3) Ahmed, Jean and Karl share a taxi home from a party. Ahmed travels 4.8 miles, Jean 5.9 miles and Karl 6.8 miles. The cost of the taxi is £8.25 and the driver is given a 50p tip. They pay in proportion to the distance travelled. How much does each of them pay? (SEG)

(4) After a bad storm, the Borough Sports Club had a repair bill of £1750 to pay. The club decided to charge each of its three sections a proportion of the bill based on the size of their membership. The numbers in the three sections were Hockey 85, Squash 209, Tennis 406. Calculate how much each section had to pay. (SEG)

(5) A private road is to be made up in front of three houses. The length of frontage of the houses is 26 m, 24.5 m and 19.5 m. The total cost of the road is going to be £17 500. How much should each household pay? Is your method fair?

3.13 Using percentages

We have already seen in Section 1.7 how to calculate a percentage. We will now look at how to use percentages.

(a) To find a percentage of a quantity, proceed as follows.

Worked example 3.14

(i) Find 12% of £64.

(ii) Find 130% of 8 m 40 cm.

(i) 12% means $\frac{12}{100} = 12 \div 100 = 0.12$.

The word 'of' means × (multiply).

Hence 12% of £64 = 0.12 × £64 = £7.68.

(ii) 130% means $\frac{130}{100} = 130 \div 100 = 1.3$.

Hence 130% of 8 m 40 cm is 1.3 × 8.4 m = 10.92 m.

(b) To express one quantity as a percentage of another, you must make sure that all quantities are measured in the same units.

Worked example 3.15

(i) Express £1.20 as a percentage of £2.

(ii) Express 120 g as a percentage of 6.4 kg.

(i) Convert both to pence, i.e. 120p as a percentage of 200p.

The required % is (120 ÷ 200) × 100 = 60%.

(ii) Convert both to g, i.e. 120 g as a percentage of 6400 g.
The required % is $(120 \div 6400) \times 100 = 1.875\%$.

Exercise 3l

(1) Find 16% of £3.50.
(2) Find 12% of £84.
(3) Find 20% of 8 m 40 cm.
(4) Find $12\frac{1}{2}\%$ of 40 cm^2.
(5) Find 120% of £75 000.
(6) Find $87\frac{1}{2}\%$ of £6.25 (to the nearest penny).
(7) Increase £8 by 8%.
(8) Increase 4.6 cm by 10%.
(9) Express £12 as a percentage of £96.
(10) Express 40p as a percentage of £3.20.
(11) Express 450 g as a percentage of 2 kg.
(12) Express 200 ml as a percentage of 4 litres.
(13) The population of a village increases from 6200 to 6800. Express the increase as a percentage of its original value.
(14) Express 8.27×10^7 as a percentage of 9.4×10^8, giving 3 significant figures in your answer.
(15) Express 400 cm^2 as a percentage of 1 m^2 (careful).
(16) Increase 8.95×10^{-2} by 15%.
(17) Find 0.05% of two million five hundred thousand pounds.
(18) Increase eighty five thousand by 18%.
(19) Express 400 ml as a percentage of 1 m^3.

Suggestions for coursework 3

1. Design a chart that you might sell to the general public which would convert well used non-metric units into metric. For example, pounds to grams, gallons to litres.
2. Investigate scoring rates in a sport which interests you.
3. Do you think that travel timetables are clearly set out? Investigate and make suggestions for alterations.
4. What quantities in life do you think most people are good at estimating? Provide evidence.
5. Look at the problem of best buys. What advice would you give people?

Miscellaneous exercise 3

1. Two lights flash, one every 60 seconds, and the other every 24 seconds. The lights are both on together at 1200 h. How many times during the next hour, including 1200 and 1300, are they on at the same time?
2. Dave, John and Fred share a taxi home from a party. Dave travels 4.2 miles, John 3.6 miles and Fred 2.2 miles. The cost of the taxi is £6.20 and the driver is given a £1 tip. They pay in proportion to the distance travelled. How much does each of them pay? (Give sensible amounts.)

3. A farmer has three fields, of areas 17 acres, 23 acres and 25 acres. He has 13 tons of fertiliser which must be spread evenly over all three fields. How much fertiliser is to be spread over each field? (SEG)
4. The instructions on a bottle of dried milk read as follows:

   'To make one pint of fresh milk add 5 tablespoons of dried milk powder to one pint of cold water and stir thoroughly. Each bottle makes 5 pints of fresh milk.'

   Rita uses the dried milk powder in the following ways:

Use	Amount
Fresh milk	5 tablespoons per pint
Tea	1 teaspoon per cup
Coffee	2 teaspoons per cup

   On average Rita makes one pint of fresh milk per week and drinks two cups of tea and two cups of coffee a day. How long, on average, will a bottle of dried milk last Rita? (1 tablespoon = 5 teaspoons.)
5. Figure 3.4 shows a gear wheel of width 21 mm fixed on to an axle which protrudes 6.54 cm on one side of the wheel and 4.2 cm on the other side. What is the total length of the axle?

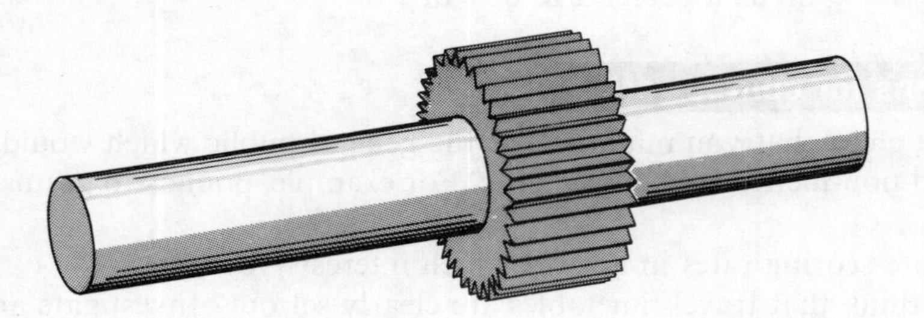

Figure 3.4

6. A large drum of oil contains 2850 litres of oil. This oil is used to fill small tins that contain 800 cm^3 of oil. How many cans can be filled from the drum?
7. A ball of string is 138.6 m long. David has been asked to cut a number of pieces of string each 2.4 m long. Calculate:
   (i) the number of pieces that he can cut;
   (ii) the length of string remaining.
8. Washco washing powder is sold in 900 g packets which cost 79p each. The instructions on the packet state:

   1 cup of powder weighs approx. 100 g ($3\frac{1}{2}$ oz)
   Quantity to use:

	Hard Water	Soft Water
Top loading automatics		
8 gallon size	3 cups	2 cups
10 gallon size	4 cups	3 cups
15 gallon size	5 cups	4 cups
Soak and handwash		
Average sink (5 gallons)	2 cups	$1\frac{1}{2}$ cups
Average bowl or bucket (2 gallons)	$\frac{3}{4}$ cups	$\frac{1}{2}$ cup

(i) What is the approximate cost of the washing powder for a 15 gallon size wash in a soft water area?

(ii) Alex lives in a hard water area and always uses the soak and handwash method, using the sink every other week for a large wash and a bowl twice every week for a small wash. On average how many weeks will one packet of Washco washing powder last Alex? (NEA)

9. In a Bowls league, each match consists of 3 rinks.
Points are awarded as follows:

4 points for every match won.
1 point for every rink won.

After each team had played 3 matches, the league table was as follows:

Team	Matches			Rinks		Points
	played	won	lost	won	lost	
Peakirk	3	3	0	8	1	20
Northborough	3	2	1	5	4	13
Glinton	3	1	2	4	5	8
Castor	3	0	3	1	8	1

The results of the next matches were as follows:

Peakirk beat Glinton by 3 rinks to 0,
Castor beat Northborough by 2 rinks to 1.

Complete the new league table.

Team	Matches			Rinks		Points
	played	won	lost	won	lost	
Peakirk	4					
Northborough	4					
Glinton	4					
Castor	4					

(MEG)

10. A grocer buys 16 kilograms of one type of tea at £2.56 per kilogram and 10 kilograms of another type at £2.34 per kilogram.
He mixes the two types of tea together and weighs them into 250 g bags.
At what price should he sell the 250 g bags to make a profit of 60% on his cost? (LEAG)

11. Taking 8 kilometres per hour to be 5 miles per hour, find:
(a) The speed in kilometres per hour equivalent to the British speed limit of 30 miles per hour.
(b) The speed in miles per hour equivalent to the French speed limit of 60 kilometres per hour.

12. At a mushroom factory, 8 people can pack 30 boxes in 20 minutes. Assuming on average everyone is working at the same rate, find:
(i) How long it would take 6 people to pack the same number of boxes.
(ii) How long it would take 5 people to pack 40 boxes.

13. The table below shows the cost of crossing the channel to France.

FARES Standard Single Fares	**E** TARIFF £	**D** TARIFF £	**C** TARIFF £	**B** TARIFF £
Cars, Minibuses and Campers				
Up to 4.00 m (13′ 1″) in length/Motorcycle combination	20 00	30 00	39 50	48 00
Up to 4.50 m (14′ 9″) in length	20 00	35 00	47 50	57 00
Up to 5.50 m (18′) in length	20 00	40 00	54 50	67 00
Over 5.50 m per extra metre or part there-of	10 00	10 00	11 00	12 00
Caravans and Trailers				
Up to 4.00 m (13′ 1″) in length	19 00	24 00	26 00	28 00
Up to 5.50 m (18′) in length	19 00	36 00	38 00	40 00
Over 5.50 m per extra metre or part there-of	10 00	10 00	10 00	10 00
Motorcycles, Scooters and Mopeds	11 00	11 00	12 00	13 00
Each Adult	11 50	11 50	11 50	11 50
Each Child (4 but under 14 yrs)	6 00	6 00	6 00	6 00

(a) Brian and Anne, with children Alison (8 years) and Julie (2 years), plan a camping holiday in June which is on Tariff C. Brian carries his camping equipment in a trailer. His car is 4.2 m long and the trailer is 2.4 m. Find the cost of the crossing.

(b) Joel and Rose, with children Jasmine (15 years) and Winston (6 years), plan a holiday for August which is on Tariff B. Their car is 5.2 m long and the caravan 5.5 m. Find the cost of the crossing.

(c) The exchange rate in June is 10.8 fr. to the £. Brian and Anne change £200 into francs. How many francs do they receive?

(d) The exchange rate in August is 11.2 fr. to the £. When Joel and Rose return to England they have 896 francs to change back to £ sterling. How much do they receive?

14. Two machines are producing components in a factory. One produces a component at the end of every 42 seconds, the other produces a component at the end of every 90 seconds. The two machines are started together; after how many seconds do they first produce a component at exactly the same time? (SEG)

15. The area of England and Wales is approximately 15 million hectares. Express this answer in square kilometres.

16. The age of the earth is thought to be about $4\frac{1}{2}$ thousand million years. Man evolved about 1 million years ago. Express this as a percentage of the life of the earth.

17. The table below is taken from the census data for 1961 and 1971 in England and Wales.

Census data, England and Wales, totals and by regions

	Population (millions)		Pensioners (millions)		Women working (%)		People in car-less households (%)		No. of households (millions)	
	1961	1971	1961	1971	1961	1971	1961	1971	1961	1971
England and Wales	46.1	48.8	6.9	8.0	42	43	49	42	14.6	16.5
Regions										
Greater London	8.00	7.45	1.20	1.19	50	50	51	46	2.66	2.65
Rest of South East	8.20	9.78	1.23	1.74	38	41	37	32	2.69	3.26
East Anglia	1.47	1.67	0.23	0.28	37	39	40	32	0.47	0.57
East Midlands	3.21	3.39	0.45	0.54	43	43	48	41	1.00	1.14
Yorks & Humberside	4.60	4.80	0.69	0.77	41	42	67	60	1.60	1.65
North	3.25	3.30	0.46	0.53	38	40	69	62	1.00	1.10
North-west	6.57	6.74	0.99	1.08	45	44	56	48	2.10	2.27
West Midlands	4.76	5.11	0.02	0.72	46	45	46	41	1.45	1.68
Wales	2.64	2.73	0.40	0.46	34	36	49	40	0.80	0.90
South-west	3.41	3.78	0.58	0.72	37	38	40	32	1.07	1.28

(*Sources*: *Census 1971*, *Census Matters*)

(i) What was the population in millions of Greater London in 1971?

(ii) What was the increase in population in millions in the East Midlands from 1961 to 1971?

(iii) In 1971 which region had the greatest percentage of women at work?

(iv) The mean household size is given by

$$\text{Mean household size} = \frac{\text{number of people} \quad \text{(Column 1)}}{\text{number of households} \quad \text{(Column 5)}}$$

What was the mean household size in Wales in 1961?

18. Look at the train timetable shown below.

London to Brighton

		(1) C	(1) C	(1) C	A	(1) CF	(1) CF	(1) CF	(1) F	A	(1) CR
London Victoria	*d*	01 00	04 00	05 00	05 47c	05 47	06 06	06 17	06 36c	07 06c	07 06
London Bridge	*d*	—	—	—	05 52	—	—	—	06 45	07 17	—
East Croydon	*d*	01 17	04 17	05 17	06 08	06 04	06 25	06 34	07 02	07 33	07 29
Brighton	*a*	02 23	05 18	06 41g	06 53	07 13	07 29	07 45	08 09	06 27	08 47
		(1) CF	(1)	(1) C	(1)	(1)	(1) C	(1) J	(1)	A	(1) C
London Victoria	*d*	07 50	—	08 32	—	09 02	09 32	—	10 02	—	10 32
London Bridge	*d*	07 25c	08 03	08 20c	08 32	08 42c	—	09 32	—	10 02	10 32g
East Croydon	*d*	08 07	08 17	08 49	08 55	09 17	09 49	09 53	10 17	10 16	10 49
Brighton	*a*	08 55	09 17	09 45	10 02	09 54	10 46	10 50	10 53	11 15	11 45
		(1) C	(1)	A	(1) C			(1) C	A	(1) C	(1) C
London Victoria	*d*	10 56	11 06	—	11 32	and at the		16 05	—	16 20	16 35
London Bridge	*d*	—	11 02c	11 02	11 32g	same minutes		16 02c	16 02	16 07b	16 22c
East Croydon	*d*	11 15	11 21	11 16	11 49	past each		16 22	16 16	16 37	16 53
Brighton	*a*	12 03	11 57	12 15	12 45	hour until		17 03	17 15	17 30b	17 46
		C	(1) C	(1) F	(1) C	(1) C	(1)	(1)	(1) C	A	(1) C
London Victoria	*d*	16 40	17 07	17 23b	17 09	17 36	18 08	18 32b	19 05	—	19 32
London Bridge	*d*	16 47b	17 02b	17 33	—	17 36c	18 02b	18 36	19 00c	19 00	19 32c
East Croydon	*d*	16 57	17 24	17 47	17 27	17 54	18 23	18 52	19 22	19 15	19 49
Brighton	*a*	18 05	18 14	18 35	18 39	18 44	19 11	19 49	20 02	20 16	20 45
		(1)	A	(1) C	A	(1) C	A	(1) C	(1)	(1) CR	(1) CF
London Victoria	*a*	20 06	—	20 32	21 00g	21 32	22 00g	22 32	23 06	23 21	23 59
London Bridge	*d*	20 02c	20 02	20 12c	21 02	21 12c	22 02	22 32c	23 02c	—	23 47c
East Croydon	*d*	20 21	20 16	20 49	21 16	21 49	22 16	22 51	23 21	23 38	00 18
Brighton	*a*	21 01	21 15	21 42	22 15	22 42	23 15	23 44	00 02	00 37	01 09

(i) What times does the 1206 from Victoria arrive at Brighton?

(ii) How many trains leave London Bridge for Brighton between 1102 and 1602 inclusive?

(iii) What is the latest time train you should catch from Victoria, to arrive at Brighton by 2100 h at the latest?

19.

To convert Imperial units into SI units multiply by			
Length	in	mm	25.4
Volume	in^3	mm^3	16 387
Density	lb/ft^3	kg/m^3	16.0185
Velocity	ft/s	m/s	0.3048

Using the above table of conversion factors, complete the following table. Write all the figures shown by your calculator.

Imperial units	SI units
17 in	mm
490 lb/ft^3	kg/m^3
in^3	275 800 mm^3
ft/s	48 m/s

20. The map below is a simplified version of how the world is divided into international time zones.

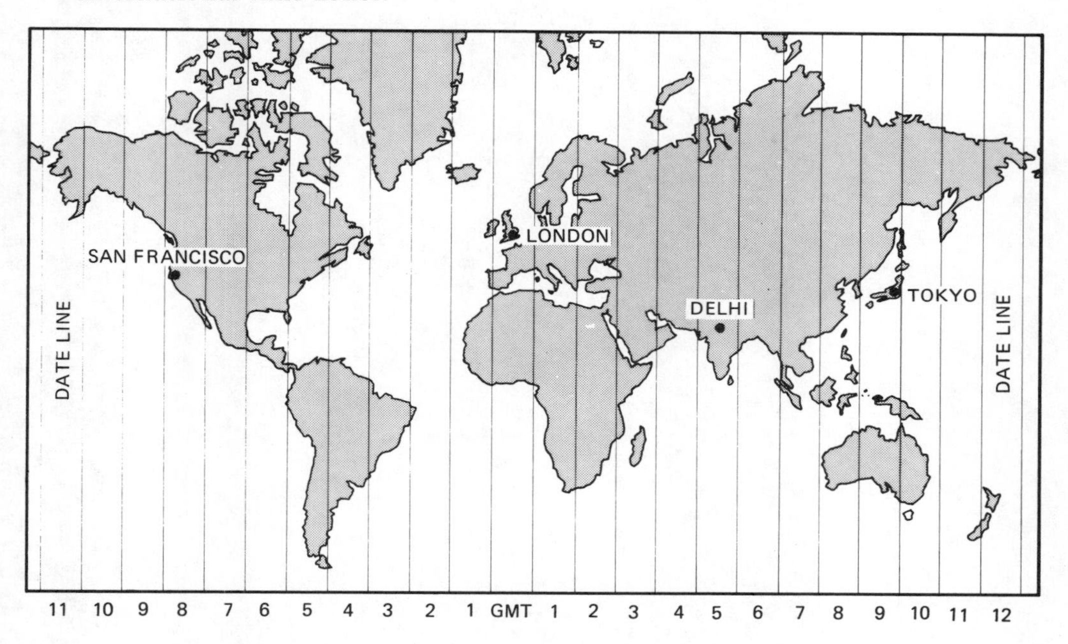

(a) When it is 5 p.m. GMT in London, what time is it in San Francisco?

(b) The Tokyo Stock Exchange closes at 3.30 p.m. Japanese time. When the London Stock Exchange opens at 9.00 a.m. GMT, for how long has the Tokyo Stock Exchange been closed?

(c) Mr Surajah is to return to Delhi on a flight lasting $8\frac{1}{2}$ hours, leaving London Heathrow at 0215 hours GMT. What time will it be in Delhi when he arrives? (NEA)

PERSONAL FINANCE AT HOME AND AT WORK

4

This unit covers the essential ideas necessary to handle money in everyday situations. You should try to master all of this unit, since the ideas are likely to be contained in any post 16 examination.

4.1 Percentages and money

In units 1 and 3, the ideas of percentages were introduced. The main area where percentages are used in everyday life is in financial transactions. To find a given percentage of an amount of money, you proceed as follows.

Worked example 4.1
Find 12% of £10.80.

This means, find $\frac{12}{100}$ of £10.80.
(The word 'of' can be replaced by $\times$.)

$$\frac{12}{100} \times \frac{£10.80}{1} = \frac{£12 \times 10.80}{100} = \frac{£129.6}{100}$$

$$= £1.29(6)$$

This would either be rounded up to £1.30, or rounded down to £1.29 depending on the policy of the shop or financial institution.

In many cases nowadays, percentages are used in describing wage settlements. The techniques are illustrated in the following example:

Worked example 4.2
Jasmin has just received an 8% pay rise. She used to earn £136 per week. What is her new weekly wage?

This type of problem can be tackled in two ways. The two methods are given below.

Method 1
Find 8% of £136, i.e.

$$\frac{8}{100} \times \frac{£136}{1} = \frac{£1088}{100} = £10.88$$

Her new weekly wage = £136 + 10.88 = £146.88.

Method 2
An increase of 8% gives a total of 108% of the original amount.
Hence we find 108% of £136, i.e.

$$\frac{108}{100} \times \frac{£136}{1} = \frac{£14688}{100} = £146.88$$

4.2 VAT

VAT is short for *value added tax*. At the time of writing this is 15% which is added on to the price of an article. When checking the price of anything you buy, always ask whether or not it includes VAT.

Worked example 4.3
Find the VAT payable on a packet of sweets costing 38p before VAT, if the current rate is 15%, and hence the retail price of the sweets.

We need to find 15% of 38p, i.e.

$$\frac{15}{100} \times \frac{38\text{p}}{1} = \frac{570}{100} = 5.7\text{p}$$

We have the problem of whether to call this 5p or 6p. If calculations are to the nearest penny, then we have the VAT = 6p.
Hence, the retail price is 38p + 6p = 44p.

4.3 Profit and loss

David went to a jumble sale, and bought a bike for £12. When he got it home, his friend saw it and said he would give him £20 for it. David decided to see what percentage profit he would make before selling it.
He found the Profit = Selling Price − Cost Price
= £20 − £12 = £8

$$\text{His percentage profit} = \frac{\text{Profit}}{\text{Cost}} \times \frac{100}{1} = \frac{8}{12} \times \frac{100}{1} = 66.7\%$$

He reckoned this was a good percentage profit, and decided to sell the bike. Do you think he made a good decision?

The Burrows family had just bought a house for £85 000. Unfortunately, Mr Burrows had to move house quickly to a new job. Houses in the area had suddenly dropped in value, and they had to sell the house for £75 000. What percentage loss did they suffer on the house?

In this case, the Loss = £85 000 − £75 000 = £10 000

$$\text{The percentage loss} = \frac{\text{Loss}}{\text{Cost}} \times \frac{100}{1} = \frac{10\,000}{85\,000} \times \frac{100}{1}$$
$$= 11.8\%$$

You will notice this time that the percentage is much smaller than in the first example, but the actual amount involved was £10 000 as opposed to £8.

You must always use percentages with great care.

4.4 Backwards percentage problems

The title of this section is to try and help you remember a certain type of % problem which people very often answer incorrectly. The situations we are looking at here are where you are given the value of a quantity *after* a percentage change, and you have to work backwards to find its value before the change took place. Although the examples given here are about money, the techniques are the same for any similar situation.

Worked example 4.4

Nadia sold her bike for £76.80, making a 20% loss on how much she paid for it. What was the original cost of the bike?

You must *not* find 20% of £76.80, and add it back on.

Since Nadia has lost 20% of the cost price, she now only has (100 − 20)%, i.e.

80% of the original cost price

Hence	80% = £76.80
÷ 80	1% = £ 0.96
× 100	100% = £96.00

The original cost price is £96.

Worked example 4.5

After a 9% pay rise, Colin's annual salary was £9156. How much did he earn before the rise?

As in the last example, you do *not* find 9% of £9156 and subtract. Colin has 100% + 9% = 109% of his original salary after the rise.

So 109% = £9156

÷ 109 1% = £ 84

× 100 100% = £8400

His original salary was £8400.

Exercise 4a

(1) Find 8% of £64 (2) Find 12% of £86 000
(3) Find $12\frac{1}{2}$% of £8.40 (4) Find 36% of £120
(5) Find 80% of 6p (6) Find $33\frac{1}{3}$% of £96.30
(7) Find 10% of 25p (8) Find 16% of £836
(9) Find 11% of £65 000 (10) Find $22\frac{1}{2}$% of £8.43
(11) Jim earns £128 a week and he has been offered a 9% pay rise. What would his new weekly wage be?
(12) The cost of a pint of beer was increased from 80p to 85p. Express this as a percentage increase.
(13) Find the VAT payable on a toy car priced at £4.30 before VAT is added.
(14) After an 8% pay rise, Ann earns £135 per week. How much did she earn before the rise?
(15) A house costs £86 000. If prices rise by 12% in a year, how much is it likely to cost a year later?
(16) In a sale, all prices were reduced by 16%. Find the cost of a book which originally sold for £6.25.

4.5 Simple interest

If a sum of money, called the Principal (P) is invested for a number of years (T) at an annual percentage interest rate of R% per annum, then the total interest earned (I) is given by the formula:

$$I = \frac{PRT}{100} \text{ (if calculated as simple interest)}$$

This formula can be rearranged using the techniques of section 2.7 to give:

$$P = \frac{100I}{RT}, \qquad R = \frac{100I}{PT}, \qquad T = \frac{100I}{PR}$$

Worked example 4.6

Find the simple interest earned on £800 deposited for 2 years 3 months at an annual percentage rate of 8.4%.

Here we have $P = 800$, $R = 8.4$, $T = 2\frac{1}{4}$ (since 3 months is $\frac{1}{4}$ year). Hence,

$$I = \frac{£800 \times 8.4 \times 2.25}{100} \quad \left[\text{Notice that } 2\tfrac{1}{4} \text{ has been written as a decimal.}\right]$$

$$= £151.20$$

Worked example 4.7
Joan has invested £600 for a certain length of time at 7.5% per annum (simple interest). At the end of this time, she found that she had £725. Find to the nearest month the length of time that the money had been invested.

We have this time, $P = 600$, $R = 7.5$, $I = 725 - 600 = 125$. Using $T = \dfrac{100I}{PR}$:

$$T = \frac{100 \times 125}{600 \times 7.5} = \frac{12\,500}{4500} = 2.78$$

Since 1 year = 12 months:

$$0.78 \text{ year} = 0.78 \times 12 = 9.36 \text{ months}$$

To the nearest month, time = 2 years 9 months.

4.6 Compound interest

In most financial calculations, interest is added each year (or more frequently) and interest is then added to the new total.

Worked example 4.8
Tim invested £1000 in a building society, which was paying interest calculated six-monthly at 3%. How much was his investment worth after 3 years if the interest rate remained the same?

After 6 months, interest = £1000 × 0.03 = £30.
The total = £1000 + £30 = £1030.
After 12 months, interest = £1030 × 0.03 = £30.90.
The total = £1030 + £30.90 = £1060.90.
After 18 months, interest = £1060.90 × 0.03 = £31.83.
The total = £1060.90 + £31.83 = £1092.73.
After 24 months, interest = £1092.73 × 0.03 = £32.78.
The total = £1092.73 + £32.78 = £1125.51.
After 30 months, interest = £1125.51 × 0.03 = £33.77.
The total = £1125.51 + £33.77 = £1159.28.
After 36 months, interest = £1159.28 × 0.03 = £34.78.
The total = £1159.28 + £34.78 = £1194.06.

I am sure you would agree that this method is rather long. If you can cope with the following formula, the method is considerably shortened. The amount of the investment $= P\left(1 + \dfrac{r}{100}\right)^n$, where P is the amount invested, r is the interest rate, and n is the number of occasions that the interest is calculated.

In worked example 4.8, we have the following:

$$P = £1000,\ r = 3\%,\ n = 6$$

$$\text{The final amount} = £1000\left(1 + \frac{3}{100}\right)^6$$

$$= £1000 \times 1.03^6$$

Using the calculator, with a power button, we get

						Display	
1000	[×]	1.03	[x^y]	6	[=]	[1194.05]	i.e. £1194.05

The answers differ by 1p!

The small error in the first method has come from repeated approximations.

Exercise 4b

(1) Find the simple interest earned on £850 if invested for 3 years at an annual percentage interest rate of 6%.

(2) Find the simple interest earned on £640 if invested for 8 months at an annual percentage interest rate of 8%.

(3) David has invested £280 for a certain length of time at an annual percentage interest rate of 6.5% (simple interest). At the end of this time, the value of his investment was £328.60. Calculate to the nearest month, the length of time the money was invested.

(4) Paula owed her father £400. He agreed to let her repay the money in three years' time, but said he would charge her simple interest, so that she would then owe him £450. What annual percentage interest rate was he charging?

(5) Jasmin had invested the money she obtained from selling her car in the building society for 5 months. Interest on the account was calculated as simple interest with an annual percentage rate of 6.8%. If she earned £100 interest, how much did she get for her car? After some further calculation, she found that she had lost 30% on the price she paid for it. How much did she originally pay for the car?

(6) Find how much £2000 becomes if invested in a savings scheme for 4 years if interest at 8% is calculated yearly on a compound basis.

(7) The value of a car depreciates by 15% each year. If it cost £5800 when new, how much is it worth after 3 years?

(8) A building society offers 4.5% interest compounded 6 monthly if money is saved for at least 3 years. Find how much an investment of £850 would be worth after 3 years.

(9) Over the last 5 years, the value of a house has appreciated on average by 15% per year. The Robinson family bought their house 5 years ago for £42 000. How much is it worth now?

(10) The price of a loaf of bread is 65p. If prices increase on average by 6% each year, how much should a loaf of bread cost in 20 years' time?

(11) The Peters family have just sold their house for £120 000. Mr Peters invests the money in a building society account which pays interest at 8% per

annum calculated 3 monthly. The family then go abroad for 2 years. How much is the investment worth when they return?

4.7 Earning a living

When you are looking at an advertisement for a job, one of the first things you are likely to want to know is how much you will earn. The main ways in which people are paid come under the headings:

(i) *Salaried*: the amount you earn each year (per annum, abbreviation p.a.) is usually stated, and this will then be paid in 12 equal monthly amounts.

(ii) *Hourly*: the amount you will earn per hour is usually stated for a certain number of hours per week. Any hours worked over this number (overtime) will usually be paid at a higher rate.

(iii) *Commission*: a person is usually paid a fixed % of sales that he or she makes. Some jobs carry a basic wage plus commission.

When considering how much you will earn in a particular job, it is also worth considering what extra benefits the job may carry, such as a car, cheap mortgages, travel subsidies, and so on.

Worked example 4.9

Tim and Nadine both work as sales assistants. Nadine gets paid £3.40 per hour for a 35 hour week, and during the year she works a total of 100 hours overtime for which she is paid time and a half. Tim gets a salary of £6000 and commission of 5% of his sales during the year. If Tim sells goods to the value of £25 000 during the year, calculate the average weekly wage that Tim and Nadine earn during the year.

In any week, Nadine earns a basic wage of $35 \times £3.40 = £119$. During the year, the overtime she works bring in:

$$£3.40 \times 1.5 \times 100 = £500$$

↑ 'time and a half'

On average each week, she earns $£510 \div 52 = £9.81$ overtime. Her average weekly wage is $£119 + £9.81 = £128.81$.

The commission earned by Tim $= £25\,000 \times \dfrac{5}{100} = £1250$.

His earnings in a full year are $£6000 + £1250 = £7250$.

His average weekly wage $= £7250 \div 52$

$= £139.42$.

(iv) *Piecework*: although this method of payment is less common now, some people (often part-time workers at home) are paid solely for each item that they produce.

Worked example 4.10

Kathy makes gloves at home. She earns 25p for each pair that she makes. If she can make 12 pairs of gloves in an hour, how many hours does she have to work to earn at least £80 per week.

In one hour, she earns 25p × 12 = 300p
= £3.

The number of hours she must work = 80 ÷ 3
= 26.7.

In other words, she must work 27 hours.

Exercise 4c

(1) Desai earns £9850 a year. How much is this a week? (Answer to the nearest penny.)

(2) Eve earns £136 per week, but averages 5 hours' overtime each week for which she is paid time and a half. If her normal hourly rate is £3.40, how much will she earn in 1 year?

(3) A salesman is paid a basic wage of £8500 a year plus 3% of all sales over £15 000. In 1988, his total sales for the year were £40 000. How much did he earn in the year?

(4) Tina assembles model cars on a production line. She is paid 15p for each car, and on a good day can assemble 22 cars an hour. How much could she expect to earn for a 36 hour week if she works at this same rate?

(5) Joan assembles plastic flowers. She is paid a basic wage of £40 per week and 3p for every flower she assembles. If in a working week she earned £112, how many flowers did she assemble?

(6) Olive earns £624 a calendar month. How much does she earn on average per week?

4.8 Income tax

Everyone is liable to pay income tax, if they earn enough money. Income tax is calculated after certain allowances have been made. These allowances include a personal allowance that everyone gets and other allowances, for example superannuation and national insurance, which are beyond the scope of this section. The remaining amount after these deductions is called your taxable income. The amount of tax you then pay is calculated according to Table 4.1 (correct at the time of writing).

Worked example 4.11

Wasim earns £158 per week. He claims the single person's allowance. How much tax will he pay each week?

Table 4.1

Taxable income	Rate of tax
£1 to £19 300	25% (standard rate)
Over £19 300	40% (higher rate)
Single person's allowance £2605	
Married person's allowance £4095	

£158 per week = £158 × 52 = £8216 per year − £2605 allowance

= £5611

Tax at 25% is £5611 × 0.25 = £1402.75.
His weekly tax payment = £1402.75 ÷ 52 = £26.98.

Worked example 4.12
Sheena earns £27 500 a year as a computer programmer. She claims allowances totalling £3050. Find her tax bill for the year, and express this as a % of her income.

Amount to be taxed = £27 500 − £3050 = £24 450.
The first £19 300 @ 25% = £19 300 × 0.25 = £4825.
£24 450 − £19 300 = £5150 @ 40%
= £5150 × 0.4 = £2060.
Her total tax bill = £4825 + £2060 = £6885.

The percentage of income $= \frac{£6885}{£27\,500} \times 100 = 25\%$.

Exercise 4d
In the following examples, calculate (a) the taxable income; and (b) the income tax paid during the year, assuming the rates of Table 4.1 apply.
(1) Tim earns £6250 per year. His total allowance is £2625.
(2) Peter earns £8375 per year. His total allowance is £3150.
(3) June earns £195 per week. Her total allowance is £2850.
(4) Wasim earns £12 800 per year. His total allowance is £3500.
(5) Sheena earns £38 000 per year. Her allowances total £5750.
(6) Tony earns £120 000 per year. His allowances come to a total of £11 250.

4.9 Money abroad

If you travel abroad, you will need to obtain the currency used in the country you are visiting. The easiest way of doing this is to buy travellers' cheques before you go, and exchange them for cash when you arrive. On your return, you then change any money left over back into pounds sterling. You will usually see the exchange rates set out in the form of a table headed 'buy' and 'sell'. The column headed 'sell' is used when travelling from this country, and the one headed 'buy'

is used when you return. You will notice the exchange rates are different in each column, which means you get slightly less for your money on the return. This is because the bank or 'bureau de change' is taking a small commission. Exchange rates are changing on a daily basis, and you should always watch them carefully to decide the best time to change money. The following example and exercise are based on the amounts shown in Table 4.2.

Table 4.2 Table of exchange rates for £1 sterling

	Sell	Buy
Austria (Schillings)	19.40	21.20
Belgium (francs)	60.30	62.50
Denmark (kroner)	10.41	11.15
France (francs)	9.85	10.36
Germany (Deutsche Marks)	2.95	3.15
Greece (drachmas)	191.20	198.50
Italy (lire)	1984.00	2036.00
Netherlands (guilders)	3.25	3.46
Spain (pesetas)	184.00	191.20
United States (dollars)	1.74	1.78

Worked example 4.13

Paul and Susan went to Greece for a week. They took £450 spending money with them which they changed on arrival into drachma. During the week they spent a total of 60 000 dr. The remaining money they changed into sterling when they returned home. How much did they actually spend during the week?

On the way out, they buy at £1 = 191.20 dr.

Note: you use the column headed 'sell', because the bank is selling to you.

So £450 = 450 × 191.20 dr = 86 040 dr.

The money they have left = 86 040 dr − 60 000 dr
= 26 040 dr.

On arrival back home, we use £1 = 198.50 dr.
So 26 040 ÷ 198.50 = 131.18.
This means they have £131.18.
The amount spent = £450 − £131.18 = £318.82.

Worked example 4.14
Thomas wanted to buy a special engine for his model railway. If he bought it in this country, it would cost £54. The engine was made in Austria and while on holiday there he found it was priced at 600 sch. What percentage does he save by buying it in Austria?

The exchange rate for schillings is £1 = 19.40 sch.
So 600 sch = £600 ÷ 19.4 = £30.92.
His saving = £54 − £30.92 = £23.08.

$$\text{The percentage saving} = \frac{£23.08}{£54} \times 100$$
$$= 42.7\%.$$

Worked example 4.15
Using the information from the 'sell' column in Table 4.2, change 200 Deutsche Marks into pesetas.

Although sterling is not directly involved in this calculation, it is easier to change into pounds sterling first. Hence:

$$200 \text{ DM} = \frac{£200}{2.95}$$

Into pesetas: $$\frac{£200}{2.95} \times 184 = 12\,475 \text{ pesetas}$$

Note: when changing from one currency into another, most banks and money changers will only change from money given in note form.

In the following exercise, use the exchange rates given in Table 4.2.

Exercise 4e
(1) Carlos went to Spain with £250 in travellers' cheques. How many pesetas can he get for his money?
(2) While on holiday in France, the Jones' family cashed cheques worth £100, £85, £146 and £83. How many francs did they receive in total?
(3) Sulim bought his mother a clock for 85 DM while on holiday in Germany. How much is this in sterling?
(4) Anita went to New York to visit some friends. She took £300 with her, which

she changed into dollars on arrival. While there, she spent \$475. On returning to this country, she changed the remaining money back into sterling. How much did her visit cost her altogether in sterling?

(5) Helen travelled from England to Spain via France. She left England with £580 sterling. In France, she spent 300 fr. The rest of the money she changed into pesetas on arriving in Spain. How many pesetas did she get for her money?

(6) Using the information in the 'sell' column of Table 4.2, find the value of the following:
(i) \$100 in lire; (ii) 400 DM into Kroner;
(iii) 400 Belgian francs into drachma;
(iv) 200 guilders into pesetas.

4.10 Borrowing money

(i) Bank or finance company loan

A bank will lend you money for a variety of things. The amount they will lend will basically depend on how much you can afford to repay each month. The normal period of repayment is usually up to 3 years, although it can be longer for certain items. You can usually calculate the amount you will have to pay each month by reading from a repayment table like that shown in Table 4.3.

Worked example 4.16

Joy wants to buy a complete sound system. She needs to borrow £675 and would like to repay the money over a two-year period. Using the information in Table 4.3, find her monthly payment, and the total amount of interest she will have to pay.

You will notice that £75 does not appear in the table, so you proceed as follows:

£600 over 24 months (2 years) is 29.98 per month

Since £75 is $\frac{1}{4}$ of 300:

the repayment for £75 over 24 months is $14.99 \div 4 = £3.75$

The total monthly repayment would be

$$£29.98 + £3.75 = £33.73$$

It follows that over 24 payments, the amount she pays back to the bank will be

$$24 \times £33.73 = £809.52$$

 The interest paid $= £809.52 - £675 = £134.52$.

Table 4.3

	12 MONTHS APR 19.7%			24 MONTHS APR 19.7%			36 MONTHS APR 19.7%			48 MONTHS APR 19.7%			60 MONTHS APR 19.7%		
Amount of Loan	Monthly Repay-ments £	Total Interest £	Total to Repay £	Monthly Repay-ments £	Total Interest £	Total to Repay £	Monthly Repay-ments £	Total Interest £	Total to Repay £	Monthly Repay-ments £	Total Interest £	Total to Repay £	Monthly Repay-ments £	Total Interest £	Total to Repay £
£300	27.52	30.24	330.24	14.99	59.76	359.76	10.86	90.96	390.96	8.83	123.84	423.84	7.63	157.80	457.80
£400	36.69	40.28	440.28	19.99	79.76	479.76	14.48	121.28	521.28	11.77	164.96	564.96	10.18	210.80	610.80
£500	45.86	50.32	550.32	24.99	99.76	599.76	18.10	151.60	651.60	14.71	206.08	706.08	12.72	263.20	763.20
£600	55.04	60.48	660.48	29.98	119.52	719.52	21.72	181.92	781.92	17.66	247.68	847.68	15.27	316.20	916.20
£700	64.21	70.52	770.52	34.98	139.52	839.52	25.34	212.24	912.24	20.60	288.80	988.80	17.82	369.20	1069.20
£800	73.38	80.56	880.56	39.98	159.52	959.52	28.96	242.56	1042.56	23.54	329.92	1129.92	20.36	421.60	1221.60
£900	82.56	90.72	990.72	44.98	179.52	1079.52	32.59	273.24	1173.24	26.49	371.52	1271.52	22.91	474.60	1374.60
£1000	91.73	100.76	110.76	49.98	199.52	119.52	36.21	303.56	1303.56	29.43	412.64	1412.64	25.45	527.00	1527.00
£1100	100.90	110.80	1210.80	54.97	219.28	1319.28	39.83	333.88	1433.88	32.37	453.76	1553.76	28.00	580.00	168.00
£1200	110.08	120.96	1320.96	59.97	239.28	1439.28	43.45	364.20	1564.20	35.32	495.36	1695.36	30.54	632.40	1832.40
£1300	119.25	131.00	1431.00	64.97	259.28	1559.28	47.07	394.52	1694.52	38.26	536.48	1836.48	33.09	685.40	1985.40
£1400	128.43	141.16	1541.16	69.97	279.28	1679.28	50.69	424.84	1824.84	41.21	578.08	1978.08	35.64	738.40	2138.40
£1500	137.60	151.20	1651.20	74.97	299.28	1799.28	54.31	455.16	1955.16	44.15	619.20	2119.20	38.18	790.80	2290.80
£1600	146.77	161.24	1761.24	79.96	319.04	1919.04	57.93	485.48	2085.48	47.09	660.32	2260.32	40.73	843.80	2443.80
£1700	155.95	171.40	1871.40	84.96	339.04	2039.04	61.55	515.80	2215.80	50.04	701.92	2401.92	43.27	896.20	2596.20
£1800	165.12	181.44	1981.44	89.96	359.04	2159.04	65.18	546.48	2346.48	52.98	743.04	2543.03	45.82	949.20	2749.20
£1900	174.29	191.48	2091.48	94.96	379.04	2279.04	68.80	576.80	2476.80	55.92	784.16	2684.16	48.36	1001.60	2901.60
£2000	183.47	201.64	2201.64	99.96	399.04	2399.04	72.42	607.12	2607.12	58.87	825.76	2825.76	50.91	1054.60	3054.60
£2100	192.64	211.68	2311.68	104.95	418.80	2518.80	76.04	637.44	2737.44	61.81	866.88	2966.88	53.46	1107.60	3207.60
£2200	201.81	221.72	2421.72	109.95	43.8.80	2638.80	79.66	667.76	2867.76	64.75	908.00	3108.00	56.00	1160.00	3360.00
£2300	210.99	231.88	2531.88	114.95	458.80	2758.80	83.26	698.08	2998.08	67.70	949.60	3249.80	58.55	1213.00	3513.00
£2400	220.16	241.92	2641.92	119.95	478.80	2878.80	86.90	728.40	3128.40	70.64	990.72	3390.72	61.09	1265.40	3665.40
£2500	229.33	251.96	2751.96	124.95	498.80	2998.80	90.52	758.72	3258.72	73.59	1032.32	3532.32	63.64	1318.40	3818.40
£2600	238.51	262.12	2862.12	129.95	518.80	3118.80	94.14	789.04	3389.04	76.53	1073.44	3673.44	66.18	1370.80	3970.80
£2700	247.68	272.16	2972.16	134.94	538.56	3238.56	97.77	819.72	3519.72	79.47	1114.56	3814.56	68.73	1423.80	4123.80
£2800	256.86	282.32	3082.32	139.94	558.56	3358.56	101.39	850.04	3650.04	82.42	1156.16	3956.16	71.28	1476.80	4276.80
£2900	266.03	292.36	3192.36	144.94	578.56	3478.56	105.01	880.36	3780.36	85.36	1197.28	4097.28	73.82	1529.20	4429.20
£3000	275.20	302.40	3302.40	149.94	598.56	3598.56	108.63	910.68	3910.68	88.30	1238.40	4238.40	76.37	1582.20	4582.20
£3250	298.14	327.68	3577.68	162.43	648.32	3898.32	117.68	986.48	4236.48	95.66	1341.68	4591.68	82.73	1713.80	4963.80
£3500	321.07	352.84	3852.84	174.93	698.32	4198.32	126.73	1062.28	4562.28	103.02	1444.96	4944.96	89.10	1846.00	5346.00

(ii) Credit cards

A credit card can be issued by a bank, a finance company or a shop. It allows the holder to spend up to an agreed maximum amount, and the holder must then pay a minimum of £5 each month or 5% of the outstanding balance. On a certain day each month, you receive a statement of your purchases, and the minimum amount you need to repay. It is important to realise that, because interest is being charged on the amount outstanding, usually at about 2% per month, you should try and repay more than the minimum if you can afford to.

Figure 4.1 shows an example of what a statement might look like.

Statement   BERKLEYKARD VISA

MR H G WELLS

Account number
4831 222 712 111   Statement date
19 OCT 84

Date	Reference	Details	Amount
		BALANCE FROM PREVIOUS STATEMENT	684.24
16 OCT		INTEREST AT 1.90% PER MONTH	13.00
19 OCT	00134	PAYMENT RECEIVED - THANK YOU	125.00CR
			572.24
22 OCT	21041	PETROL	14.00
23 OCT	33011	CHEMIST WHOLESALER	24.00
25 OCT	24102	BRITISH RAIL	48.00
30 OCT	46891	HARRODS	137.00
5 NOV	50334	FIREWORK FACTORY	27.00

Credit limit	Present balance
£1000	822.24
Minimum payment	**To reach us by**
£41.00	12 NOV 84

Figure 4.1

Worked example 4.17

David uses a well-known credit card quite regularly. During August, he made the following purchases: petrol £8, theatre tickets £15, British Rail £8.70, petrol

£12, books £16, petrol £12. His balance on the previous statement was £141.80. If interest is charged at 2% per month on the outstanding balance, draw up a statement of his account for the month of August.

The statement would appear as follows:

	Amount
BALANCE FROM PREVIOUS STATEMENT	141.80
INTEREST AT 2.00% PER MONTH	2.84
PETROL	8.00
TICKETS	15.00
BRITISH RAIL	8.70
PETROL	12.00
BOOKS	16.00
PETROL	12.00
	216.34
MINIMUM PAYMENT 5% OF BALANCE	10.00

(rounded down to the nearest £)

Exercise 4f
Use the information in Table 4.3 to find
(a) the monthly payments; (b) the total interest paid for the following purchases:

(1) Some kitchen furniture costing £1800 over 3 years.
(2) A car costing £2800 over 2 years.
(3) A radio-controlled helicopter costing £825 over 2 years.
(4) A garden shed costing £875 over 4 years.
(5) A settee costing £650 over 12 months.
(6) A holiday costing £1460 over 2 years.
(7) A bike costing £350 over 24 months.
(8) A caravan costing £6200 over 5 years.
(9) Vivek has just receive his credit card bill. The outstanding balance *before* last month's payment was £120 (including the interest). During the month, he made purchases totalling £85. What is the minimum payment required for this month's statement, if last month he sent the credit card company £35?
(10) June has just received her credit card statement from a well-known electrical store. The previous balance was £84, before interest at 2.6% per month was charged. She used the card during the month to buy a portable television costing £185. What will be the minimum payment required (rounded down to the £) on next month's statement.

(iii) Hire purchase
The third most common method of buying an item is by using hire purchase, or HP. This is usually arranged by a shop or mail order catalogue. Sometimes

you have to a pay a deposit, and then the rest is paid in equal instalments. The interest rate often turns out to be quite high and you may find it cheaper to find other ways of raising the money.

Worked example 4.18

A video recorder priced at £315 can be bought for a deposit of £80 and 12 monthly instalments of £25.80.

(i) Find the total cost of buying the video recorder this way.

(ii) Find the interest you have paid and express this as a % of the cash price.

(i) Total cost = £80 + 12 × £25.80
= £80 + £309.60 = £389.60.

(ii) The interest = £389.60 − £315 = £74.60.

The percentage interest $= \frac{74.60}{315} \times 100 = 23.7\%$.

Exercise 4g

Find the cost of the articles illustrated. In each case, find the extra amount you pay by buying it on HP instead of cash.

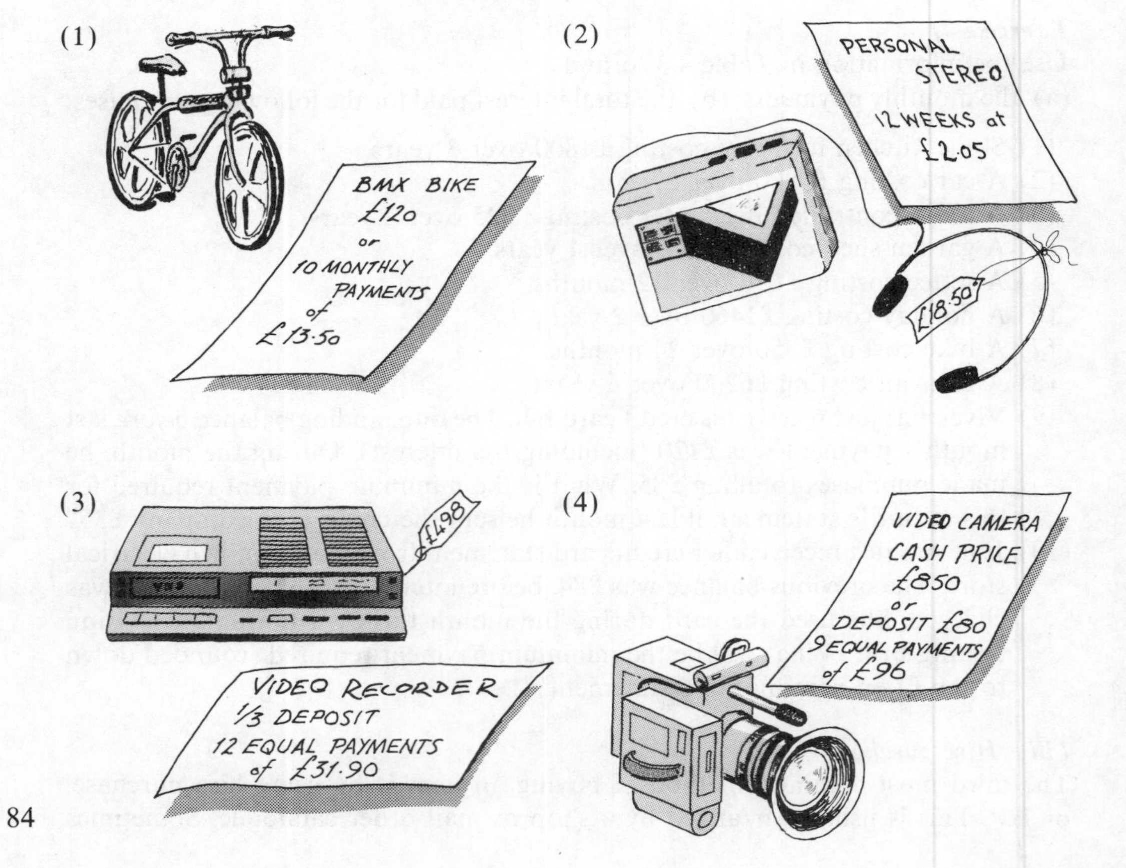

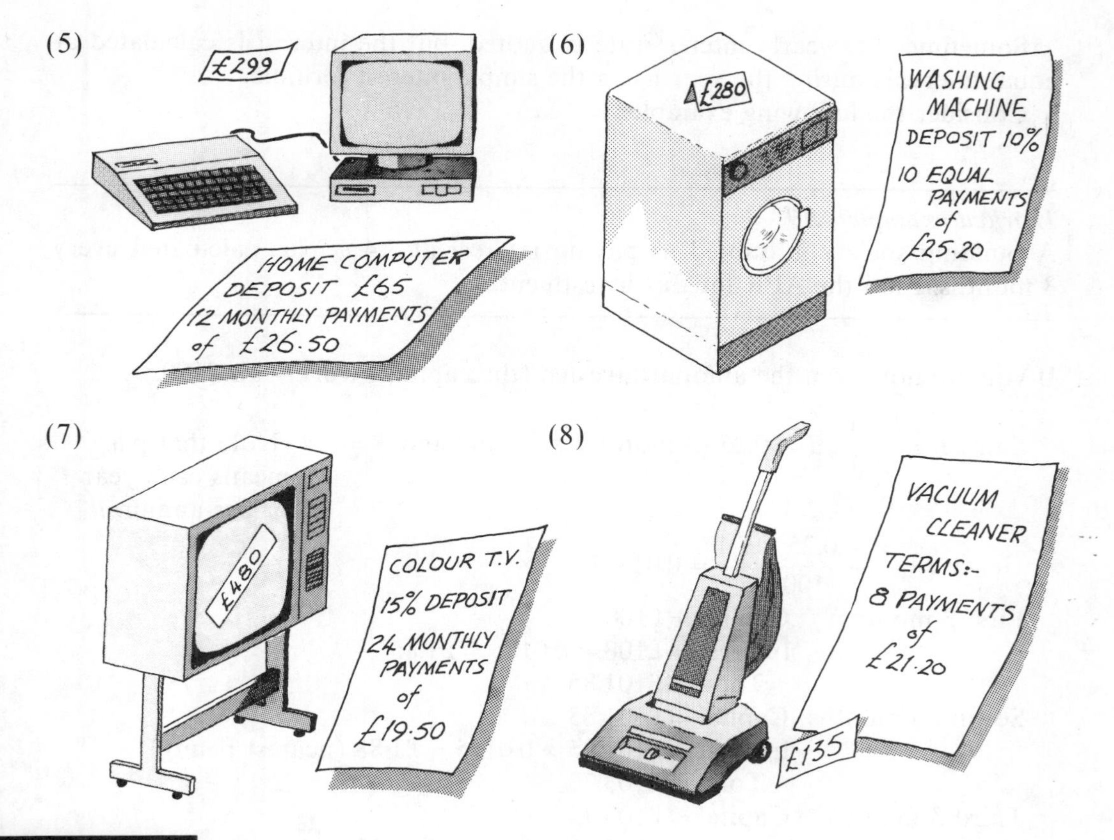

4.11 APR

The abbreviation APR stands for the *annual percentage rate* of interest charged on some forms of loan or credit agreement. It is a useful quantity in that it helps you to compare more easily different ways of raising finance. The problem usually is that interest may be charged monthly (credit cards), or three-monthly, or even weekly.

For example, a credit card may state that the interest rate is $1\frac{3}{4}\%$, charged monthly.

The annual rate of interest is not $12 \times 1\frac{3}{4} = 21\%$. To calculate the APR, we use the rules applied to compound interest.

If N is the number of times per year that interest is charged (e.g. $N = 12$, if it is monthly), and r is the % rate for the period that it is charged, then the APR is given by the formula:

$$\text{APR} = \left[\left(1 + \frac{r}{100}\right)^N - 1\right] \times 100$$

For our credit card of $1\frac{3}{4}\%$ per month:

$$\begin{aligned}\text{APR} &= \left[\left(1 + \frac{1.75}{100}\right)^{12} - 1\right] \times 100 \\ &= [1.0175^{12} - 1] \times 100 = [1.231 - 1] \times 100 \\ &= 23.1\%\end{aligned}$$

Sometimes the yearly interest rate is quoted, but the interest is calculated at equal intervals during the year using the simple interest formula.

Consider the following example.

Worked example 4.19

A building society is quoted as paying interest at 7.4% p.a., calculated every 3 months. Find the APR for this investment.

If you are not given the amount invested (the capital), work with £100.

Since $I = \dfrac{PRT}{100}$, $T = 0.25$ (3 months is $\frac{1}{4}$ year) and $R = 7.4$ (Note that p.a. means each year or per annum.)

Hence $I = \dfrac{P \times 0.25 \times 7.4}{100} = 0.0185P$

First 3 months: Capital = £100
Interest = £100 × 0.0185 = £1.85
Total = £101.85

Second 3 months: Capital = £101.85
Interest = £101.85 × 0.0185 = £1.88 (nearest penny)
Total = £103.73

Third 3 months: Capital = £103.73
Interest = £103.73 × 0.0185 = £1.92
Total = £105.65

Fourth 3 months: Capital = £105.65
Interest = £105.65 × 0.0185 = £1.95
Total = £107.60

The total interest paid = £7.60

Since the initial capital was £100, we have that the APR = 7.6%.

Exercise 4h

Find the APR from the following information:

(1) 2.1% charged monthly
(2) 1.75% charged monthly
(3) 5% charged 3 monthly
(4) 0.6% charged weekly
(5) 8% p.a. charged quarterly
(6) 6.5% charged monthly

4.12 Buying a house

For most people, buying a house is likely to involve borrowing a large sum of money. The money borrowed is called a *mortgage*. If you cannot keep up the repayments on the house, you will be forced to sell and repay the amount owing. A mortgage can be obtained from a building society, a bank or, in some cases, through a local authority or the firm you work for.

Normally, you will have to find a deposit on the house as well, although some institutions will lend you 100% of the purchase price. The money is usually repayed over a period of about 25 years, but seeing that most people will move before 25 years have elapsed, it is the monthly payment that influences how much you borrow. Interest rates tend to change considerably nowadays, and it is advisable to bear in mind when taking out a mortgage, that the payments may soon increase.

Table 4.4 shows typical monthly repayments at an interest rate of 11%.

Table 4.4

Number of years	15	20	25	30	35
Monthly repayment per £1000	£11.25	£10.50	£9.90	£9.60	£9.40

Worked example 4.20

Janice and Peter are trying to buy a flat which costs £46 000. Peter earns £180 a week on average, and Janice earns a salary of £9800. The building society will lend them $3\frac{1}{2} \times$ Peter's annual income, plus $1 \times$ Janice's salary. Find the size of the mortgage they can obtain, the corresponding lowest monthly payment, and the deposit they need to find, using the information in Table 4.4.

Peter earns $52 \times £180 = £9360$ a year, so they will lend

$3\frac{1}{2} \times £9360 = £32\,760$ on Peter's wages
$+1 \times £9800 = £\ 9\,800$ on Janice's salary

£42 560

The deposit required $= £46\,000 - £42\,560 = £3440$.
The monthly repayment is lowest over 35 years.
The monthly repayment $= 42.56 \times £9.40$
$= £400.06$

Exercise 4i

(1) A building society offers a 20-year mortgage for monthly repayments of £11.50 per £1000 borrowed.

(a) What are the monthly repayments on a loan of £45 000?

(b) What is the total amount of money paid out by the borrower over the 20 years?

(2) Frances and David are trying to buy their first house. Frances earns £6500 a year, and David earns £14 000 per year. The building society has offered them 4 × their joint salary over a period of 30 years. Use the information given in Table 4.4 to find the monthly repayment they will have to find.

(3) The Patel family have saved £4200 for the deposit on a house. The building society will offer a 95% mortgage. What is the maximum price they can pay for a house? What would the monthly repayments on this mortgage be if they borrowed the money over a 20-year period, assuming the figures in Table 4.4 apply?

4.13 Running a car

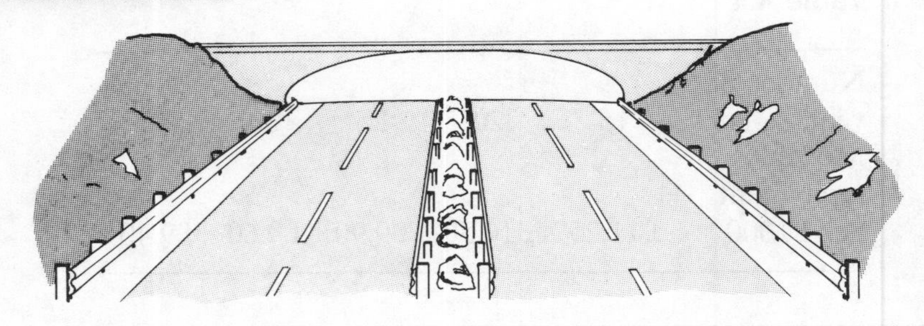

When calculating the cost of running a car, there are several factors that you need to take into account.

(1) *Road tax*

At the time of writing, this is £100 per year.

(2) *Insurance*

This depends on a number of things, such as type of car, age of driver, area in which you live, whether it is comprehensive or third party. See section 4.14 for calculations about this.

(3) *The cost of repairs*

This needs to be averaged out over a year, and it may include obtaining an MOT certificate.

(4) *Petrol*

(5) *Depreciation*

This is often not included by people when calculating the running cost of a car.

Worked example 4.21

The Depak family have just bought a nearly new Ford Sierra for £4800. The costs of running the car for a year are as follows: road tax: £100; insurance: £240;

repairs: about £150; petrol: 12 000 miles at an average of 30 m.p.g. using 4 star at £1.75 a gallon. Assuming the value of the car depreciates by 15% a year, estimate how much per week it costs to run the car.

The amount of petrol used = 12 000 ÷ 30 = 400 gallons.

The cost = 400 × £1.75 = £700

Depreciation = 0.15 × £4800 = £720

The total cost is £720 petrol
£700 depreciation
£100 tax
£240 insurance
£150 repairs
―――
£1910

The cost per week = £1910 ÷ 52 = £36.73.

Exercise 4j

(1) Toby is a student and runs a very old car. The tax is £100 and he manages to get third party insurance for £190 per year. He travels about 5000 miles each year on 4 star which costs roughly £1.90 per gallon. His repair bills including the MOT for the year come to £180 and he is not worried about depreciation. How much on average a week does it cost Toby to run his car? [Assume the car averages 30 m.p.g.]

(2) Janice has just bought a new car for £5200. The road tax is £100, and her comprehensive insurance policy costs £280. She is expecting to travel about 14 000 miles a year using 4 star petrol at £1.78 per gallon. Assuming no repair bills, and a depreciation of 10% on the price, estimate the monthly cost to Janice of running the car. [Assume the car averages 30 m.p.g.]

4.14 Insurance

People take out insurance to cover themselves in case something goes wrong. This may be insurance on a car, a house, or against the loss of an item that would be expensive to replace. The variations in the premiums (repayments) charged by different companies are often not great, but it is usually worth while shopping around.

Worked example 4.22

The rates of car insurance charged by the PROODENTIAL insurance company are given in Table 4.5. For drivers over the age of 25, there is a 30% reduction.

A no claims bonus scheme also offers percentage reductions on the full premium of 20% after 1 year, 40% after 2 years, and 60% after 3 years, together with a further reduction of £10, if you are prepared to pay the first £100 of any damage in an accident.

Table 4.5 Comprehensive insurance cover

District	Insurance group of car 1	2	3	4	5
1	320	430	560	720	1050
2	330	440	570	730	1080
3	350	460	590	750	1100
4	360	470	600	770	1200

(i) Find the annual premium to insure a car in group 3 if you live in district 2 and have 40% no claims bonus, assuming you are prepared to pay the £100 excess and you are over 25.

(ii) David is 23 years old, and has just bought a Ford Escort which is in group 4. How much will his basic premium be if he does not want to pay the excess? This is the first time he has taken out a car insurance policy, and he lives in district 4.

(i) The basic premium = £570.
Over 25 gives a reduction of 30%, i.e. 0.3 × £570 = £171. His no claims bonus is 40% of £570 = £570 × 0.4 = £228. A further deduction of £10 for the excess gives us

$$£570 - £171 - £228 - £10 = £161$$

(ii) The basic premium is £770.
Since David is under 25, he must pay the full amount. It is his first insurance policy, so there is no no claims bonus. The answer is unfortunately still £770.

Exercise 4k

(1) The Helpful Insurance Company quotes the following rates for insuring a house and its contents

£1.35 per £1000 (or part) insured for the building

32p per £100 (or part) for contents

An extra 20% is payable on contents if you live within a 15 mile radius of central London.

(i) What would it cost to insure a house valued at £72 000 outside London?

(ii) What would the premium be on a flat valued at £135 000 in central London?

(iii) David and June want to insure the contents of their house in central London, for £12 600. If they pay the premium in monthly instalments, how much will they have to pay each month?

(iv) Mr and Mrs Jones are retiring to Brighton. They have bought a bungalow for £97 000, and they also need to insure the contents for £7500. What will be the annual premium?

(2) Use the information given in Table 4.5 to find the cost of the following car insurance:
   (i) Peter is 36 years old, lives in district 2 and wants to buy a car in group 5. He is not prepared to pay the £100 excess. What is his annual premium?
   (ii) Joan is 22 years old, living in district 1. She has a car in group 2 and a full 60% no claims bonus. She is not prepared to pay the £100 excess, so what is her annual insurance premium?
   (iii) Mrs Davies is 60 years old. She has been driving for 31 years and never had an accident. Her car is in category 5 and she lives in district 3. What is the cheapest comprehensive policy she can obtain?

4.15 Balancing the books

You may find yourself one day in the position of having to draw up some accounts for a club or society that you belong to. Without making things too complicated, the following example should show you how to tackle the problem.

Worked example 4.23

The following list shows payments and receipts made by Sheena who is treasurer of the local hockey team's social club.

			£
Jan	1	Balance carried forward	130.65
	3	Paid the milk bill	8.70
	7	Paid laundry bill	4.60
	8	Collected subs	15.00
	12	Bought 2 new hockey balls	9.70
	12	Paid pitch hire fees	12.60
	15	Collected subs	15.00
	18	Purchased stationery	3.85
	19	Umpires fees paid	23.00
	22	Collected subs	14.00
	25	Bought 2 new shirts	23.80
	29	Collected subs	15.00

Draw up a balanced account of these figures, including the balance carried forward to February.

The account is laid out under 2 headings, namely credit (receipts) and debit (payments).

Credit			Debit		
Jan		£	Jan		£
1	balance b/f	130.65	3	milk	8.70
8	subs	15.00	7	laundry	4.60
15	subs	15.00	12	hockey balls	9.70
22	subs	14.00	12	hire fees	12.60
29	subs	15.00	18	stationery	3.85
			19	umpires	23.00
			23	shirts	23.80
				balance c/f Feb.	103.40
	Total	189.65		Total	189.65

You should notice that the initial balance b/f (brought forward) is counted as a credit, and the final balance carried forward to the next month counts as a debt. The amount carried forward at the end of the month is calculated so as to make each side of the account balance.

Exercise 4l

The following list shows the payments made during October by Steve who is treasurer of the Seagulls football team. Draw up a balanced account of these figures, including the balance carried forward to November.

			£
October	1	Balance brought forward	85.20
	4	kit washing	2.80
	5	first aid kit	5.60
	7	collected subs	12.00
	10	referee's fees	5.60
	14	collected subs	12.00
	17	kit washing	2.80
	23	new football	18.00
	28	collected subs	12.00
	29	referee's fees	4.80

Suggestions for coursework 4

1. Decide on an item that you would like to buy costing between £300 and £500. Compare the different makes available on the market, and the prices charged by different retailers. Find out possible ways of raising the money and decide which is best for you.
2. Try to find out what you need to earn each week in your area if you are about to leave home and live on your own. You should try and draw up a budget to cover all the items you are likely to need to buy, the expenses you will incur in travel, heating, lighting, etc. Do you think you can afford it? If you have a friend living in a different area, try and compare the cost of living in that area with yours.

3. Over a period of a few weeks, look at the exchange rates for certain currencies against the pound. Try to illustrate your results on a graph. Do they all move together or at different times? Is it possible to predict any changes from your graphs?
4. Plan a holiday to somewhere exotic which may take some time to save for. It should include the cost of trips when you arrive, and an estimation of how the prices may increase while you are saving.

   This particular project can include calculations of distances etc. It is particularly suitable for diagrams, but remember these should be mathematical, and not just nice pictures.
5. Try and find for a selection of different types of shop, the cost of an article to the shopkeeper, and how much he sells it for. You can then see what percentage profit he makes on average. What factors do you think influence this percentage mark up?

Miscellaneous exercise 4

1. During the year, Jasmin received £1.50 per week pocket money for 14 weeks, and the rest of the year £1.80 per week. Calculate:
   (i) Her average weekly pocket money during the year.
   (ii) The percentage of the yearly amount she saved, if on average she spent £1.20 per week.
2. A DIY shop sells paint at a discount of 20% off the normal price. VAT at 15% is then added after the discount has been taken off. I buy four tins of paint. The normal prices of them are £4.65, £2.70 and £0.95. Copy and complete this bill.

	£	p
2 tins of paint @ £4.65		
1 tin of paint @ £2.70		
1 tin of paint @ £0.95		
Total normal price		
Discount @ 20%		
Selling price without VAT		
VAT @ 15%		
Amount to pay		

(O and C)

3. A couple plan to paint the outside of their house. To reach the bedroom windows they will need a ladder. To reach the ground floor windows they

will need a pair of steps. They calculate that if there is no rain the work will taken them at least 13 days.

A local shop hires equipment. Part of the shop's price list is shown below.

ITEM	Hire charge per week or part of week
Ladder (up to 32 ft)	£10.20
Ladder (up to 24 ft)	£7.60
Steps	£5.00
Electric paint stripper	£3.00

(a) Calculate the cost of hiring a ladder (up to 24 ft), steps, and an electric paint stripper for 2 weeks. Set out your calculations as a bill from the shop, as shown below:

HIRE CHARGES

ITEM	No. of weeks	Charge per week	Charge
Ladder	2		
Steps			
Paint stripper			

(b) What will the total cost be if the work takes longer than expected and lasts for 19 days?

The shop also sells equipment and the prices of the items the couple will use are Ladder £37.60, Steps £12.75, Paint Stripper £9.60.

(c) Say which of the items you would advise them to buy instead of hiring, giving reasons for your selection. (O and C)

4. When Harry Lime goes abroad he cashes Eurocheques made out to his account at his English bank. These are the charges made by his bank.

When each Eurocheque is received in the UK, commission of 1.25% is added to the sterling value as a payment to the foreign bank for the work it undertakes. A transaction charge of 29p per cheque is also added to the sterling value.

On his last visit to Europe he cashed the following four cheques.

Switzerland	250 Swiss francs
Italy	50 000 lira
Italy	78 000 lira
France	4000 French francs.

The exchange rates at the time of his visit were as follows.

Switzerland	2.94 francs to the £
Italy	2 380 lira to the £
France	10.7 francs to the £

How much will be charged to his English bank account for the four cheques altogether? (NEA)

5. A manufacturer sells articles to a wholesaler and makes a profit of $33\frac{1}{3}\%$ on the cost of manufacture. The wholesaler then sells to a retailer, making a profit of 20% on the price paid by the wholesaler. The retailer sells each article for £15, making 25% profit on the price paid to the wholesaler. Calculate:
   (i) the price per article paid by the retailer,
   (ii) the cost of each article to the wholesaler,
   (iii) the cost of manufacture of each article.
   The manufacturing cost per article rises by £1.50. The manufacturer reduces his profit to 30% of the cost of manufacture. Calculate:
   (iv) the new selling price per article if the wholesaler and the retailer maintain the same profit margins as before.
6.

32ND ISSUE CERTIFICATES

52%

NO TAX, NO RISK, NO HASSLE, NO FEAR OF FALLING INTEREST RATES.

The interest that you can earn from National Savings Certificates is TOTALLY free from Income Tax and Capital Gains Tax. You don't even have to declare it on your Income Tax return.

The 32nd Issue offers a guaranteed return of 52% after five years. This is equivalent to 8.75% a year over the five years.

You can invest from £25 to £5,000, in addition to any other Issues you already hold. Each member of your family can invest up to the full amount in their own names.

For full details ask at your bank or post office.

We guarantee freedom from tax, with high performance and absolute safety. There is nothing to touch National Savings Certificates.

NATIONAL SAVINGS

The above advertisement contains the following statement.

'The 32nd Issue offers a guaranteed return of 52% after five years. This is equivalent to 8.75% a year over the five years.'

(a) Investigate the truth of this statement by completing the following table to show the year by year growth of an initial investment of £100 at 8.75% a year.

	Amount at end of year
Year 1	
Year 2	
Year 3	
Year 4	
Year 5	

From this table, write down, correct to one place of decimals, the total percentage increase over the five years.

(b) Given that National Savings Certificates are bought in multiples of £25, find the minimum amount of money which would have to be invested initially in order to produce a total of at least £1000 at the end of the five years. (NEA)

7. Mr Spender received his bank statement on the first of April and was horrified to discover that his account was overdrawn. Here is a copy of his bank statement.

SHIRE BANK Mr S. M. Spender				STATEMENT OF ACCOUNT Acc. No. 30124276 31 March 1988	
Details		Payments	Receipts	Date	Balance
Balance Forward				1 March	57.60
Counter Credit			387.56	2 March	445.16
Cheque No. 157		50.00		2 March	395.16
Deetown Rates	DDR	42.80		7 March	352.36
British Gas	DDR	63.00		11 March	289.36
Mersey Building Society	STO	98.50		15 March	190.86
Cheque No. 158		100.00		22 March	90.86
N. Electric	DDR	26.00		22 March	64.86
Cheque No. 159		150.00		22 March	85.14 DR*

*Bank current account charges now payable

Abbreviations: STO Standing Order DDR Direct Debit
DR Overdrawn Balance

(a) If an account is overdrawn the bank charges for every withdrawal made during the month are as follows:

32p for each cheque and standing order
20p for each direct debit

(i) What will Mr Spender pay in bank charges?
(ii) What will his balance be after the bank charges are included in the statement?

(b) The bank had included the direct debit for Deetown rates by mistake since the rates were payable every month *except* March and April.
(i) How much was Mr Spender's yearly rates bill?
(ii) If the rateable value of Mr Spender's house is £400, calculate at what rate in the £ the rates are levied.

(c) In fact Mr Spender had his cheque book stolen on March 21st. He informed the bank immediately so should not have cheques 158 and 159 debited to his account. Write out his correct balance column putting right the bank's errors with the cheques and the rates.

(d) What percentage of Mr Spender's monthly pay of £387.56 was spent on gas and electricity? Give your answer correct to one decimal place.

(MEG)

8. Mr and Mrs Dent went on holiday to the USA. They went by air and returned by sea.
(a) Their journey to the USA took them 11 hours and they travelled at an average speed of 460 miles per hour.
(i) How many miles did they travel?
(ii) If the fare was £340 per person what was the cost, to the nearest penny, in pence per mile?
(b) On the return journey they travelled 3100 miles and the journey took them 4 days and 16 hours.
(i) What was their average speed in miles per hour?
(ii) If the fare was £320 per person what was the cost, to the nearest penny, in pence per mile?
(c) Before leaving England Mr and Mrs Dent changed £560 into dollars at 1.42 dollars to the pound. During the holiday they spent 705 dollars. On their return they changed their remaining dollars into pounds and pence.
If they received 1.40 dollars to the pound, how much did they receive in pounds and pence? (LEAG)

9. A man and his wife are planning a short holiday. The travel agent's brochure price is £165 per person for a one week holiday (7 days) with bed and breakfast. There is a supplement of £3 per person per day for an evening meal.
(i) If they book for the week including the evening meal what is the total charge for the two of them?
(ii) They can fly to the resort from one of two airports.
(a) From Cardiff: There is a supplement of £18 per person. In addition they will have to drive to and from the airport in Cardiff, a distance of 80 miles each way, and pay £8 to garage the car while they are away. Their car travels 24 miles on a gallon of petrol which costs £1.90 per gallon.
(b) From Gatwick: There is no supplement on the holiday, but the car journey is 130 miles each way and the charge for garaging the car is £10.

Calculate the total cost for (a) Cardiff, (b) Gatwick.
Say which airport you would advise them to use, giving reasons for your answer. (O and C)

10. John Brown has just been promoted and he will move to a country area from the city in which he now lives. He decides to change his car and decides to work out the cost of his car insurance.

His insurance company has supplied him with a table of basic premiums for different areas and different groups (of car). The figures in bold type relate to a 'comprehensive' policy and the figures in italics relate to a 'third party, fire and theft' policy.

	Group											
Area	1		2		3		4		5		6	
1	**252**	*154*	**284**	*168*	**333**	*188*	**391**	*214*	**445**	*237*	**623**	*331*
2	**274**	*164*	**308**	*176*	**360**	*202*	**433**	*228*	**482**	*252*	**680**	*352*
3	**284**	*172*	**318**	*183*	**377**	*210*	**441**	*239*	**504**	*265*	**706**	*370*
4	**307**	*181*	**344**	*197*	**406**	*224*	**477**	*255*	**550**	*284*	**770**	*398*
5	**332**	*193*	**374**	*208*	**441**	*242*	**518**	*273*	**588**	*307*	**823**	*429*
6	**365**	*213*	**411**	*234*	**485**	*266*	**570**	*300*	**647**	*337*	**906**	*472*

(a) He is presently living in Area 5, drives a car in group 2 and insures his car for third party, fire and theft. What is his basic premium?
(b) John receives a 60% no claims discount and a further 12.5% discount for restricting the driving to himself or his wife. Find his net premium.
(c) John will be moving to a house in area 2 and will buy a larger car which will be in group 6. What will be his new basic premium for a third party, fire and theft policy?
(d) As John will be buying a new car he decides to see whether he can afford the cost of comprehensive insurance. In his new home, with the new car, what would be this basic premium?
(e) Because John will also agree to be responsible for the first £200 of damage he will receive a 15% discount in addition to the discounts given above. What will be the net premium for this comprehensive policy?
(f) Explain the difference in cover between a 'comprehensive' policy and a 'third party, fire and theft' policy. (SEG)

11. Mark Mullin is a retail butcher. On 31 March 1986, the end of his accounting year, the following information was available:

	£
Freehold premises at cost	50 000
Equipment at cost	16 000
Provision for equipment depreciation	7 500
Stock in trade	1 800
Trade debtors	300

Expenses paid in advance	40
Trade creditors	1 350
Drawings during year	9 000
Bank overdraft	150
Net profit for year	13 400
Capital (1 April 1985)	54 740

(a) Prepare the balance sheet of Mark Mullin's business as at 31 March 1986 paying particular attention to presentation.
(b) Calculate the amount of Mullin's working capital at 31 March 1986. Express Mullin's profit as a percentage on capital invested for the year ended 31 March 1986 and name two other pieces of information which would enable you to say whether or not the profit was satisfactory.

(SEG)

12. My Bank Account is with the Unity Bank which makes no charges on my Account provided that I do not overdraw. It pays me no interest on the money in my Account.

The Loamshire Bank would charge me 10p for each cheque that I draw on an Account, but it would pay me 5% interest on the 'minimum balance' each month. The 'minimum balance' is the smallest amount that there is in the Account during the month.

Looking back over the past year I see that the minimum balance in my Account was never less than £60. On average I drew 42 cheques each month.
(i) Calculate the interest that would be paid to me by the Loamshire Bank in a month when the minimum balance was £60.
(ii) Calculate the total charges the Loamshire Bank would make for the 42 cheques drawn in a month.
(iii) Use your answers to (i) and (ii) to calculate how much I would pay to the Loamshire Bank in the month if I had my Account with them.

I decide to change my Account to the Loamshire Bank, but to put extra money into the Account so that I shall 'break even' in an average month when I draw 42 cheques.
(iv) Calculate how much extra I must put into the Account. (O and C)

13. The repayment table overleaf shows you how much your monthly repayments would be on any bank loan from £200 to £5000, over a period of between 1 and 3 years.
(a) What would be the total payable on a bank loan of £1500 over 12 months?
(b) Mrs Weller has made 5 monthly payments of £92.50 off a bank loan of £1000. How much has she yet to pay?
(c) Vijay wants to borrow £2000 for a motor bike and can only afford to pay off the bank loan at a maximum of £80 per month. What is the total payable on this loan?
(d) Instead of borrowing money from the bank, Vijay could afford a 10% deposit and 36 payments of £75 on Hire Purchase. How much more would it cost him?

Amount of loan £	200	500	1000	2000	5000
Repayment term 12 months at 11% p.a. flat					
Total Payable £	222.00	555.00	1110.00	2220.00	5550.00
Monthly Repayment £	18.50	46.25	92.50	185.00	462.50
Repayment term 24 months at 11% p.a. flat					
Total Payable £	244.08	610.08	1219.92	2440.08	6100.08
Monthly Repayment £	10.17	25.42	50.83	101.67	254.17
Repayment term 36 months at 11% p.a. flat					
Total Payable £	266.04	664.92	1329.84	2660.04	6649.92
Monthly Repayment £	7.39	18.47	36.94	73.89	184.72

(NEA)

14. Mrs Morris is a very careful driver and she has a reliable 10-year old car. To run the car each year she must pay the following:

Road Fund Tax	£100
MOT Test Fee	£9.50
Insurance	£174
Servicing and repairs	£120

Mrs Morris' car travels 17 miles on each gallon of petrol and, on average, she travels 9000 miles each year.

(a) Calculate the total cost of the petrol she buys in a year if she pays, on average, £1.76 per gallon.

(b) Find the total cost of a year's motoring for Mrs Morris.

Her friends suggest that she should sell her car and buy a new, smaller car that would average 47 miles on every gallon. Her 10-year old care is quite valuable and she could sell it for £3500, but the new car will cost £4950.

(c) If she sells her old car and buys the new car, how much money must she find?

She must borrow this money from her Bank and repay it with interest over 12 months, repaying the Bank a total of £1536. The new car will not require an MOT test, but the other costs (Road Fund Tax, etc.) will be unchanged.

(d) How much will Mrs Morris spend on petrol in a year if she buys the new car?

(e) If Mrs Morris buys the new car, work out what she spends altogether in the first year that she owns it, including repayment to her Bank, petrol, and other costs.

(f) Would you advise Mrs Morris to buy the new car? Give reasons for your answer.

15. (a) KAMIKAZE Video Cassette Recorder
CASH PRICE: £500
HIRE-PURCHASE TERMS: Deposit 20% of Cash Price plus 24 monthly instalments of £22.50.
MAINTENANCE: 1st year, free.
2nd year, £8.50,
3rd year, £8.50.

Calculate:
(i) the deposit,
(ii) the total of the 24 instalments,
(iii) the total hire-purchase price,
(iv) the net cost over 3 years if the buyer who bought on HP took out maintenance, and were to sell the set for £120 at the end of 3 years.

(b) A rental company offers a similar Video Cassette Recorder for the all-in charges shown below:
1st year: £18.00 per month,
2nd year: £17.50 per month,
3rd year: £17.00 per month.

Calculate the amount that would be paid over 3 years by this method.

16. During the course of a year a motorist drove 28 000 km. For a quarter of this distance he drove in England and used a car with average petrol consumption of 14 litres per 100 km. For the remainder of this distance he drove in France and used a more economical car with an average petrol consumption of 8 litres per 100 km. Calculate
(i) the total number of litres of petrol used,
(ii) his average petrol consumption for the whole year giving the answer in litres per 100 km.

His total expenditure for the year on petrol was £930. Given that the average price he paid for petrol in England was 34p per litre, calculate the total amount spent on petrol in France.

Assuming the average rate of exchange to be £1 = 10.8 francs, calculate the average price paid for petrol in France, giving the answer in francs per litre to the nearest tenth of a franc.

Calculate how much the motorist would have saved had he driven the more economical car in England as well as in France.

17. I am considering the purchase of a new car. Two suppliers advertise the same model, but under different terms.

CASH PRICE CARS	CREDITABLE CARS
0% interest Minimum deposit £670 12 monthly payments of £225 **NO INTEREST CHARGES**	Your old car (any age) pays the deposit Deposit (your old car) £670 24 monthly payments of £140

(i) Using the figures for CASH PRICE CARS calculate the total cash price of the car.
(ii) For CREDITABLE CARS calculate the total amount paid in the 24 monthly instalments.
(iii) CASH PRICE CARS will allow me £250 part exchange for my old car against the deposit for the new one; the rest of the deposit must be paid in cash. What is the total I shall pay if I buy the car from this supplier? (Exclude the £250 I receive in part exchange for my old car.)
(iv) Say, with reasons, which supplier you would advise me to buy the car from.
(v) Give one reason why you might suggest that I buy from the other supplier instead, if my present car is worth only £250 as part exchange value.

(O and C)

BASIC GEOMETRY AND CONSTRUCTION

5

In this unit we look at properties and shapes of commonly occurring geometrical figures formed by straight lines and circles.

5.1 Angles and straight lines

An *angle* is a measure of rotation or turn. The unit of measure is *degrees* and there are 360° in one complete rotation.

Worked example 5.1

Arif's lunch break starts at 12.40 p.m. (Figure 5.1) and finishes at 1.30 p.m. If he is well known as a clock watcher, through how many degrees will he see the minute hand move during his lunch break?

Figure 5.1

The minute hand has moved through 50/60ths of a complete turn, therefore the angle through which the hand has turned is

$$\frac{50}{60} \times 360^\circ = 300^\circ$$

Worked example 5.2

Figure 5.2 shows a representation of two roads which meet at the junction

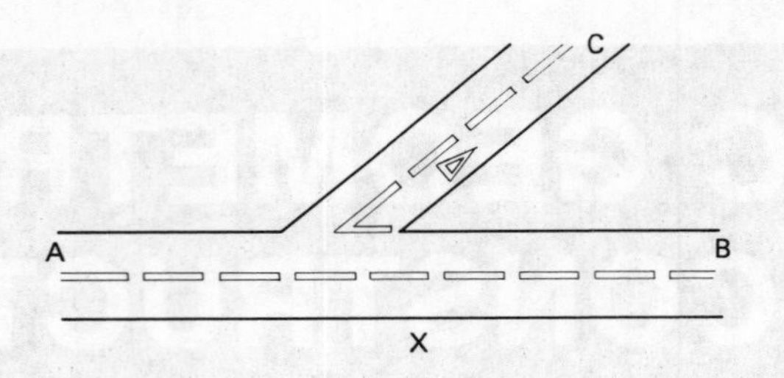

Figure 5.2

labelled X. A motorist travelling from A to B passes the junction at X and realises that he really wants to go to C, so he turns round to go back to the junction. Through how many degrees has he turned? If he then turns through an angle of 140° to go in the direction of C, what is the angle CXB?

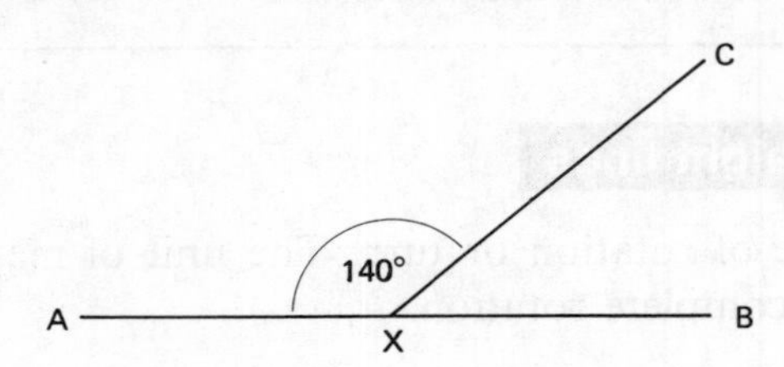

Figure 5.3

The figure can be simplified into a line diagram as shown in Figure 5.3. In turning around to head back towards the junction, the motorist turns through half a turn which is 180°. At the junction, he turns through 140° which is marked on the line diagram. The required angle CXB is the difference between the 180° and 140°.

So using $\angle$ as a shorthand abbreviation for 'angle':

$$\angle CXB = 180° - 140° = 40°$$

This example illustrates the property that angles on a straight line add up to 180°.

Using this property, it can be shown that when two straight lines intersect, the angles opposite to each other are equal.

Worked example 5.3

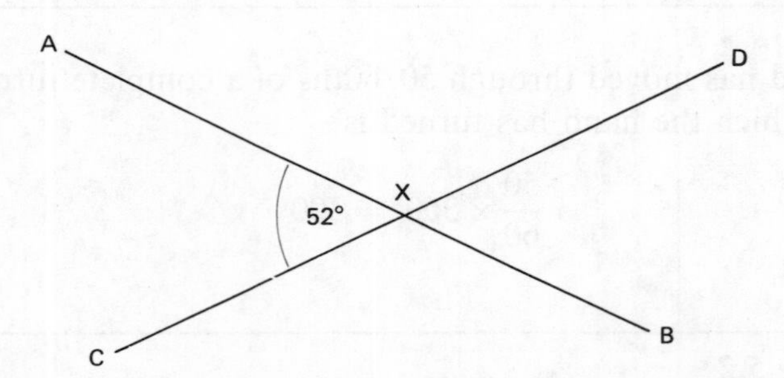

 Figure 5.4

In Figure 5.4, $\angle AXC = 52°$. Using the property that angles on a straight line add up to 180°, determine the value of $\angle DXB$.

$$\angle AXC + \angle AXD = 180° \text{ (angles on the straight line CD)}$$

so $$\angle AXD = 180° - 52°$$

$$= 128°$$

but $$\angle AXD + \angle DXB = 180° \text{ (angles on the straight line AB)}$$

so $$\angle DXB = 180° - 128°$$

$$= 52°$$

5.2 Types of angles

There are different types of angles dependent on their size.

1. An angle equal to 90° (a quarter of a turn) is called a *right angle.*
2. An angle less than a right angle is called an *acute angle.*
3. An angle more than 90° but less than 180° (half a turn) is called an *obtuse angle.*
4. Angles greater than 180° but less than 360° are called *reflex angles.*

Angles which are equal on a diagram are usually marked with the same symbol, and 90° angles are marked as shown in Figure 5.5.

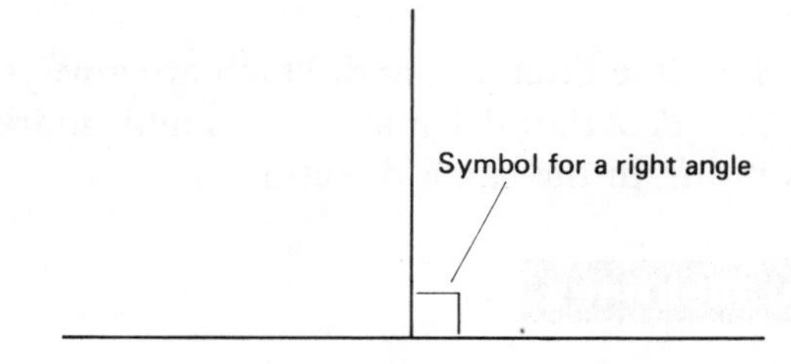

Figure 5.5

5.3 Bearings

True bearings are a measure of angle based on the points of the compass. These are measured in a clockwise direction from North and are always given as 3 figures, i.e. a bearing of 8° is written as 008°. The four points of the compass are North, South, East and West.

Worked example 5.4

From Leeds, Newcastle is due North, Sheffield due South, Hull due East and Bradford is due West. Draw a diagram to illustrate this information. What is the bearing of Bradford from Leeds? What is the bearing of Leeds from Sheffield? Pickering is North East of Leeds, draw the appropriate bearing on the diagram and write down the bearing of Pickering from Leeds.

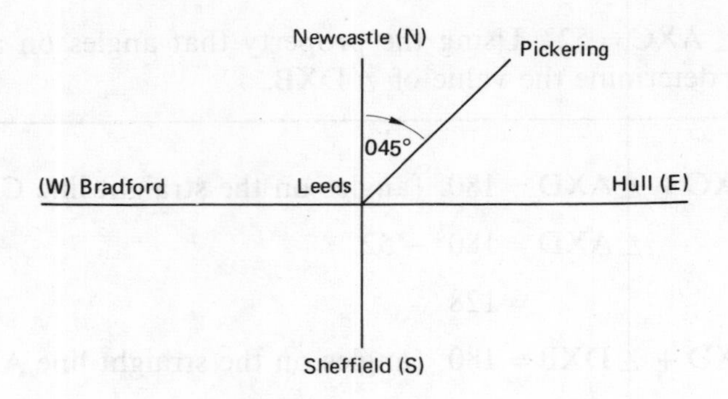

Figure 5.6

The four points North, East, South and West are on bearings of 000°, 090°, 180°, and 270° respectively.

As Bradford is due West of Leeds, the bearing of Bradford from Leeds is 270°.

Leeds is due North of Sheffield, therefore the bearing of Leeds from Sheffield is 000°. Pickering is North East of Leeds, and as the direction suggests, it is half way between North and East therefore the bearing of Pickering from Leeds is 045°.

Conversely, Leeds is South West of Pickering. As a bearing due south is 180°, and a further turn to the direction South West is 45° (half of 90°), the bearing of Leeds from Pickering is 225°.

5.4 Parallel lines

Lines which point in the same direction and which are always the same distance apart are said to be *parallel*. Parallel lines are usually marked with the same number of arrows, pointing in the same direction.

5.5 Angles and parallel lines

If a straight line crosses a pair of parallel lines, then a number of angles are formed. (This line is sometimes referred to as a *tranversal.*)

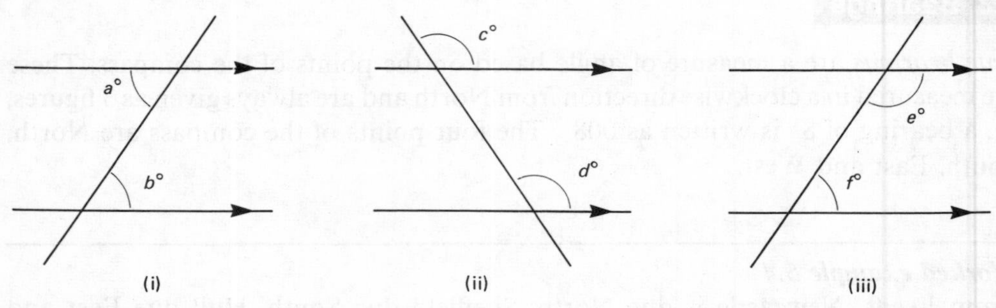

Figure 5.7

The following are the properties of such angles formed and refer to Figure 5.7.

(i) $a = b$. a and b are on opposite sides of the transversal and are said to be *alternate angles* between parallel lines.

(ii) $c = d$. c and d are on corresponding sides of the two parallel lines and are said to be *corresponding angles* between parallel lines.

(iii) $e + f = 180°$. e and f are *interior angles* between the two parallel lines.

Worked example 5.5

Figure 5.8 shows two supports AB and BC for a bridge. If RBS is parallel to AP and QC, ∠BAP = 32° and ∠ABC is a right angle. Find ∠BCQ.

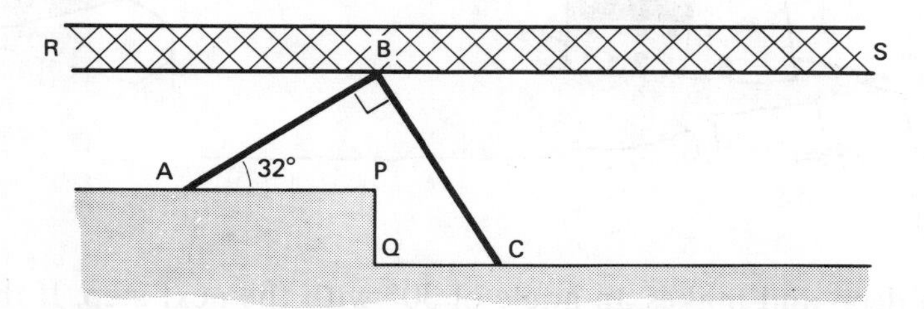

Figure 5.8

Now ∠RBA = ∠BAP = 32° (alternate angles between two parallel lines)

As ∠RBA + ∠ABC + ∠SBC = 180° (angles on a straight line)

∠SBC = 180° − 90° − 32°

= 58°

So ∠BCQ = ∠SBC = 58° (alternate angles)

Alternatively, ∠RBC + ∠BCQ = 180° (interior angles)

so ∠BCQ = 180° − ∠RBC = 180° − (90° + 32°)

= 58°

Exercise 5a

(1) How many degrees are there in 3/8ths of a revolution?

(2) Arif starts work in the afternoon at 1.30 p.m. If he has a tea break at 3.05, through how many degrees has the minute hand moved during this time?

(3) A screw has 6 threads to the centimetre (Figure 5.9). Through what angle must it be turned in order to penetrate 1 cm into a piece of wood?

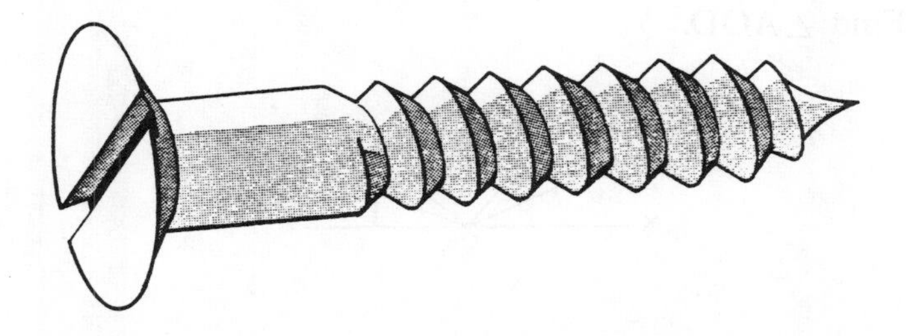

Figure 5.9

(4) A lighthouse has a circular stairway to the top (Figure 5.10). Each step is

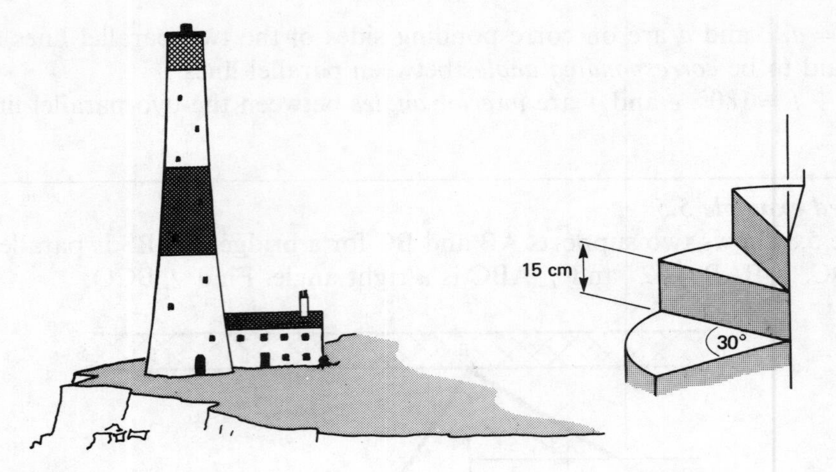

Figure 5.10

15 cm deep and makes an angle of 30° with the next step. If the lighthouse keeper has to turn through 10 complete revolutions to get to the top, how high has he climbed? A keeper at a similar lighthouse has to climb a height of 21.6 metres. How many steps does he climb?

(5) In Figure 5.11, find x.

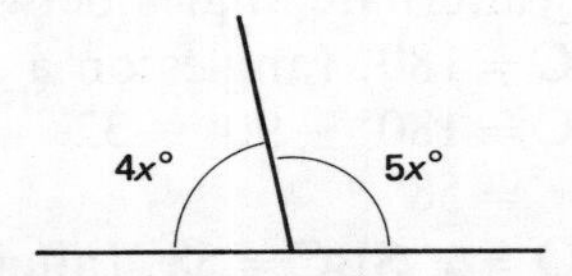

Figure 5.11

(6) In Figure 5.12, XY is a straight line. c is twice as great as a, and b is three times as great as c. Find the angles a, b and c.

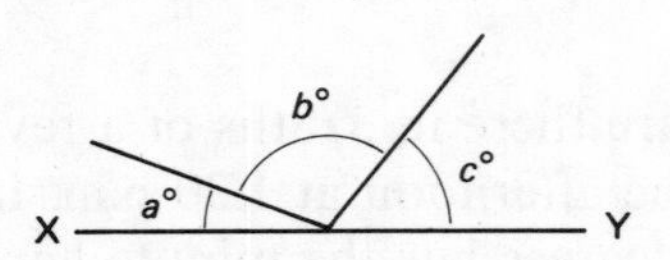

Figure 5.12

(7) In Figure 5.13, XY is a straight line. AO bisects ∠BOX and DO bisects ∠COY. Find ∠AOD.

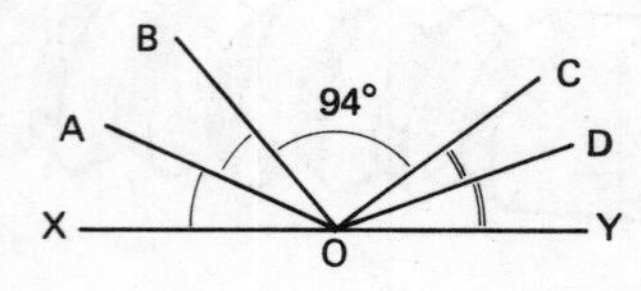

Figure 5.13

(8) How many degrees are there in 2/5ths of a right angle?

(9) In Figure 5.14, find x.

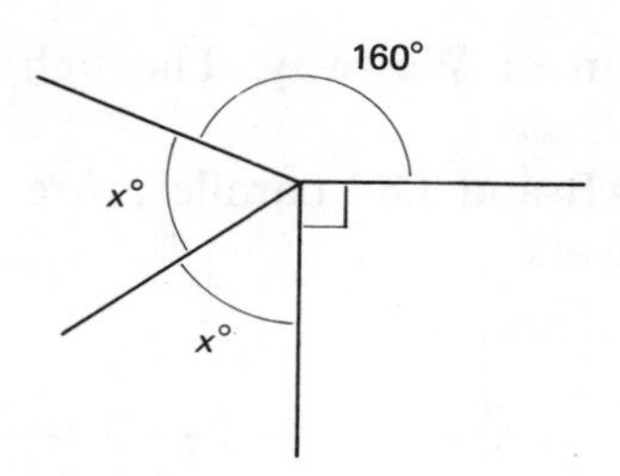

Figure 5.14

(10) In Figure 5.15, find the value of the largest angle.

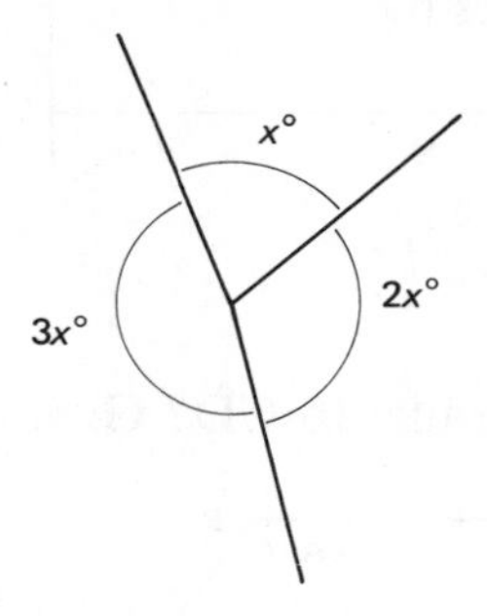

Figure 5.15

(11) B lies to the north of A and the bearing of C from A is South East. What is the obtuse angle BAC?

(12) Figure 5.16 shows a motorway interchange. The three routes off the motorway, Westway, New Easton and South Eastleigh, are in the directions West, North East and South East respectively. A motorist travelling northbound on the motorway wishes to travel to South Eastleigh. Through how many degrees will he turn? Travelling from Westway, Rebecca wishes to travel south on the motorway. Through how many degrees will she turn?

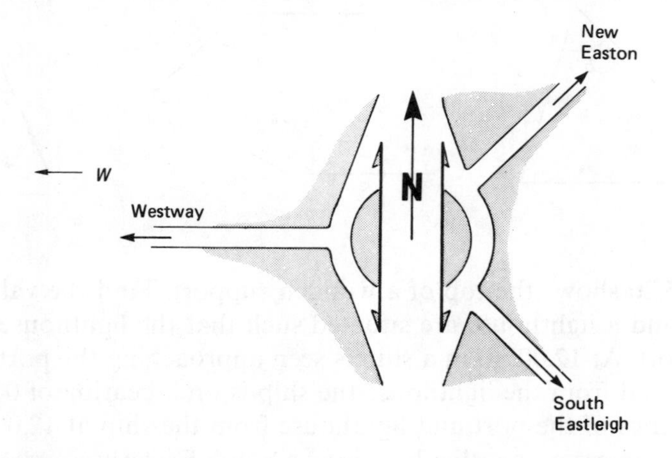

Figure 5.16

Travelling from New Easton, Graham is undecided where he wants to go and does one complete circuit of the roundabout. On the second circuit he leaves in the direction of Westway. Through how many degrees has he turned?

(13) In Figure 5.17, are AB and DC parallel? Are AD and BC parallel? Give reasons for your answers.

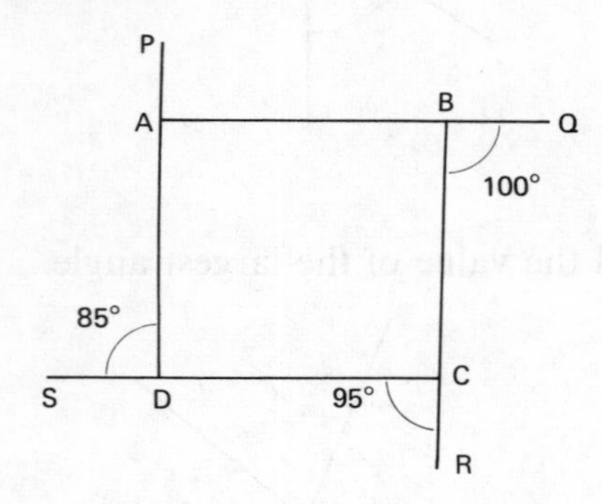

Figure 5.17

(14) In Figure 5.18, is AB parallel to CD? Give a reason for your answer.

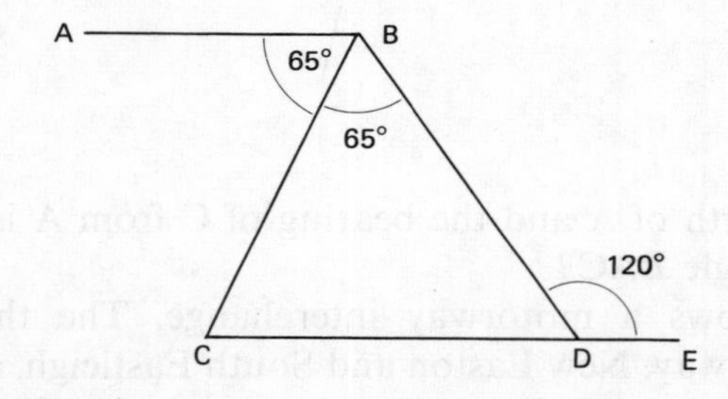

Figure 5.18

(15) In Figure 5.19, find the marked angles *a*, *b*, *c*, *d*, *e* and *f*.

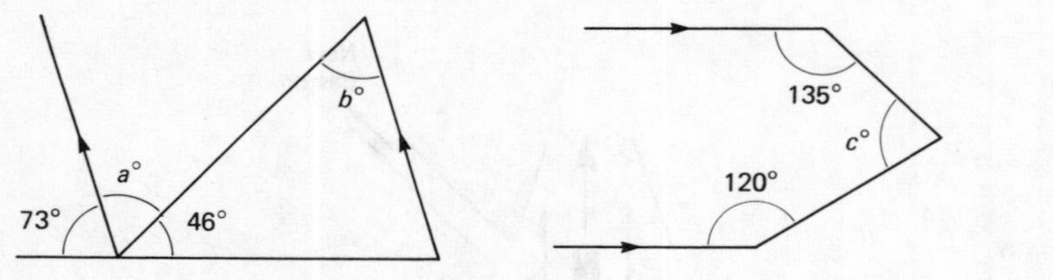

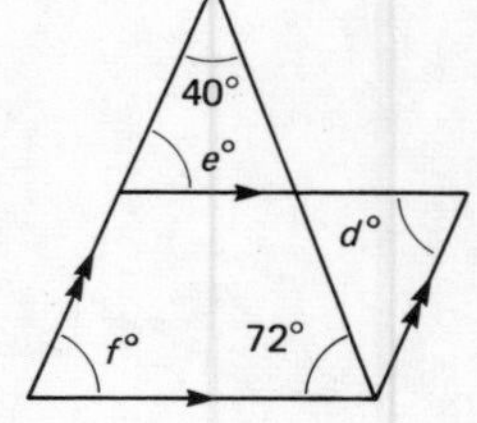

Figure 5.19

(16) Figure 5.20 shows the top of a wooden support. Find the value of *x*.

(17) A port and a lighthouse are situated such that the lighthouse is due South of the port. At 12.00 noon a ship is seen approaching the port on a bearing of 134°, and from the lighthouse the ship is on a bearing of 062°. What are the bearings of the port and lighthouse from the ship at 12.00 noon?

At the same time, another boat is on a South Easterly course away from the lighthouse. If the bearing of this boat is 165° from the port, through how many degrees will the boat have to turn in order to change direction towards the port?

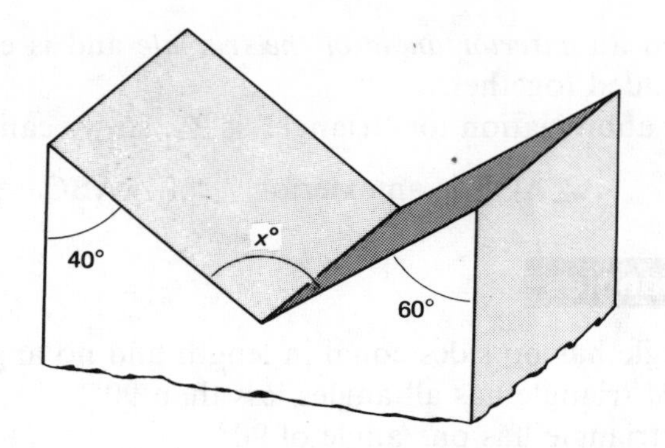

Figure 5.20

5.6 Triangles

Look at Figure 5.21.

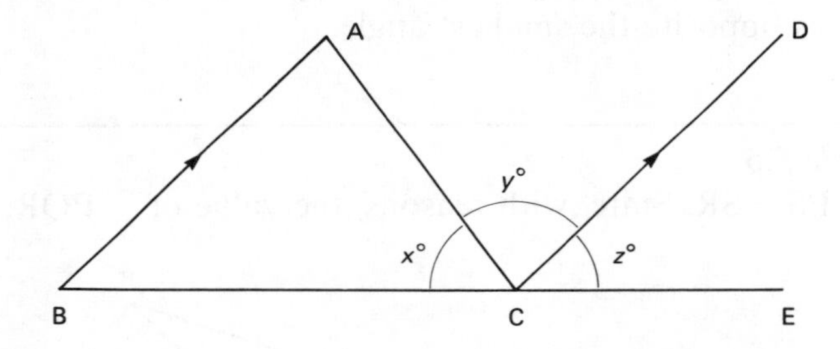

Figure 5.21

From the previous sections on angles and parallel lines,

$$\angle \text{BAC} = y^\circ \text{ (alternate angles)}$$

and

$$\angle \text{ABC} = z^\circ \text{ (corresponding angles)}$$

So the three angles of the triangle ABC are x, y and z whose sum is given by

$$x^\circ + y^\circ + z^\circ$$

But these three angles lie on the straight line BCE, so

$$x^\circ + y^\circ + z^\circ = 180^\circ$$

Hence the sum of the angles of a triangle equals 180°.

Removing the parallel line CD from the figure, we have Figure 5.22.

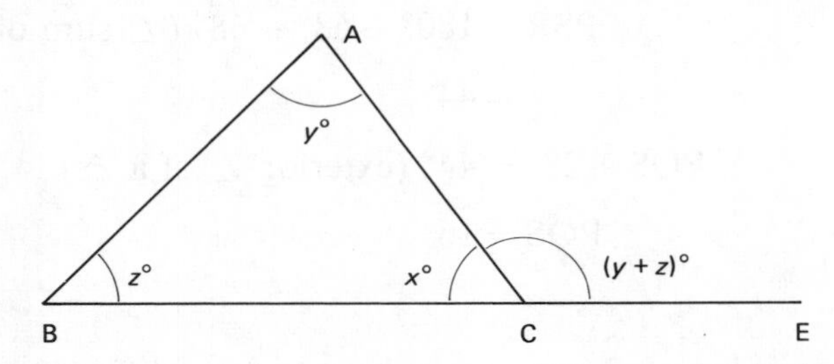

Figure 5.22

$\angle$ACE is called an *exterior angle of the triangle* and is equal to the interior opposite angles added together.

The shorthand abbreviation for 'triangle' is $\triangle$, so we can write:

$$\angle ACE \text{ is an exterior } \angle \text{ of } \triangle ABC$$

5.7 Types of triangles

1. A *scalene* triangle has no sides equal in length and no angles equal in size.
2. An *acute-angled* triangle has all angles less than 90°.
3. A *right-angled* triangle has one angle of 90°.
4. An *obtuse-angled* triangle has one angle greater than 90°.
5. An *isosceles* triangle has two sides equal. The angles opposite the equal sides are also equal.
6. An *equilateral* triangle has all sides and all angles equal.

Note: In any triangle, the largest side is opposite the largest angle, and the smallest side is opposite the smallest angle.

Worked example 5.6

In Figure 5.23 PS = SR. State, with reasons, the value of $\angle$PQR.

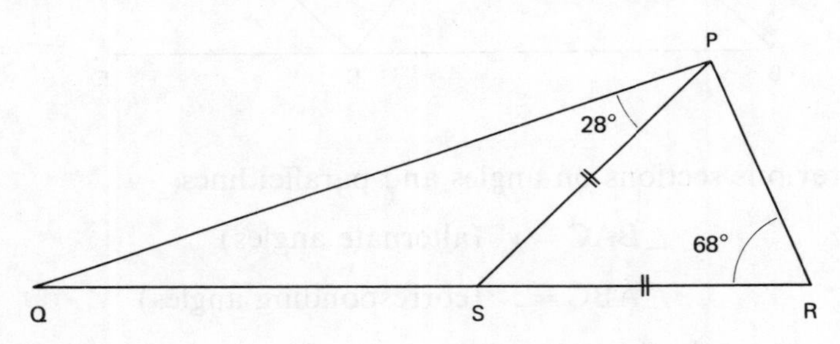

Figure 5.23

Note that sides which are equal in length are usually marked with equivalent number of notches.

As PS = SR, $\triangle$PSR is an isosceles triangle

so $\quad \angle SPR = \angle PRS = 68°$.

$$\angle PSR = 180° - 68° - 68° \ (\angle \text{ sum of } \triangle)$$

$$= 44°$$

Now $\quad \angle PQS + 28° = 44°$ (exterior $\angle$ of a $\triangle$)

so $\quad \angle PQS = 16°$

Exercise 5b

(1) A piece of wire of length 30 cm can be bent to form different triangles. Write

down possible lengths of the three sides if the triangle is:
(i) scalene; (ii) isosceles; (iii) equilateral.

(2) In Figure 5.24, find x.

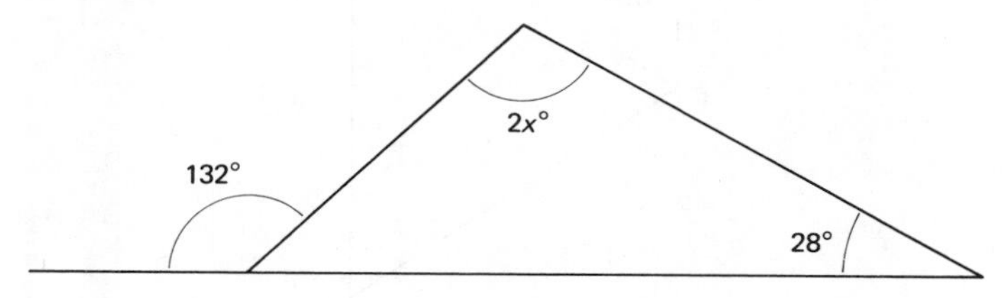

Figure 5.24

(3) An exterior angle of a triangle is $148°$ and the interior opposite angles are each $x°$. Find x.

(4) A triangle ABC has an obtuse angle at A and angle $B = 45°$. Sketch such a possible triangle and name the shortest side.

(5) In Figure 5.25, if $\angle QPS = \angle RPS$ and ST is parallel to RP, state giving your reasons, a triangle which is isosceles.

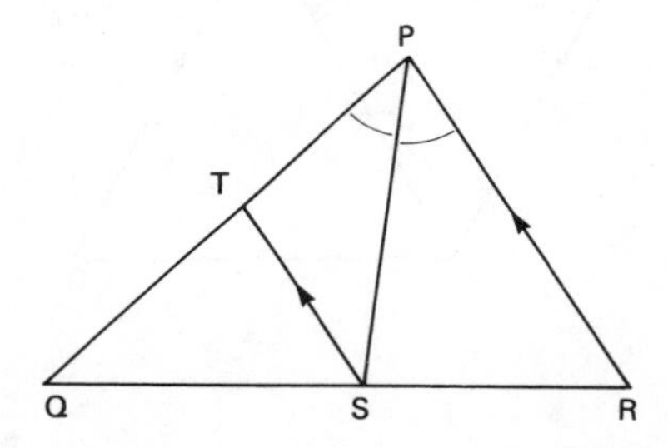

Figure 5.25

(6) Figure 5.26 shows three points A, B and C such that C is due North of A and the bearing of B from A is 040°. If $\angle CBA = 90°$, find:

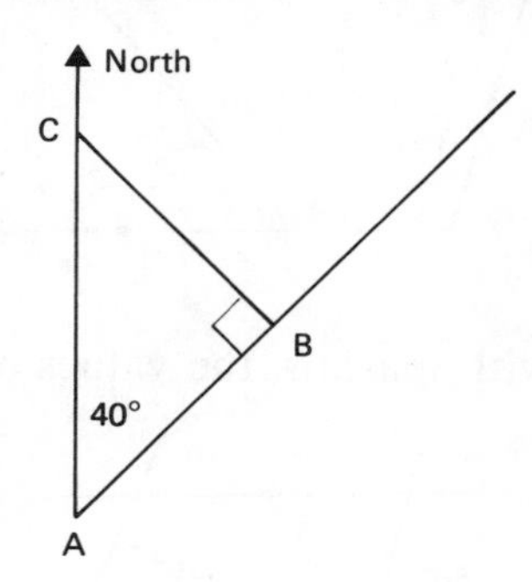

Figure 5.26

(i) the bearing of A from B,
(ii) the bearing of C from B.

(7) In Figure 5.27, the three points A, B and C form an equilateral triangle. If the bearing of A from B is 050°, find
(i) the bearing of C from B,

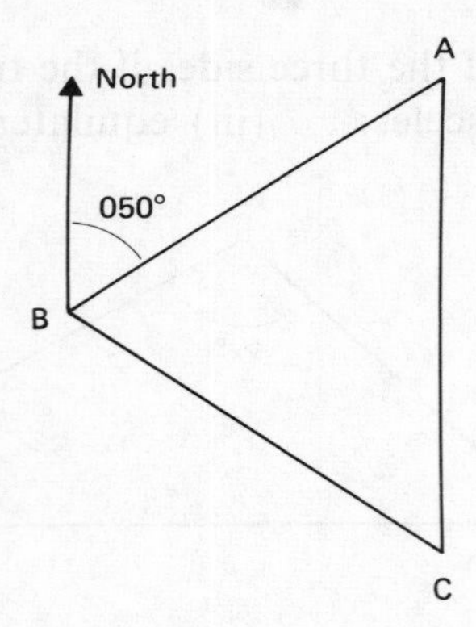

Figure 5.27

(ii) the bearing of B from C,

(iii) the bearing of C from A.

(8) In Figure 5.28, AB is parallel to DC and CD = AC = BC. Find

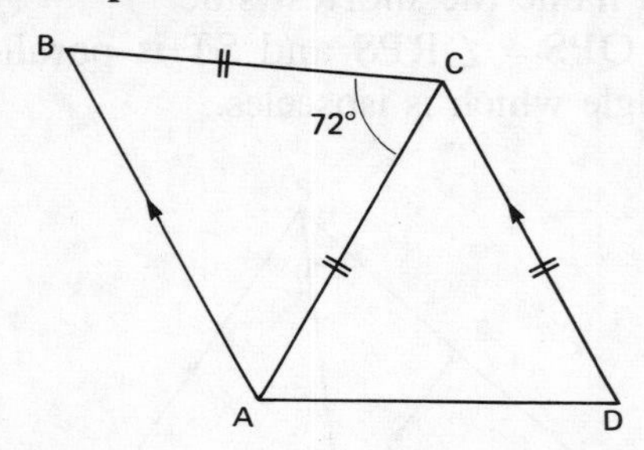

Figure 5.28

(i) ∠BAC, (ii) CAD.

(9) In Figure 5.29, find *a*, *b*, *c*, *d* and *e*.

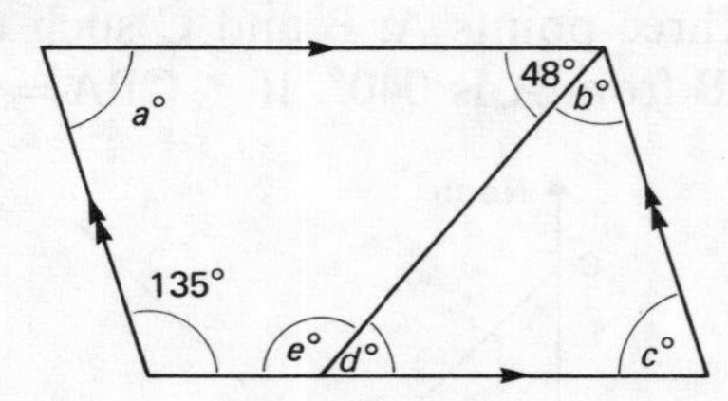

Figure 5.29

(10) In Figure 5.30, state, with reasons, the values of *a*, *b* and *c*.

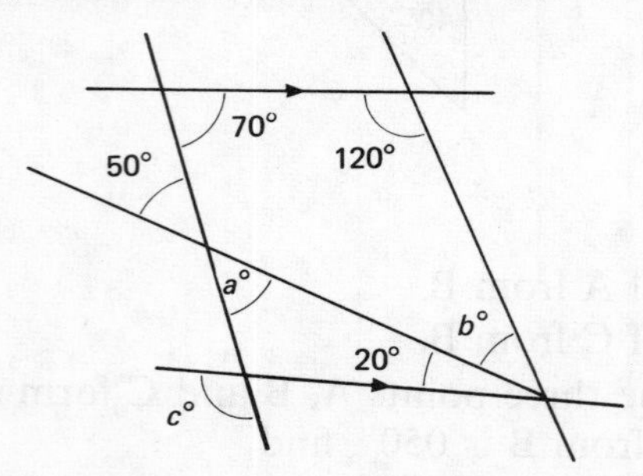

Figure 5.30

5.8 Polygons

A *polygon* is a closed plane figure whose sides are straight lines. The meeting point of two sides is called a *vertex*. In the previous section we discussed the triangle, which is a polygon with the minimum number of sides of 3.

A polygon is *regular* if all sides and angles are equal, otherwise it is *irregular*. The regular 3-sided polygon is called an *equilateral* triangle.

A 4-sided polygon is called a *quadrilateral.* Like all polygons, the quadrilateral can be divided into triangles. Figure 5.31 shows that the quadrilateral can be divided into two triangles and the sum of the six angles shown $= 2 \times 180° = 360°$.

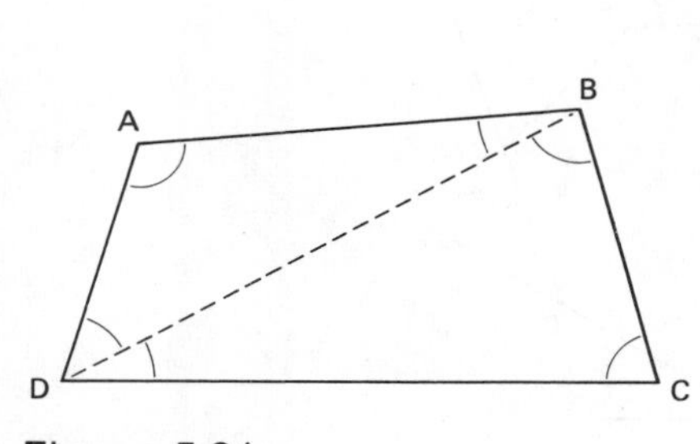

Figure 5.31

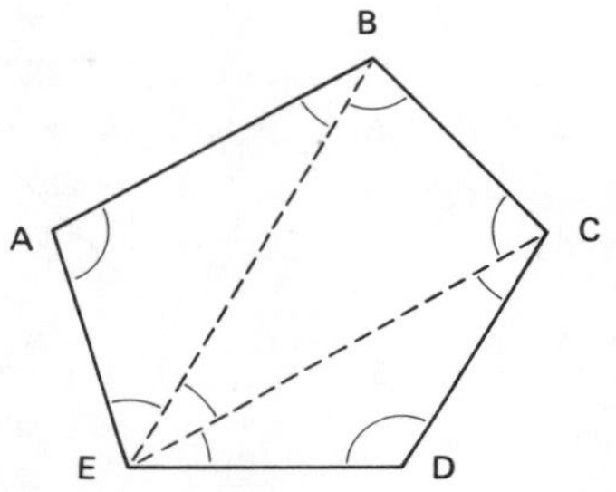

Figure 5.32

Carrying out a similar process with a 5-sided polygon, a *pentagon*, we can see that three triangles are formed (Figure 5.32). Therefore the sum of the interior angles is

$$3 \times 180° = 540°$$

Table 5.1 shows the sum of the interior angles of the first four polygons. It is left as an exercise to continue the table of polygons with more sides.

Table 5.1

Sum of Interior Angles of Polygons			
Name of polygon	Number of sides	Number of triangles	Sum of interior angles
TRIANGLE	3	1	180
QUADRILATERAL	4	2	360
PENTAGON	5	3	540
HEXAGON	6	4	720
⋮	⋮	⋮	⋮

The table shows that there is a pattern in that there are two fewer triangles than there are sides in a polygon. As each triangle has an angle sum of 180°, we can say that for a polygon with n sides, the sum of the interior angles is

$$180 \times (n-2)^\circ$$

So the sum of the interior angles of the 8-sided polygon, an *octagon*, is

$$180 \times (8-2)^\circ = 1080^\circ.$$

5.9 Exterior angles of a polygon

In any polygon we can extend each side in order to form exterior angles at each vertex, marked $e_1, e_2, e_3, e_4, e_5, \ldots$ in Figure 5.33.

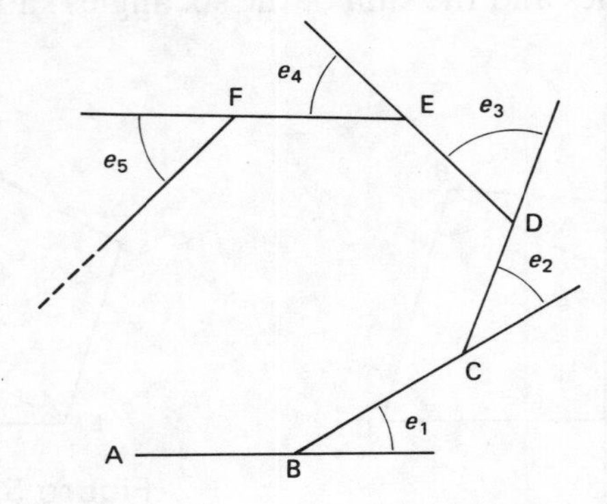

Figure 5.33

Each angle denotes the amount of turn at each vertex. If you start at A and walk around the perimeter of the polygon until you again return to A, you will have turned through one complete rotation. Therefore the number of degrees turned through is 360°, which is equivalent to the sum of the exterior angles.

For a regular polygon with n sides, the value of each exterior angle is

$$\frac{360^\circ}{n}$$

Note: the interior angle + exterior angle = 180°.

Worked example 5.7
Each interior angle of a regular polygon is 162°. How many sides has the polygon?

If the interior angle = 162°, then the exterior angle
$= 180^\circ - 162^\circ$
$= 18^\circ.$

The number of sides $= 360/18$
$= 20$ sides.

Worked example 5.8
Figure 5.34 shows a pentagon BCDEF where FB is parallel to DC and the three angles marked are equal.

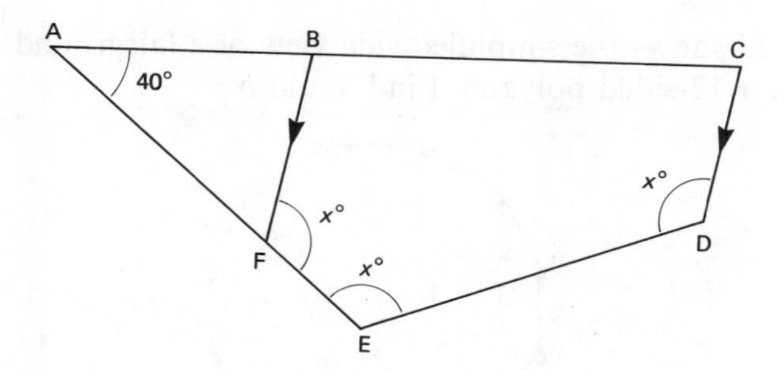

Figure 5.34

(i) Find x.
(ii) If $\angle FAB = 40°$, what is the value of $\angle FBC$?

(i) The sum of the interior angles of a pentagon

$$= 540°.$$

Now $\angle$'s FBC and BCD add up to 180° (interior $\angle$'s)

so $$3x = 540° - 180°$$
$$= 360°$$

therefore $$x = 120°.$$

(ii) $$\angle AFB + \angle BFE = 180°$$

so $$\angle AFB = 180° - 120°$$
$$= 60°.$$

Using the angle sum of a triangle:

$$\angle ABF = 180° - (60° + 40°)$$
$$= 80°$$

so $$\angle FBC = 100°.$$

Exercise 5c

(1) Find the size in degrees of each interior angle of a regular polygon with 5, 8 and 10 sides.
(2) The interior angle of a regular polygon is twice the exterior angle. How many sides has the polygon?
(3) Explain why the statement that 'each external angle of a regular polygon is 16°' cannot be true.
(4) A polygon is divided into 6 triangles by drawing all possible diagonals through one of its vertices. How many sides has it?
(5) How many diagonals has a polygon with 4, 5, 6 and 7 sides? Can you suggest how many diagonals a polygon with 12 sides would have?
(6) Each exterior angle of a regular polygon is 24°. How many sides has the polygon?

(7) Figure 5.35 shows the simplified side view of a fairground big wheel which is a regular 12-sided polygon. Find a and b.

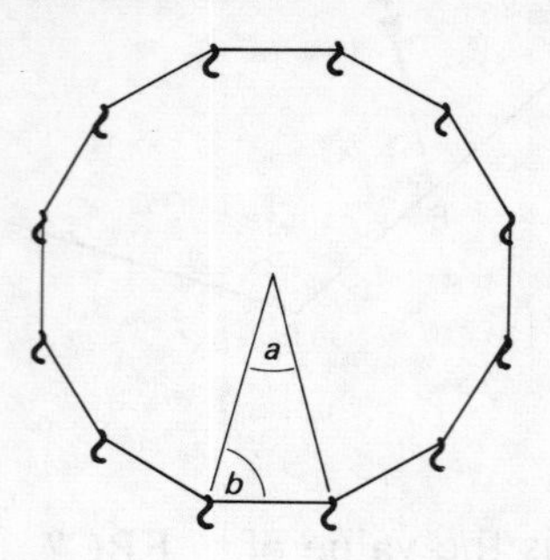

Figure 5.35

(8) A regular hexagon and a regular pentagon are placed side by side as shown in Figure 5.36. Find the value of x.

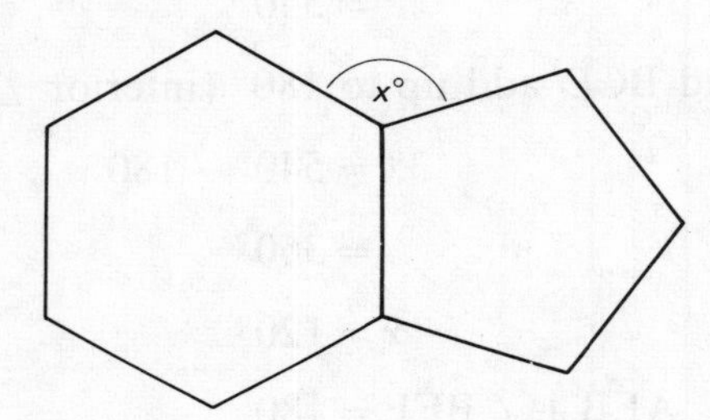

Figure 5.36

(9) In Figure 5.37, find x.

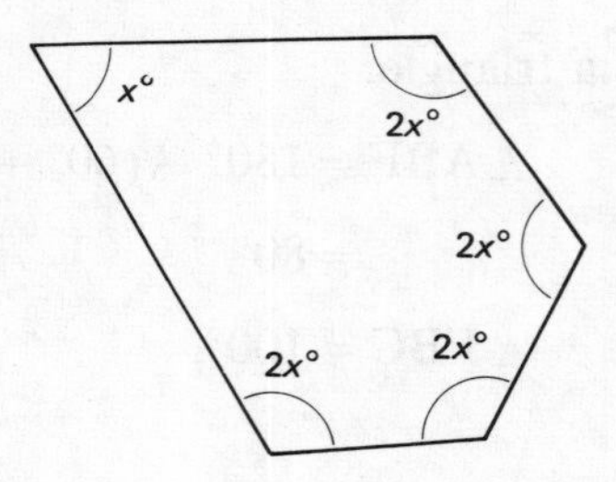

Figure 5.37

(10) In Figure 5.38, state with reasons, the value of x.

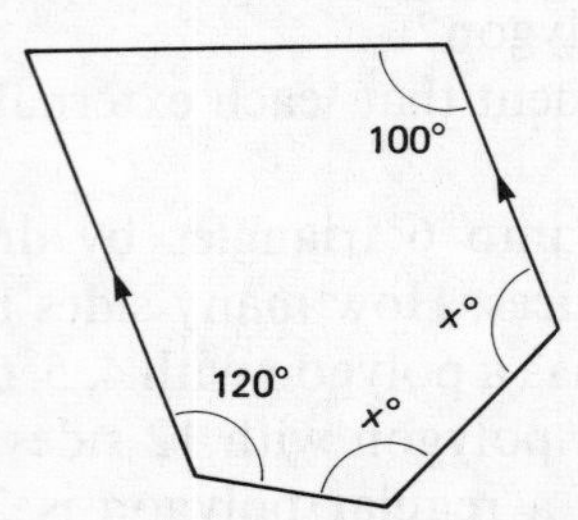

 Figure 5.38

5.10 Symmetry

For any plane figure, if a line can be drawn so that the figure is divided into two parts which are exactly the same, then the line is a *line of symmetry*.

Any plane figure can be rotated through 360° about a point such that at least once through the rotation, its position will appear to be the same. The number of times that the figure's position appears to be the same is the *order of rotational symmetry*. Since a rotation of 360° brings any figure back to its original position, the order of rotational symmetry is at least 1 for all plane figures.

If a figure appears to be the same after rotating through 180°, it is said to have *point symmetry*.

5.11 Symmetry and simple plane figures

Plane figure	Lines of symmetry	Order of rotational symmetry	Properties
(i) ISOSCELES TRIANGLE	1	1	Two sides equal Two angles equal
(ii) EQUILATERAL TRIANGLE	3	3	Three sides equal All 3 angles equal to 60°
(iii) SQUARE	4	4	Four equal sides Four angles of 90° Diagonals bisect each other at right angles
(iv) RECTANGLE	2	2	Opposite sides equal and parallel Four angles of 90° Diagonals bisect each other
(v) PARALLELOGRAM	0	2	Opposite sides equal and parallel Opposite angles equal Diagonals bisect each other
(vi) RHOMBUS	2	2	All four sides equal Opposite sides parallel Diagonals bisect at right angles Diagonals bisect vertices Opposite angles equal

(continued)

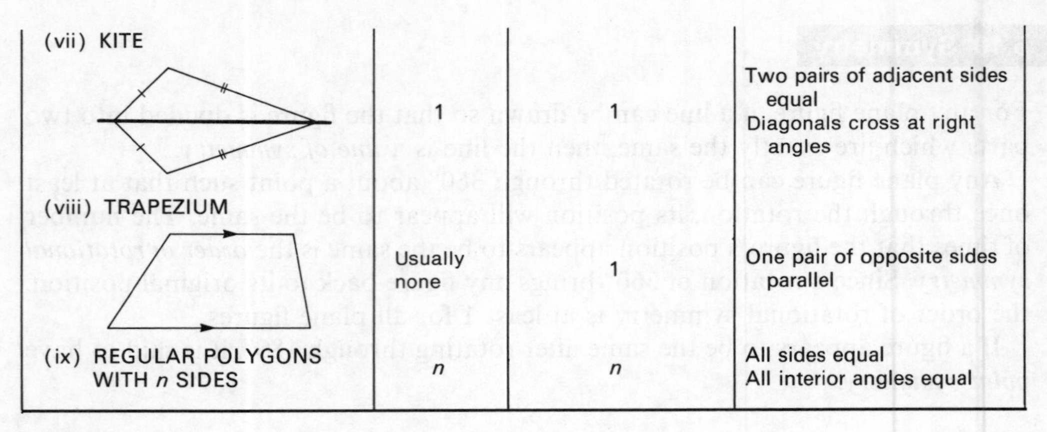

(vii) KITE	1	1	Two pairs of adjacent sides equal Diagonals cross at right angles
(viii) TRAPEZIUM	Usually none	1	One pair of opposite sides parallel
(ix) REGULAR POLYGONS WITH n SIDES	n	n	All sides equal All interior angles equal

Figure 5.39

5.12 Tessellations

Patterns which completely cover a plane are called *tessellations.* For a shape or group of shapes to tessellate, there must be no overlapping and no gaps.

When one shape is repeated the tessellation is said to be a *simple* tessellation.

When more than one shape is used in the repeated pattern, it is said to be a *compound* tessellation.

Worked example 5.9

How big is the angle gap when you try to tessellate regular pentagons?

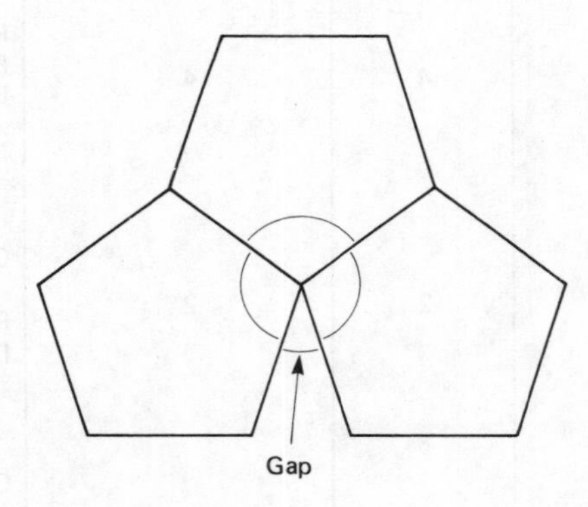

Figure 5.40

The interior angle of a regular pentagon is 108°, therefore we can only place three vertices together (see Figure 5.40) to make a total angle of

$$3 \times 108° = 324°$$

The angle gap is therefore

$$= 360° - 324°$$

$$= 36°.$$

Exercise 5d

(1) Figure 5.41 shows a kite. Find the angles *a*, *b* and *c*.

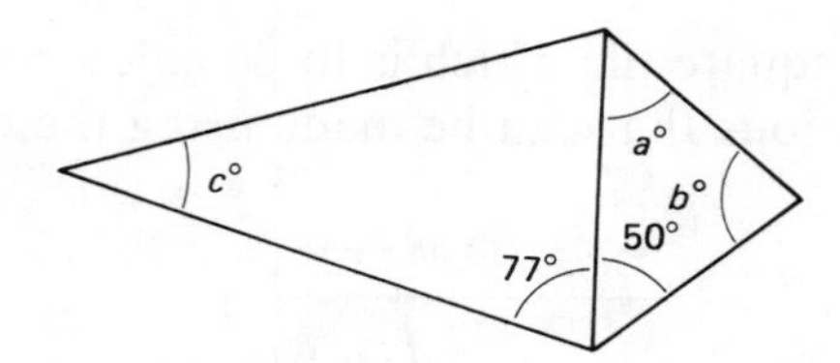

Figure 5.41

(2) Name the shape ABCD shown in Figure 5.42. Show that triangle DEC is an isosceles triangle.

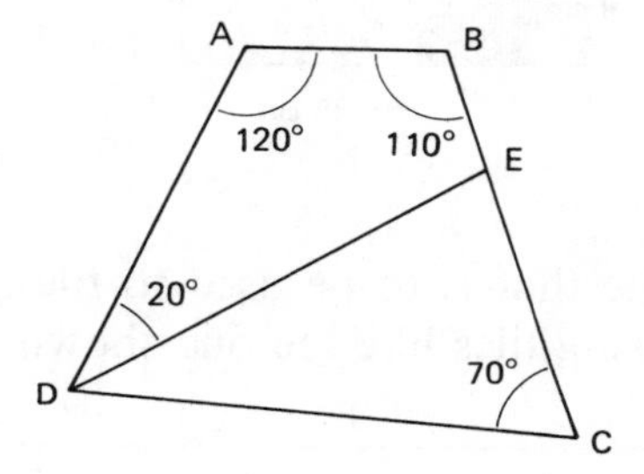

Figure 5.42

(3) Which of the letters in the sign shown in Figure 5.43 have at least one line of symmetry and also have point symmetry?

Figure 5.43

(4) ABCD is a quadrilateral which is symmetrical about the diagonal AC. If AB and BD are each 5 cm and the size of angle BCD is 120°, find the other three angles of the quadrilateral.

(5) Name the only three regular shapes which tessellate.

(6) What indoor pastime uses irregular shaped pieces to tessellate?

(7) A rectangular entrance hall is to be tiled with the two shapes shown in Figure 5.44. Show how these tiles would fit together to tessellate across the floor.

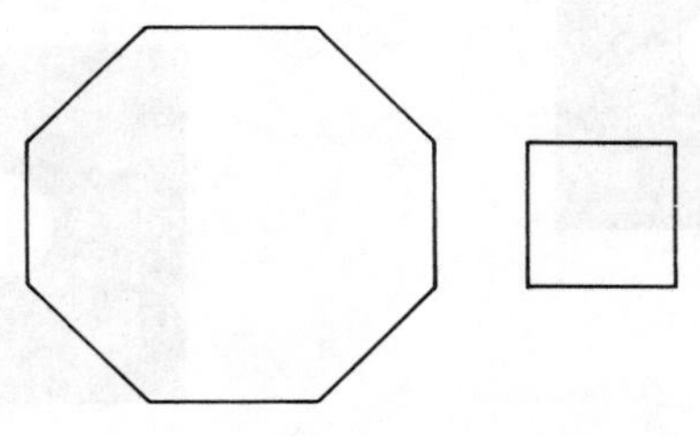

Figure 5.44

Certain other shaped tiles would be needed to fill in the corners and down the sides. What shapes would be required? Calculate the angles of one such shape.

(8) Figure 5.45 shows a square tile which is to be coloured red and blue. Sketch two different tessellations that can be made using these patterned tiles.

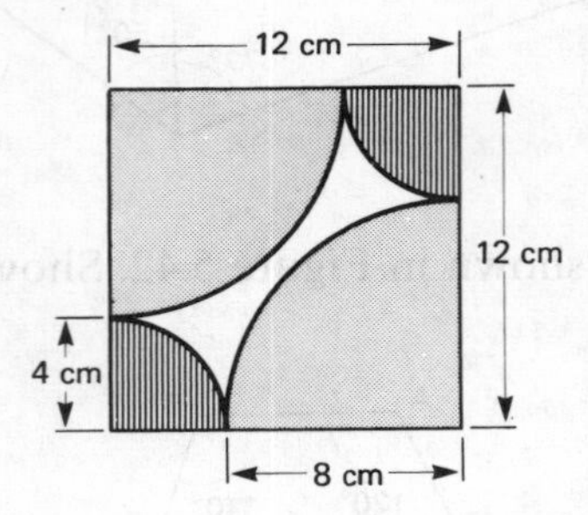

Figure 5.45

(9) Figure 5.46 shows a tile that is to be used to tile a floor. Name the shape. Sketch a tessellation using tiles like the one shown.

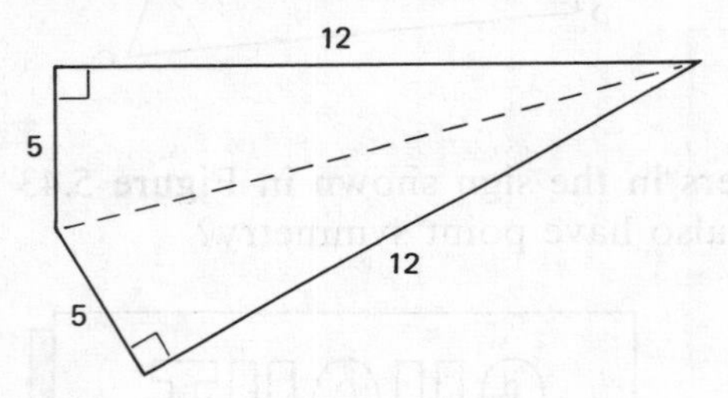

Figure 5.46

5.13 Congruency and similarity

Two figures which have the same shape and size, even though they may be in different orientations, are said to be *congruent*. One simple way of looking at the idea of congruency is to imagine placing one figure on top of the other. If there is an exact fit, the shapes are congruent.

By turning shape A shown in Figure 5.47 clockwise through 90°, you should

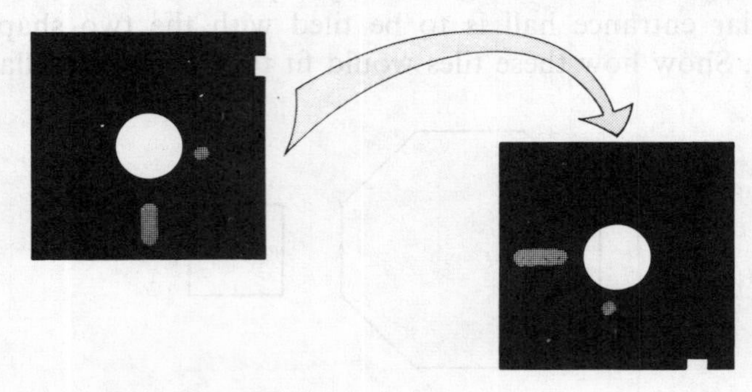

 Figure 5.47

be able to see that it can be placed exactly on top of shape B. A and B are said to be congruent.

The two triangles shown in Figure 5.48 are not congruent as they are not equal in area, however they are *similar* in shape and we can say that one triangle is an *enlargement* of the other.

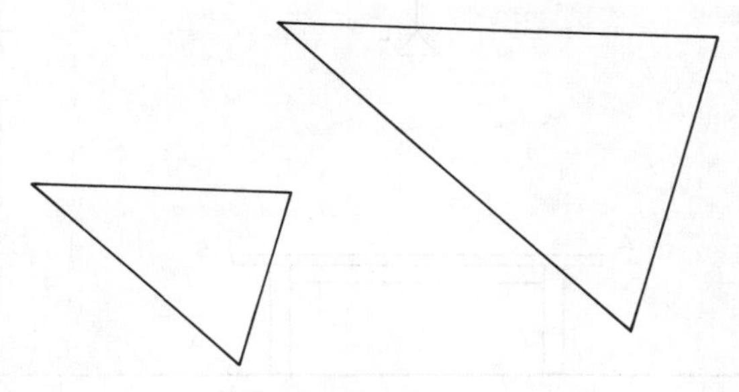

Figure 5.48

Joining corresponding vertices in the two triangles and extending the three lines until they meet gives a focus or *centre of enlargement* marked O in Figure 5.49.

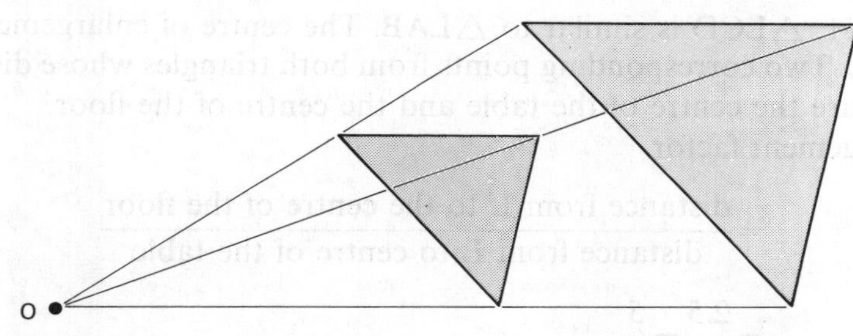

Figure 5.49

Using a ruler and measuring from O to corresponding points on the two triangles, you will find the distances to the larger triangle are twice the distances to the smaller triangle. This factor is known as the *enlargement factor* and lengths measured on the larger triangle are *twice* the corresponding lengths on the smaller triangle.

Note that we can also consider that the smaller triangle has lengths *half* those of the larger triangle. It is not necessary to have a centre of enlargement for two figures to be similar. Figure 5.50 is an example of two similar shapes. The larger shape is a rotation followed by an enlargement of the smaller figure.

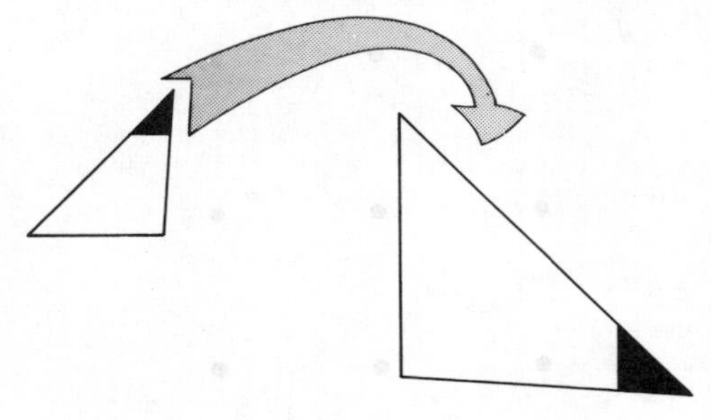

Figure 5.50

Worked example 5.10

A light is positioned 1.5 m immediately above the centre of a table which is 1 m high and 1.5 m long. Find the length of the shadow cast.

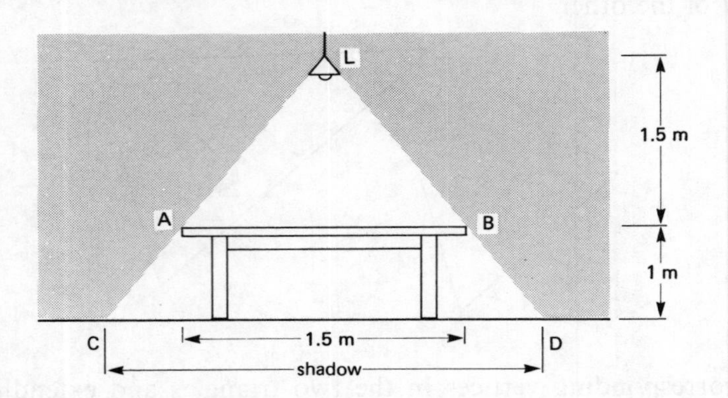

Figure 5.51

In Figure 5.51, △LCD is similar to △LAB. The centre of enlargement for both triangles is L. Two corresponding points from both triangles whose distance from L we know are the centre of the table and the centre of the floor.

The enlargement factor

$$= \frac{\text{distance from L to the centre of the floor}}{\text{distance from L to centre of the table}}$$

$$= \frac{2.5}{1.5} = \frac{5}{3}$$

The shadow on the floor corresponds to the table length in the two triangles, therefore the length of the shadow

$$= 1.5 \times \frac{5}{3} = 2.5 \text{ m}$$

Exercise 5e

(1) Using the dots shown in Figure 5.52 as the vertices of a triangle, draw two non-right-angled triangles which are congruent.

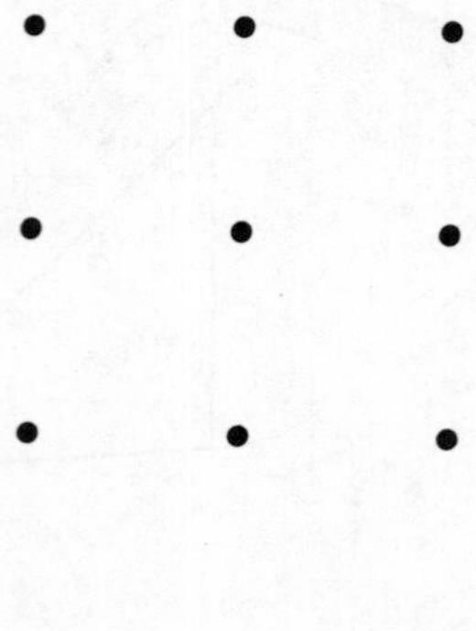

Figure 5.52

(2) In Figure 5.53, parallel lines are marked. Name two triangles which are congruent. Name two triangles which are similar, but not congruent.

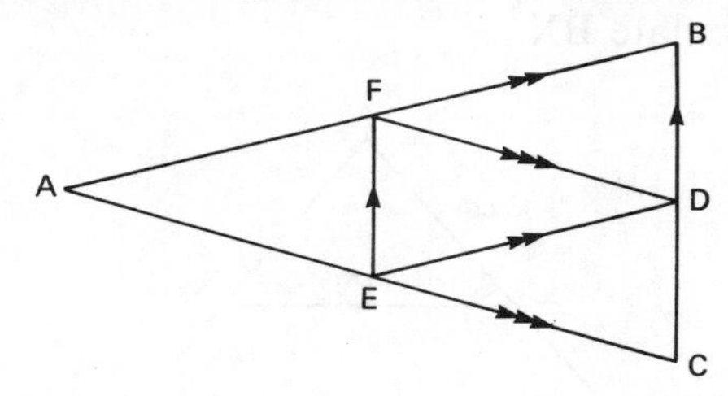

Figure 5.53

(3) In Figure 5.54, which triangles are similar? Justify your answer.

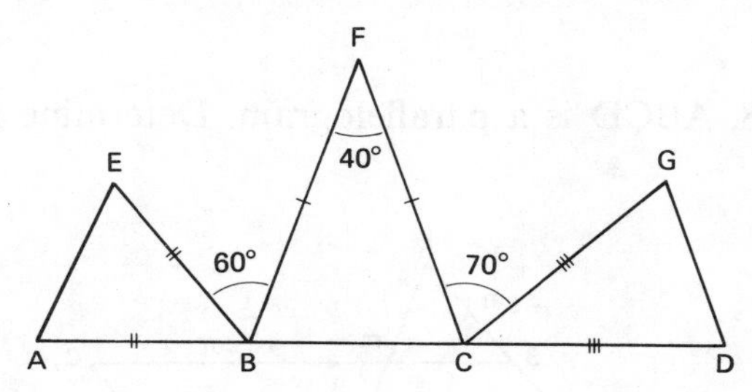

Figure 5.54

(4) In Figure 5.55, ABCD is a rectangle and $\angle DEF = 90°$. Name two similar triangles.

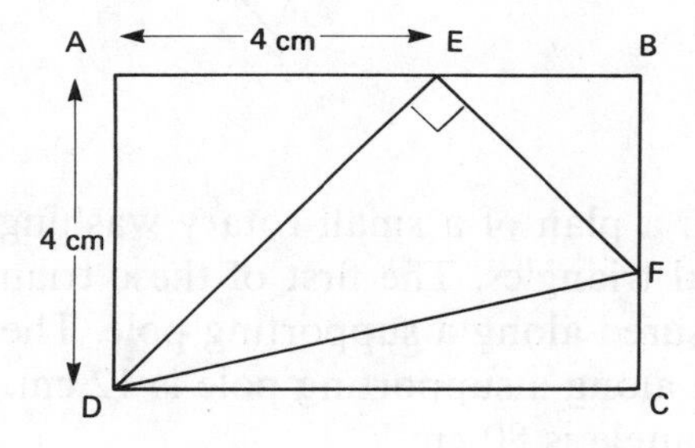

Figure 5.55

(5) To measure the approximate height of a tree, Paula made use of the sun's shadow. She measured the shadow cast by a metre stick and found it to be 0.8 m. At the same time, she measured the shadow cast by the tree and found it to be 19.4 m (see Figure 5.56). How tall was the tree?

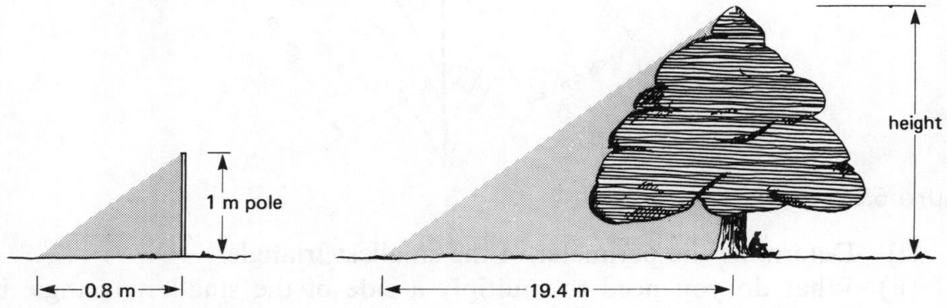

Figure 5.56

(6) An overhead projector transparency is 25 cm by 25 cm. What factor of enlargement must be used to fill a screen 1.5 m by 1.5 m?

(7) In Figure 5.57, calculate BX.

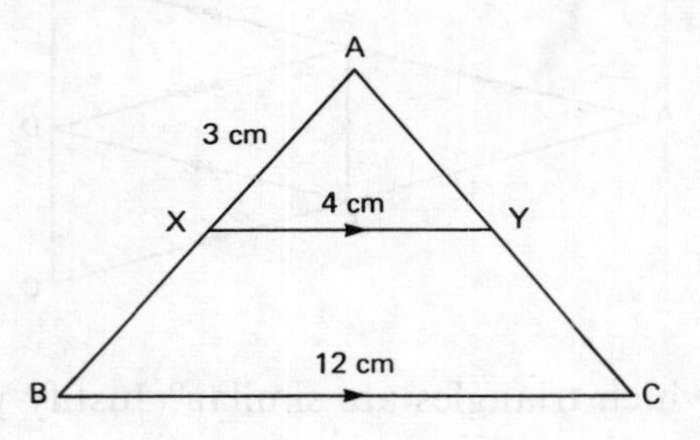

Figure 5.57

(8) In Figure 5.58, ABCD is a parallelogram. Determine the lengths BF, GD and DH.

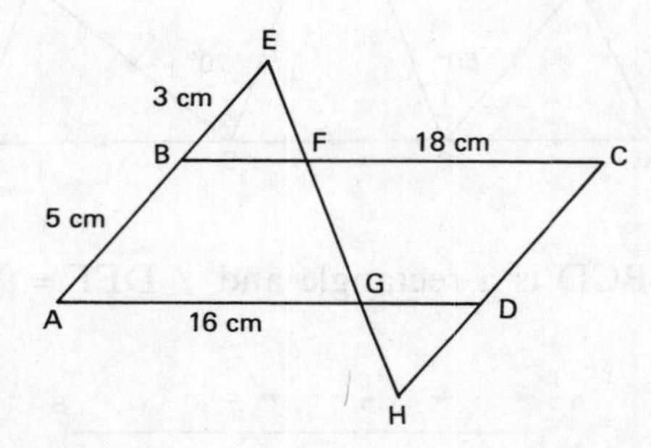

Figure 5.58

(9) Figure 5.59 shows a plan of a small rotary washing line. The line is made up of 4 equilateral triangles. The first of these triangles is 48 cm from the central pole, measured along a supporting pole. The distance between each triangle measured along a supporting pole is 12 cm. The length of one side of the smallest triangle is 80 cm.

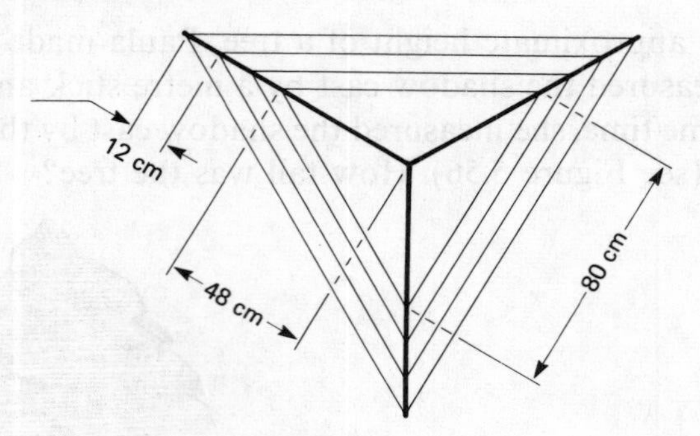

Figure 5.59

(i) Determine the perimeter of the smallest triangle.

(ii) What do you need to multiply a side of the smallest triangle by to obtain the length of a side of the next triangle in size?

(iii) What is the perimeter of the *second* smallest triangle?
(iv) Determine whether or not the line is capable of holding 15 m length of washing.

(10) In Figure 5.60, ABCD is a rectangle, and P, Q and R are the mid-points of CD, DA, and AC respectively.

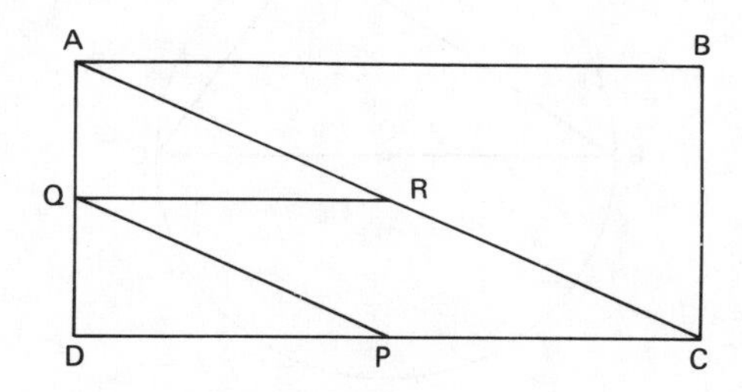

Figure 5.60

(a) Name:
(i) a triangle congruent to $\triangle ABC$,
(ii) a triangle similar to $\triangle ABC$, but not congruent to it.

(b) What type of quadrilateral is:
(i) CPQR
(ii) ACPQ?

5.14 The circle

Figure 5.61 shows the component parts of a circle.

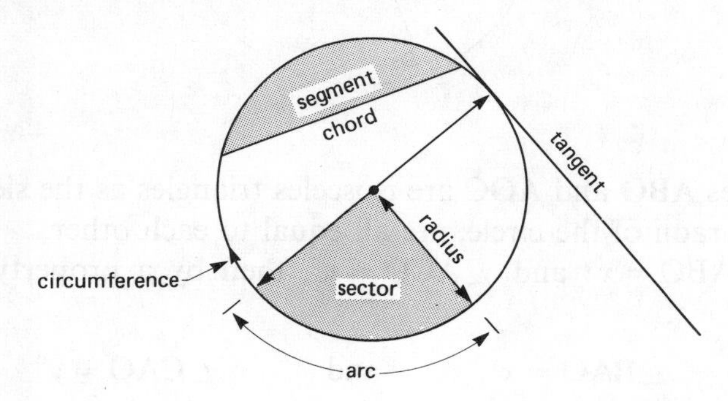

Figure 5.61

The perimeter of the circle is the *circumference*. Any part of the circumference is called an *arc*.

A *radius* is a line from the centre of the circle to the circumference.

A *chord* is any line joining two points on the circumference.

A *diameter* is a chord which passes through the centre of the circle.

A *tangent* is a line drawn from a point outside the circle which touches the circle at one point.

A *sector* is an area of the circle bordered by two radii (plural of radius), and the circumference.

A *segment* is an area of the circle bordered by a chord and an arc.

Angles in a semicircle

Figure 5.62 shows a triangle, ABC, with the point A on the circumference of the circle and BC a diameter. What is the value of $\angle$ BAC?

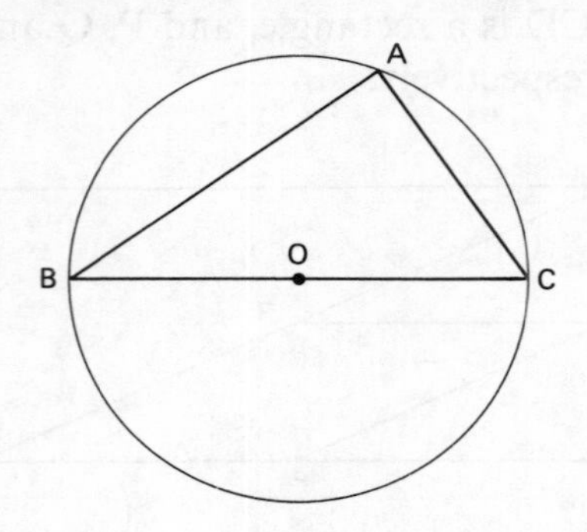

Figure 5.62

To determine this value we need to turn our attention to Figure 5.63. It is the same diagram as the previous figure except that there is an added line, AOD.

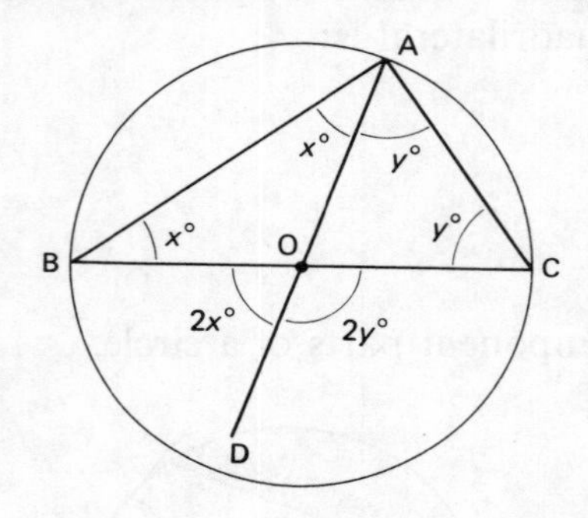

Figure 5.63

Now triangles ABO and AOC are isosceles triangles as the sides OA, OB and OC, which are radii of the circle, are all equal to each other.

If we let $\angle \text{ABO} = x^\circ$ and $\angle \text{ACO} = y^\circ$, then by a property of an isosceles triangle:

$$\angle \text{BAO} = x^\circ \qquad \text{and} \qquad \angle \text{CAO} = y^\circ$$

and by the property of exterior angles of a triangle:

$$\angle \text{BOD} = 2x^\circ \qquad \text{and} \qquad \angle \text{COD} = 2y^\circ$$

Now $2x^\circ + 2y^\circ = 180^\circ$ as these two angles lie on a straight line, so

$$x^\circ + y^\circ = 90^\circ$$

But the diagram shows that

$$\angle \text{BAC} = x^\circ + y^\circ$$
$$= 90^\circ$$

So the angle in a semicircle $= 90^\circ$

Worked example 5.11

In Figure 5.64, BC is a diameter of the circle and $\angle ABD = 142°$. Find x.

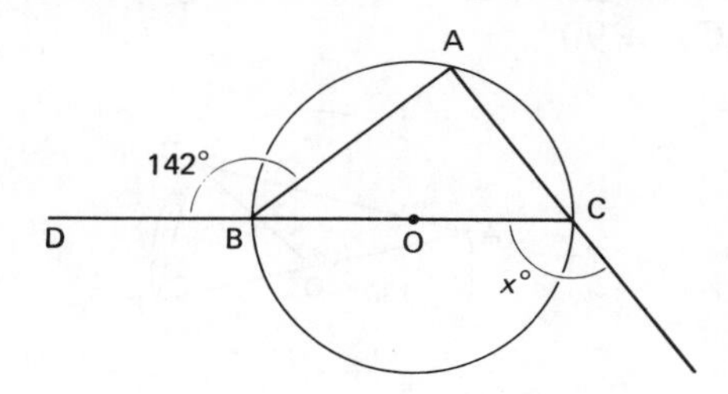

Figure 5.64

	$\angle ABC = 180° - 142°$	(angles on a straight line)
	$= 38°$	
and	$\angle BAC = 90°$	(angle in a semicircle)
so	$\angle ACB = 180° - 90° - 38°$	(angle sum of a triangle)
	$= 52°$	
so	$\angle BCE = 180° - 52°$	(angles on a straight line)
	$= 128°$	
therefore	$x = 128.$	

5.15 Tangent and radius

From any point outside a circle, not one, but *two* tangents can be drawn to a circle. Where a tangent meets the circle, the angle between the tangent and the radius is a *right angle.*

Figure 5.65 shows two tangents from a point P, to a circle. The quadrilateral PAOB is distinctly kite-like shaped, and it can be shown that this is the case as the two triangles are congruent to each other and PO is a line of symmetry.

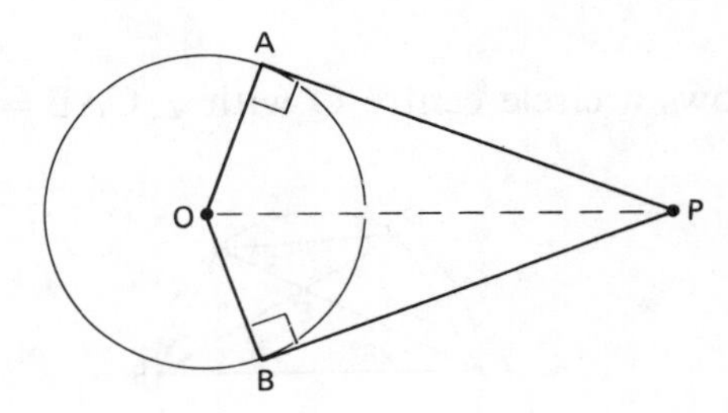

Figure 5.65

This property leads to the conclusion that the two tangents drawn from one point to a circle are equal in length.

Exercise 5f

(1) In Figure 5.66, O is the centre of the circle and $\angle AOC$ is 130°. Find
(i) $\angle CAO$, (ii) $\angle OBC$.
Show that $\angle ACB = 90°$.

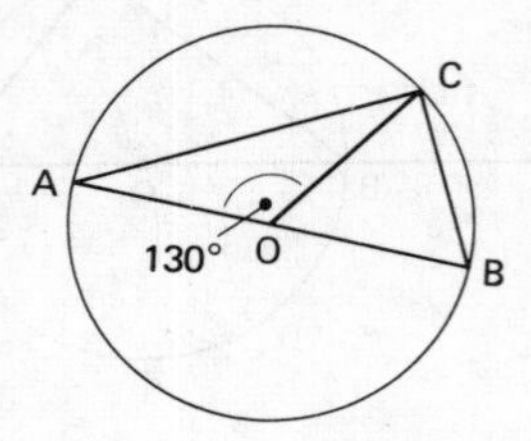

Figure 5.66

(2) In Figure 5.67, DC is a diameter, AB is parallel to DC, and angle ABD = 28°. Find x.

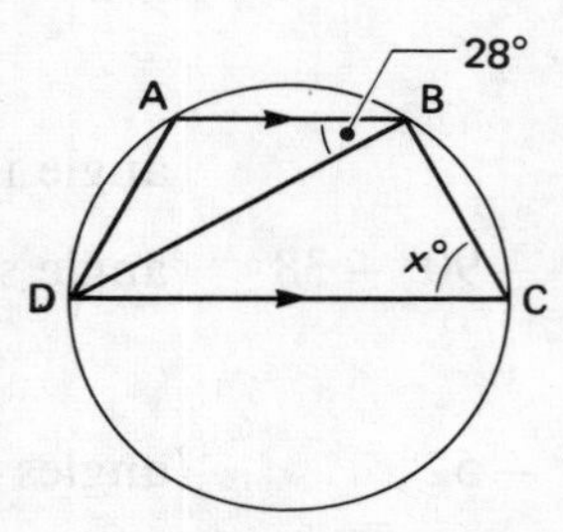

Figure 5.67

(3) In Figure 5.68, O is the centre of the circle. $\angle BAO = 25°$, and AB = BO. Find
(i) $\angle CBO$, (ii) $\angle COB$, (iii) $\angle COE$.

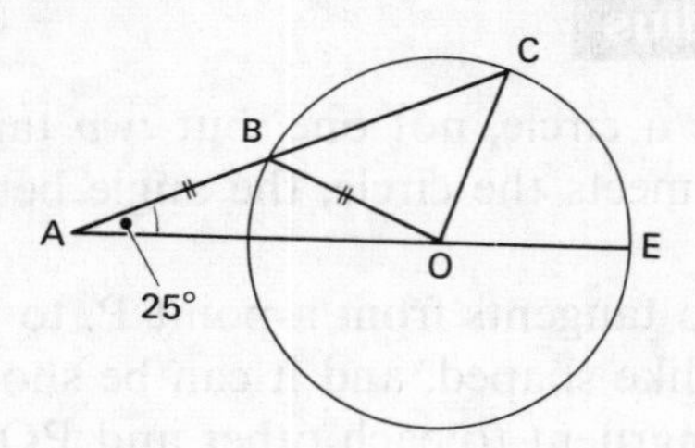

Figure 5.68

(4) Figure 5.69 shows a circle centre O with $\angle CAB = 28°$ and $\angle DAC = 34°$.

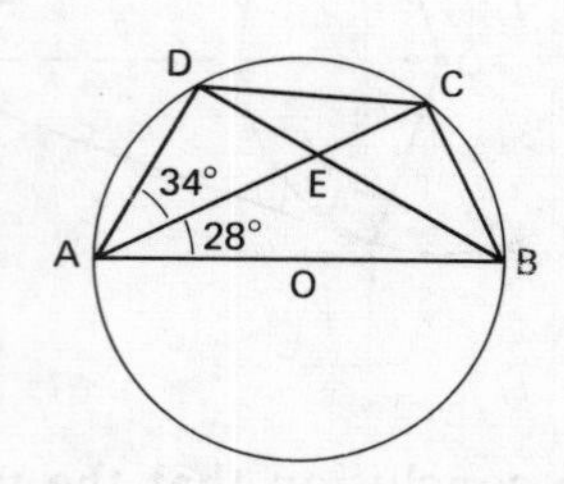

 Figure 5.69

Find
(i) ∠ABC, (ii) ∠DBC.
What shape is △AEB?

(5) In Figure 5.70, TA and TB are tangents, and O is the centre of the circle. Write down the value of the angles *a*, *b*, *c* and *d*.

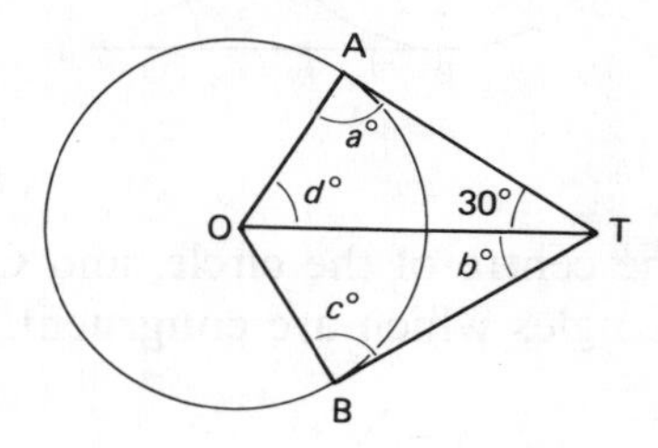

Figure 5.70

(6) In the three diagrams of Figure 5.71, TA and TB are tangents and O is the centre of the circle. State, with reasons, which of the diagrams have been incorrectly marked.

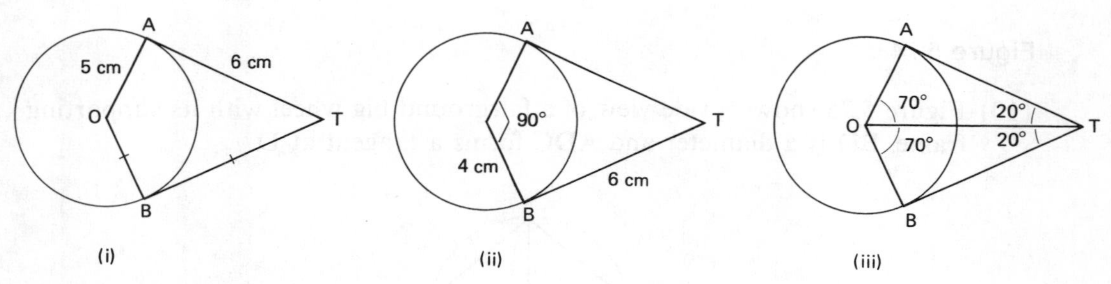

Figure 5.71

(7) In Figure 5.72, PT is a tangent to the circle centre O, and angle APT = 15°. Find, giving reasons:
(i) ∠TOP, (ii) ∠ATP.

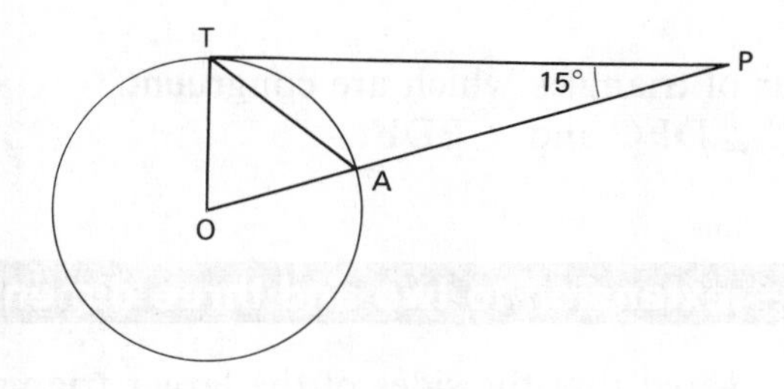

Figure 5.72

(8) Figure 5.73 shows a circle centre O, and PT is a tangent touching the circle at P. If OR is parallel to PT and angle OTP = 32°, find:
(i) ∠QOS, (ii) ∠QSO, (iii) ∠SQR.

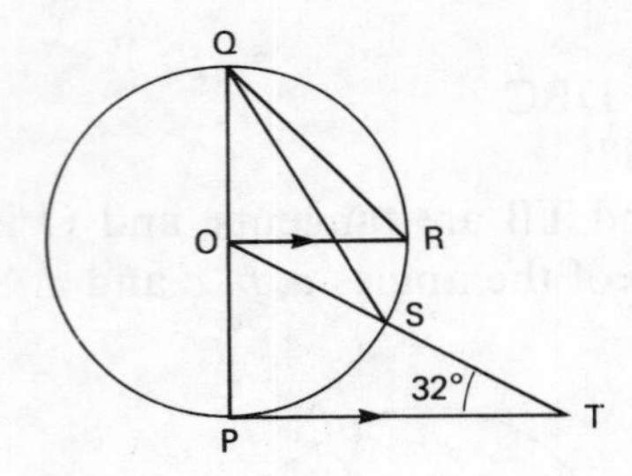

Figure 5.73

(9) In Figure 5.74, O is the centre of the circle, and CB and CD are tangents. State three pairs of triangles which are congruent.

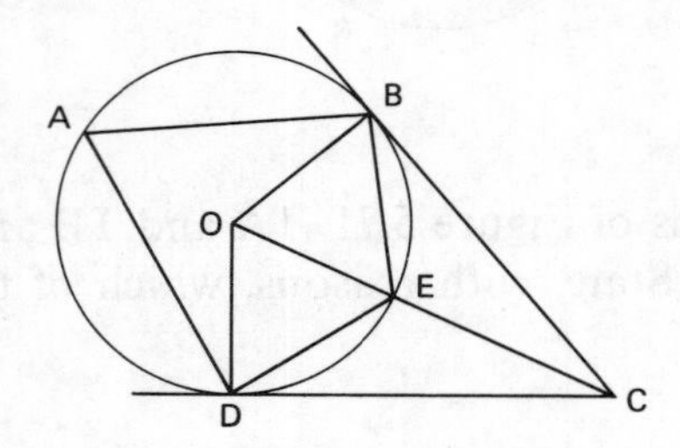

Figure 5.74

(10) Figure 5.75 shows a side view of a fairground big wheel with its supporting frame. BD is a diameter and ADC forms a tangent at D.

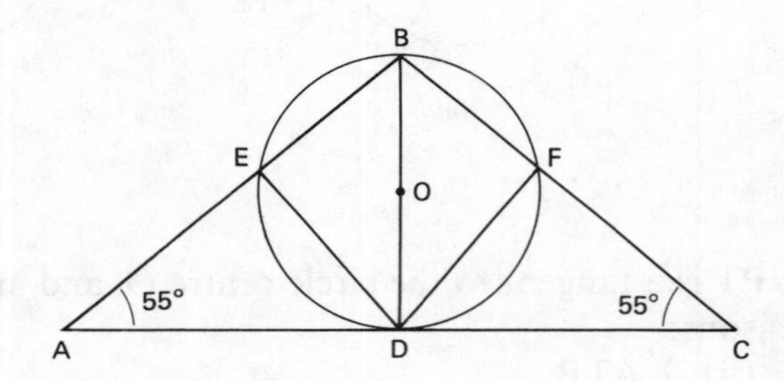

Figure 5.75

(i) State any pair of triangles which are congruent.
(ii) Find $\angle$ ABC, $\angle$ DFC and $\angle$ BDF.

5.16 Enlargement — extension to two and three dimensions

In Figure 5.49, we discovered that the sides of the larger triangle were twice the size of the smaller triangle and the enlargement factor was 2. How many triangles congruent to the smaller one can we fit inside the larger triangle? Figure 5.76 shows, in fact, that there are four.

Let us look at another example:

If we take a square of side 1 cm and a square of side 3 cm, the enlargement

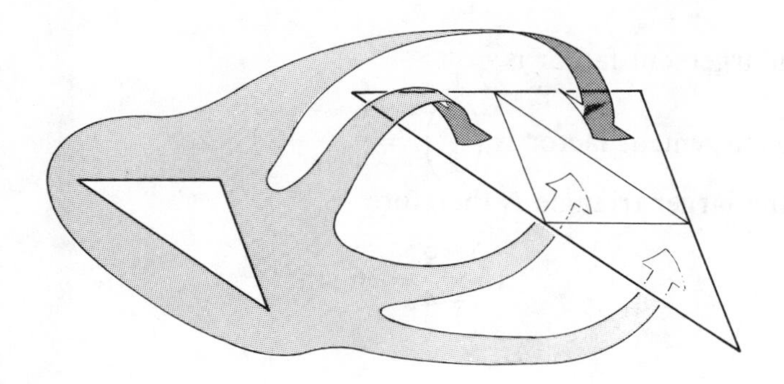

Figure 5.76

factor is 3. In terms of area, however, we can fit 9 of the smaller squares in the larger square (see Figure 5.77).

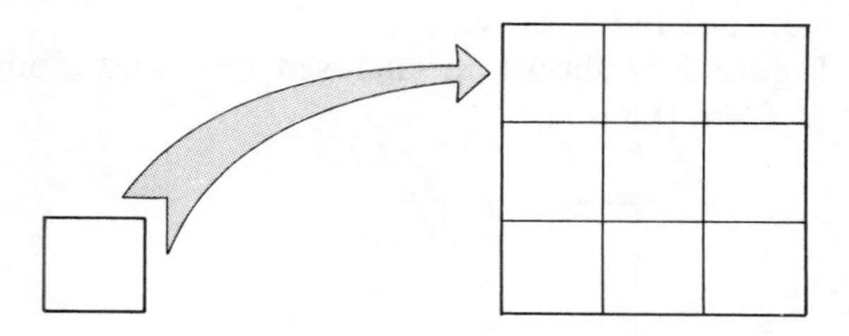

Figure 5.77

So for a *length* enlargement factor of 3, the equivalent *area* enlargement factor is 9.

Length enlargement factor	*Area enlargement factor*
$\times 2$	$\times 4$
$\times 3$	$\times 9$

In general, if the length enlargement factor is x, the area enlargement factor is x^2.

Extending to three dimensions, the *volume* enlargement factor is x^3, i.e. in Figure 5.78 the length enlargement factor is 3 so the volume enlargement factor is 3^3 which is 27. The figure shows that, using a little imagination, we can place 27 of the smaller 1 cm^3 cubes inside the larger cube.

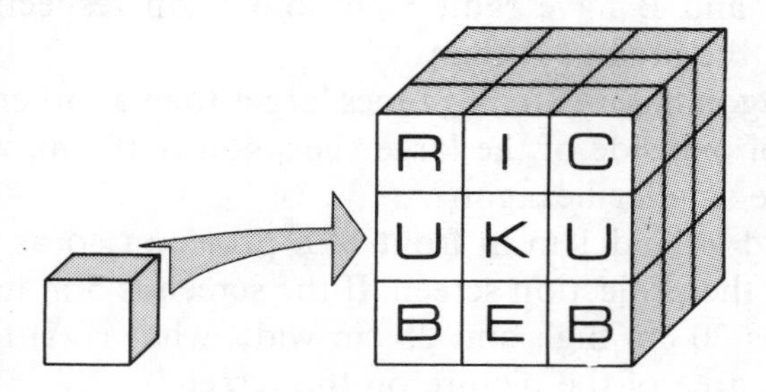

Figure 5.78

Worked example 5.12

The larger of two similar triangles is $1\frac{1}{2}$ times taller than the smaller one. If the area of the smaller triangle is 16 cm^2, find the area of the larger triangle.

The length enlargement factor is $\frac{3}{2}$.

So the area enlargement factor is $\left(\frac{3}{2}\right)^2 = \frac{9}{4}$.

The area of the larger triangle is therefore

$$16 \times \frac{9}{4} = 36 \text{ cm}^2$$

5.17 Nets

A *net* is the end result of unfolding and opening out a geometric solid so that all faces of the solid lie on one flat surface.

As an example, Figure 5.79 shows an enclosed box, and along side it a net which can be used to form the box.

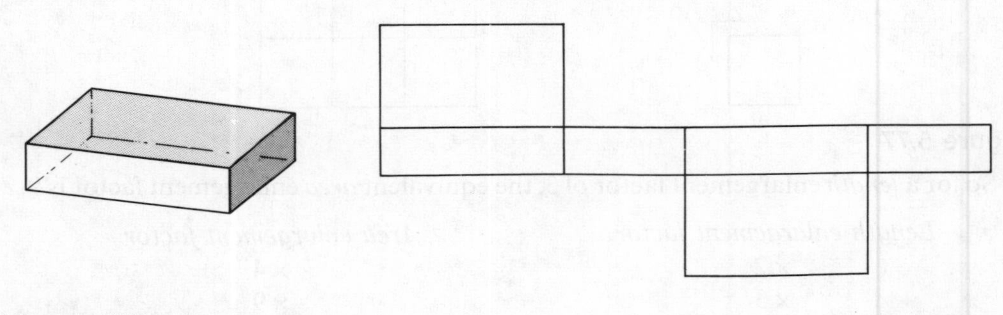

Figure 5.79

Do not cut this Figure out of the book, but make a copy and try it for yourself!

Exercise 5g

(1) Two circles A and B have radii 3 cm and 5 cm respectively. How many times larger is the area of B than A?

(2) A regular hexagon has an area $6\frac{1}{4}$ times larger than another regular hexagon. If the length of one side of the larger hexagon is 10 cm, what is the length of a side of the smaller hexagon?

(3) A piece of card is held 1 m in front of a projector so as just to blank out the picture on the projection screen. If the screen is 5 m from the projector, and the card is 20 cm high and 25 cm wide, what is (i) the height, (ii) the width, (iii) the area of the picture on the screen?

(4) The local Q-less DIY superstore is promoting its own brand of paint this month by offering free sample pots for customers to try out. The pots are identical to the 1.6 litre cans of paint but are only 1/4 of the height. How many ml of paint are there in a sample pot? The quantity of paint in the sample pot is sufficient to paint a surface area of 0.5 m². What surface area would a 1.6 litre can of paint cover?

Figure 5.80

(5) Varnworth VIth Form College 1st XI won the U19 Challenge Cup. The cup has to be returned after a year but the college are given a replica to keep. The cup is 30 cm tall and the replica is 15 cm tall, and both are made of the same material. If the larger cup weighs 640 g, what is the weight of the replica?

(6) Four equal cubes are such that the sum of their surface areas is the same as the surface area of a cube with sides of length 6 cm. What is the length of an edge of each of the smaller cubes?

(7) Which of the nets shown in Figure 5.81 cannot be formed into a cube?

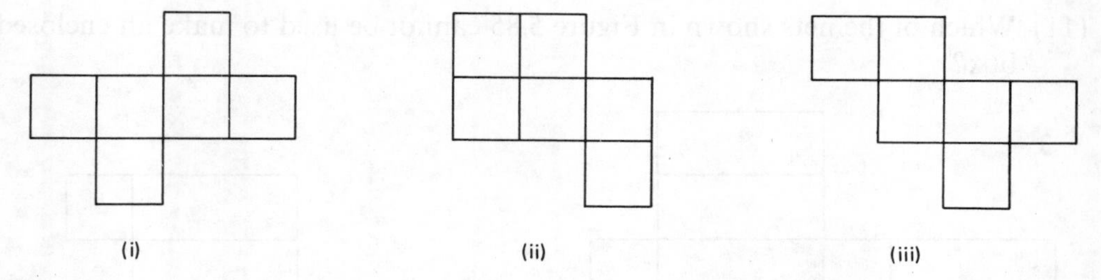

Figure 5.81

(8) Figure 5.82, not drawn to scale, shows a net for a square based pyramid. Write down the lengths ZA, ZD, BX, BC. State a pair of congruent triangles.

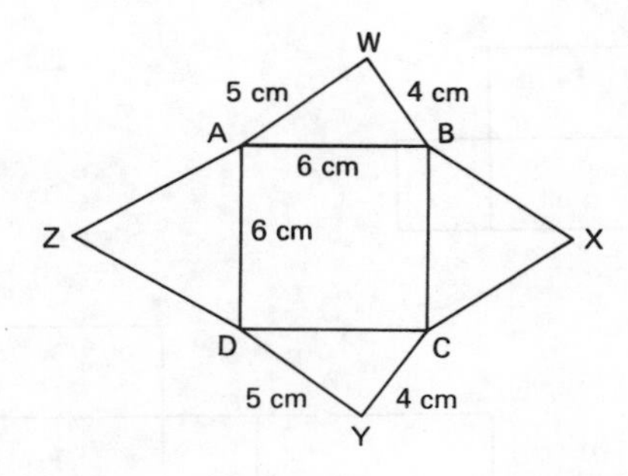

Figure 5.82

(9) In Figure 5.83, another square needs to be added to the net so that when you cut it out and fold it up you will get a cube. Copy the figure and add one more square to complete the net.

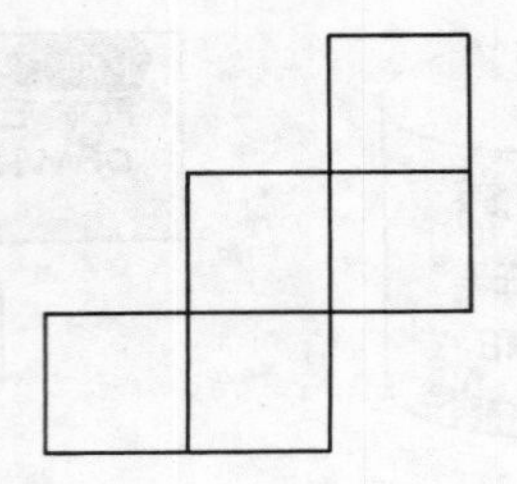

Figure 5.83

(10) Which of the nets shown in Figure 5.84 cannot be used to make a triangle based pyramid?

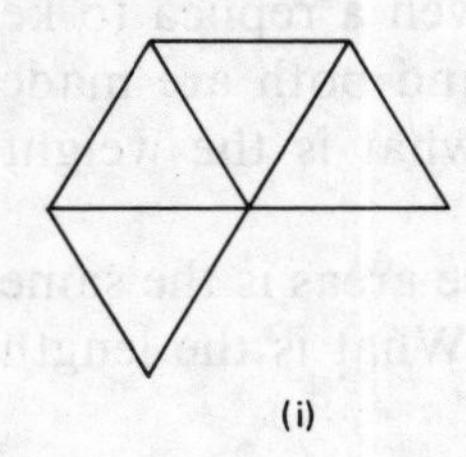
(i)

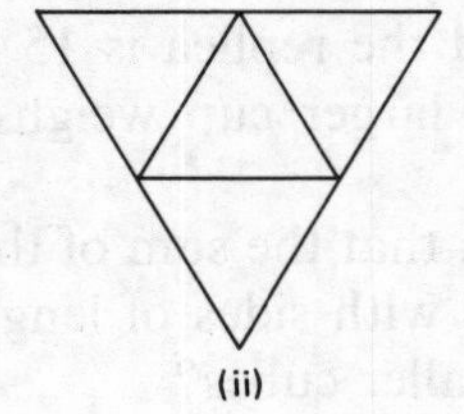
(ii)

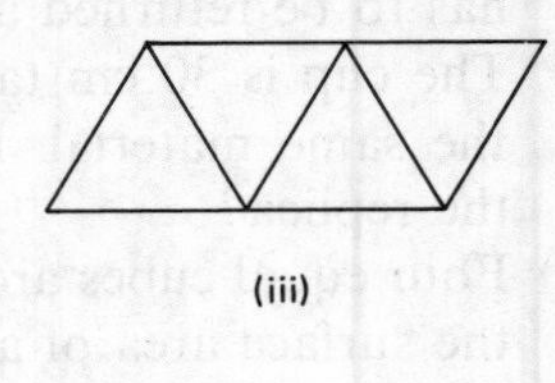
(iii)

Figure 5.84

(11) Which of the nets shown in Figure 5.85 cannot be used to make an enclosed box?

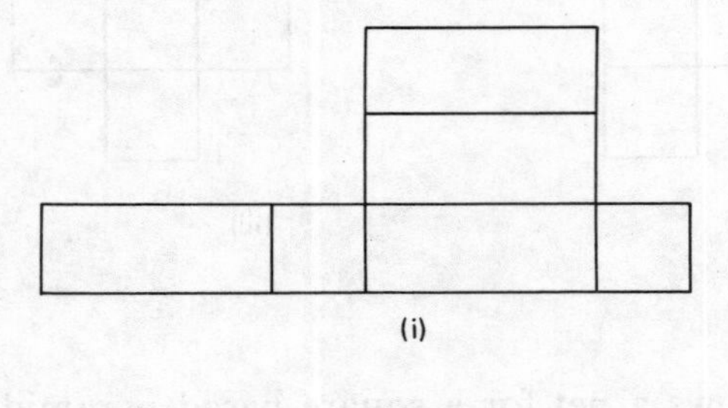
(i)

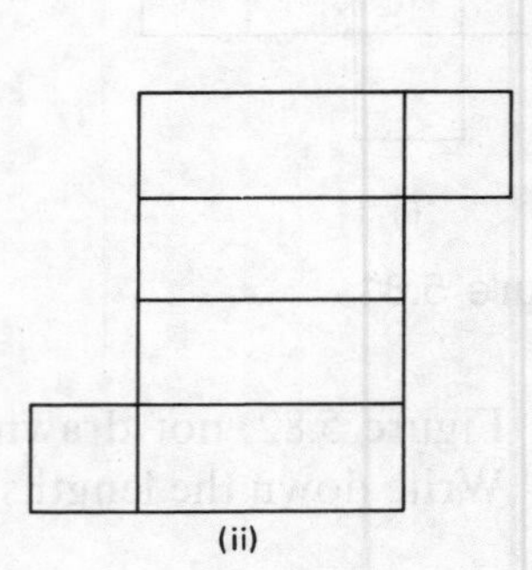
(ii)

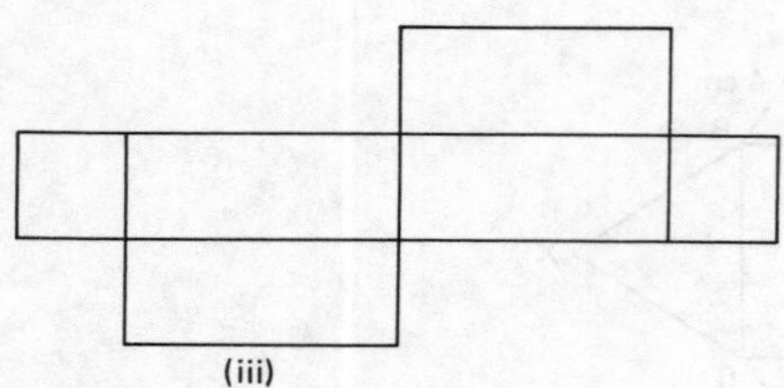
(iii)

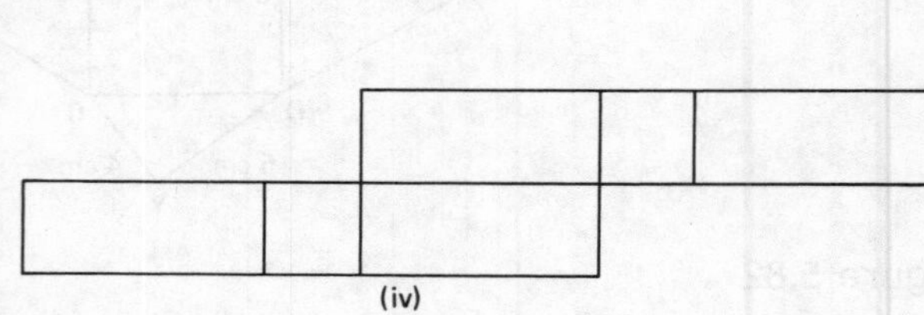
(iv)

Figure 5.85

(12) Only one of the three nets drawn can be used to construct the die shown in Figure 5.86. Determine which is the correct net.

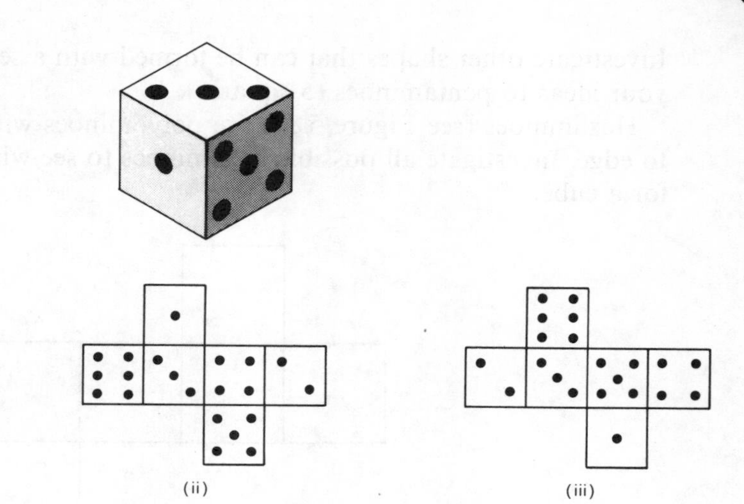

Figure 5.86

Suggestions for coursework 5

1. Figure 5.87 shows a circle with 4 points A, B, C and D marked on the circumference. All possible pairs of points are joined to form chords. Count the number of lines and the number of regions formed. Extend this investigation for 5, 6, 7 ... points on the circumference.

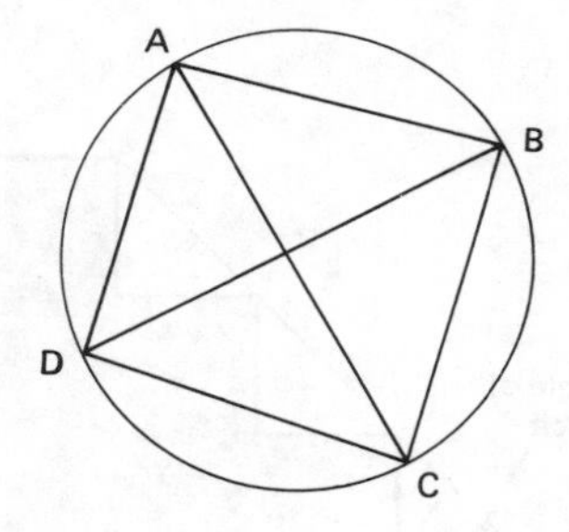

Figure 5.87

2. Squares which are joined together edge to edge to form flat shapes are called *polyominoes*. Polyominoes with four squares are called tetraminoes (see Figure 5.88).

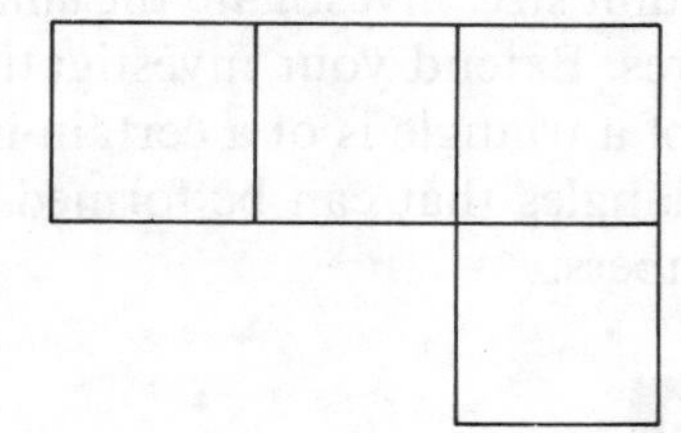

Figure 5.88

Investigate how many different tetraminoes can be designed. Investigate the number of different ways of constructing a 4×4 square, using four tetraminoes.

Investigate other shapes that can be formed with a set of tetraminoes. Extend your ideas to pentaminoes (5 squares).

Hexaminoes (see Figure 5.89) are polyominoes with 6 squares joined edge to edge. Investigate all possible hexaminoes to see which can be used as a net for a cube.

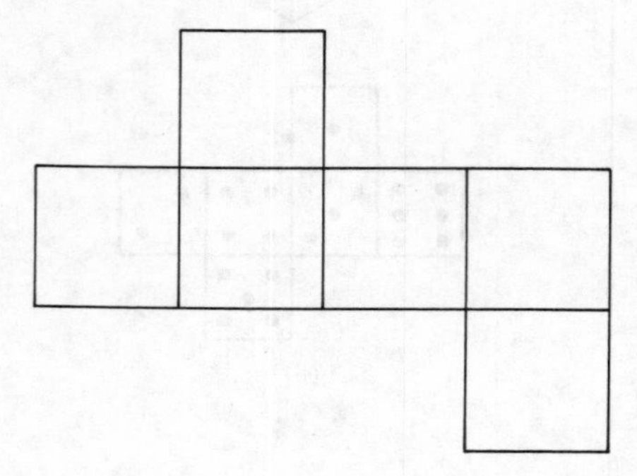

Figure 5.89

3. The steepness of stairs or steps can be measured by the angle of pitch (see Figure 5.90). This angle is too shallow if less than 25° and too steep if it is greater than 42° for a private stairway (i.e. at home), or greater than 38° for a common stairway in a block of flats, office or school. Investigate the pitch of stairs or steps in buildings of your choice (this idea could be extended to ladders!).

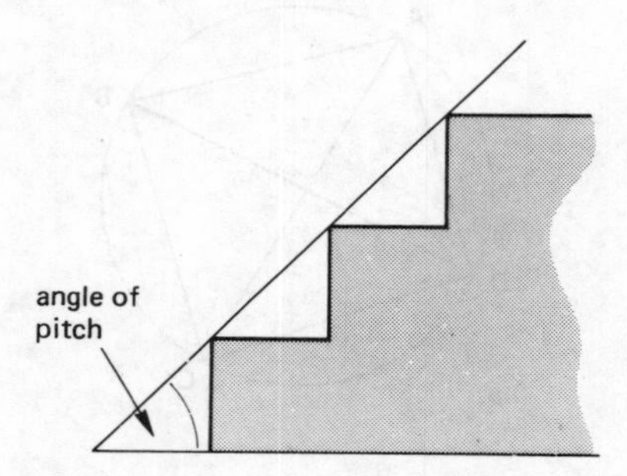

Figure 5.90

4. Starting with 5 squares of unit size, investigate the smallest square frame which can fit around these squares. Extend your investigations to 6, 7, ... squares.
5. Given that the perimeter of a triangle is of a certain integer length, investigate the number of possible triangles that can be formed if the length of the sides should also be whole numbers.

Miscellaneous exercise 5

1. Figure 5.91 shows the location of three places: Anywhere, Bear Inn and Combe Pass. Write down the bearing of Combe Pass from Bear Inn and the bearing of Anywhere from Bear Inn.

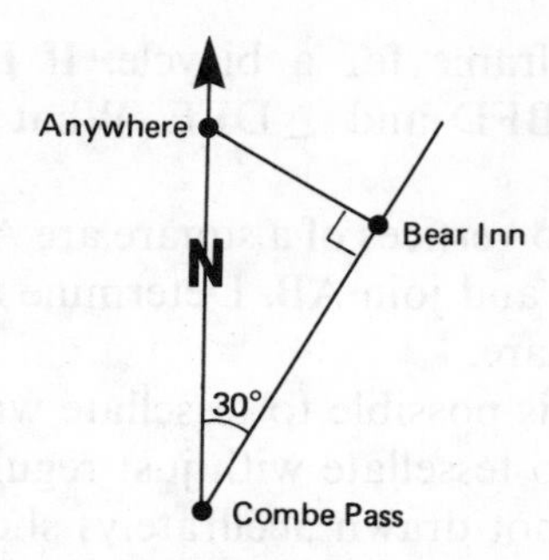

Figure 5.91

2. Figure 5.92 shows the side view of a jewellery box with the lid partially open. Calculate the value of x.

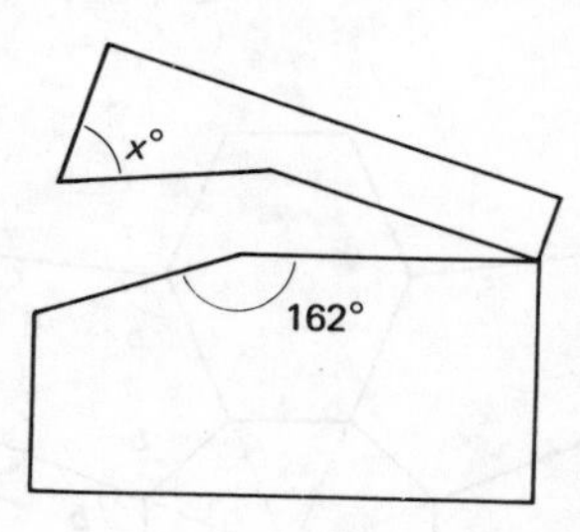

Figure 5.92

3. The isosceles triangle with a 90° angle and two 45° angles can be said to have one of its angles twice one of its other angles ($90° = 2 \times 45°$). Find the other isosceles triangle which has the same property.
4. Which of the following is not an interior angle of a regular polygon? Justify your answer.
   (i) 160°, (ii) 162°, (iii) 165°, (iv) 166°.
5. Two of the interior angles of an irregular polygon are 90°. What are the exterior angles corresponding to these two interior angles? If all the remaining interior angles are 144°, what is the value of the corresponding exterior angle to one of these interior angles? How many sides has the polygon?
6. The four angles of a quadrilateral are in the ratio 3:4:5:6. Calculate the angles and draw a sketch of the quadrilateral. What name do we give to this type of quadrilateral?

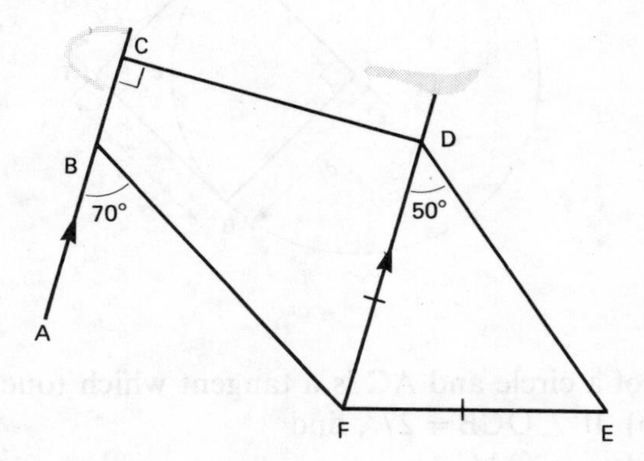

Figure 5.93

7. Figure 5.93 shows a frame for a bicycle. If DF = FE, ∠ABF = 70° and ∠FDE = 50°, find ∠BFD and ∠DFE. What name is given to the shape BCDF?
8. The co-ordinates of two vertices of a square are A(1, 1) and B(8, 0). Plot these points on graph paper and join AB. Determine the co-ordinates of the other two vertices of the square.
9. Explain briefly why it is possible to tessellate with regular hexagonal shapes but it is not possible to tessellate with just regular pentagons.
10. Figure 5.94 (which is not drawn accurately) shows part of a tiled floor in a church in Corfu. Three kinds of tiles are used. Some are regular octagons. Some are hexagons with two lines of symmetry. Some are pentagons with one line of symmetry. Calculate the angles *a*, *b*, *c*, *d*. Give reasons for your answers.

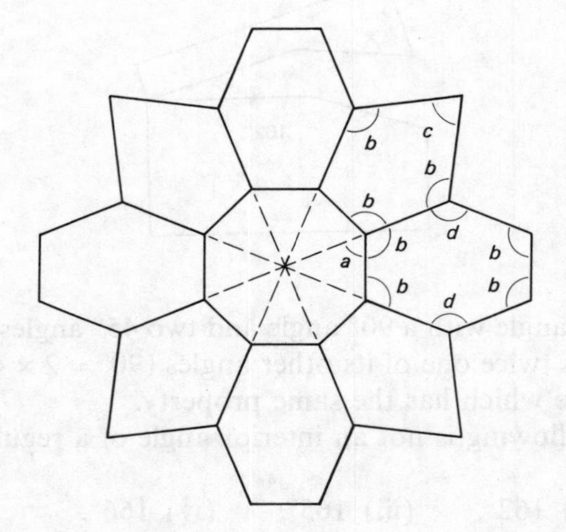

Figure 5.94

11. In Figure 5.95, TA and TB are tangents and O is the centre of the circle. Find *a*, *b*, *c* and *d*.

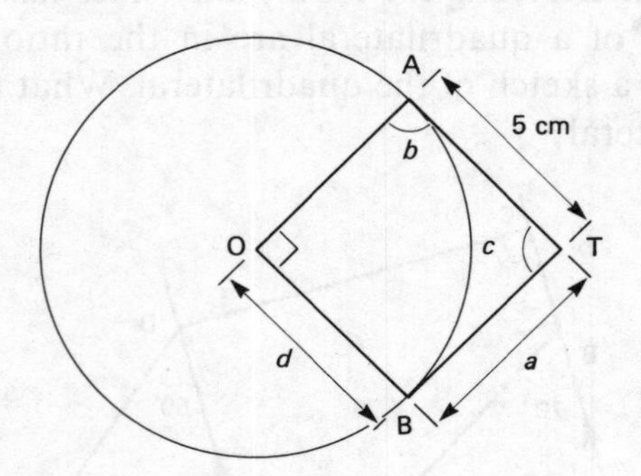

Figure 5.95

12. O is the centre of a circle and AC is a tangent which touches the circle at C (see Figure 5.96). If ∠OCB = 27°, find
(i) ∠OBC, (ii) ∠OAC.

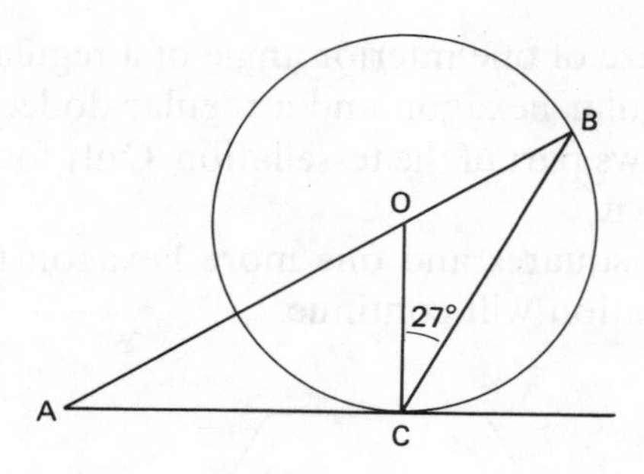

Figure 5.96

13. Two regular hexagons have their sides in the ratio 5:2. If the area of the smaller hexagon is 16 cm^2, what is the area of the larger hexagon?
14. A small photograph is to be enlarged to three times its dimensions (see Figure 5.97).

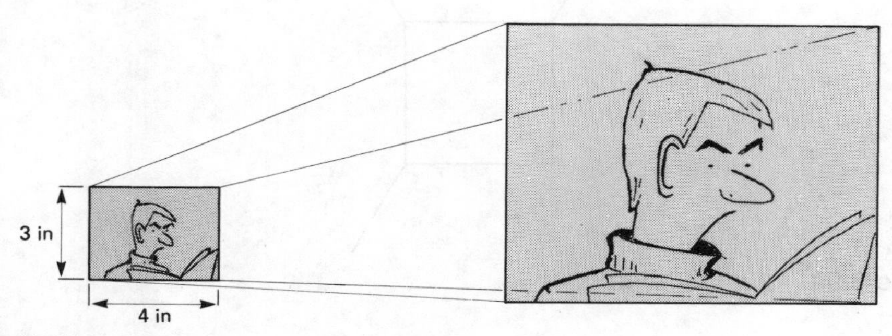

Figure 5.97

(a) Calculate the area of the small photograph.
(b) Write down the dimensions of the enlargement.
(c) What is the perimeter of the enlargement?
(d) Calculate the area of the enlargement.
(e) How many times bigger is the area of the enlargement than that of the small photograph?

15. Figure 5.98 shows a semicircle with AB as diameter. O is the centre of the circle. A tangent is drawn from C, a point on AB produced. The tangent touches the semicircle at T and angle TAO = 34°.

Calculate the angles marked $x°$, $y°$, $z°$ in the diagram, giving a reason for each step in your calculation. (MEG)

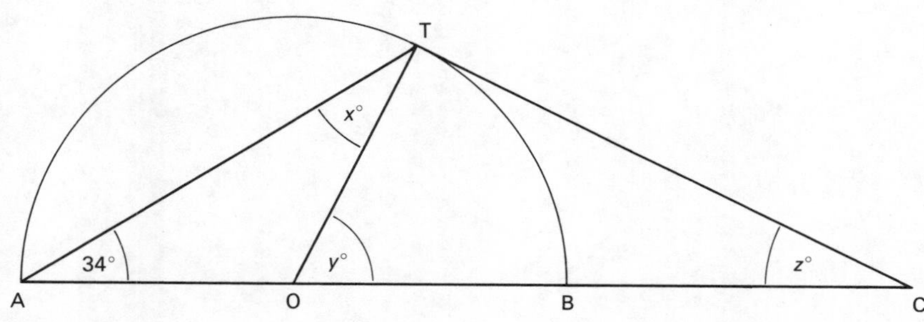

Figure 5.98

16. (a) Calculate the size of one interior angle of a regular dodecagon (12 sides).
(b) A square, a regular hexagon and a regular dodecagon can be tessellated. Figure 5.99 shows part of the tessellation. Only four sides of the dodecagon have been drawn.
Add two more squares and one more hexagon to the diagram to show how the tessellation will continue. (MEG)

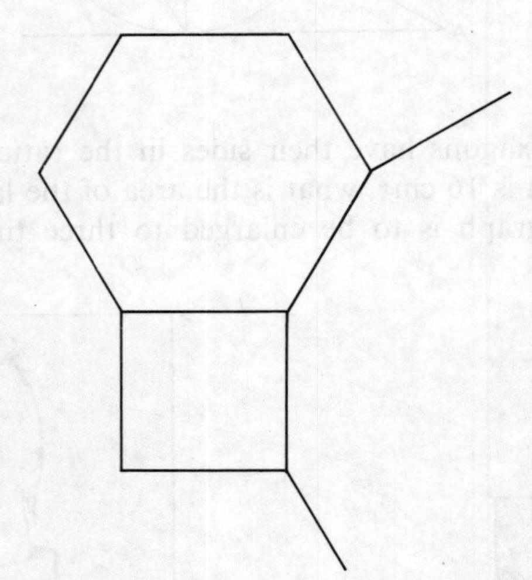

Figure 5.99

MENSURATION

6

The purpose of this unit is to consider ways of finding perimeters, areas and volumes associated with shapes most commonly encountered in everyday life. Also, to look at shapes which are not regular. Some examples are included that involve costings associated with these quantities.

6.1 The definition of area

Area is a measure of the extent covered by a region. Figure 6.1 shows a region R drawn on a square grid composed of 1 cm squares. Each square measures 1 cm by 1 cm, and we say its area is 1 square centimetre, written 1 cm^2. By counting, the number of complete squares shown unshaded is found to be 20. The shaded parts around the edge have to be estimated. This is not a particularly easy exercise, and at best, $\frac{1}{4}$, $\frac{1}{2}$ and $\frac{3}{4}$ squares can be used. The total of the shaded part was found by one student to be $10\frac{3}{4}$ squares, that is 10.75 cm^2 (check the accuracy of this estimation).

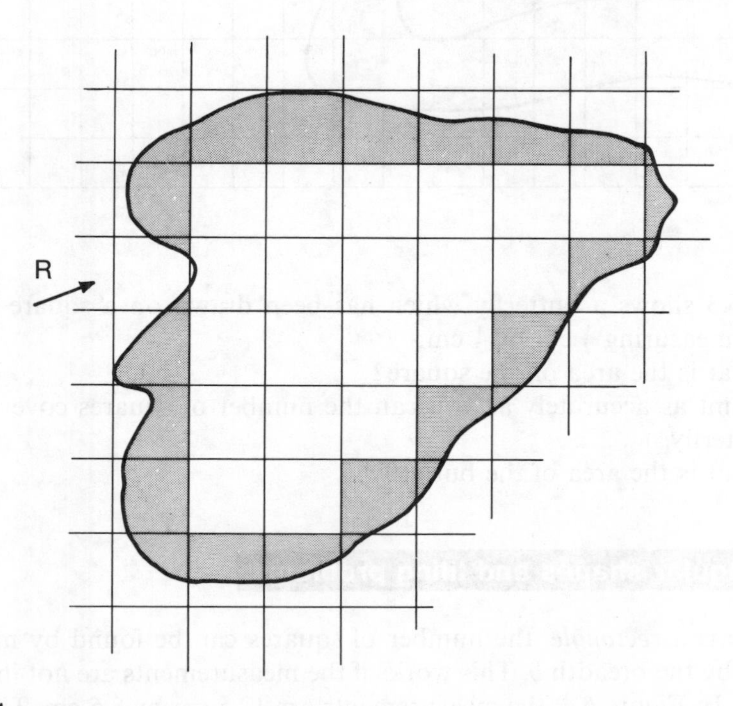

Figure 6.1

Hence the total area of R is $20 + 10.75 = 30.75\ \text{cm}^2$.

If the square grid does not measure 1 cm by 1 cm, you must find the area of each square and then multiply by the number of squares. See example 2 in Exercise 6a.

Exercise 6a

(1) Find approximations to the regions shown in Figure 6.2. Assume the squares are each $1\ \text{cm}^2$.

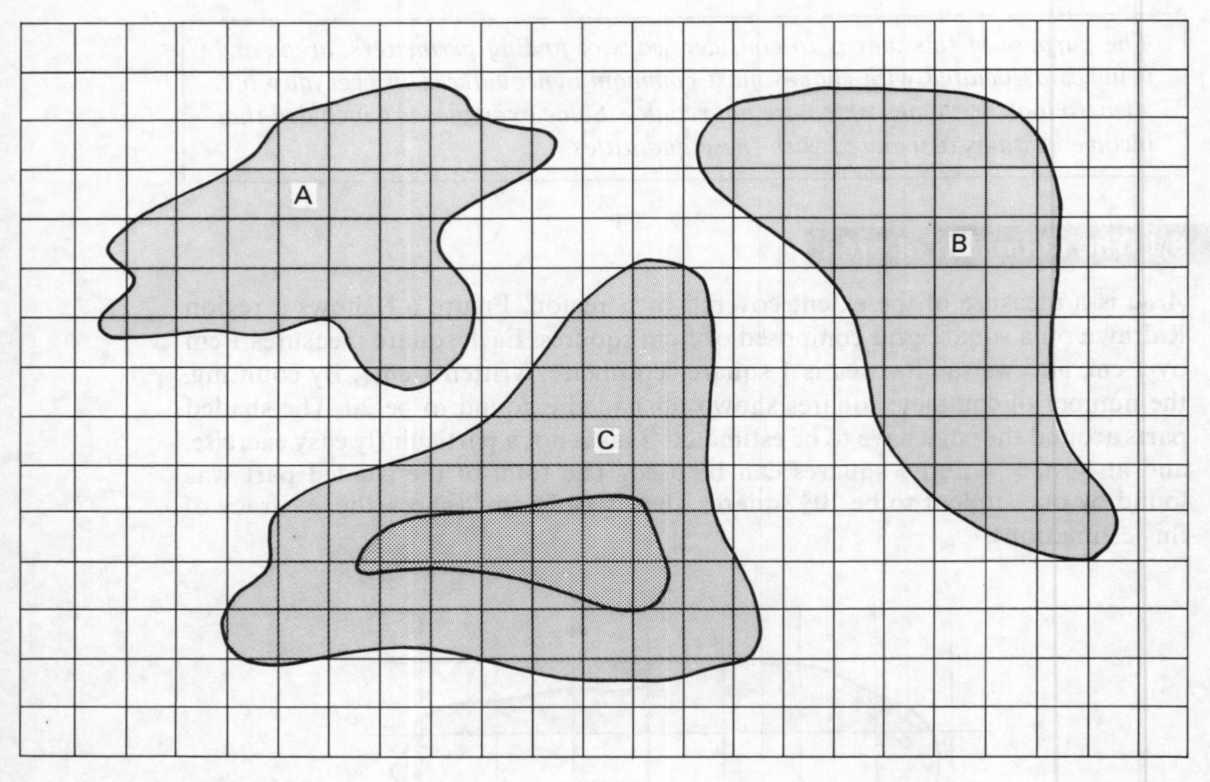

Figure 6.2

(2) Figure 6.3 shows a butterfly which has been drawn on a square grid with squares measuring $\frac{1}{2}$ cm by $\frac{1}{2}$ cm:
- (i) What is the area of one square?
- (ii) Count as accurately as you can the number of squares covered by the butterfly.
- (iii) What is the area of the butterfly?

6.2 Rectangular shapes and tiling problems

If the shape is a *rectangle*, the number of squares can be found by multiplying the length l by the breadth b. This works if the measurements are not in complete centimetres. In Figure 6.4, the measurements are 12.5 cm by 8.5 cm. The number

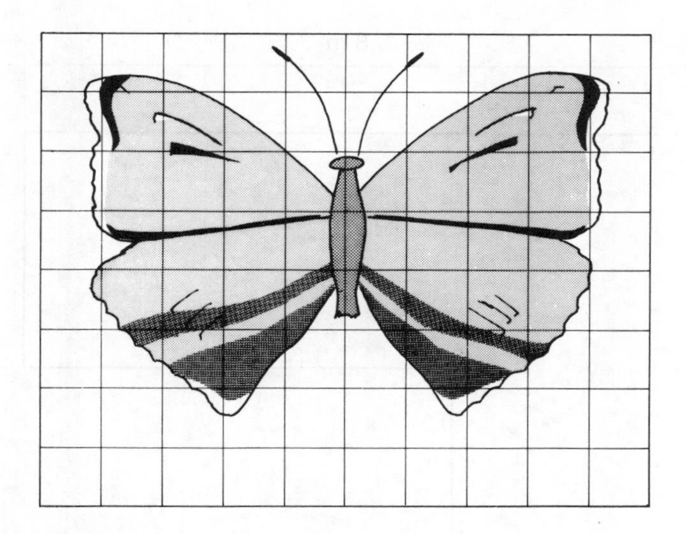

Figure 6.3

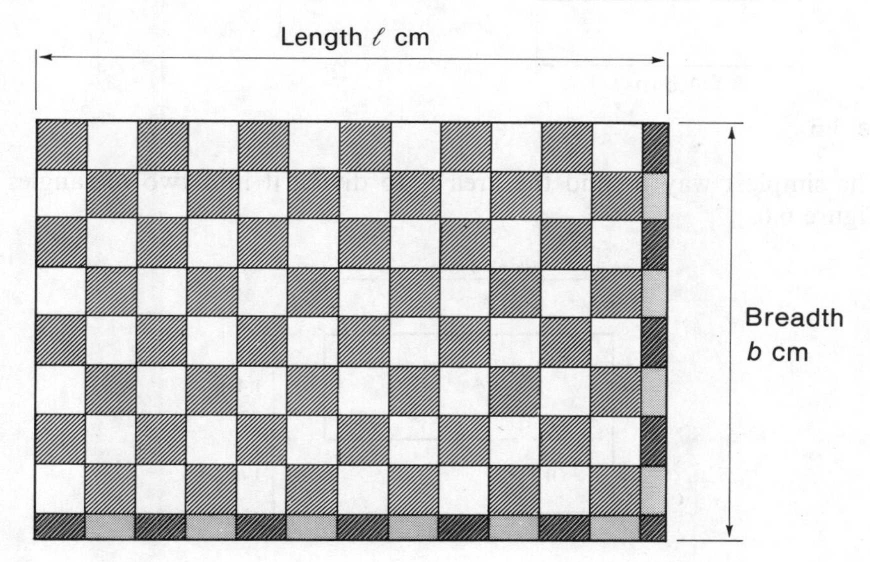

Figure 6.4

of squares in the pattern is given by $12.5 \times 8.5 = 106.25$ (this can be checked by counting).

Hence the area of a rectangle = length × breadth, or by formula

$$\text{Area} = lb$$

Worked example 6.1

Figure 6.5 shows the plan of a small room which is to be covered with tiles measuring 30 cm × 30 cm. Find:

(a) the area of the floor;
(b) the least number of tiles required to tile the floor.

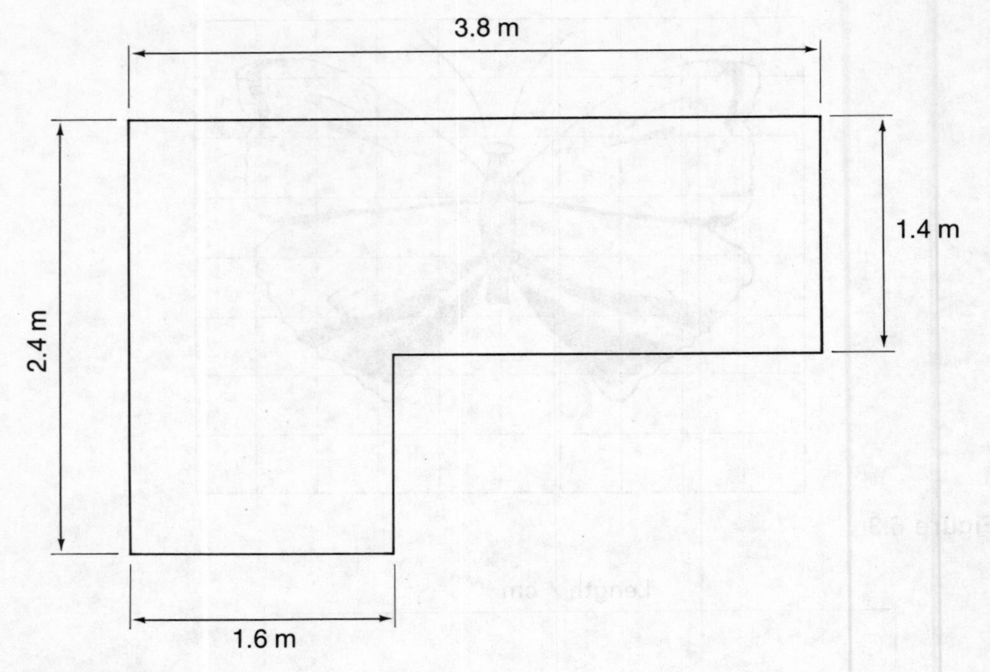

Figure 6.5

(a) The simplest way to find the area is to divide it into two rectangles as in Figure 6.6.

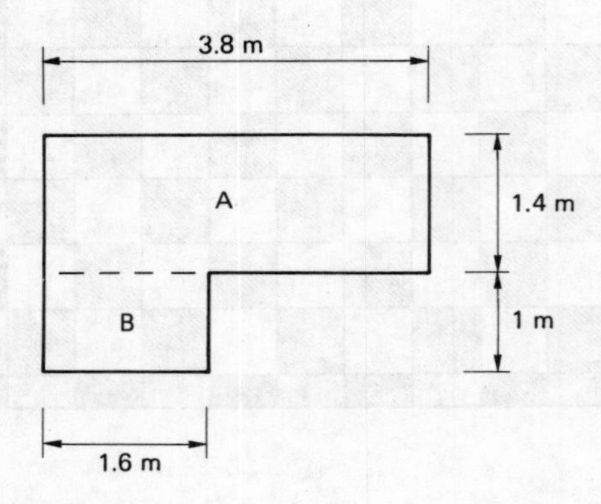

Figure 6.6

$$\text{Area of A} = 3.8 \times 1.4 = 5.32 \text{ m}^2$$
$$\text{Area of B} = 1 \times 1.6 = 1.6 \text{ m}^2$$
$$\text{The total area} = 5.32 + 1.6 = 6.92 \text{ m}^2.$$

(b) Tiling problems are often not quite as easy as they seem. The area of the floor is $6.92 \times 100 \times 100$ cm^2 = 62 900 cm^2. The area of each tile is $30 \times 30 = 900$ cm^2.

Is it true that the number of tiles = 62 900 ÷ 900 = 77 (nearest whole number)?

A look at Figure 6.7 shows two ways of laying the tiles in practice. It would appear that the minimum number of tiles required is 83. (Can you find a different number?).

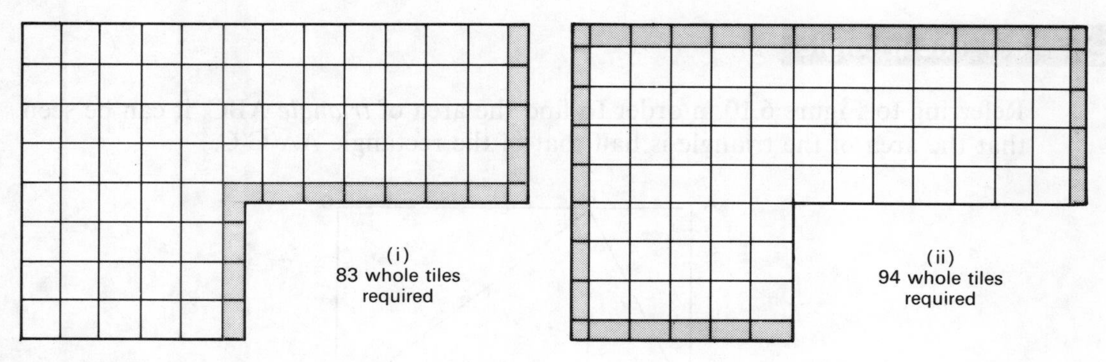

Figure 6.7

Exercise 6b
Figure 6.8 shows two plans of different kitchens. They are to be covered with square tiles each measuring 30 cm by 30 cm. Decide how you would lay the tiles in each case to use as few tiles as possible. (Assume the tiles are all the same colour.)
Investigate how the problem alters if more than one coloured tile is used.

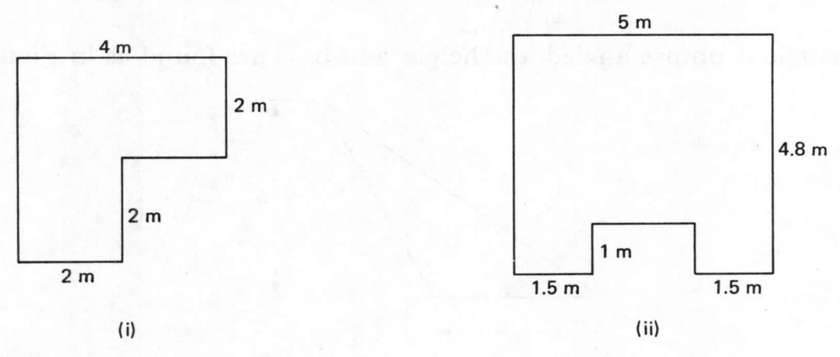

Figure 6.8

6.3 The parallelogram

In order to find the area of the *parallelogram* ABCD shown in Figure 6.9, it can be seen that it is the same as the area of the rectangle AB′C′D.

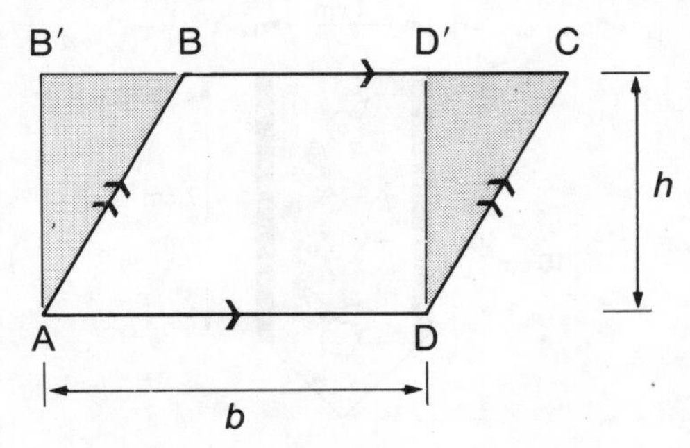

Figure 6.9

Hence the area of the parallelogram = base × height
$= bh$

6.4 The triangle

Referring to Figure 6.10 in order to find the area of *triangle* ABC, it can be seen that the area of the triangle is half that of the rectangle AA′C′C.

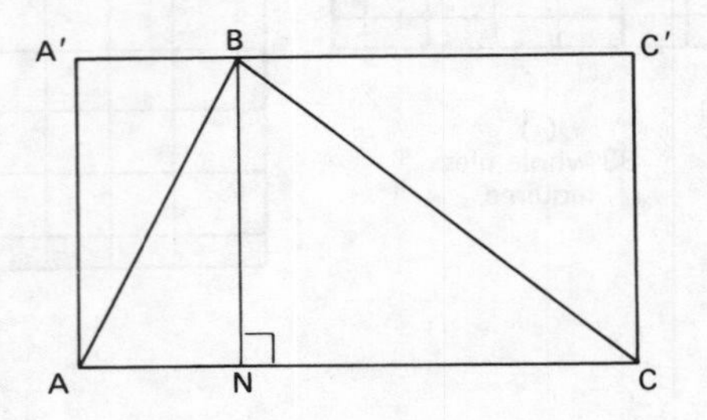

Figure 6.10

Hence the area of the triangle $= \frac{1}{2}(AC \times AA')$.
But AA′ is the same as BN, the height of the triangle.
Hence, the area of a triangle $= \frac{1}{2}(\text{base} \times \text{height})$

$$= \frac{1}{2}bh$$

If the triangle is obtuse angled, the height and base are found as in Figure 6.11.

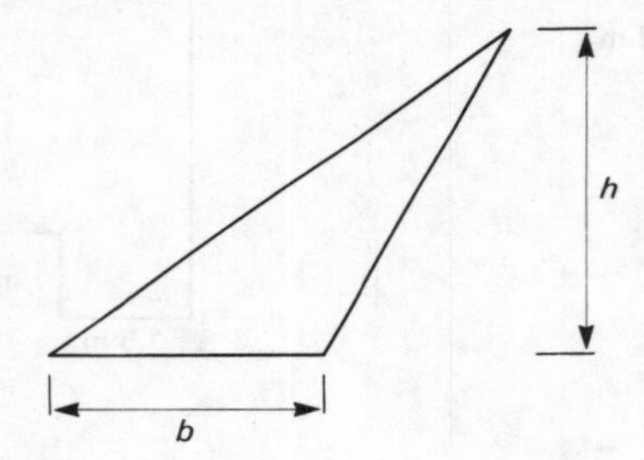

Figure 6.11

Worked example 6.2

Find the area of the bookmark shown in Figure 6.12.

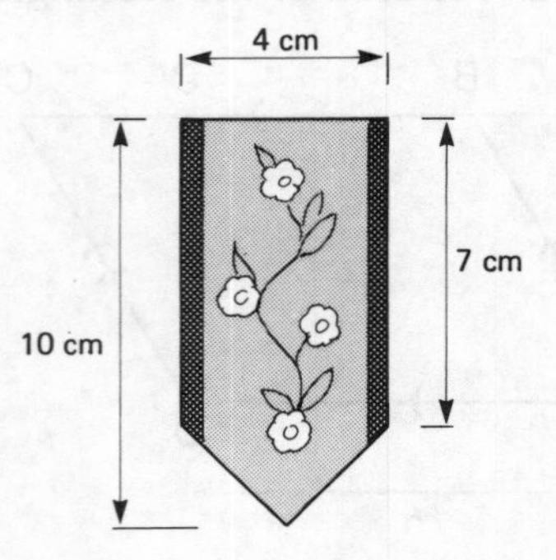

Figure 6.12

The shape consists of a rectangle and a triangle joined together. These two shapes are shown with measurements in Figure 6.13.

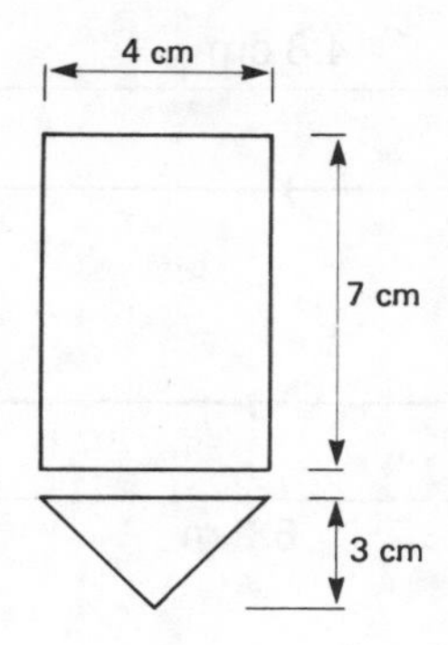

Figure 6.13

The area of the rectangle $= 4 \times 7$
$= 28 \text{ cm}^2$
The area of the triangle $= \frac{1}{2} \times 4 \times 3$
$= 6 \text{ cm}^2$
Hence the area of the bookmark $= 28 + 6 = 34 \text{ cm}^2$

6.5 The trapezium

A *trapezium* is a quadrilateral with two parallel sides (see Figure 6.14). If A_1 and A_3 are halved as shown, a rectangle PQRS can be made with the same area as the trapezium ABCD. The sides of the rectangle are h and the average of a and b, that is $\frac{1}{2}(a+b)$.

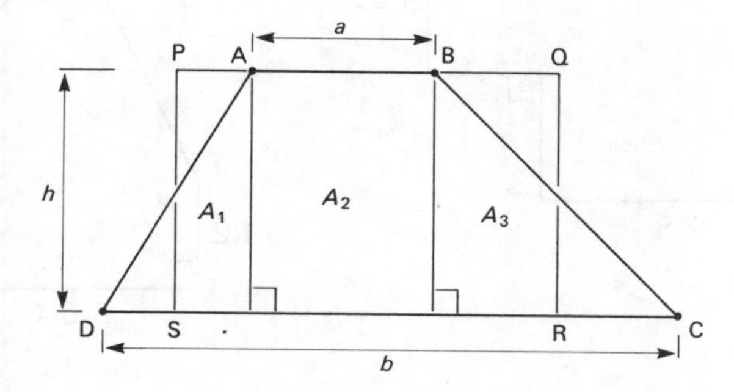

Figure 6.14

The area of the rectangle $= h \times \frac{1}{2}(a+b) = \frac{1}{2}h(a+b)$.
Hence the area of the trapezium $= \frac{1}{2}h(a+b)$.

Worked example 6.3
PQRS is a trapezium, with PQ parallel to SR. If PQ = 6.2 cm and SR = 4.8 cm, find the area of the trapezium, if the distance between PQ and SR is 1.6 cm.

The trapezium is shown in Figure 6.15. Using the formula from Section 6.5:

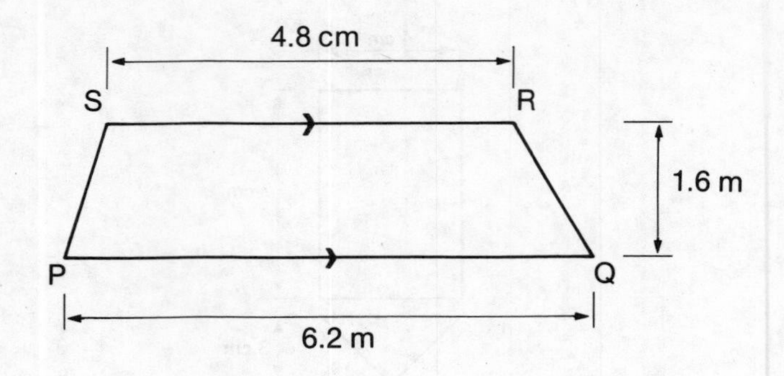

Figure 6.15

$$a = 4.8,\ b = 6.2,\ h = 1.6$$

Hence the area of the trapezium

$$= \tfrac{1}{2} \times 1.6 \times (4.8 + 6.2)$$
$$= 0.8 \times 11 = 8.8 \text{ cm}^2$$

Exercise 6c

Find the areas of the following shapes. All measurements are in cm.

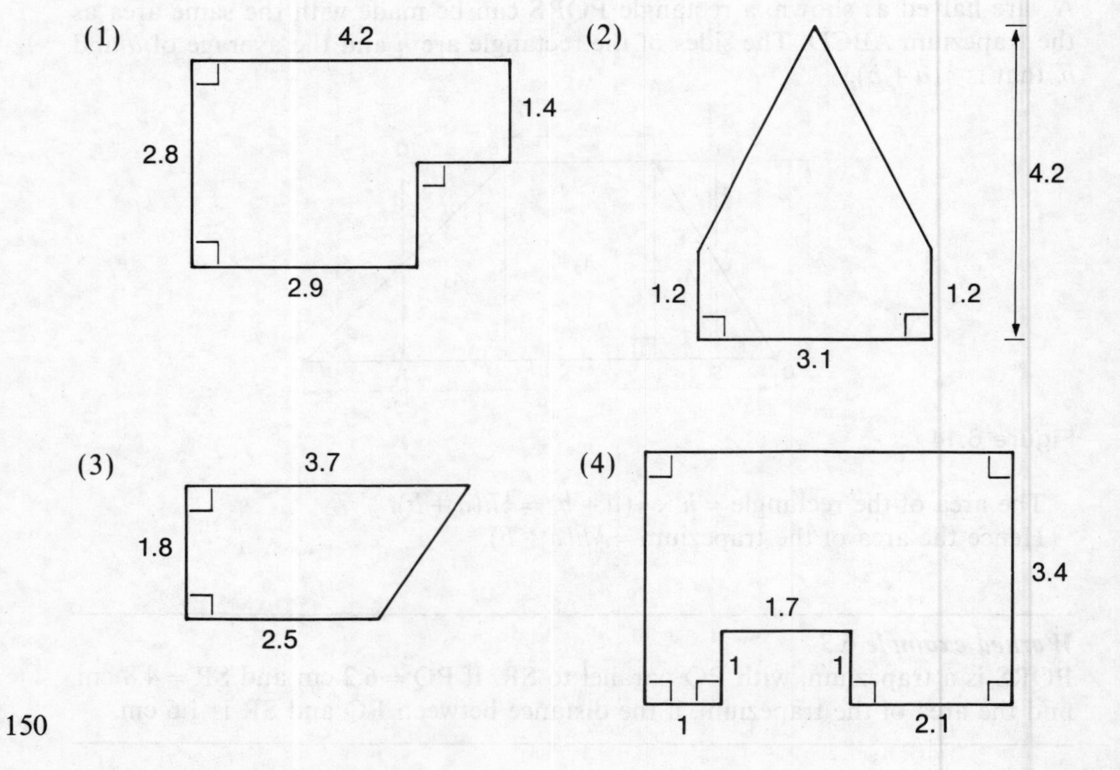

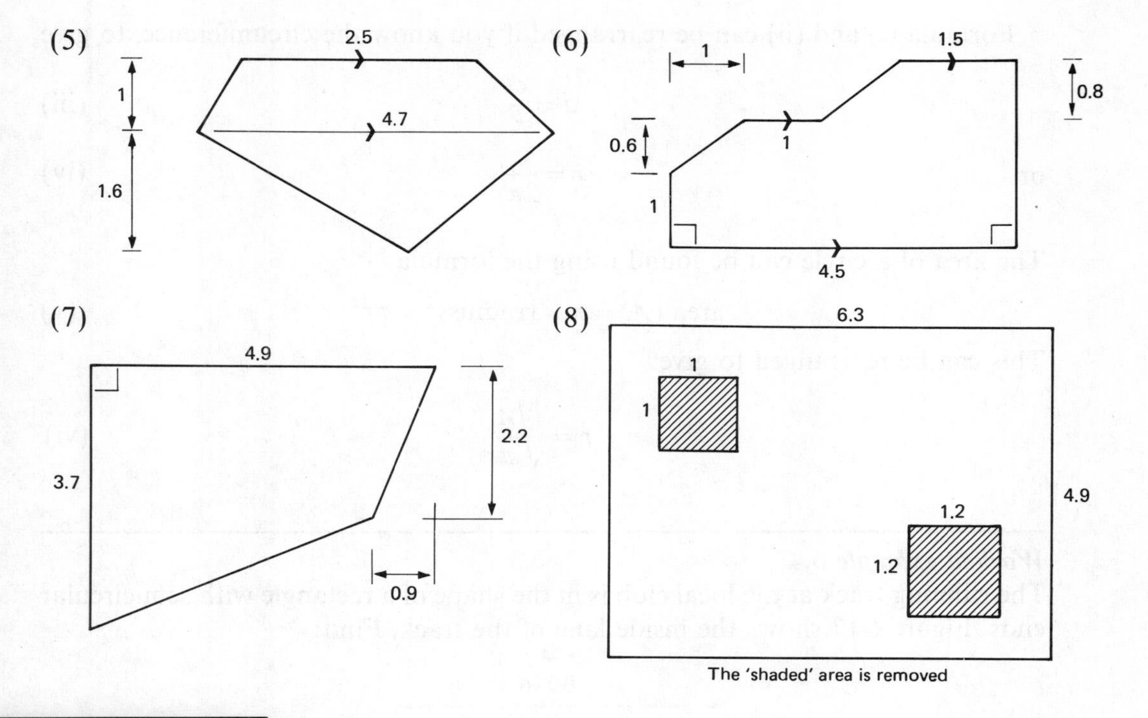

6.6 The circle

The perimeter or circumference of a *circle* can be easily found using the formula:

$$\text{Circumference } (c) = \pi \times d = \pi d \qquad \text{(i)}$$

Since the diameter is twice the radius, $d = 2r$, the formula becomes

$$c = 2 \times \pi \times r = 2\pi r \qquad \text{(ii)}$$

Various values are taken for π depending on how accurate an answer is required.

If you have a $\boxed{\pi}$ button on your calculator, it will probably give you $\pi = 3.141\,592\,654$!

For an approximate answer, take $\pi = 3$.

For the accuracy required in this book $\pi = 3.14$.

If you want an approximate fraction, $\pi = 3\frac{1}{7} = \frac{22}{7}$.

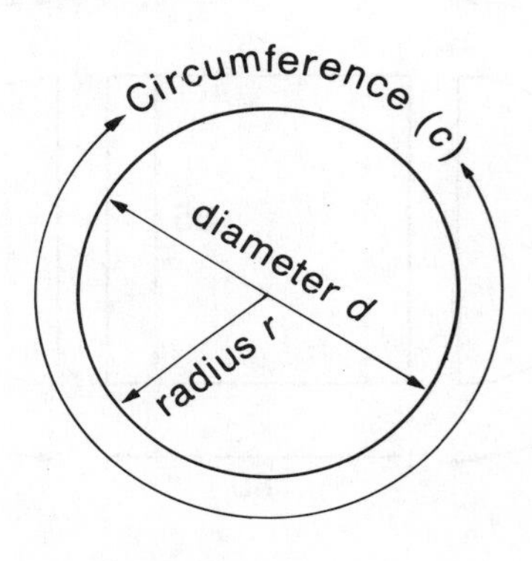

Figure 6.16

Formula (i) and (ii) can be rearranged if you know the circumference, to give

$$d = \frac{c}{\pi} \qquad \text{(iii)}$$

or

$$r = \frac{c}{2\pi} \qquad \text{(iv)}$$

The area of a circle can be found using the formula

$$\text{area } (A) = \pi \times (\text{radius})^2 = \pi r^2 \qquad \text{(v)}$$

This can be rearranged to give

$$r = \sqrt{\frac{A}{\pi}} \qquad \text{(vi)}$$

Worked example 6.4

The running track at the local club is in the shape of a rectangle with semicircular ends. Figure 6.17 shows the inside lane of the track. Find:

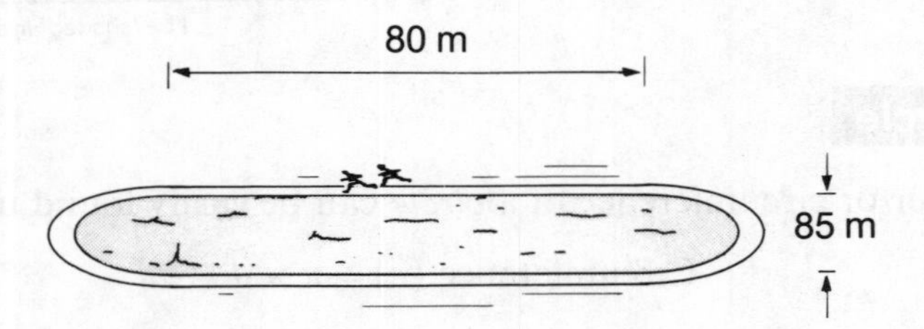

Figure 6.17

(a) the distance round the inside lane;
(b) the area of the grass in the centre.

The shape can be divided into a rectangle with two semicircles, one at each end (see Figure 6.18).

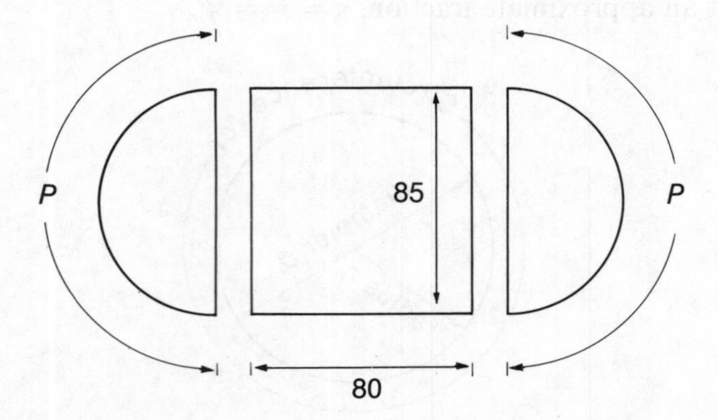

 Figure 6.18

(a) The total perimeter of the two semicircles $2P$ is just the circumference of a circle.

Hence $2P = \pi \times 85 = 266.9$ m.

The total distance round the track $= 2 \times 80 + 266.9 = 426.9$ m.

(b) The area of the grass is the area of a circle of radius 42.5 m and a rectangle of area 80×85 m^2.

Hence the total area $= 3.14 \times 42.5^2 + 80 \times 85$ m$^2 = 12\,500$ m^2 (3 sig. figs).

Worked example 6.5

A rough estimate of the circumference of the earth is 25 000 miles. Find the value of the diameter.

Using formula (iii) in Section 6.6, we have

$$d = \frac{c}{\pi} = \frac{25\,000}{3.14} = 7962$$

The diameter would be 7962 miles.

Worked example 6.6

A small circular metal coin has an area of 2.6 cm^2. What is the diameter of the coin?

Using formula (vi) in Section 6.6, we have

$$r = \sqrt{\frac{A}{\pi}} = \sqrt{\frac{2.6}{3.14}}$$

By calculator, we have

					Display
2.6	÷	3.14	=	√	0.909959

Take the radius as 0.91 cm.

The diameter $= 2 \times 0.91$ cm $= 1.82$ cm.

Exercise 6d

(1) Find the circumference of a circle if the diameter is 8 cm.

(2) Find the radius of a circle of circumference 20 cm.

(3) Find the radius of a circle if its area is 100 cm^2.

(4) The diameter of a bicycle wheel is 50 cm. If the wheel rotates 20 times, find the distance covered by the wheel along the ground, assuming it does not slip.

(5) Figure 6.19 shows an oval of a model railway track.

  (i) Find the distance travelled by an engine if it travels once round the oval.

  (ii) If it takes 8 seconds to travel the distance, estimate the speed of the train in km/h!

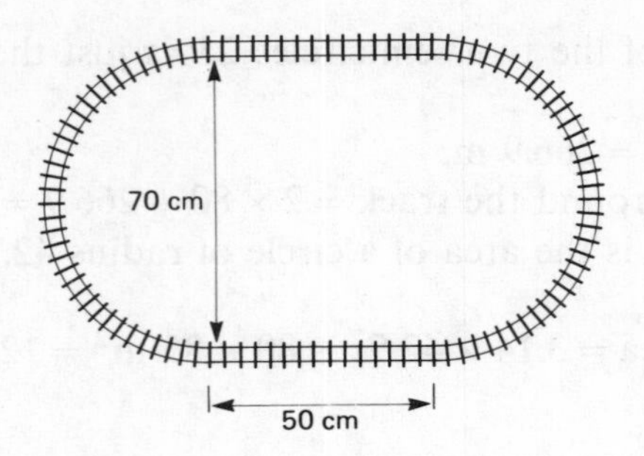

Figure 6.19

(6) Metal discs of radius 2 cm are stamped out from a strip of material 4 cm wide and 100 cm long. What percentage of the metal is wasted?
(7) Find the diameter of a circle if its area is 6.4×10^{-2} mm^2.
(8) The diameter of the cycle wheel of a bike is 65 cm.
  (a) Calculate the circumference of the wheel.
  (b) Tanya is riding the bike in a gear for which the wheel makes 4 complete revolutions for every 1 complete revolution of the pedals. How many complete revolutions of the pedals are needed for Tanya to make a journey of 2 km if she stays in this gear.

6.7 Surface area (painting and covering problems)

There are many everyday situations where it is required to paint or cover some form of 3-dimensional solid. In order to do this, we often need to know the surface area of the solid. To help us find the surface area, the *net* of the solid may be helpful.

Worked example 6.7
A cube of side 4 cm is to be coated with an expensive anti-corrosion compound. The manufacturer claims that a £20 tin of this compound will cover a total area of 2800 cm^2. Find the total cost of coating 1000 of these cubes. Give your answer to the nearest £10.

The net of the cube is shown in Figure 6.20. It consists of 6 squares, giving a total surface area of $6 \times 4 \times 4 = 96$ cm^2.

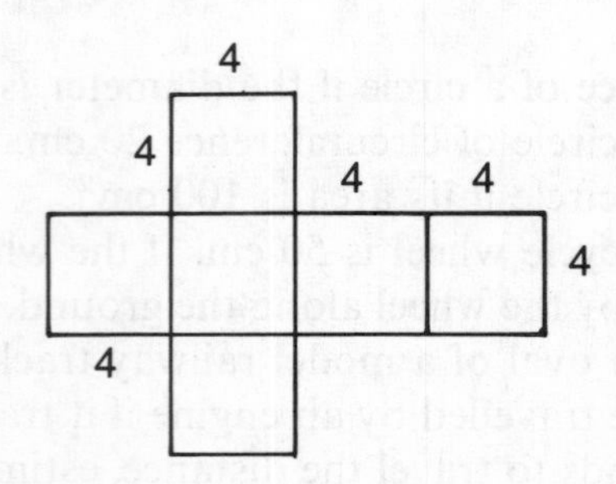

 Figure 6.20

The total area of 1000 such cubes is $1000 \times 96 = 96\,000 \text{ cm}^3$.
The number of tins required $= 96\,000 \div 2800 = 34.29$.
Hence number of tins is 35.
The total cost $= £35 \times 20 = £700$.
The exact cost if part tins are allowed is $£34.29 \times 20 = £690$ (to the nearest £10).

6.8 Volume of a cuboid

A *cuboid* is a solid with rectangular faces. A cubic centimetre is the volume of a cube measuring 1 cm × 1 cm × 1 cm. To find the number of cubes that can be fitted into the shape would give us the volume of the cuboid. See Figure 6.21.

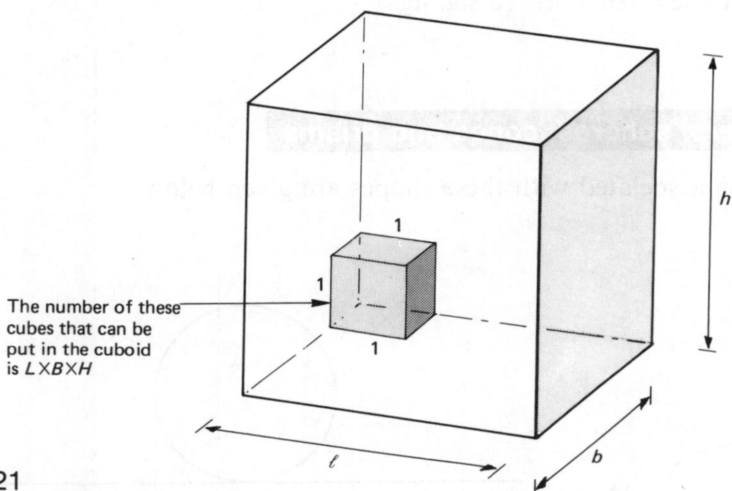

Figure 6.21

Hence the volume V of a cuboid $= l \times b \times h$
that is $V = lbh$

Compound shapes can be treated in similar ways to compound areas, see the following example.

Worked example 6.8

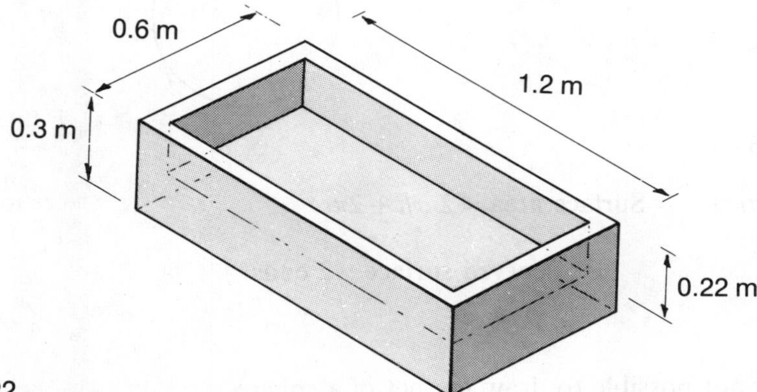

Figure 6.22

A concrete flower container is in the shape of a cuboid measuring 1.2 m × 0.6 m × 0.3 m. The centre part is 0.22 m deep and the concrete is 4 cm thick at the sides. Find the volume of concrete needed to make the container.

As different units are given in the problem, it is first necessary to change to a common unit, cm being the best probably.

The outside has a volume $120 \times 60 \times 30 = 216\,000$ cm^3.

The inside has a volume $112 \times 52 \times 22 = 128\,128$ cm^3.

Hence the volume of concrete $= 216\,000 - 128\,128 = 87\,872$ cm^3.

If the volume in m^3 is required, divide by 10^6.

Hence volume $= 0.088$ m^3 (2 sig. figs).

6.9 Cylinders, spheres, cones and prisms

The formulae associated with these shapes are given below.

Cylinder

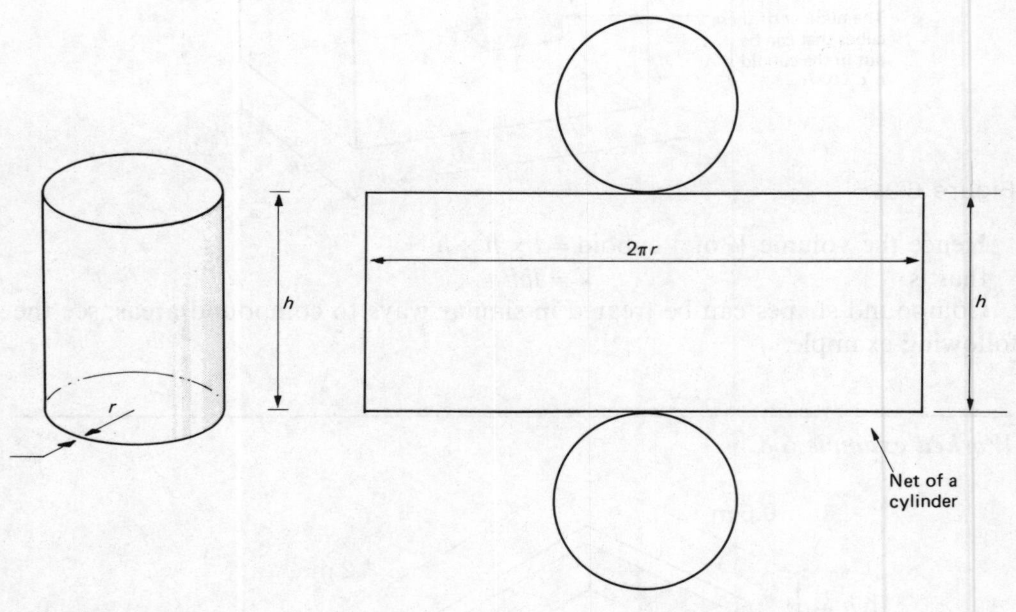

Figure 6.23

Volume $= \pi r^2 h$ Surface area $= 2\pi rh + 2\pi r^2$

↑ curved surface ↑ 2 ends

Sphere

 It is clearly not possible to draw the net of a sphere.

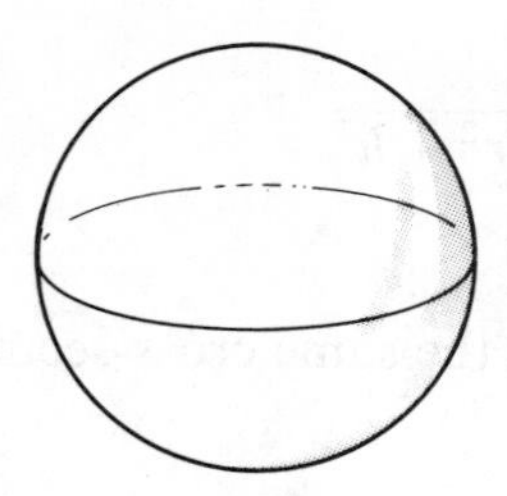

Figure 6.24

$$\text{Volume} = \frac{4\pi r^3}{3} \qquad \text{Surface area} = 4\pi r^2$$

Pyramid

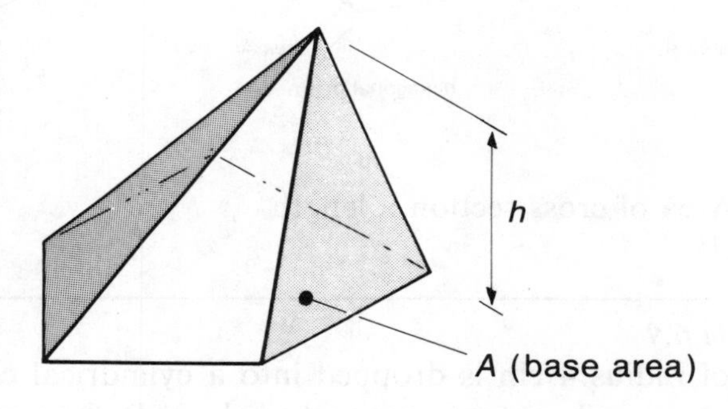

Figure 6.25

$$\begin{aligned}\text{Volume} &= \tfrac{1}{3}\text{ base area} \times \text{height} \\ &= \tfrac{1}{3}Ah\end{aligned}$$

Cone

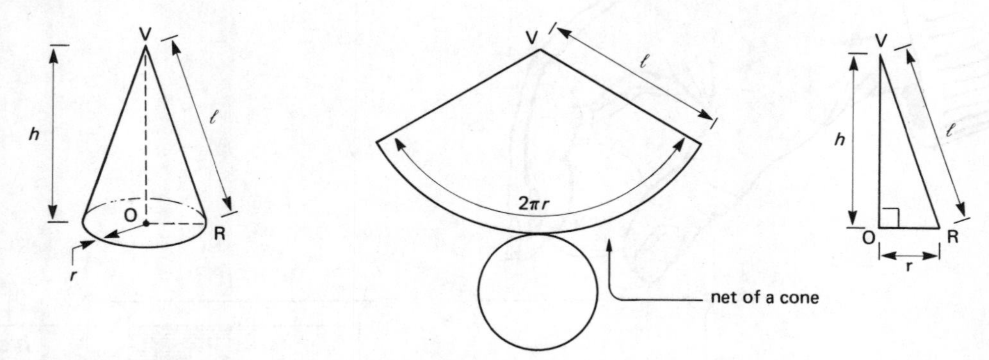

Figure 6.26

$$\text{Volume} = \tfrac{1}{3}\pi r^2 h$$

$$\text{Surface area} = \underset{\substack{\uparrow \\ \text{curved surface}}}{\pi r l} + \underset{\substack{\uparrow \\ \text{base}}}{\pi r^2}$$

Pythagoras's theorem
gives $l^2 = r^2 + h^2$, hence $l = \sqrt{r^2 + h^2}$

Prism

A prism is any solid which has the same cross-section throughout. For examples, see Figure 6.27.

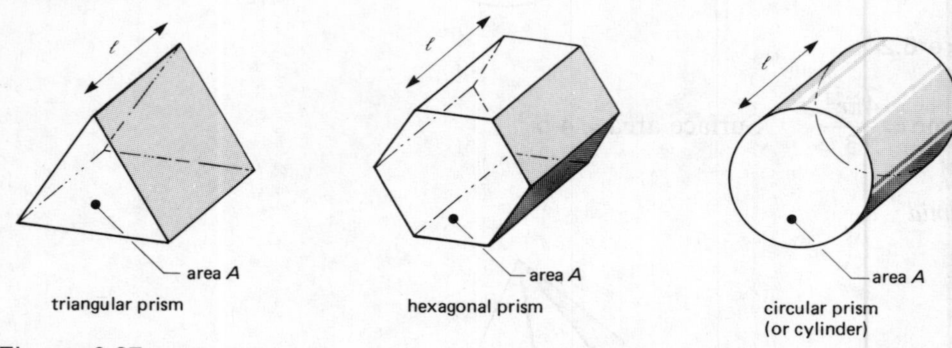

Figure 6.27

The volume = Area of cross-section × length
$= Al$

Worked example 6.9
A lead sphere of radius 4 cm is dropped into a cylindrical container of radius 10 cm, containing enough water to cover the sphere. By how much does the level of water rise?

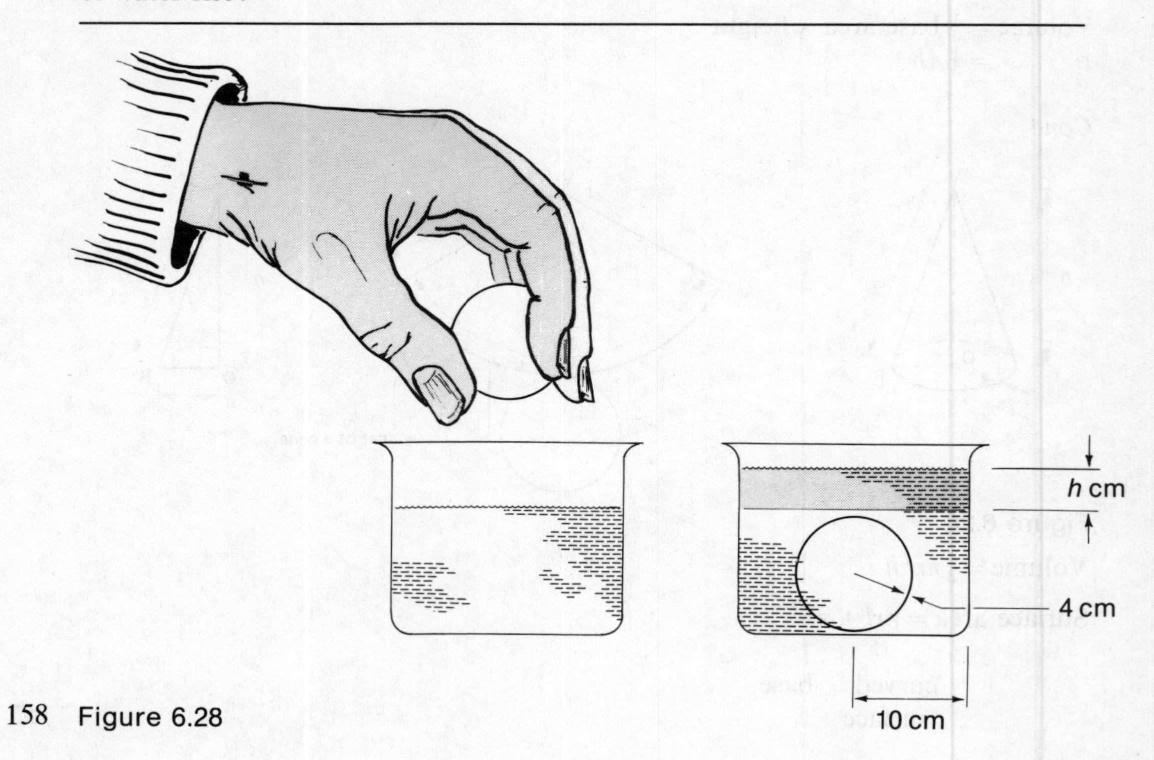

Figure 6.28

Figure 6.28 shows the sphere displacing some water. The volume of this displaced water must equal the volume of the sphere.

Volume of the sphere $\frac{4}{3} \times 3.14 \times 4^3 = 267.9\ \text{cm}^3$

Volume of the cylinder $= 3.14 \times 10^2 \times h = 314h\ \text{cm}^3$

We have the equation $314h = 267.9$.

Hence $h = \dfrac{267.9}{314} = 0.85\ \text{cm}$

Worked example 6.10

The diameter of the dome of St Paul's Cathedral is roughly 40 m. Find the volume of air that the dome holds in litres. Give your answer in standard form.

Figure 6.29

W have to make an assumption that the dome is shaped like half of a sphere.

Its volume then will be $\frac{1}{2}$ of $\dfrac{4\pi r^3}{3} = \dfrac{2\pi r^3}{3}$.

Since the diameter is 40 m, then $r = 20$ m.

Hence the volume $= \dfrac{2 \times 3.14 \times 20^3}{3} = 16\,750\ \text{m}^3$

$= 1.675 \times 10^3\ \text{m}^3 = 1.675 \times 10^4 \times 10^6\ \text{cm}^3$

$= 1.675 \times 10^{10}\ \text{cm}^3$

Now 1 litre $= 10^3\ \text{cm}^3$.

$\therefore$ Volume of air $= 1.675 \times 10^{10} \div 10^3 = 1.675 \times 10^7$ litres

Exercise 6e

(1) Find the volume of a cylinder of radius 4 cm and height 8 cm.

(2) Find the volume of a small ball of radius 4 cm.

(3) Find the volume of water which would cover a field of area 1 hectare if 0.5 cm of rain falls on it.

(4) Find the total surface area of a solid cylinder of radius 10 cm and height 8 cm.

(5) A small sphere of radius 2 cm is dropped into a cylinder of radius 3 cm which contains water. What will be the rise in the level of the water?

(6) Find the total surface area of a solid hemisphere of radius 10 cm.

(7) A metal cube of side 10 cm is melted down and cast into small spheres of radius 1 cm. How many of these spheres can be made?

(8) Find the volume of a paper weight made in the form of a pyramid with base area 12 cm^2 and height 3 cm.

(9) A cuboid measures 8 cm by 6.4 cm by 3.2 cm. What is the surface area of the cuboid?

(10) A triangular prism has a cross-section area of 12 cm^2 and length 1 m. What is the volume of the prism?

6.10 Density

The mass of a unit volume of a given substance, is called its *density*.

We have $\text{Density} = \dfrac{\text{Mass}}{\text{Volume}}$, or $D = \dfrac{M}{V}$

This can be arranged to $M = DV$ or $V = \dfrac{M}{D}$.

The units of density are g/cm^3 or kg/m^3.

Worked example 6.11

A roll of copper wire is made of wire 200 m long with a circular cross-section of diameter 1.2 mm. Calculate the volume of the coil.

If the density of copper is 8.8 g/cm^3, calculate the mass of the copper wire in kilograms.

Care must be taken with the units. Since the density is given per cubic centimetre, it is best to calculate the volume in cubic centimetres.

Using $V = \pi r^2 h$:

$$V = 3.14 \times 0.06^2 \times 20\,000$$
$$= 226 \text{ cm}^3.$$

$$\text{Hence the mass} = \frac{226 \times 8.8}{1000} \text{ kg}$$
$$= 1.989 \text{ kg}.$$

Exercise 6f

Complete the following table:

	Substance	Volume (cm^3)	Density (g/cm^3)	Mass (g)
(1)	aluminium	41	2.7	
(2)	gold	100		1932
(3)	lead		11.37	45.48
(4)	sand	30	2.63	
(5)	glass	30		75
(6)	ice		0.92	73.6
(7)	olive oil	24	0.92	
(8)	air	200		0.258
(9)	hydrogen		0.000 09	11.34

Suggestions for coursework 6

1. Try to find out as much as you can about π, and the different places it occurs. Try and devise a method for finding π by accurate drawing.
2. In Figure 6.30 a square has been divided into four regions by two straight lines. Investigate how these two lines can be drawn so that the areas of A, B, C, D are all equal. Look at other shapes that your method might work for.

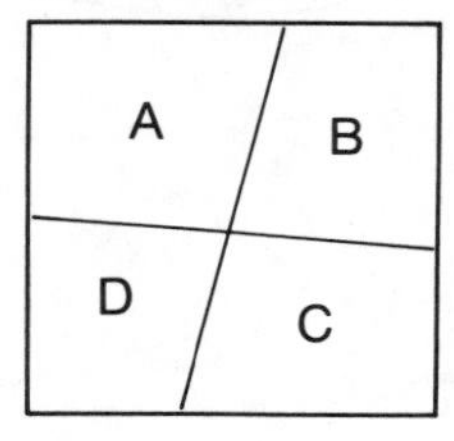

Figure 6.30

3. Figure 6.31 shows a metal scoop. The container part is made from a flat sheet of metal measuring 10 cm by 8 cm. The front slopes at an angle of 45°. Use a graphical method to find the largest volume that the scoop could hold. Look at other problems similar to this.

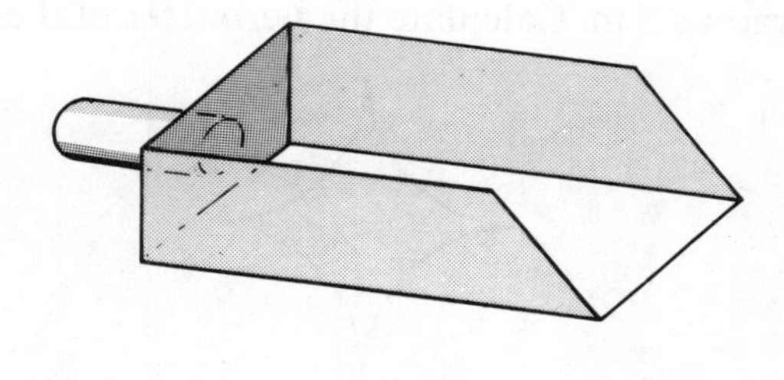

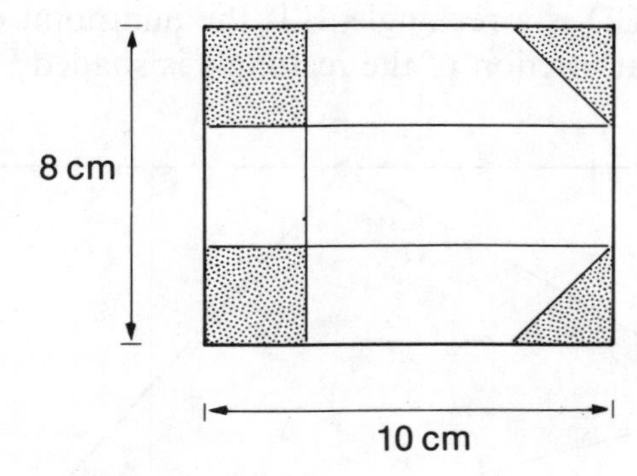

Figure 6.31

Miscellaneous exercise 6

1. Figure 6.32 shows a rectangular allotment garden 30 m by 10 m. It contains

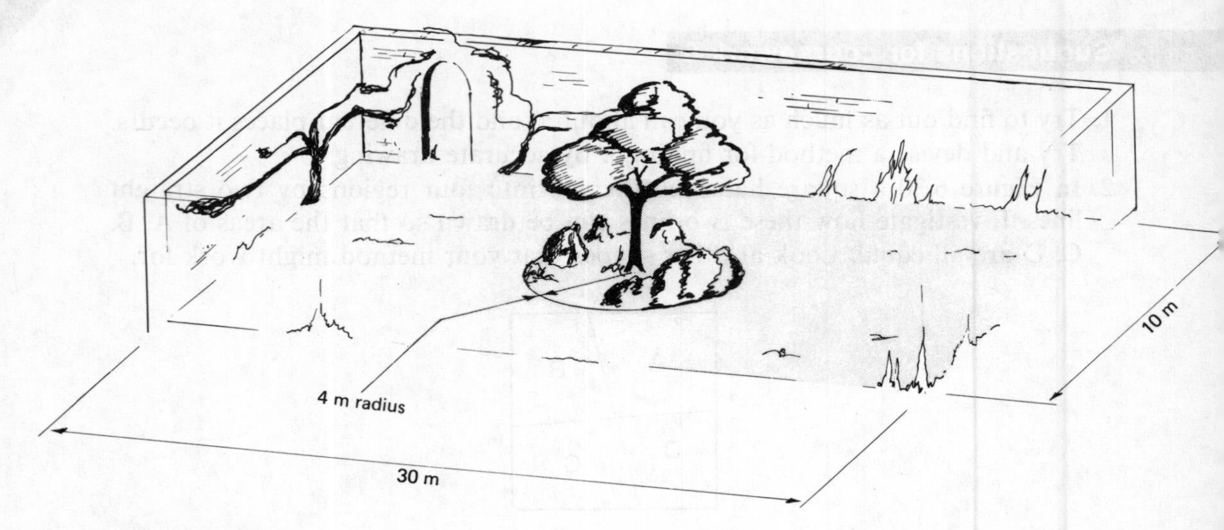

Figure 6.32

a fruit section in the shape of a circle of diameter 8 m. In the rest of the garden, vegetables are grown.

(a) Find the area of the circular fruit section.

(b) Find the area of the vegetable section. (MEG)

2. Figure 6.33 shows the plan of a flower bed. It is formed from four quarter circles, each of radius 5 m. Calculate the perimeter of the flower bed ($\pi = 3.1$). (MEG)

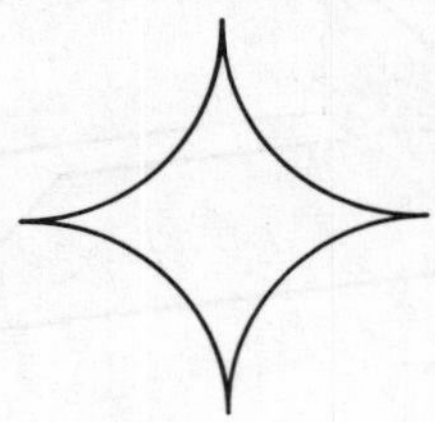

Figure 6.33

3. In Figure 6.34, ABCD is a rectangle, E is the mid point of AD and F is the mid point AB. What fraction of the rectangle is shaded?

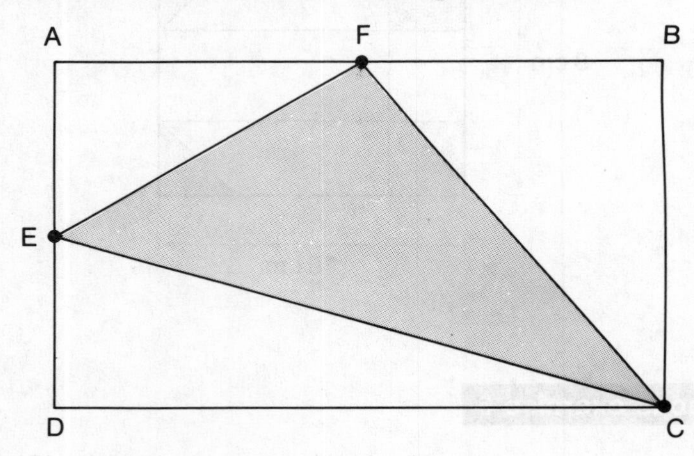

 Figure 6.34

4. In Figure 6.35, ABCD is a square. S and T are the middle points of BC and CD respectively. Write down the area of each of the following as a fraction of the area of the whole square ABCD.

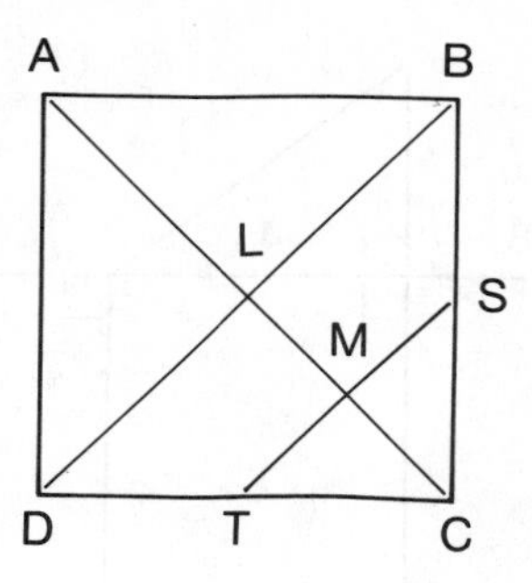

Figure 6.35

(a) Triangle ALB.
(b) Triangle STC.
(c) Triangle SMC.
(d) Trapezium BLMS.
(e) Trapezium BSTD.

5. Find the area of the label on a tin of baked beans (Figure 6.36) 12 cm tall with a radius of 4 cm. Take $\pi = 3.14$. (SEG)

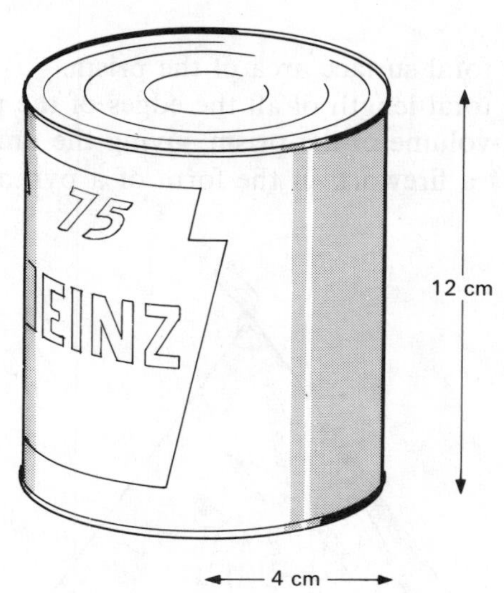

Figure 6.36

6. A cardboard box with internal dimensions of 60 cm by 48 cm by 45 cm contains packets of butter each measuring 10 cm by 9 cm by 8 cm. How many packets of butter does the box contain, assuming that no space is wasted? (SEG)

7. An iron bar of rectangular cross-section 20 mm by 15 mm is reduced by corrosion to 19 mm by 14 mm.
(a) Calculate the original area of the cross-section.

(b) Calculate the cross-sectional area after corrosion.

(c) Calculate the percentage decrease in cross-sectional area. (LEAG)

8. Figure 6.37 shows the net of a solid prism. All lengths are given in centimetres.

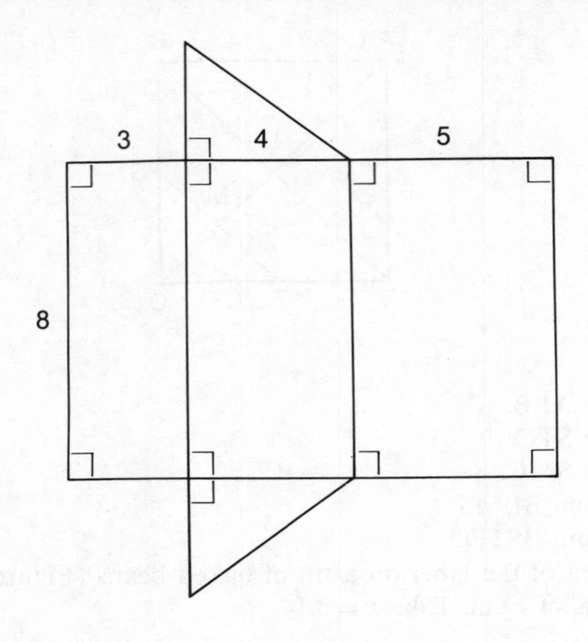

Figure 6.37

(a) Calculate the total surface area of the prism.

(b) Calculate the total length of all the edges of the prism.

(c) Calculate the volume of the prism, giving the units of your answer.

9. Figure 6.38 shows a firework in the form of a pyramid of height 6 cm and radius 2.5 cm.

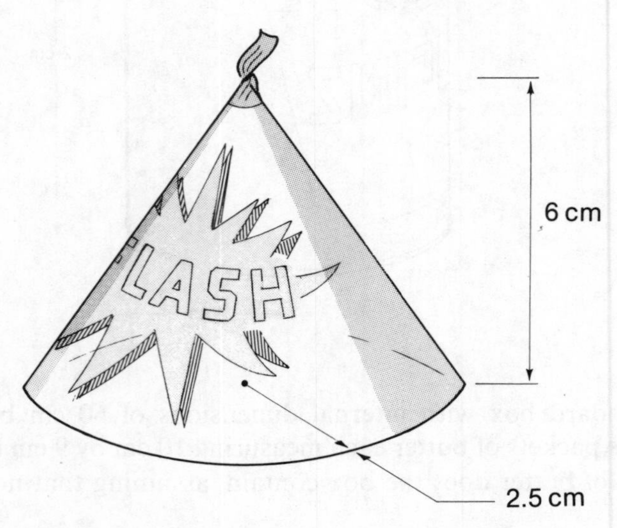

Figure 6.38

(a) Find the volume of the firework.

(b) Find the surface area of the label on the firework.

10. A metal ingot is in the form of a solid cylinder of length 7 cm and radius 3 cm.
 (a) Calculate the volume, in cm^3, of the ingot. The ingot is to be melted down and used to make cylindrical coins of thickness 3 mm and radius 12 mm.
 (b) Calculate the volume, in mm^3, of each coin.
 (c) Calculate the number of coins which can be made from the ingot, assuming that there is no wastage of metal. (MEG)
11. A child's circular paddling pool has a radius of 2 ft 4 in and is 1 ft 6 in deep. What is the volume of water in the pool when it is half full? Take $\pi = \frac{22}{7}$.
12. Figure 6.39 represents six equal cylindrical tins packed into a rectangular box ABCD, the height of the box being exactly equal to the height of the tins. The radius of each tin is 7.0 cm.

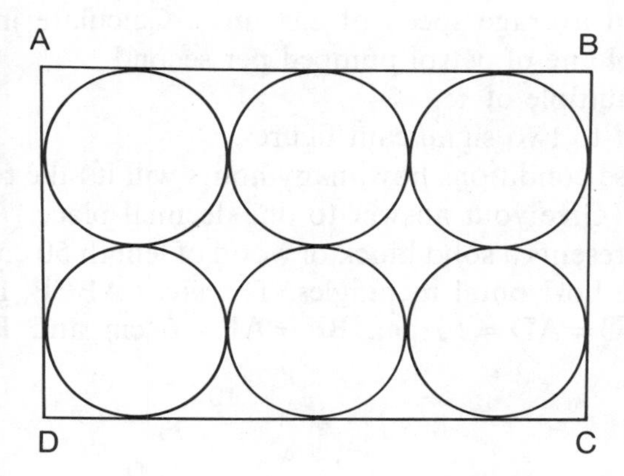

Figure 6.39

 (a) Calculate
  (i) the length of AB;
  (ii) the length of AD;
  (iii) the area ABCD.
 (b) Calculate
  (i) the area of the circular top of a tin;
  (ii) the total area of all 6 tops of the tins.
 (c) Calculate the space occupied by the six tins as a percentage of the space in the box. Give your answer correct to one decimal place.
13. The tank on a petrol tanker is assumed to be cylindrical of radius 2 m and length 12 m.
 (a) Find the capacity of the tank in litres.
 (b) A small car holds 40 litres of petrol when full. If the car can run 10 miles on 1 litre of petrol, how far could the car travel if it used all the petrol that the tanker can hold?
14. A petrol storage tank consists of two hemispheres (Figure 6.40) of radius 3.00 m joined by a circular cylinder also of radius 3.00 m, whose length is 2.00 m.
 (i) Calculate the capacity of the tank in cubic metres
  (a) as a multiple of π;
  (b) correct to two significant figures.
 (ii) Petrol can be pumped out through a cylindrical outflow pipe of radius

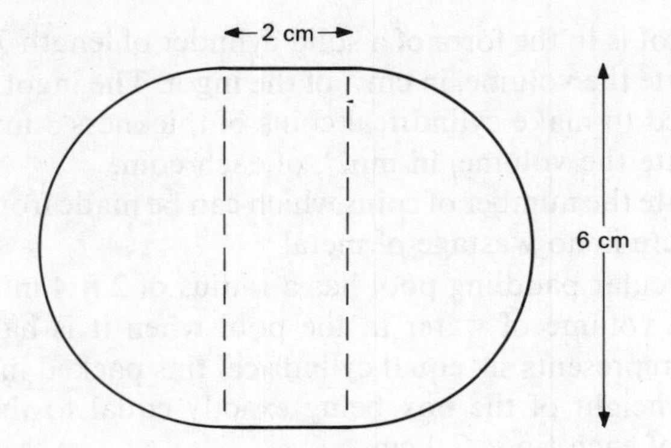

Figure 6.40

6 cm at an average speed of 2.00 m/s. Calculate in cubic metres the average volume of petrol pumped per second
(a) as a multiple of π;
(b) correct to two significant figures.
(iii) Under these conditions how many hours will it take to empty a full tank of petrol? (Give your answer to one decimal place.)

15. Figure 6.41 represents a solid block of wood of length 50 cm. The faces ABCD and EFGH are horizontal rectangles. The faces ABFE, BCGF and ADHE are vertical. BC = AD = 10 cm, BF = AE = 6 cm and FG = EH = 18 cm.

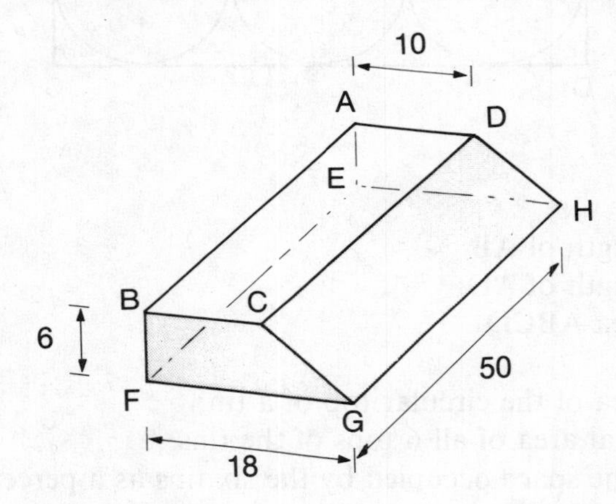

Figure 6.41

Calculate
(i) CG;
(ii) the volume of the block;
(iii) the *total* surface area of the block.

16. A company buys steel rods of circular cross-section. Each rod is of radius 12 mm and of length of 9 m.
(a) Write down the radius in metres.
(b) Calculate the volume of steel in one rod. Give your answer in cubic metres to 3 significant figures.
(The density of steel is 7800 kg/m^3.)
(c) Calculate the mass of one of these rods in kilograms to 2 significant figures.

The company later decides to buy steel from another firm which sells it in rods of circular cross-section with radius 15 mm.

(d) Calculate the length, in metres to 2 decimal places, of one of these rods given that it has the same volume as one of the original rods.

17. [In this question take π to be 3.142 and give each answer correct to three significant figures.]

A solid silver sphere has a radius of 0.7 cm.

(a) Calculate

(i) the surface area of the sphere;

(ii) the volume of the sphere.

(b) A silversmith is asked to make a solid pyramid with a vertical height of 25 cm and a square base. To make the pyramid, the silversmith has to melt down 1000 of the silver spheres.

Assuming that none of the silver is wasted, calculate the total surface area of the pyramid.

18. (In this question, take π to be 3.14. Give each answer correct to 1 decimal place.)

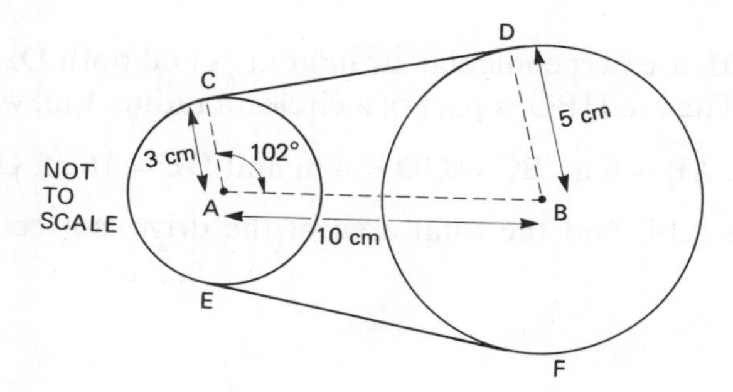

Figure 6.42

A manufacturer wishes to fit a continuous belt round two circular gear wheels of radii 3 cm and 5 cm. The centres, A and B, of the two wheels are 10 cm apart.

(a) The manufacturer has calculated that the angle BAC is 102°. Find:

(i) the length of belt in contact with the smaller wheel, centre A,

(ii) the length of belt in contact with the larger wheel, centre B.

(b) Find the length of belt CD, not in contact with either wheel.

(c) Find the total length of belt needed.

The ratio of the gear wheels is $1:1\frac{2}{3}$. This means that when the larger wheel rotates once, the smaller one rotates $1\frac{2}{3}$ times.

(d) If the smaller wheel rotates 10 times, how many times does the larger wheel rotate?

(e) After tests, the manufacturer decides that this gear ratio does not meet his requirements. He wishes to keep the larger wheel the same size, but alter the size of the smaller wheel so that it rotates 15% faster than before. Find:

(i) the new gear ratio in the form $1:n$,

(ii) the new radius of the smaller wheel. (MEG)

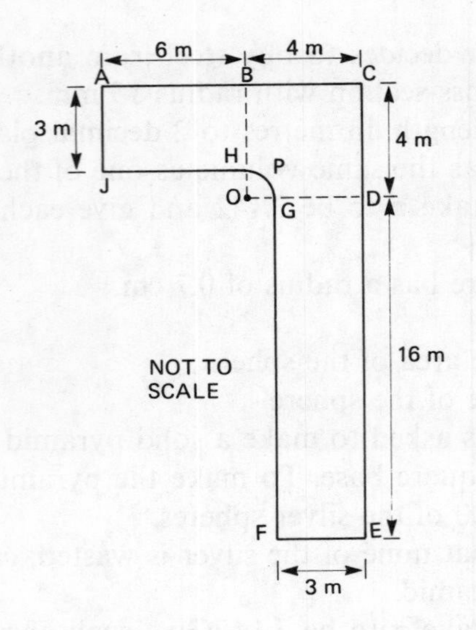

Figure 6.43

19. Figure 6.43 shows the plan of a concrete drive that Mr Fraser intends to lay in his garden.

    ABC and CDE are perpendicular straight lines and both DEFG and ABHJ are rectangles. The arc HPG is part of a circle, of radius 1 m, with centre at O.

    AJ = EF = 3 m. AB = 6 m, BC = CD = 4 m and DE = 16 m. OB = OD = 4m.

    Taking π to be 3.14, find the total area of the drive, correct to the nearest square metre. (NEA)

7

TRIGONOMETRY OF RIGHT-ANGLED TRIANGLES

The ideas of trigonometry of right-angled triangles are developed and applied to two- and three-dimensional situations in this unit.

7.1 Right-angled triangles

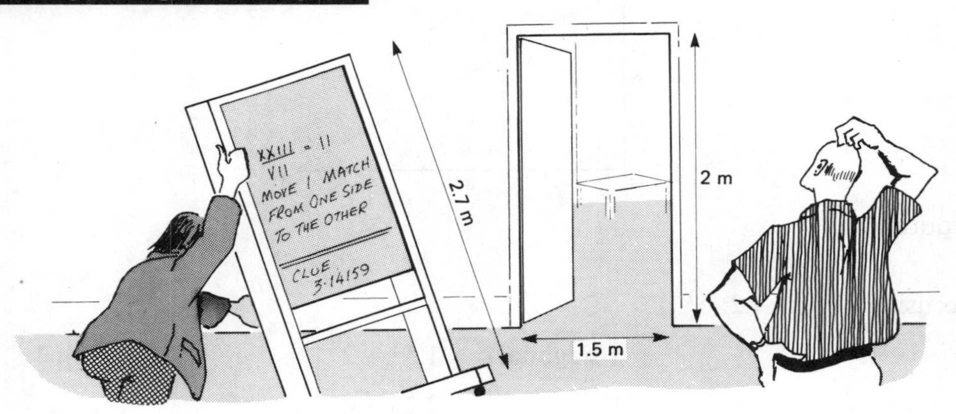

Figure 7.1

Moving a roller blackboard from one classroom to another can be rather a tricky task. Without dismantling the board in Figure 7.1, or turning it on its side, is it possible to get it through the opening? As the height of the board is greater than the height of the door opening, the board needs to be 'angled'. We therefore need to work out the length of the diagonal of the door opening as this will give us the maximum permissible length which will go through.

The right-angled triangle shown in Figure 7.2 will help us to find the diagonal length which is the side opposite the 90° angle. The general name given to this side in a right-angled triangle is the *hypotenuse.*

The property that we will use was developed by *Pythagoras* and is known as *Pythagoras's theorem.*

In any right-angled triangle the square on the hypotenuse is equal to the sum of the squares on the other two sides.

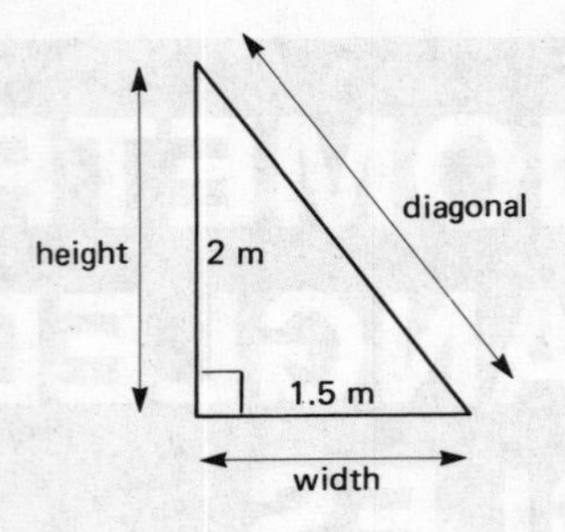

Figure 7.2

What this means is if you square the lengths of the two shorter sides of a right-angled triangle and add them together, the result will be equal to the square of the length of the hypotenuse.

If we label the sides of a right-angled triangle with the letters a, b and c as shown in Figure 7.3, then

$$a^2 + b^2 = c^2$$

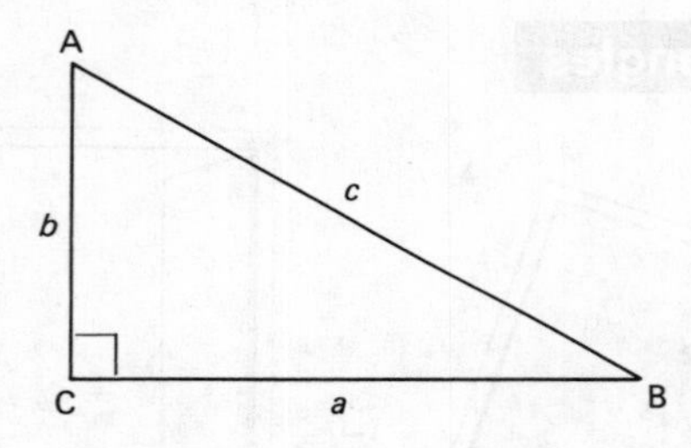

Figure 7.3

So using Figure 7.2:

$$\begin{aligned}(\text{diagonal})^2 &= 1.5^2 + 2^2\\ &= 2.25 + 4\\ &= 6.25\end{aligned}$$

The length of the diagonal therefore $= \sqrt{6.25} = 2.5$ m.

Clearly a case for either putting the board on its side or dismantling it!

Worked example 7.1

A Young Enterprise group in the first year sixth embarks on a designer jewellery scheme. The aim is to produce earrings in the shape of isosceles triangles (see Figure 7.4). Determine the amount of material in cm^2 required to make a set of earrings.

Using the symmetrical properties of an isosceles triangle, the perpendicular bisector of the base passes through the vertex of the triangle.

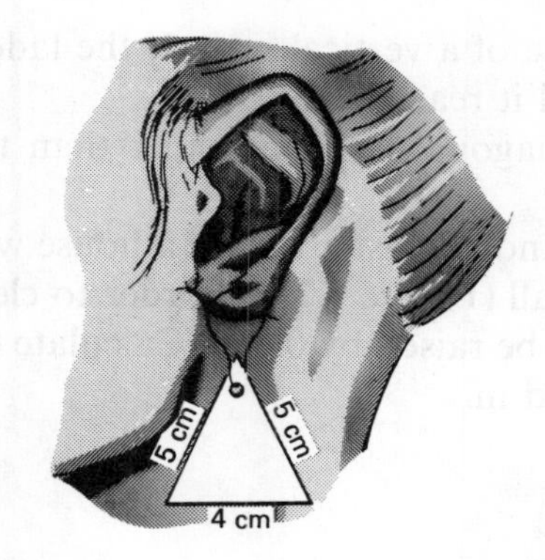

Figure 7.4

Figure 7.5

By drawing this line (Figure 7.5), we now have two equal right-angled triangles. Using Pythagoras's theorem on one of these triangles we have:

$$(\text{height})^2 + 2^2 = 5^2$$

$$(\text{height})^2 + 4 = 25$$

so $$(\text{height})^2 = 21$$

therefore $$\text{height} = 4.58 \text{ cm}$$

$$\text{The area of one earring} = \frac{1}{2} \times 5 \times 4.58 \text{ cm}^2$$

$$= 11.45 \text{ cm}^2$$

So the area of material required for two earrings

$$= 2 \times 11.45 \text{ cm}^2$$

$$= 22.9 \text{ cm}^2$$

Exercise 7a

(1) Which of the triangles in Figure 7.6 are incorrectly labelled? Justify your answers.

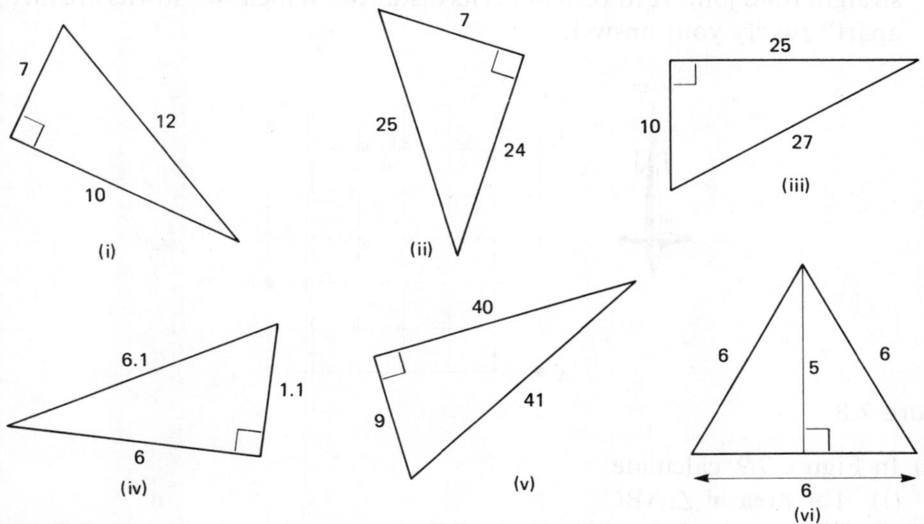

Figure 7.6

(2) The foot of a ladder is 2.5 m from the base of a vertical wall. If the ladder is 6.5 m in length, how far up the wall will it reach?

(3) A plank of wood is 84 cm long and its diagonal is 1 cm longer than this measurement. How wide is the plank?

(4) A ladder of length 6 m stands on level ground against a wall of a house with its base 3 m out from the bottom of the wall (Figure 7.7). In order to clean the window, the top of the ladder needs to be raised by 0.5 m. Calculate the distance by which the base must be pushed in.

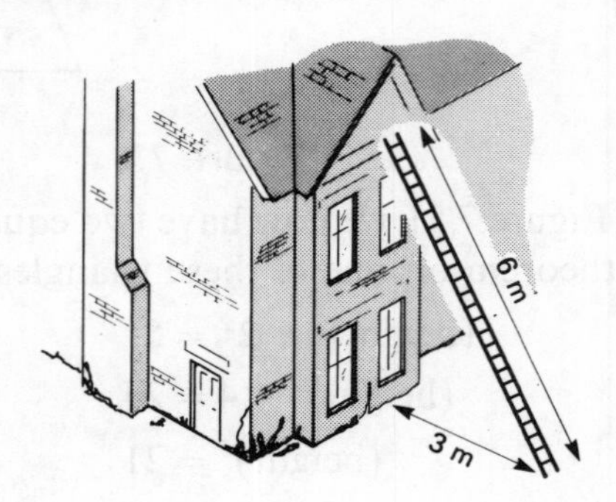

Figure 7.7

(5) The following table shows the distances in km, East and North of three towns B, C and D from A.

Town	B	C	D
Distance East of A in km	6	4	0
Distance North of A in km	2	6	5

Copy the grid shown in Figure 7.8 and plot the positions of C and D. A straight road joins A to B. What is its distance? Which two towns are furthest apart? Justify your answer.

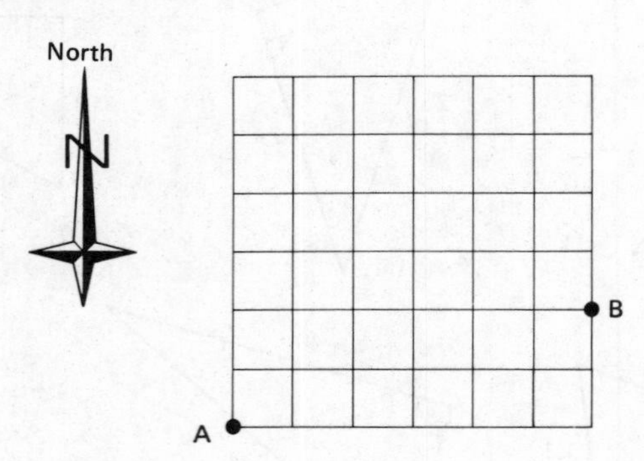

Figure 7.8

(6) In Figure 7.9, calculate:
(i) The area of $\triangle ABC$.
(ii) The length of the sides of $\triangle ABC$.

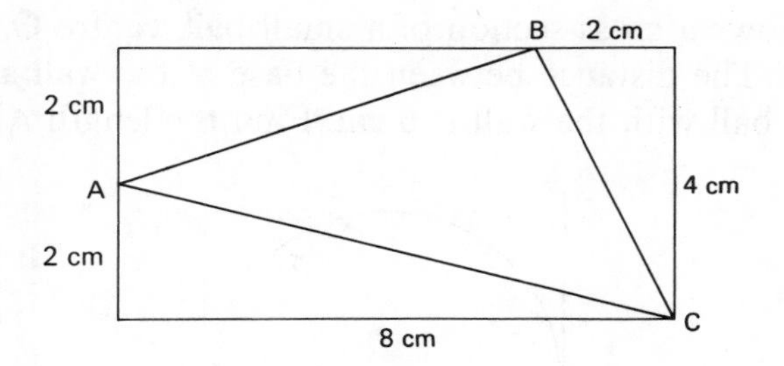

Figure 7.9

(7) To support a pole carrying a CB aerial, four wires are used from a point on the pole 12 metres from the ground to 4 pegs in the ground, each 5 metres from the foot of the pole. What is the length of one wire? If an extra metre is required to secure each wire, how much wire is required?

(8) Two straight roads run East–West and North–South through a town. Fatima and Iftekar leave the intersection of these two roads at the same time, Fatima cycling East at 12 km/hour and Iftekar walking North at 5 km/hour. How far apart are they after 30 minutes?

(9) Figure 7.10 shows the cross-section of a greenhouse belonging to Mr Potter. Mr Potter, who is 1.8 m tall, wishes to install a light from the apex of the roof. What is the maximum length for the cable and lamp in order that Mr Potter should not bang his head?

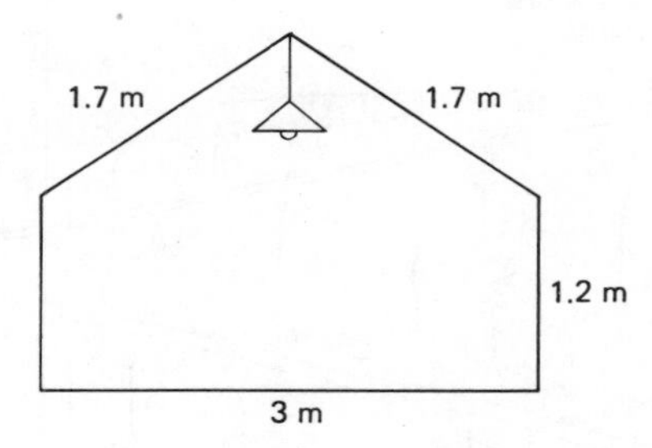

Figure 7.10

(10) Figure 7.11 shows a ball of radius 5 cm resting between the two sides of a V-shaped trough. If the distance from the base of the trough to the point of contact with the ball is 12 cm, determine the height of the top of the ball above the base of the trough.

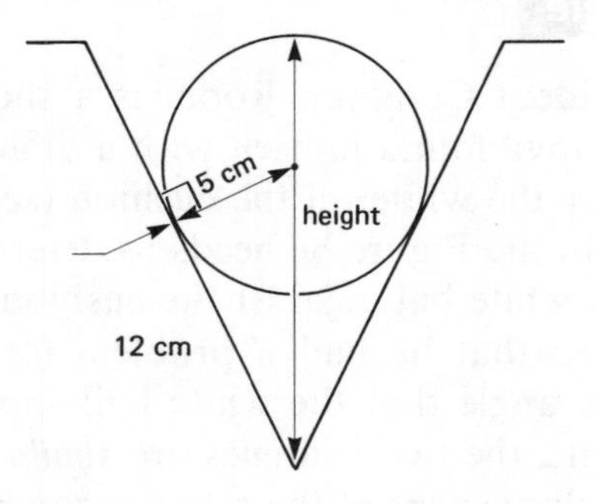

Figure 7.11

(11) Figure 7.12 shows a cross-section of a small ball, centre O, resting against a vertical wall. The distance between the base of the wall and the point of contact of the ball with the wall is 6 cm. Find the length AB.

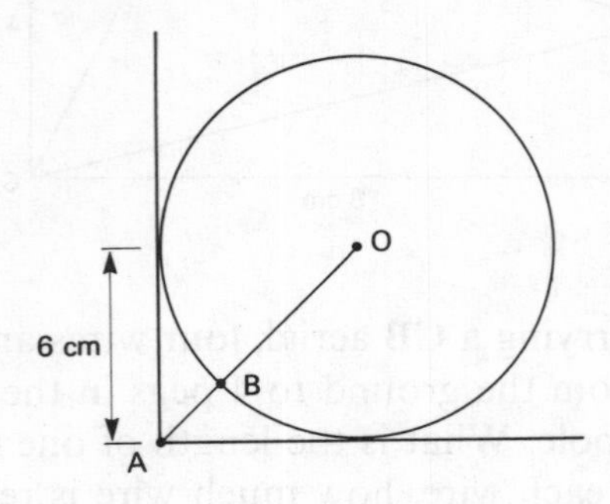

Figure 7.12

(12) From a height of 170 m above the surrounding countryside, Varnworth Castle enjoys fine views. The path up to the castle is a climb up two paths starting at a point immediately beneath the castle. The first stage is a gradual climb of some 250 m to a resting and viewing platform. Using the information given in Figure 7.13, determine the height of this viewing platform above the starting point. Determine also the total walking distance to the castle.

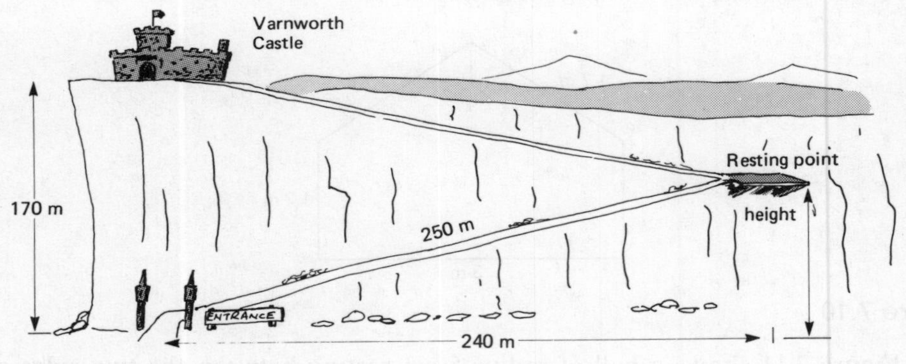

Figure 7.13

7.2 Tangent of an angle

Pride of place in the Student Common Room is a snooker table. During one particular lunch break, Floyd found himself with a difficult shot. In order to hit the blue ball he must play the white off the cushion (see Figure 7.14). Working out the distances shown in the Figure he needs to determine where, and at what angle, he should play the white ball against the cushion.

He immediately noticed that he had a problem involving two right-angled triangles and because the angle that the white ball should hit the cushion was equal to the rebound angle, the two triangles are *similar*.

 As one triangle is an enlargement of the other, a comparison of corresponding

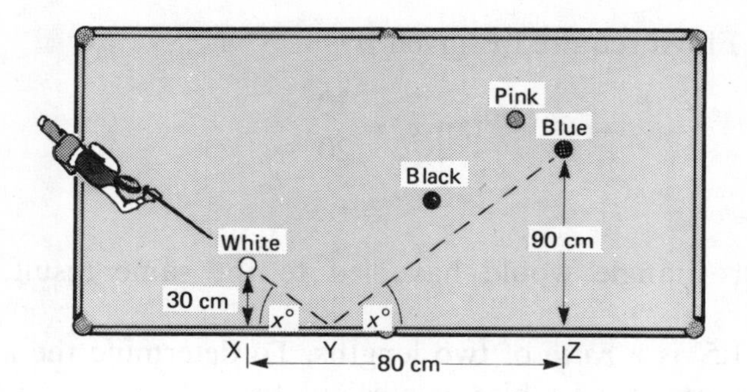

Figure 7.14

sides gave a multiplying factor of 3. The sides of the larger triangle were therefore 3 times longer than the smaller triangle.

So $$\mathrm{YZ} = 3 \times \mathrm{XY}$$

and $$\mathrm{XY} + \mathrm{YZ} = 80 \text{ cm}$$

These two equations gave

$$4\mathrm{XY} = 80 \text{ cm}$$

so $$\mathrm{XY} = 20 \text{ cm (and YZ} = 60 \text{ cm)}$$

With this information, Floyd was able to determine the angle using the *tangent ratio* of a right-angled triangle.

In Section 7.1 we gave the general name for the longest side of a right-angled triangle – the *hypotenuse*. For the rest of this unit it will be useful to refer to the other two sides as the *opposite* and the *adjacent*. To determine which is which, besides the right angle, there is usually one angle either given or that we are interested in finding in right-angled triangles. Figure 7.15 shows a representation of the smaller triangle in Floyd's problem with the angle $x°$ marked. The *opposite*, by its position, is opposite the angle and the *adjacent* is one of the sides which makes up the angle $x°$ along with the hypotenuse.

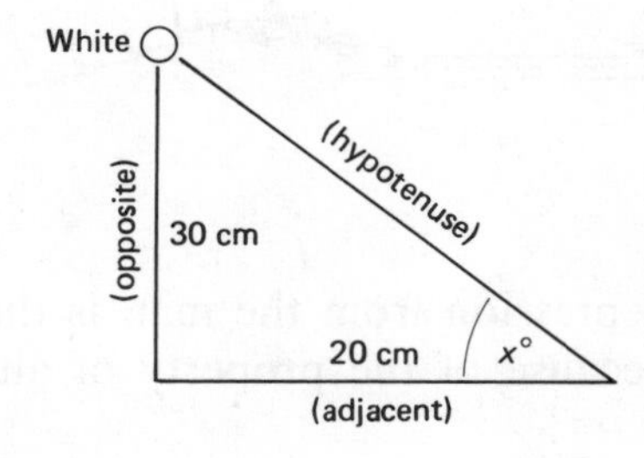

Figure 7.15

The tangent of an angle is defined as:

$$\tan x° = \frac{\text{opposite}}{\text{adjacent}}$$

so the angle that Floyd required is given by

$$\tan x^\circ = \frac{30}{20}$$

$$= 1.5$$

[*Note*: the larger triangle would have led to the same result, i.e. $\tan x^\circ = 90/60 = 1.5$.]

This value of '1.5' is a *ratio* of two lengths. To determine the angle, we need to use either trigonometrical tables or a calculator.

Floyd used a calculator and tapped the following keys:

1.5 [INV] [TAN]

or [TAN⁻¹]

The screen display showed the figures 56.309932 which is the angle in degrees. The angle that Floyd required to the nearest tenth of a degree was

$$56.3^\circ$$

7.3 Angles of elevation and depression

Figure 7.16 illustrates what is meant by the terms *angle of elevation* and *angle of depression.* Looking upwards from a horizontal line, the man on the clifftop can see a seagull, and looking downwards from a horizontal line he can see the boat. The angles formed are angle of *elevation* and angle of *depression* respectively.

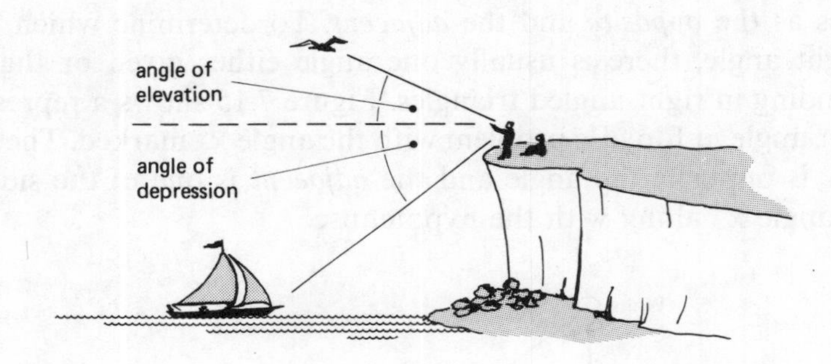

Figure 7.16

Note that the angle of depression from the man is the same as the angle of elevation from the boat, because of the property of alternate angles between parallel lines.

Worked example 7.2

Francis needs to fall a tree of unknown height in his garden. It is situated 20 m from his house. In order to determine the tree's height, he measures the angle of elevation of the top of the tree at a distance of 40 m from the base of

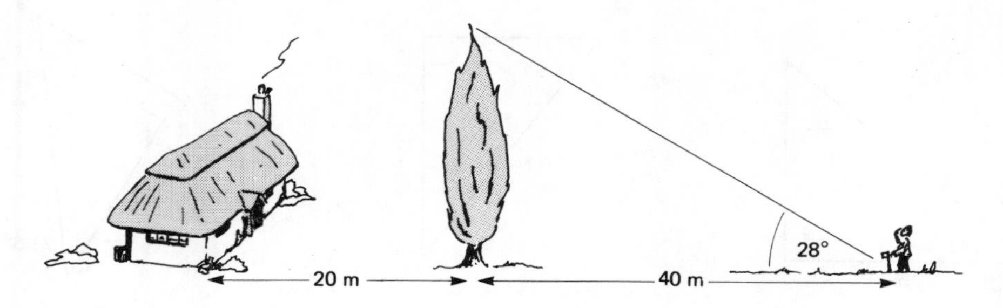

Figure 7.17

the tree. The angle was found to be 28° (see Figure 7.17). If the tree falls in the direction of the house, will it cause any damage?

Using the tangent ratio:

$$\tan 28^\circ = \frac{\text{height}}{40}$$

Rearranging the equation:

$$40 \times \tan 28^\circ = \text{height}$$

so the height of the tree = 40 [×] 28 [TAN] [=] Display [21.268]

$$= 21.3 \text{ m}$$

Evasive action may therefore be necessary.

Exercise 7b

(1) In Figure 7.18, determine the size of the angles marked, to the nearest tenth of a degree.

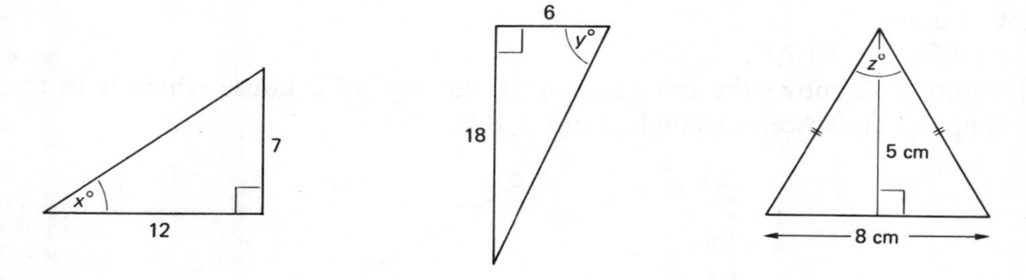

Figure 7.18

(2) In Figure 7.19, use the tangent ratio and Pythagoras's theorem to determine the lengths of the other two sides of each triangle correct to 1 decimal place.

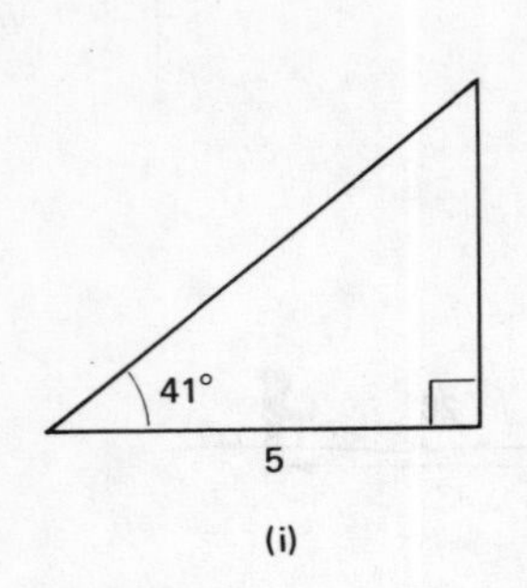

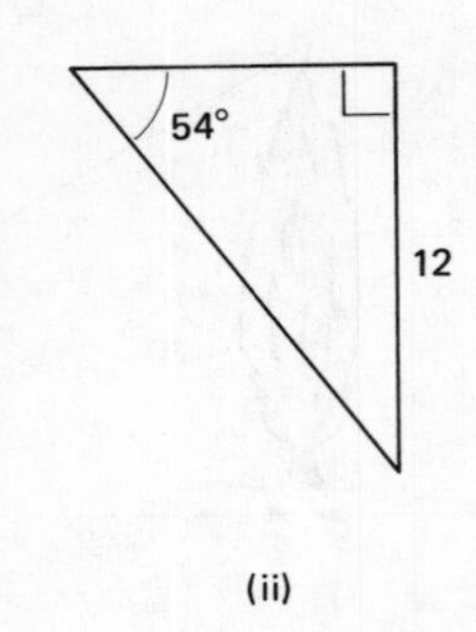

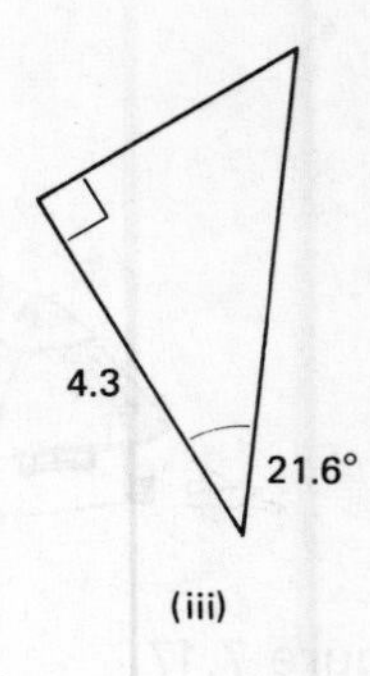

Figure 7.19

(3) Figure 7.20 shows a triangle ABC inscribed in a circle with BC as a diameter. If AC = 15 cm and ∠ ACB = 31°, find AB and the radius of the circle.

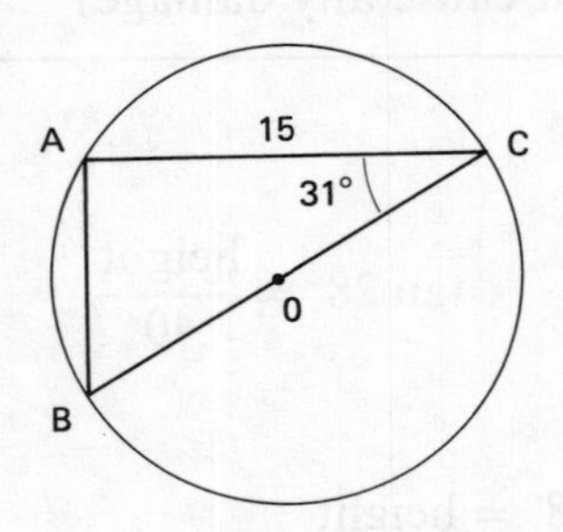

Figure 7.20

(4) In Figure 7.21, AB = 5 cm, ∠ ABC = 90° and ∠ BAD = ∠ DAC = 30°.

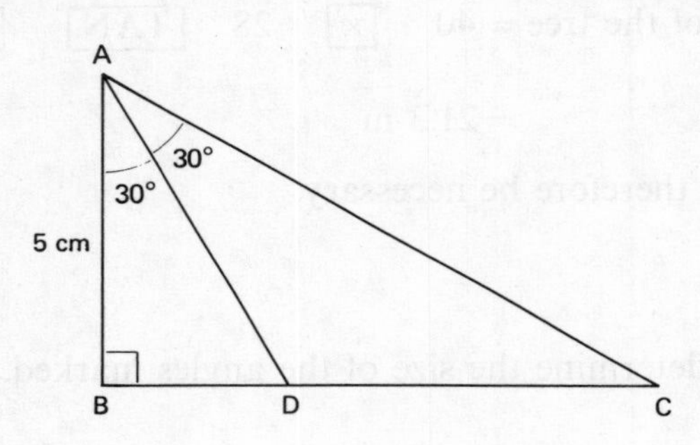

Figure 7.21

Calculate:
(i) BD, (ii) AC.

(5) Figure 7.22 shows the cross-section of the roof of a house which is in the shape of an isosceles triangle. Find ∠ BAC.

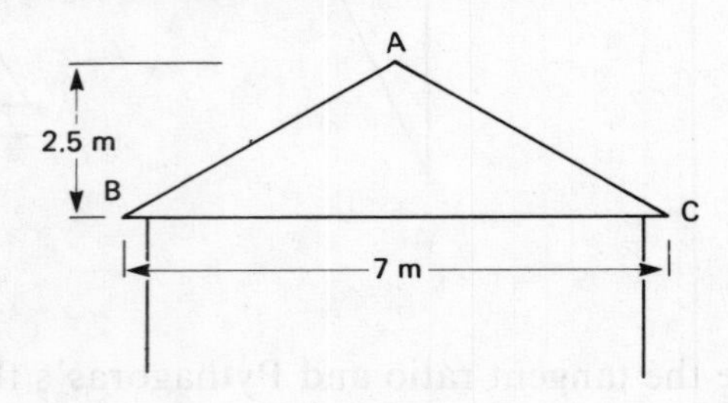

 Figure 7.22

(6) From a first floor window (Figure 7.23) the top of a lamp-post is seen at an angle of elevation of 21.8°. If the height of the window above the ground is 4 m and the distance of the window from the lamp-post is 20 m, calculate the height of the lamp-post.

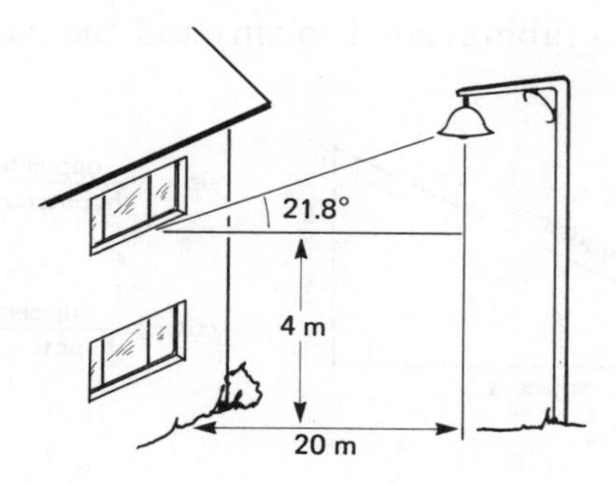

Figure 7.23

(7) From the top of a cliff which is 50 m high, Josie looks directly out to sea and sees a windsurfer at an angle of depression of 18°. She also observes a swimmer at an angle of depression of 26°. Given that Josie, the swimmer and the windsurfer are in the same vertical plane, what is the distance between the swimmer and the windsurfer?

(8) In Figure 7.24, AB = 5 cm, BC = 12 cm and AC = CD. Find $x°$.

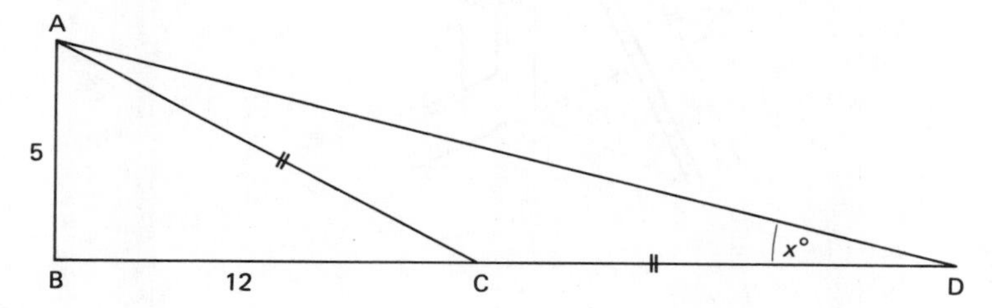

Figure 7.24

(9) Figure 7.25 shows a ball stuck in a V-shaped drain. An angle and a length are marked on the diagram. Determine the radius of the ball.

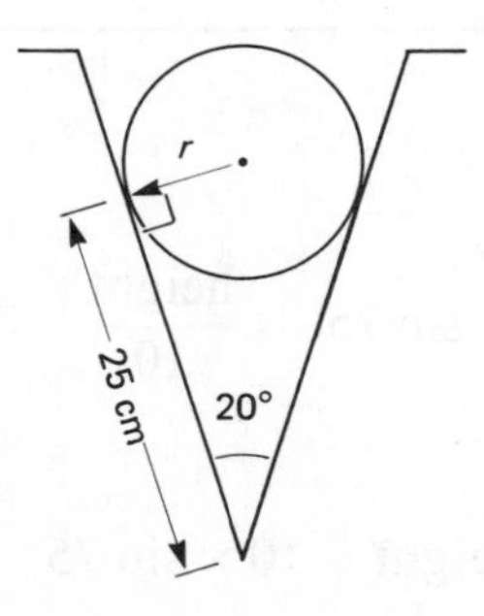

Figure 7.25

7.4 Sine and cosine of an angle

Although the tangent ratio is a useful method to determine the angles and sides of a right-angled triangle, on its own it may not be sufficient and two further ratios are required.

These are the *sine ratio* (abbreviated to sin) and the *cosine ratio* (abbreviated to cos).

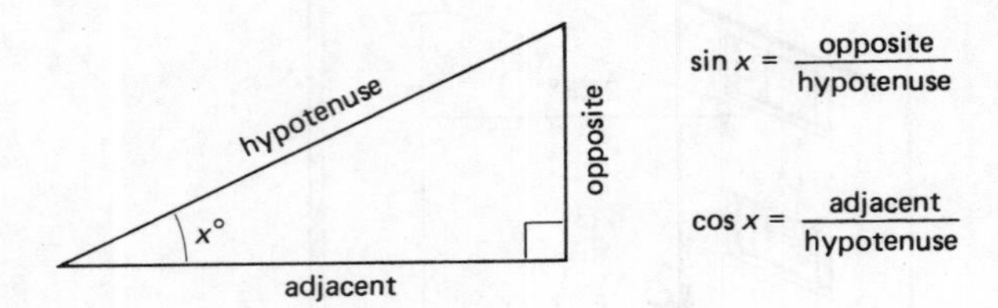

Figure 7.26

Worked example 7.3

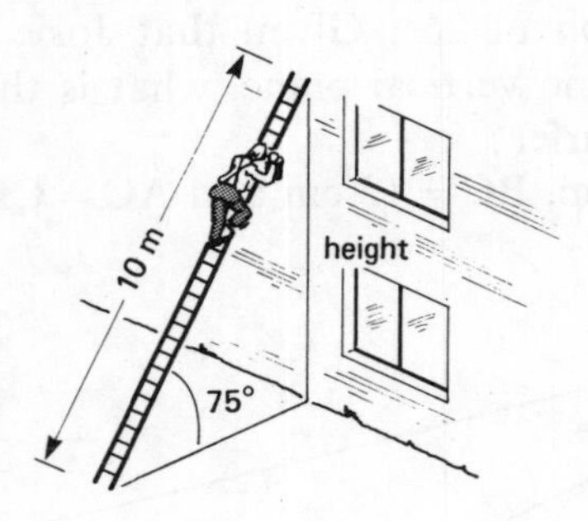

Figure 7.27

An external decorator has a ladder which extends to a length of 10 m. On the grounds of safety, when the ladder is fully extended the decorator places the ladder at an angle of inclination of no more than 75° to the horizontal. Find the point of contact of the top of the ladder against a vertical wall when the ladder is inclined at its maximum angle.

Using the sine ratio:

$$\sin 75° = \frac{\text{height}}{10}$$

so

$$\text{height} = 10 \times \sin 75°$$

$$= 9.66 \text{ m}$$

Using a calculator:

					Display
10	[×]	75	[SIN]	[=]	[9.659...]

Worked example 7.4

Figure 7.28

In order to paint the exterior when the ladder is at its maximum angle of elevation, the decorator finds that he can safely reach the wall with his brush provided he is within 4 m of the top of the ladder. Find the decorator's maximum safe reach.

We require to calculate the 'adjacent' side of the right-angled triangle so we use the cosine ratio.

$$\cos 75^\circ = \frac{\text{distance}}{4}$$

so

$$\text{distance} = 4 \times \cos 75^\circ$$

$$= 1.04 \text{ m}$$

Using a calculator:

					Display
4	[×]	75	[COS]	[=]	[1.0352]

Worked example 7.5

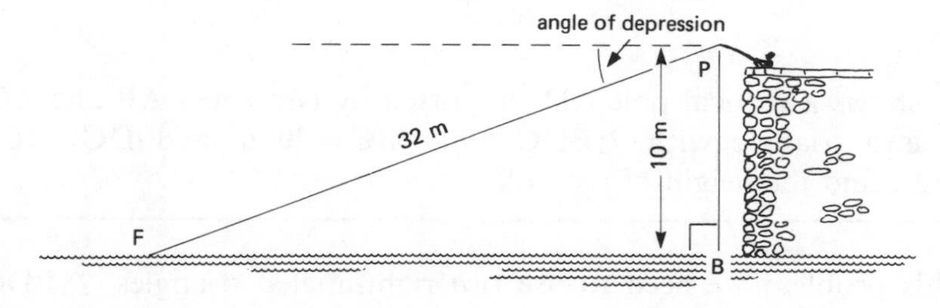

Figure 7.29

Rizwan was fishing from the parapet of a bridge. If the height of the parapet is 10 m above the water and the amount of fishing line he has let out is 32 m, determine the angle of depression of the line. You may assume that the line is taut (see Figure 7.29).

Using the right-angled triangle shown in Figure 7.29, we use the property of alternate angles between parallel lines to calculate $\angle$ PFB. This is equal to the angle of depression.

Using the sine ratio:

$$\sin \angle \text{PFB} = \frac{10}{32}$$

so $$\sin \angle \text{PFB} = 0.3125$$

therefore $$\angle \text{PFB} = 18.2°$$

Using a calculator:

0.3125 [INV] [SIN] Display [18.209]

(or [SIN^{-1}])

7.5 Three-dimensional problems

In solving problems of this nature, it is necessary to find suitable right-angled triangles within the diagrams.

Worked example 7.6

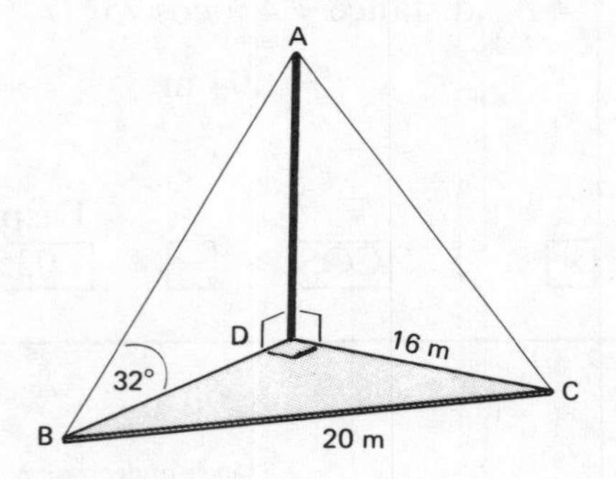

Figure 7.30

Figure 7.30 shows a vertical pole AD supported by two ropes AB and AC. The base BDC is a triangle with $\angle$ BDC = 90°, BC = 20 m and DC = 16 m. If $\angle$ ABD = 32°, find the height of the pole.

To solve this problem we need to use *two* right-angled triangles, $\triangle$BDC and $\triangle$ABD (see Figures 7.31 and 7.32).

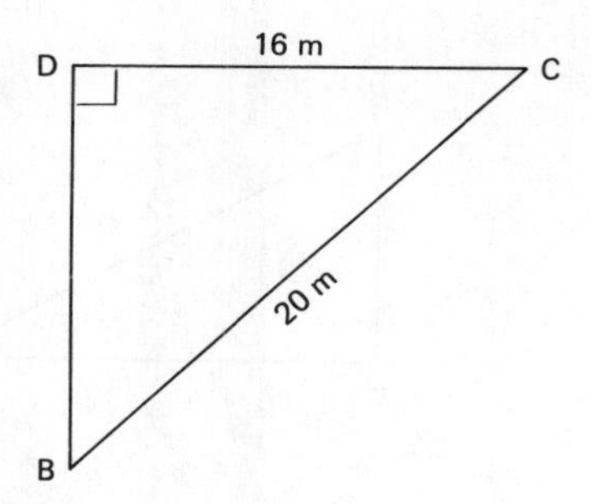

Figure 7.31

Figure 7.32

Using Pythagoras's theorem on $\triangle BDC$:

$$BD^2 + 16^2 = 20^2$$

so
$$BD^2 + 256 = 400$$

therefore
$$BD = \sqrt{144} = 12 \text{ m}$$

Using this information in $\triangle ABD$:

$$\tan 32^\circ = \frac{AD}{12}$$

so
$$AD = 12 \times \tan 32^\circ$$
$$= 7.50 \text{ m}$$

Using a calculator:

					Display
12	[×]	32	[TAN]	[=]	[7.498]

Exercise 7c

(1) In Figure 7.33, determine the size of the angles marked, to the nearest tenth of a degree.

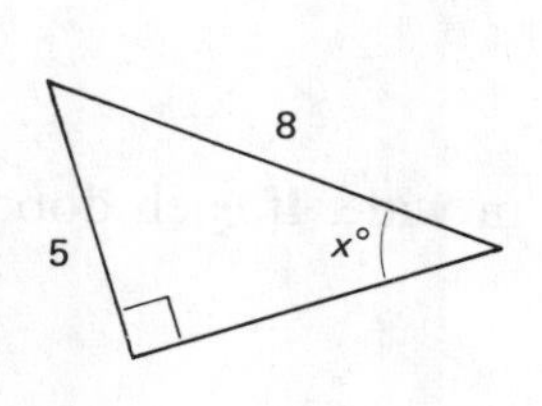

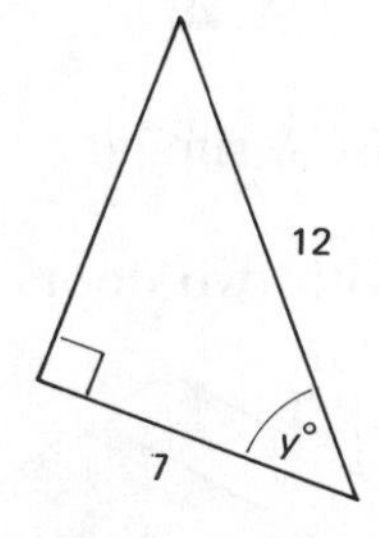

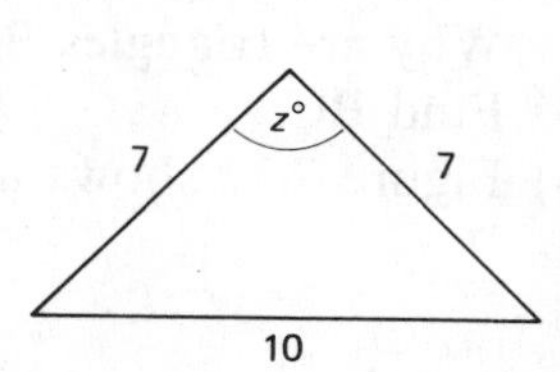

Figure 7.33

(2) In Figure 7.34, determine the lengths of the other two sides of each triangle correct to 1 decimal place.

(3) A ladder of length 7 m rests against a vertical wall (Figure 7.35). If the base of the ladder is 3 m from the wall, what is the angle of elevation of the ladder? If the ladder is extended by 2 m but the base remains in the same position, determine the new angle of elevation.

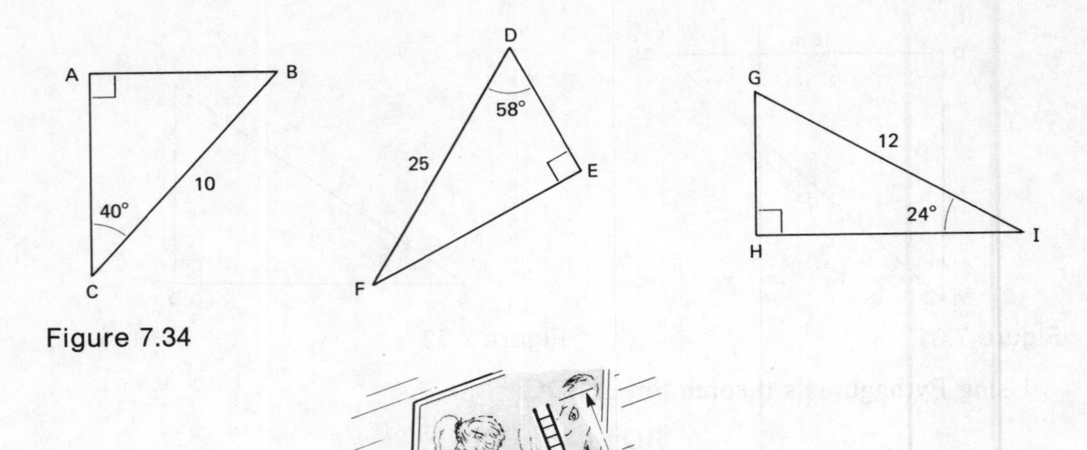

Figure 7.34

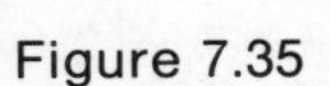

Figure 7.35

(4) An escalator slopes at an angle of elevation of 25°. If the difference in height between the two floors is 5 m, find the length of the escalator.

(5) Using Figure 7.36, write down the sine of ∠ADB.

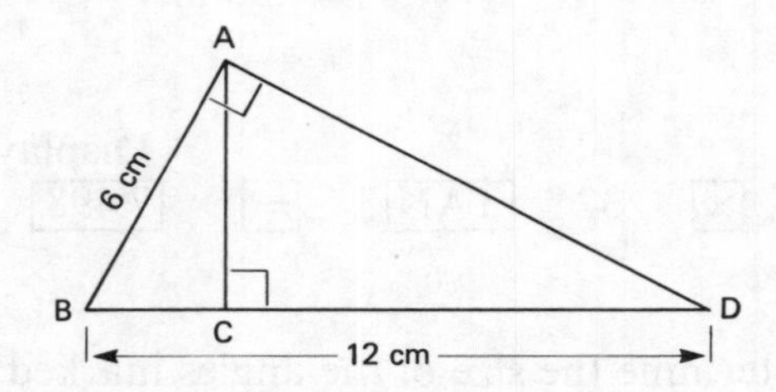

Figure 7.36

Which other angle is equal to ∠ADB?
Find the length of AD.
Why are triangles BAD and BCA similar?
Find BC.

(6) Figure 7.37 shows a garage with two doors each 1.5 m wide. If each door

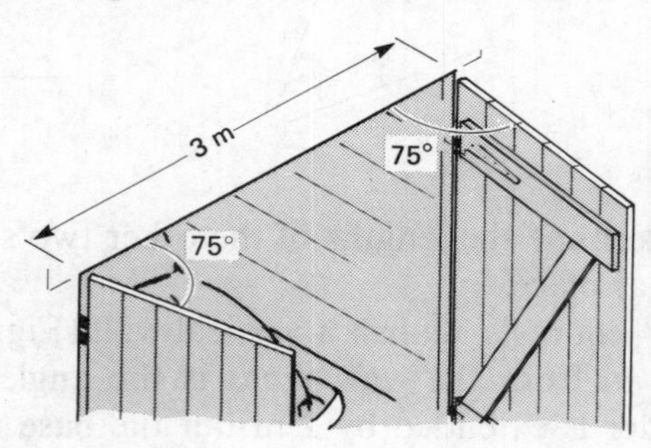

 Figure 7.37

can only be swung on its hinges through 75°, how wide a vehicle can enter (or leave) the garage?

(7) Figure 7.38 shows the framework of a crane. If $\angle DBC = 90°$, $\angle BDC = 20°$, $DC = 8$ m, $AC = 4$ m and $\triangle ABC$ is isosceles, find the lengths BD and AB.

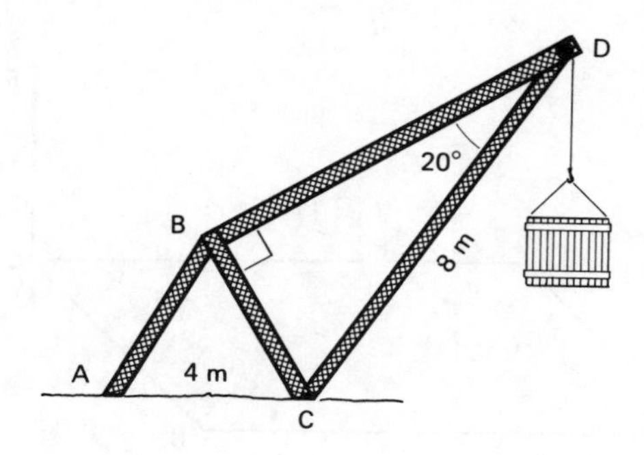

Figure 7.38

(8) Figure 7.39 shows the course of a boat which starts at Arbor and travels a distance of 15 km on a bearing of 075° to Bowtin. It then changes course to 060° and travels a further 12 km to Chartor. Find how far North and East Bowtin is from Arbor. Find how far North and East Chartor is from Bowtin. How far North and East of Arbor is Chartor? What is the bearing of Chartor from Arbor?

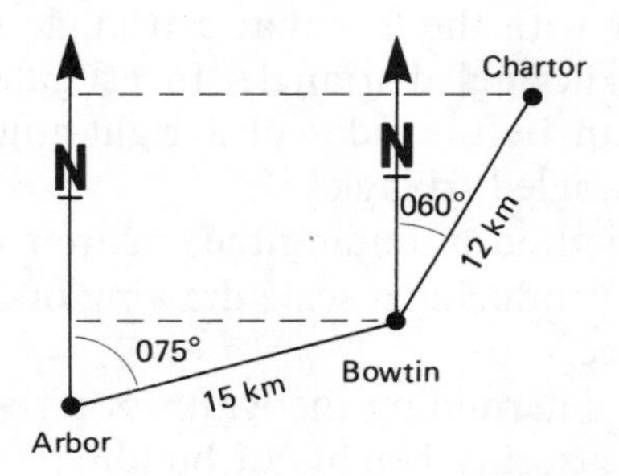

Figure 7.39

(9) Figure 7.40 shows a cube of side 1 cm.

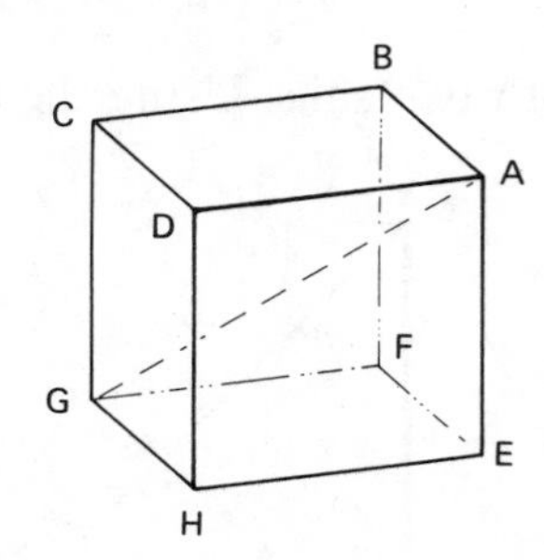

Figure 7.40

(a) Calculate the length of AH.
(b) Calculate the length of AG.

(c) Calculate ∠HAG.
(d) Explain why ∠HAC is 60°.

(10) Figure 7.41 shows a pole of height 12 m, situated at the corner of a square field ABCD. The angle of elevation of the top of the pole from the corner A is 18°.

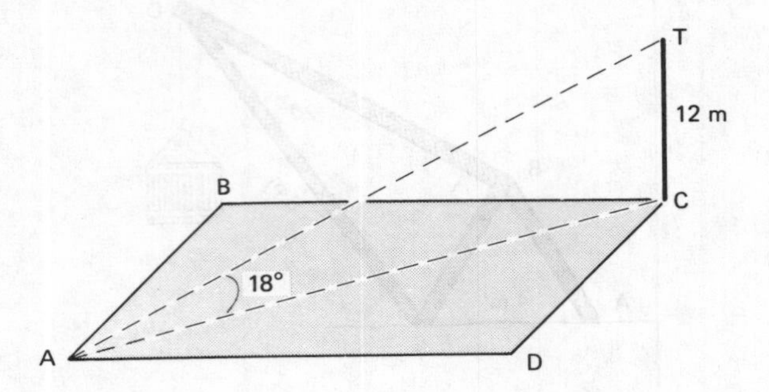

Figure 7.41

(a) Calculate the length of the diagonal AC.
(b) Write down the value of ∠CAD.
(c) Find AD.
(d) Find the area of the field.

Suggestions for coursework 7

1. You should be familiar with the fact that a triangle with sides of length 3, 4 and 5 units forms a right-angled triangle. Investigate other combinations of three integers which can be the sides of a right-angled triangle. Investigate everyday uses of right-angled triangles.
2. Devise a geometrical method of finding the square root of any integer.
3. Investigate methods of producing a scale drawing of the elevation of the front of your school or college.
4. Investigate methods of determining the width of a river at a given point.
5. Devise a method of measuring heights of buildings or landmarks. Test your method and compare results with the correct measurements.

Miscellaneous exercise 7

1. Figure 7.42 shows a boat with two sails. Using the information contained on

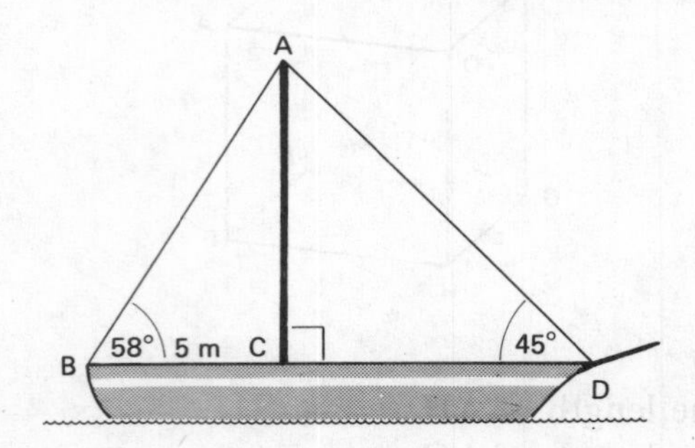

 Figure 7.42

the figure, determine the height of the mast above the deck. Write down the length BD.

2. Figure 7.43 shows a ball attached to a 1 m pole. The length of the chord and the ball is 95 cm. A skittle of height 10 cm is placed 30 cm from the base of the pole. Is it possible to swing the ball and hit the skittle? Justify your answer.

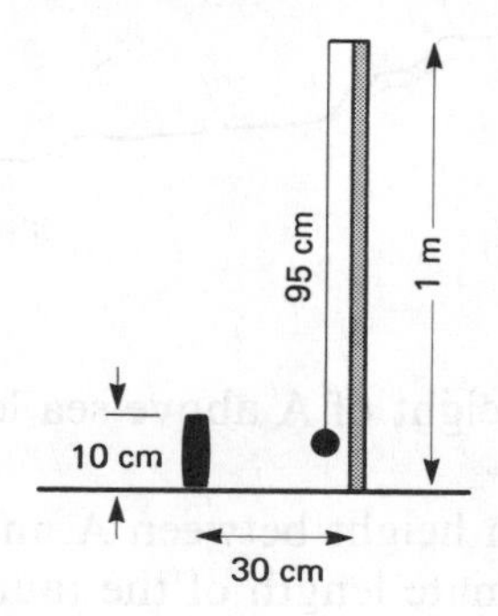

Figure 7.43

3. A theodolite is used to measure the height of a tree (Figure 7.44). The distance between the tree and the theodolite is 100 m and the height of the theodolite is 1.5 m. The angle of elevation of the top of the tree measured by the theodolite is 15°. Calculate the height of the tree.

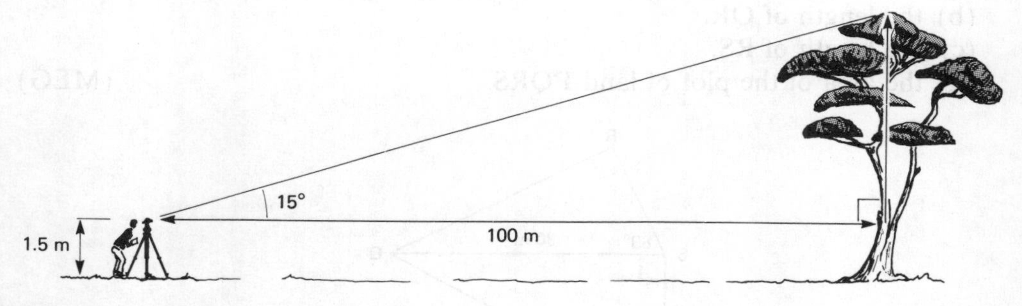

Figure 7.44

4. In Figure 7.45, AC = 12 m, AD = 15 m, AE = 21 m, ∠ABC = 55° and ∠ACD = 90°. Calculate (i) CD, (ii) ∠AEC, (iii) AB.

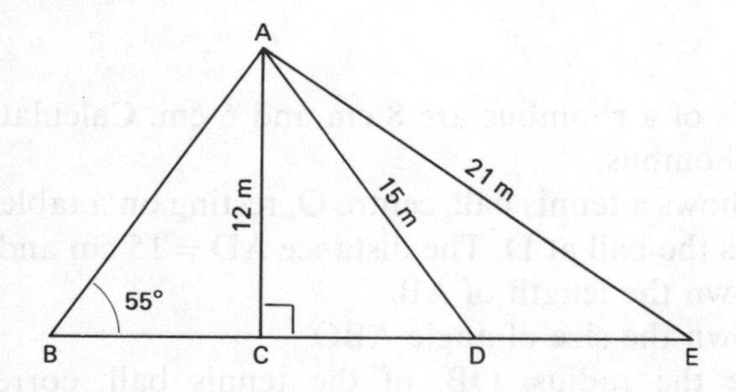

Figure 7.45

5. In Figure 7.46, A and B are points at the feet of two identical pylons on a mountain to be connected by a cable car. The horizontal distance between

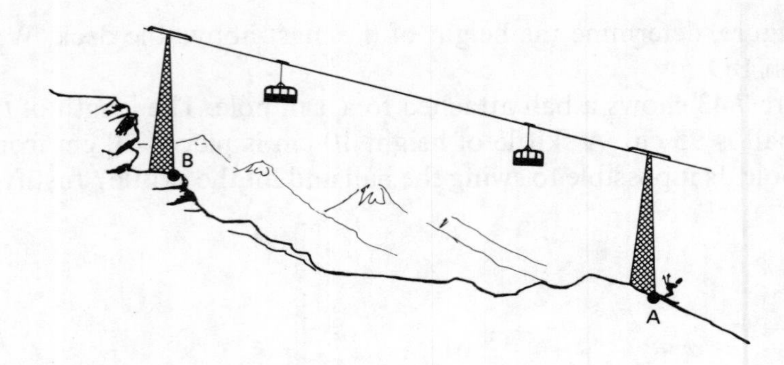

Figure 7.46

A and B is 420 m. The height of A above sea level is 2340 m and the height of B is 2490 m.

(a) Find the difference in height between A and B.
(b) What is the approximate length of the taut cable joining A and B?
(c) A skier is at point A. What is the angle of elevation of point B?

(SEG)

6. In Figure 7.47, PQRS represents a horizontal plot of land. PS = 20 m, SQ = 30 m, $\angle$ PSQ = 90°, $\angle$ QSR = 63° and $\angle$ QRS = 90°. Calculate, correct to two significant figures:
(a) the length of PQ,
(b) the length of QR,
(c) the length of RS,
(d) the area of the plot of land PQRS. (MEG)

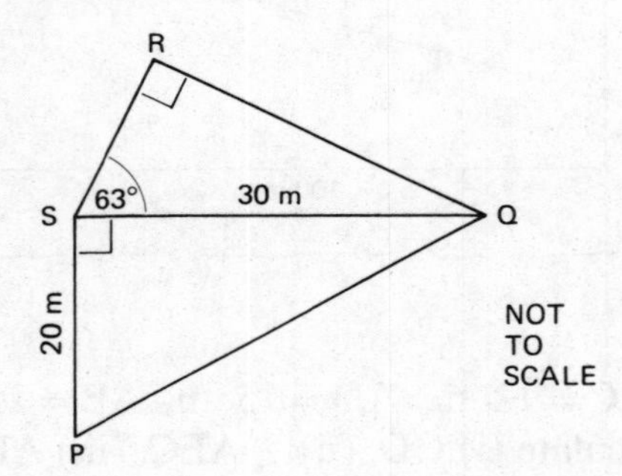

Figure 7.47

7. The diagonals of a rhombus are 8 cm and 6 cm. Calculate the length of the sides of the rhombus.
8. Figure 7.48 shows a tennis ball, centre O, resting on a table top, ABC. A ruler, ADE, touches the ball at D. The distance AD = 15 cm and angle DAB = 24°.
(a) Write down the length of AB.
(b) Write down the size of angle ABO.
(c) Calculate the radius, OB, of the tennis ball, correct to the nearest millimetre. (MEG)
9. At 3.00 p.m. a ship is observed at A, 6 km due South of a lighthouse, L. The ship maintains a constant speed on a bearing of 060° and reaches a point B,

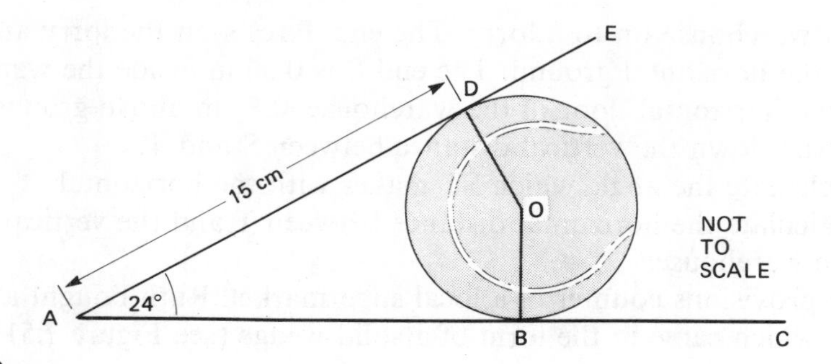

Figure 7.48

due East of the lighthouse, at 3.40 p.m. C is the point on the ship's path which is nearest to L.

(a) Draw a sketch showing the positions of A, L, B and C and calculate the distance LC, correct to 2 significant figures.
(b) Find the distance AC.
(c) Find the time at which the ship reaches C. (MEG)

10. Airport A is 400 km due East of airport B. An aircraft leaves A at 21 55 hours to fly to B. Its speed over the ground is 320 km/h (see Figure 7.49).

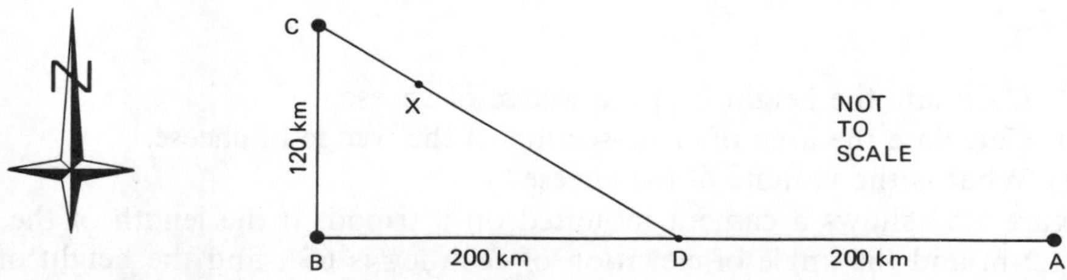

Figure 7.49

(a) (i) Calculate the time the aircraft takes to fly 400 km.
(ii) At what time is the aircraft expected to arrive at B?
(b) When the aircraft is at D, halfway between A and B, it is diverted to airport C because of fog at airport B. Airport C is 120 km due North of B.
(i) Calculate the bearing of C from D.
(ii) Calculate the distance from D to C.
(iii) The point on the aircraft's path nearest to B is X. Calculate the distance of X from B. (MEG)

11. In Figure 7.50, ST represents a ramp, 3.50 m long, used when goods are moved

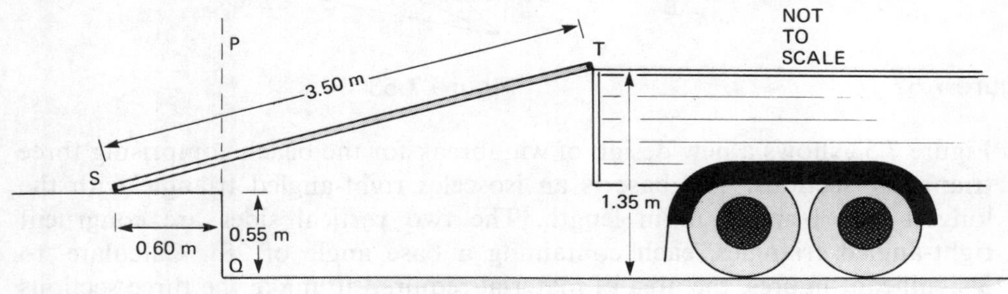

Figure 7.50

from a warehouse on to a lorry. The end T rests on the lorry and is 1.35 m above the horizontal ground. The end S is 0.60 m inside the warehouse and is on the horizontal floor of the warehouse, 0.55 m above ground level.

(a) Write down the vertical distance between S and T.

(b) Calculate the angle which ST makes with the horizontal.

(c) Calculate the horizontal distance between T and the vertical wall PQ of the warehouse. (MEG)

12. At the provisions counter of a local supermarket. Ruth bought a quantity of cheese which came in the form of a solid wedge (see Figure 7.51).

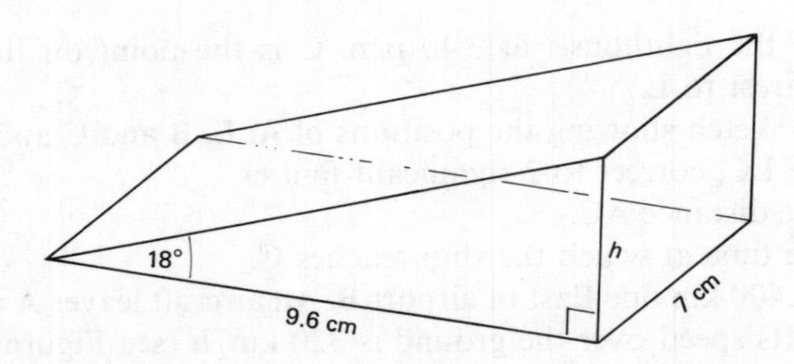

Figure 7.51

(i) Calculate the height h of the wedge of cheese.

(ii) Calculate the area of cross-section of the wedge of cheese.

(iii) What is the volume of the cheese?

13. Figure 7.52 shows a camera mounted on a tripod. If the length of the legs is 1.2 m and the angle of elevation of each leg is 65°, find the height of the central support (marked 'height' on the diagram), if the base of the camera is to be 1.2 m from the ground.

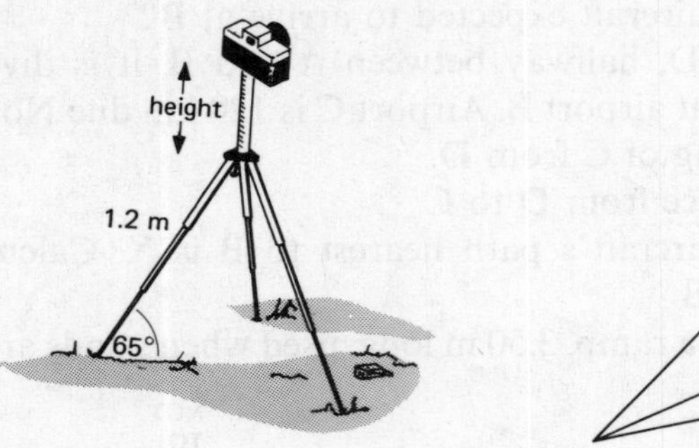

Figure 7.52

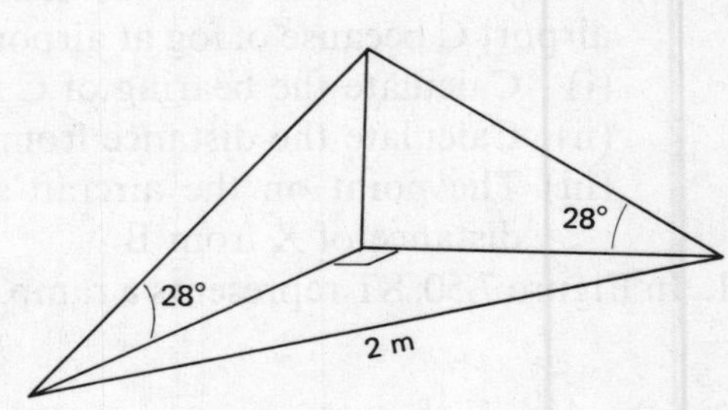

Figure 7.53

14. Figure 7.53 shows a new design of windbreak for the beach comprising three triangular sections. The base is an isosceles right-angled triangle with the longest side being 2 m in length. The two vertical sides are congruent right-angled triangles, each containing a base angle of 28°. Calculate, to 3 significant figures, the area of material required to make the three sections of the windbreak.

USING GRAPHS

8

Graphs are an important visual representation of information. This unit shows you how to interpret and draw graphs for a variety of situations.

8.1 Co-ordinates

Co-ordinates in the conventional mathematical sense are numbers which are used to identify uniquely a point or a position on a graph. A horizontal and a vertical line, called *axes*, can be drawn through the *origin*, and uniform scales marked along both of these lines. Co-ordinates can then be written as two numbers in brackets separated by a comma. The first number represents a measurement horizontally from the origin, and the second number represents a vertical measurement from the origin. Direction has to be taken into account and horizontal measurements to the *left* of the origin are *negative*, and likewise vertical measurements *below* the origin are also *negative*.

Worked example 8.1

On the grid in Figure 8.1(i), O represents the origin. Write down the co-ordinates of the points A and B, and determine the distance between them.

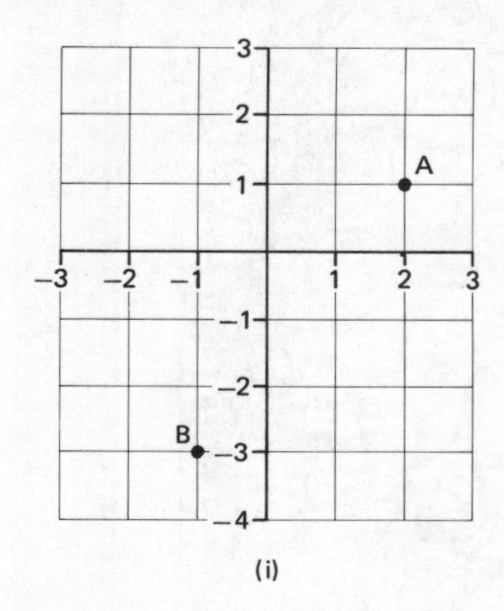

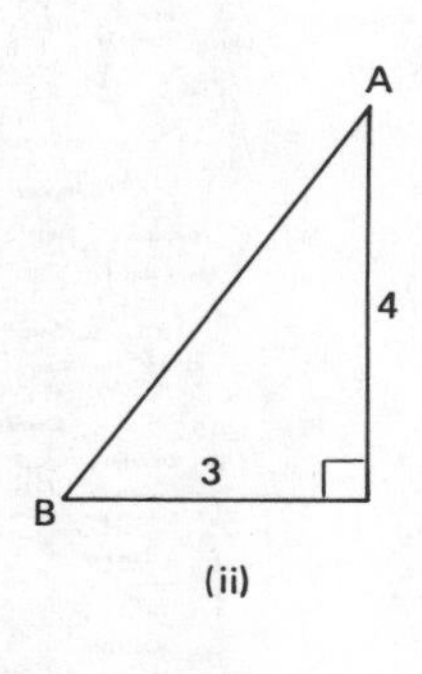

Figure 8.1

A is 2 units to the right and 1 unit up from the origin so the co-ordinates of A are:

$$(2, 1)$$

B is 1 unit to the left and 3 units below the origin so the co-ordinates of B are:

$$(-1, -3)$$

The distance between these two points can be determined by using Pythagoras's property of a right-angled triangle with AB as the hypotenuse. The measurements on the other two sides are the horizontal and vertical *differences* between the two points (see Figure 8.1(ii)).

So $$AB^2 = 3^2 + 4^2$$

$$= 25$$

Therefore $$AB = 5 \text{ units}$$

This representation of a co-ordinate is not unique and neither is it solely confined to using numbers. Theatre seats are often identified using a combination of letters and numbers. So, for example, the seat C24 would perhaps identify the 24th seat along the third row from the front.

8.2 Reading maps

Those students who have participated in the Duke of Edinburgh Award Scheme will be familiar with a practical use of co-ordinates when they have needed to determine map references for a route plan. Figure 8.2 shows an extract from an Ordnance Survey map. Each square is identified by the co-ordinates of the bottom left-hand corner of the square, and this is referred to as the *grid reference* of the square. The two co-ordinates are written as one 4-figure number without brackets. So the grid reference of the square containing the village of Calton is 1050 and the grid reference of the square containing Cheshire Wood is 1153.

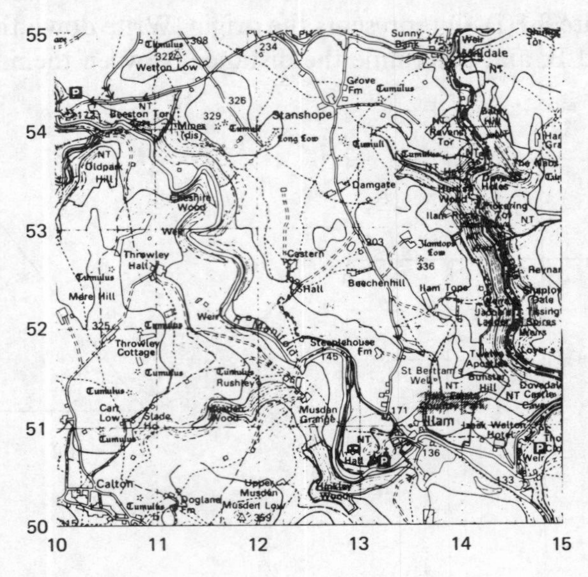

 Figure 8.2

We can identify points on the map by dividing each side of a square into *tenths*. The co-ordinates of the point would then be estimated by eye and the resulting grid reference written down as a 6-figure number.

Worked example 8.2
Find the 6-figure grid reference of Hamtops Low.

The grid reference of the square containing Hamtops Low is 1352. Figure 8.3 shows this square and if we draw a horizontal and vertical line through Hamtops Low we can estimate that the co-ordinates of the intersection are (13.7, 52.7). So the 6-figure grid reference is 137527.

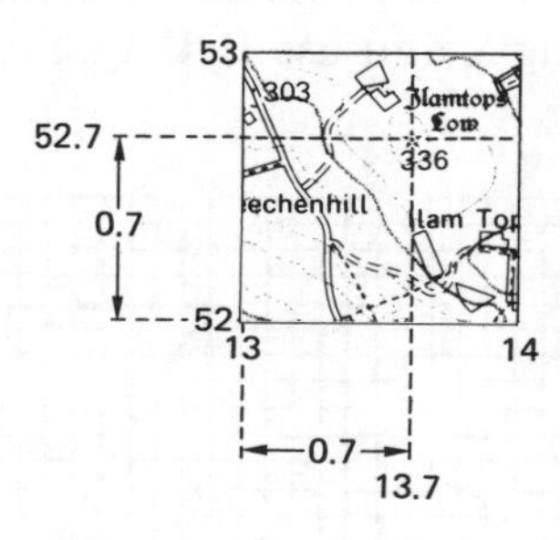

Figure 8.3

Exercise 8a

(1) Figure 8.4 shows a partially completed game of 'Noughts & Crosses'. If 'X' is to play, write down the co-ordinates of the squares which would ensure a winning position.

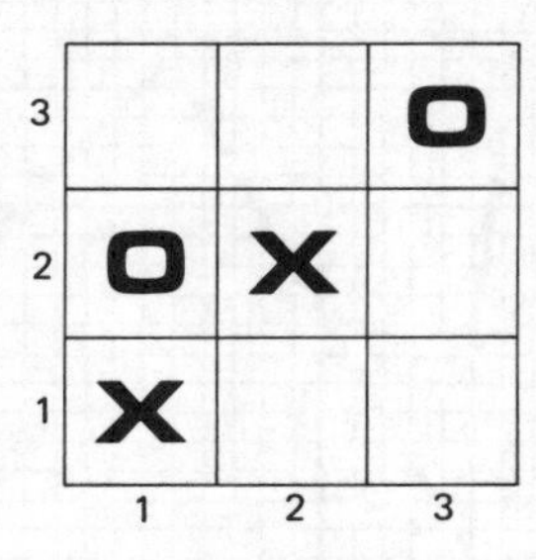

Figure 8.4

(2) Figure 8.5 shows a side view of a plan to put a dormer extension into a roof. Unfortunately some of the co-ordinates are missing. Given that the plan is symmetrical, determine the missing co-ordinates A, B, C and D.
(3) The area of a rectangle ABCD is 12 square units. If the co-ordinates of A are (4, 7) and B are (1, 7), give both possible co-ordinates of C and D.
(4) By using a suitable grid, plot the points A(1, 1), B(0, 3), C(4, 5) and D(5, 3). Join the points in the order ABCD to form a quadrilateral. What is the name of the quadrilateral formed? By using Pythagoras, estimate the lengths of the

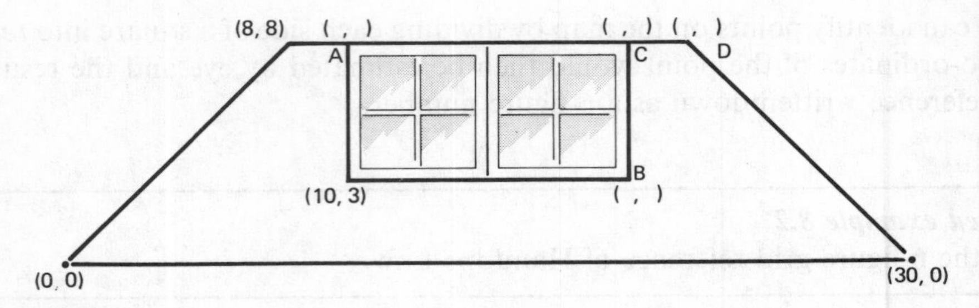

Figure 8.5

sides of this quadrilateral to one decimal place. Hence estimate the area of this shape.

(5) Figure 8.6 shows three triangles A, B and C. Determine which of the three triangles is isosceles. Write down the co-ordinates of the midpoints of the sides of this isosceles triangle, and use Pythagoras to determine the length of the shorter side.

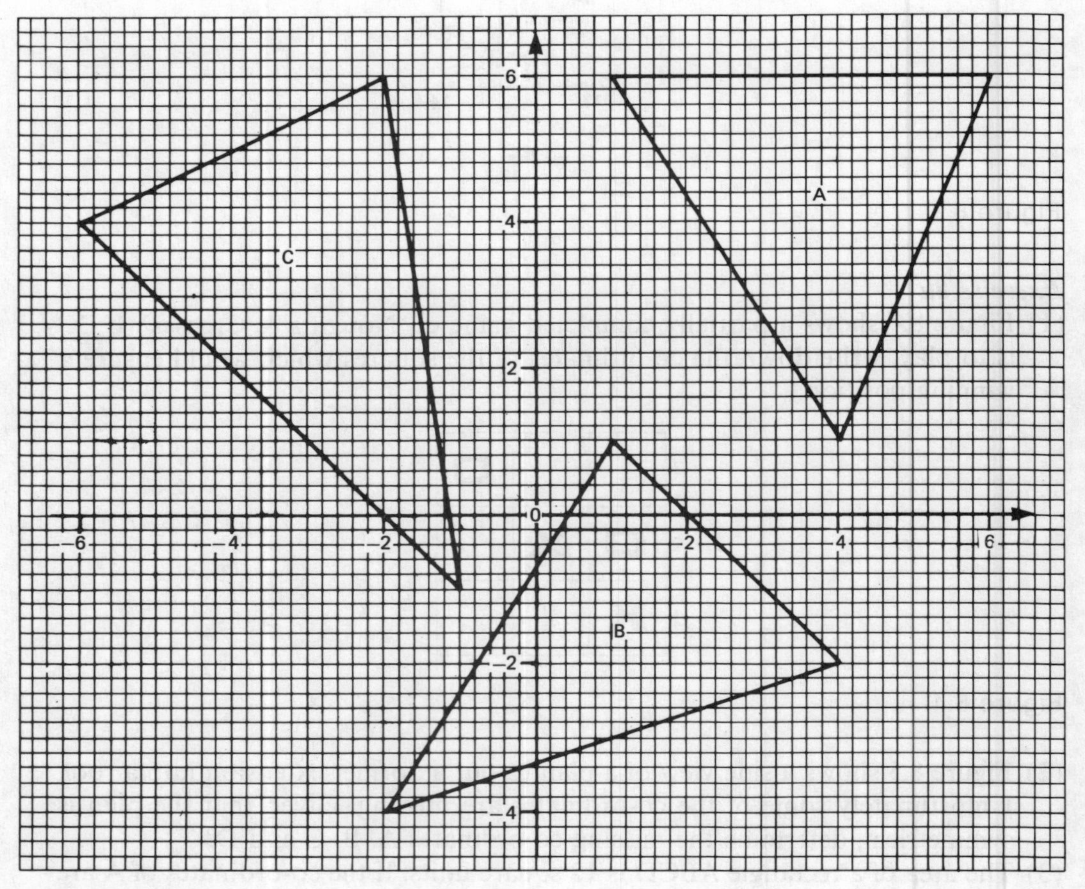

Figure 8.6

(6) Three vertices of a square are given by:

A(2, 3) B(7, 0) D(5, 8)

Using a suitable grid, plot these points and determine the co-ordinates of the fourth vertex, C. Measure the length of AB and write down the area of ABCD.

(7) As Majid was walking into the theatre, he picked up a ticket from the floor which was numbered J18. Assuming that the rows were lettered alphabetically starting with A, and that there are the same number of seats in each row, what is the minimum number of seats in the theatre? The lady who claimed the ticket took up her seat which was the third seat from one end of the row and the third row from the back of the theatre. How many people can be seated in the theatre?

(8) A counter is placed on a grid at the origin (see Figure 8.7). A fair coin is now tossed 6 times. If a head is uppermost, the counter is moved one unit to the right. If a tail is uppermost, the counter is moved one unit upwards. Draw this grid and mark *all* possible positions of the counter after 6 tosses of the coin, with an X. Now place a ring around all the crosses which are more than 5 units from the origin.

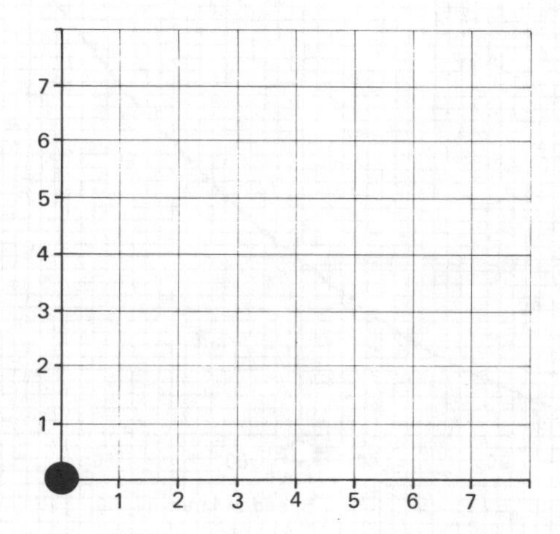

Figure 8.7

(9) Using Figure 8.2,
 (a) write down the 4-figure references of the square which contains:
  (i) Throwley Cottage, (ii) Stanshope, (iii) Hinkley Wood
 (b) Write down the 6-figure grid references of the following:
  (i) St Bertram's Well, (ii) Damgate, (iii) Grove Farm
 (c) What is situated at:
  (i) 112503, (ii) 129508, (iii) 114548, (iv) 129549, (v) 144508?

8.3 Interpretation of graphs

A graph can be used to present a visual picture of a relationship between two sets of data. In order to obtain information from a graph we need to know for both axes:

(i) the units used;
(ii) the scale used.

To interpret a graph, it is therefore vital that it is suitably labelled.

Worked example 8.3

Figure 8.8 shows the stopping distance of a car on a dry road for different speeds. From the graph, estimate the stopping distance of a car travelling at 50 km/h. If a car takes 74 metres to stop, at what initial speed must it have been travelling?

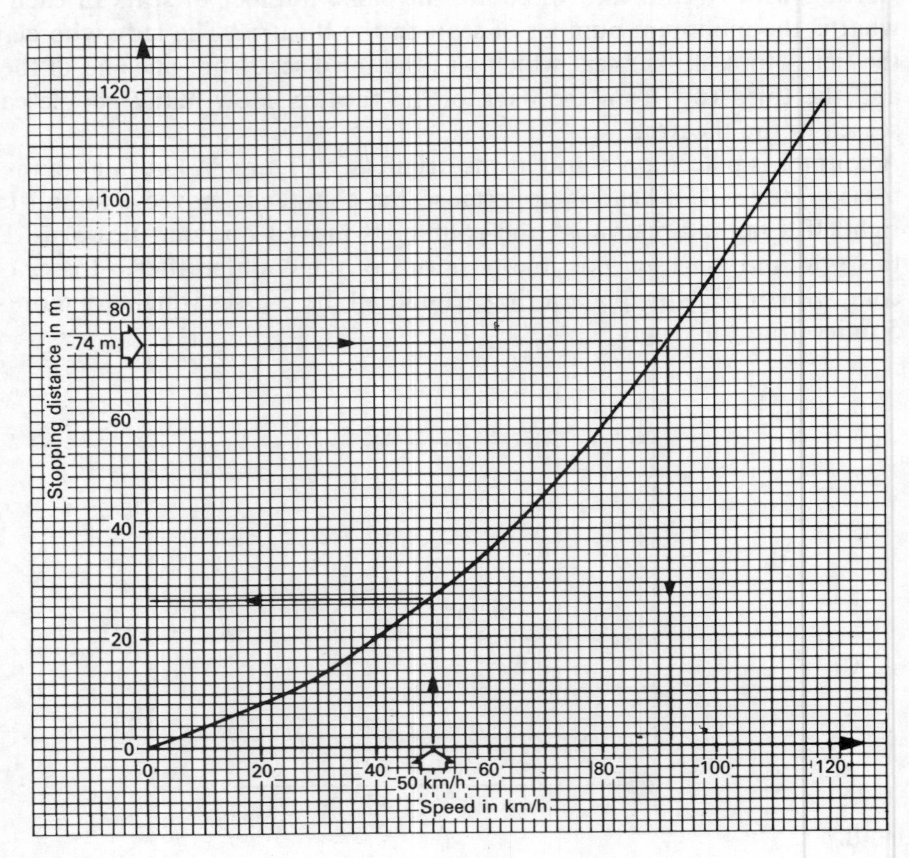

Figure 8.8

When the road is wet, the stopping distance for any speed doubles. From the graph, estimate the speed on a wet day which has the same stopping distance as a speed of 90 km/h on a dry day.

To find the stopping distance of a car travelling at 50 km/h we need to start at the axis showing the scale of km/h. At the point on the scale which is equivalent to 50 km/h, a line is drawn to the graph. At the point where this line meets the graph, another line is drawn to the other axis. The value of 27 m is then read off the scale marked on this axis.

To find the initial speed of a car which takes 74 metres to stop, the same process is repeated except that we need to start at the axis showing distance in metres. The estimated speed from the graph is 92 km/h. Both estimates are shown marked on the figure.

 In order to find the speed on a wet day, we need to first find the stopping distance for the car on a dry day travelling at a speed of 90 km/h. From the graph, this is 72 m. Now as the day is wet, this distance is double what it should be on a dry day. Therefore on a dry day, the stopping distance would have been 36 m. From the graph this would be equivalent to 60 km/h. So the speed on the wet day is 60 km/h.

8.4 Conversion graphs

In Unit 3 we looked at conversion rates. A *conversion graph* can be used to show a comparison between two different units which increase or decrease at the same rate. The resulting graph is a straight line which may or may not pass through the origin.

Worked example 8.4

Marina S. Tate, a second year Sixth Form student, has just passed her driving test and has purchased a car. The cost of running this car per week is shown in Figure 8.9. From the graph, find out how much it costs her to travel a total of 150 kilometres in her first week. How many kilometres has she travelled in the second week if she estimates the cost for the week at £24.00? In the third week she travels 200 kilometres. What is the average cost per kilometre? It is now proving to be expensive so she decides to do without the car for the next four weeks. What will be the total running costs for the car during this four week period?

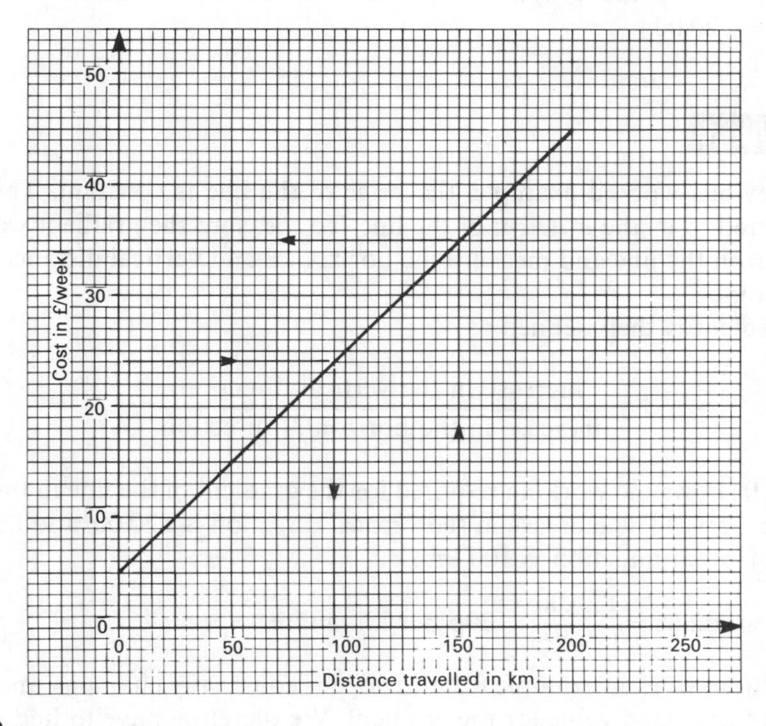

Figure 8.9

From the graph, the cost of running the car when the distance travelled is 100 km

$$= £35.00.$$

The number of kilometres travelled in one week when the total running cost is £24.00

$$= 95 \text{ km}.$$

To find the average cost per kilometre we need to determine the total cost for a week in which Marina has travelled 200 kilometres.

So the cost of running the car when the distance travelled is 200 kilometres

$$= £45.00$$

So the average cost per kilometre

$$= \frac{£45.00}{200}$$

$$= 22.5 \text{ pence}$$

During the four week period when Marina is not driving the car, the total number of kilometres travelled is 0. The cost per week is therefore £5.00 (the value where the graph crosses the vertical axis) so the total cost is $£5.00 \times 4 = £20.00$.

8.5 Gradient

In the previous worked example, the *slope* of the line is constant. This slope is often referred to as the *gradient* of the line. To measure the gradient we take *any* two points on the line and measure the horizontal and vertical distances between the two points.

The gradient is then defined as:

$$\frac{\text{increase in the vertical direction}}{\text{increase in the horizontal direction}}$$

Figure 8.10 shows the working required for the gradient of the line in the previous worked example. The increase in the vertical direction is £20.00, and the increase in the horizontal direction is 100 km.

$$\text{So the gradient} = \frac{£20.00}{100 \text{ km}} = 20\text{p per kilometre.}$$

Note that in worked example 8.3, the slope is constantly changing and therefore there is not one fixed value for the gradient. We therefore have to find a gradient at a particular point and this technique can be seen in worked example 8.10.

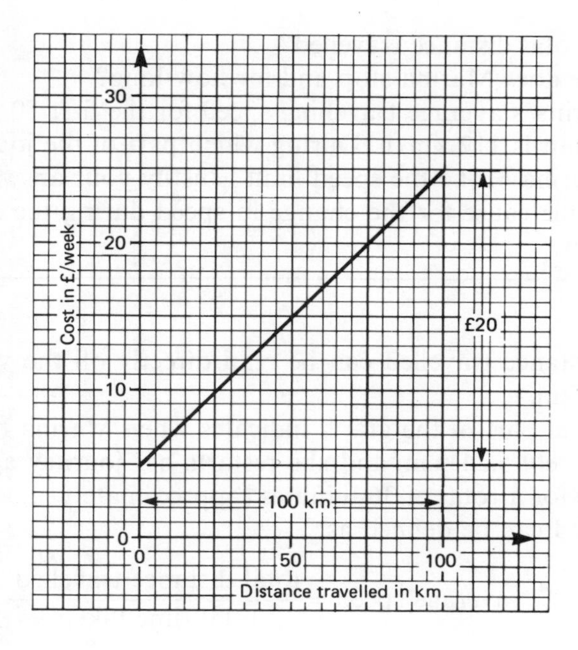

Figure 8.10

8.6 Distance–time graphs

Worked example 8.5

With her new found freedom, Marina decides to visit some friends for a weekend. She joins the M1 at 4.00 p.m. on a Friday to travel North. The graph (Figure 8.11) shows her journey on the motorway. The horizontal scale is in hours and the vertical scale shows how far Marina has travelled from her starting point on the motorway in kilometres. It is known as a *distance–time graph.*

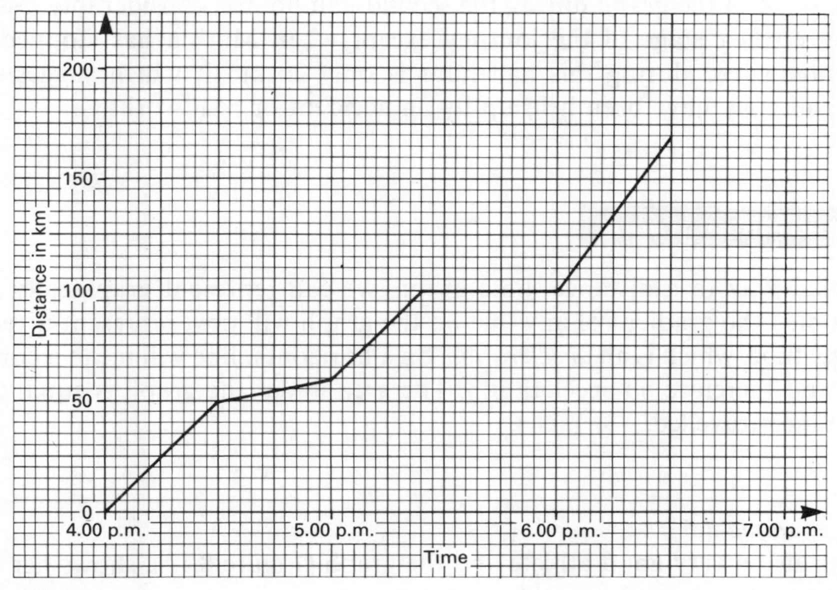

Figure 8.11

(i) What is the total distance travelled?
(ii) At what time does Marina stop and for how long?
(iii) What is Marina's average travelling speed for the first 50 km?
(iv) The speed limit is 112 km/h. During which part of the journey can you be sure that Marina broke the speed limit? Justify your answer.
(v) Give a possible cause for the change in speed during the second half hour of the journey.

(i) The total distance travelled can be read directly off the vertical axis. This value is 170 km.
(ii) The horizontal part of the graph indicates when Marina has stopped. This is approximately 5.24 p.m. and she restarts her journey again at 6.00 p.m. So she stops for a total of 36 minutes (approximately!).
(iii) Average speed can be defined as:

$$\text{average speed} = \frac{\text{total distance travelled}}{\text{total time taken}}$$

But it can be seen that the *gradient* of any line on a distance–time graph is also a distance divided by a time. So the gradient of the line representing the first party of Marina's journey $= \frac{50 \text{ km}}{1/2 \text{ h}} = 100$ km/h.
(iv) The fastest part of her journey is where the graph is steepest, which is during the last stage. The gradient of this line gives an average speed of $\frac{70\text{km}}{1/2 \text{ h}} = 140$ km/h, which is greater than the maximum speed limit. So at some stage during this part of the journey she will have broken the speed limit.
(v) The slope of the graph during the second half hour is considerably less than at any other stage of the journey except when Marina has stopped. The average speed during this half hour is only 20 km/h which suggests that Marina may have been held up by an accident or roadworks.

8.7 Speed–time graphs

In a similar way to a distance–time graph, we can represent graphically speed against time. The gradient of any line on the graph will now represent a velocity divided by time, which is an *acceleration*. So a horizontal line will mean a constant velocity and therefore *no* acceleration. The area under the graph is equal to the total distance travelled.

Worked example 8.6
Marina is also a very good athlete and the graph (Figure 8.12), shows her speed during training measured over 14 seconds.

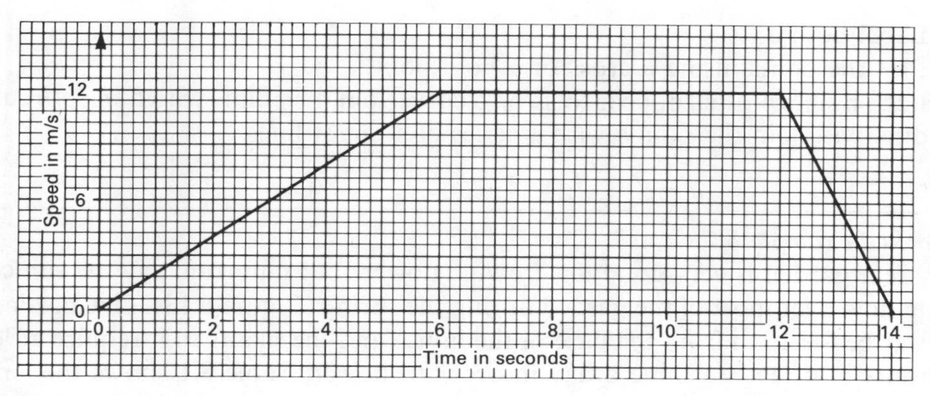

Figure 8.12

(i) Describe what happens during the 14 seconds.
(ii) How far has Marina run in the first 12 seconds?

(i) For the first six seconds, Marina *accelerates* to a speed of 12 m/s.

$$\left(\text{The acceleration is } \frac{\text{increase in speed}}{\text{time}} = \frac{12 \text{ m/s}}{6 \text{ s}} = 2 \text{ m/s}^2\right)$$

She then travels for 6 seconds at 12 m/s and finally in the last two seconds she *decelerates* until she stops.

$$\left(\text{The deceleration is } \frac{-12 \text{ m/s}}{2 \text{ s}} = -6 \text{ m/s}^2\right)$$

(ii) The distance travelled during the first 12 seconds is the area under the graph

$$= \underbrace{\frac{1}{2} \times 6 \times 12}_{\text{first 6 seconds}} + \underbrace{6 \times 12 \text{ m}}_{\text{from 6 to 12 seconds}}$$

$= 108$ m.

8.8 Drawing graphs

It is sometimes necessary to draw a graph in the first instance before the relevant information can be obtained from the data. The following seven points may help you to carry out this task.

1. Two sets of data will normally be presented in tabular form. The *first* set is used for the *horizontal* co-ordinates. The *second* set is used for the *vertical* co-ordinates.
2. Determine the *range* for both sets of co-ordinates. That is the maximum and minimum values.
3. If a scale is given, make sure you use it, otherwise choose scales which will allow you to fit extreme values on the graph.
4. Draw your horizontal and vertical axes in a position which will allow you to plot all the points.

5. Label both axes.
6. Plot the points using a cross or a bold point.
7. If the points plotted appear to lie on a curve, join the points with a *smooth* curve; otherwise join the points with a series of straight lines.

Worked example 8.7

My rear garden is very good for growing flowers but unfortunately the sun casts a shadow over most of the garden during the spring (Figure 8.13). One particular flower bed is 10 m from the base of the house and the plants in it require at least 5 hours of direct sunshine a day in order to flower. Table 8.1 shows the length of the shadow over the garden measured from the base of the house for hourly intervals during a spring day.

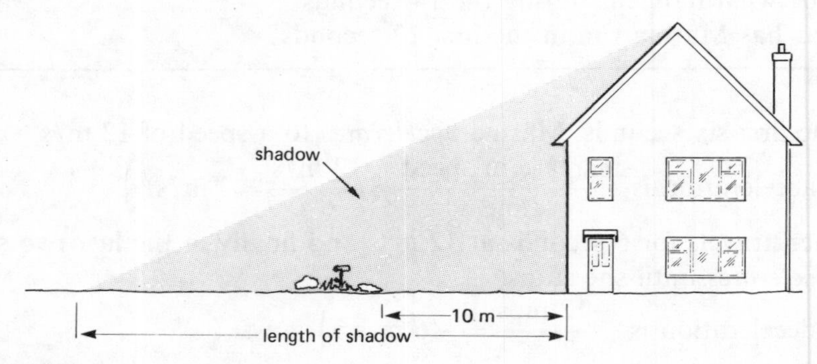

Figure 8.13

Table 8.1

Time (hs)	10.00	11.00	12.00	13.00	14.00	15.00	16.00
Length of shadow (m)	22	13.5	9	7.5	9	13.5	22

(i) Using scales of 2 cm to 1 hour and 2 cm to 5 m, plot the points and join them with a smooth curve.
(ii) From your graph, determine the number of hours of direct sunshine on all of the flower bed.
(iii) Use your graph to determine what should be the minimum distance of the flower bed from the base of the house in order for all the plants to receive 5 hours of direct sunlight.

(i) The graph is shown in Figure 8.14.
(ii) To find the number of direct hours of sunshine on the flower bed, we need to draw a horizontal line between the curve where the length of the shadow

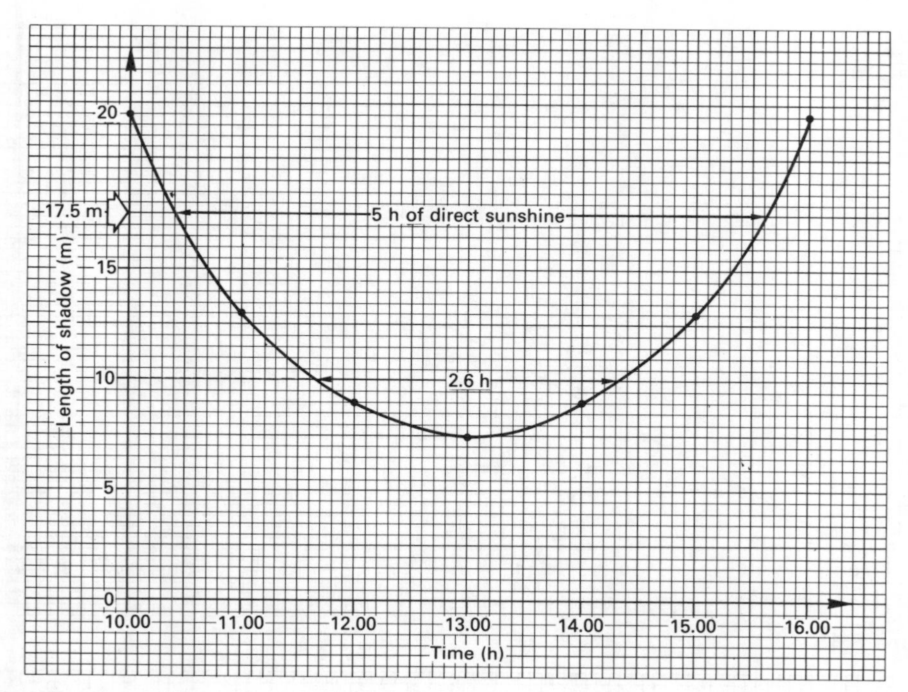

Figure 8.14

is 10 metres. Using the horizontal scale the length of this line, 5.2 cm, converts to 2.6 hours or just over two and a half hours.

(iii) To find the minimum distance from the base of the house to a flower bed which would have five hours of direct sunlight requires the use of a ruler. We need to find the two points on the curve where the horizontal distance between them is 10 cm (or 5 hours on the scale). These two points can be found by moving a ruler up and down to find the correct position. The distance can then be read from the vertical axis. From the graph it can be seen to be 17.5 m.

Exercise 8b

(1) Figure 8.15 is a conversion graph for changing gallons into litres, and litres into gallons.
  (a) Use the graph to complete these statements.
    (i) 20 litres is equivalent to .... gallons.
    (ii) .... litres is equivalent to 6 gallons.
  (b) Estimate how many gallons is equivalent to 60 litres. (MEG)

(2) Julian and Christine are twins. Each month their mother recorded their weights and the results are shown in Figure 8.16.
  (a) What was Julian's weight when he was age 14 years?
  (b) How old was Christine when she weighed 25 kg?
  (c) What was their age the first time they were the same weight?
  (d) What did each weigh the second time they were the same weight?
  (e) Estimate how much heavier Christine was than Julian when they were aged 12 years. (MEG)

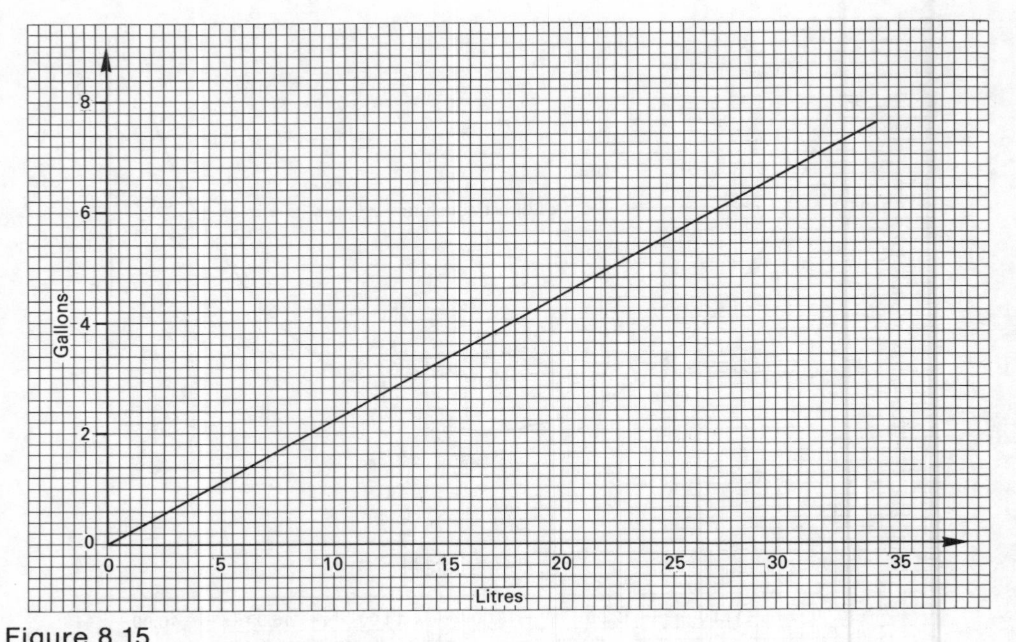

Figure 8.15

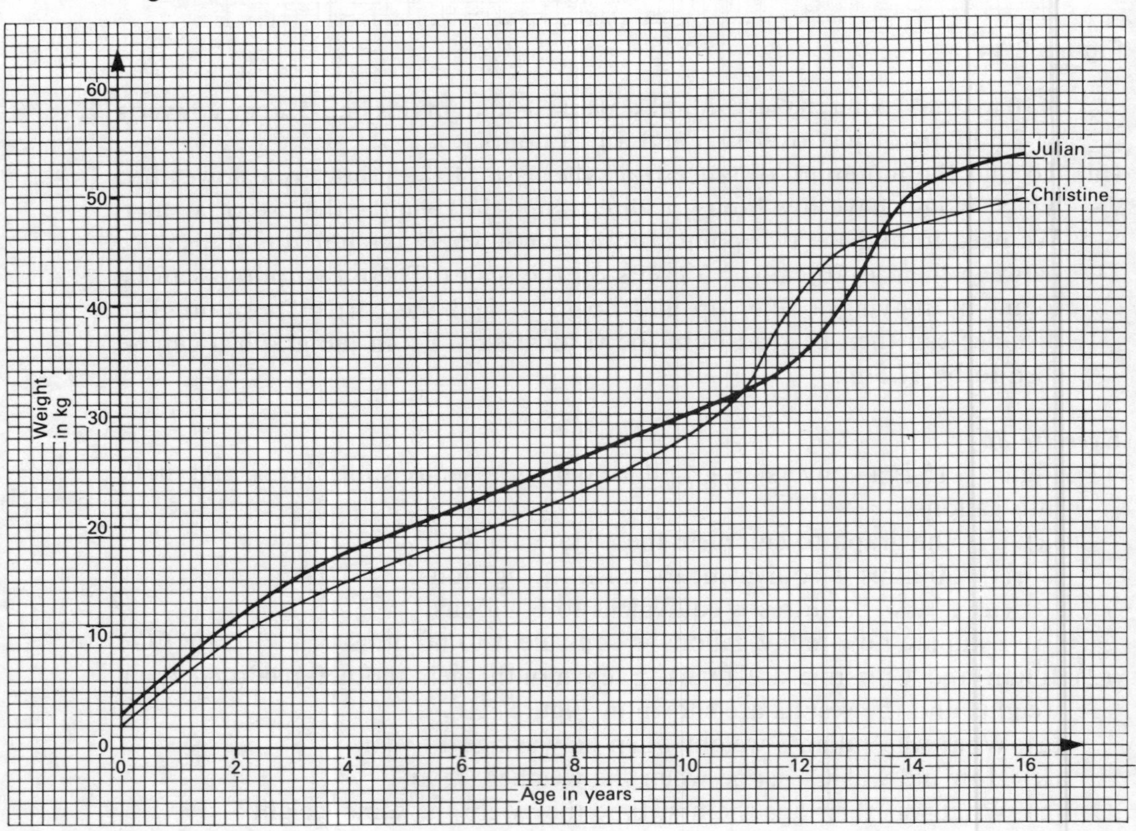

Figure 8.16

(3) Mr Turner lives at a seaside resort. He has been made redundant from his job.

He thinks that there are several ways that he can earn his living and decides to divide his time between them. One is to make model lighthouses from local stone and sell them to summer visitors.

He estimates that the number of lighthouses it will be worth his while to make in his first year will depend on the price he charges for them. Figure 8.17 shows this.

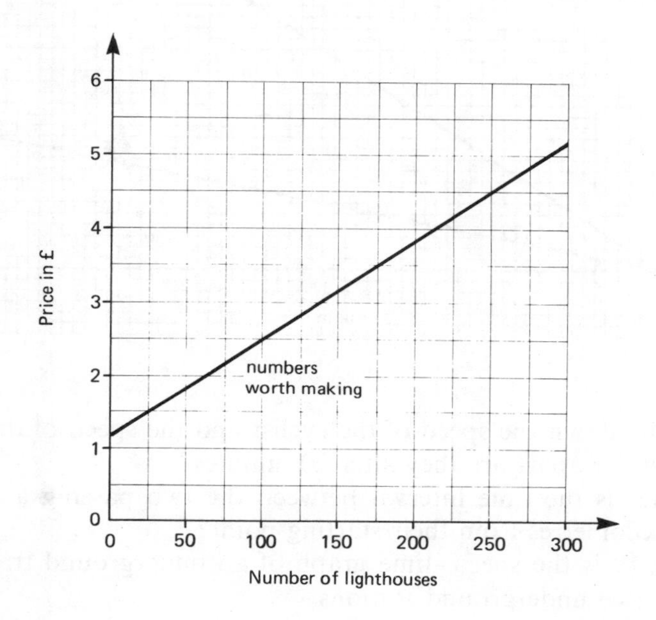

Figure 8.17

(a) At what price would it be worth making 250 lighthouses?

(b) How many should be made if he decides to charge £2.50?

A friend warns him, however, that the number of lighthouses he can expect to sell also depends on the price he charges. The numbers for some prices are shown in Table 8.2.

Table 8.2

Number of lighthouses	50	100	200	250
Price (in £) at which he can expect to sell them	6	5	3	2

(c) Copy Figure 8.17 and plot the points representing these values and join them with a straight line.

(d) For what number of lighthouses do the graphs cross? (MEG)

(4) On the first stage of the Tour de France cycle race, a cyclist and a support car start at the same time and travel along the same road for 1 hour. Figure 8.18 shows the distance-time graph for both the cyclist and the support car.

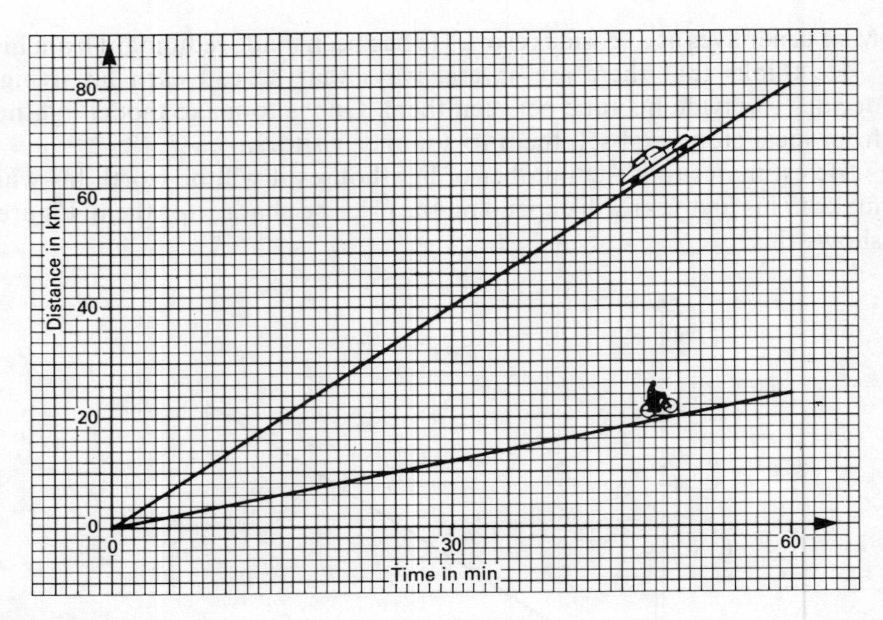

Figure 8.18

(i) Write down the speed of the cyclist and the speed of the support car.
(ii) How far apart are they after 20 minutes?
(iii) What is the time interval between the two passing a service station, 20 kilometres from their starting point?

(5) Figure 8.19 is the speed–time graph of an underground train as it travels between two underground stations.

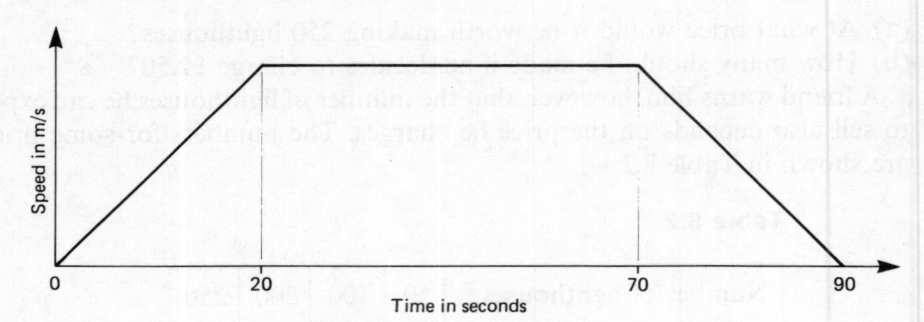

Figure 8.19

Given that the total distance travelled is 1400 metres, find:
(i) the maximum speed of the train,
(ii) the acceleration of the train during the first 20 seconds,
(iii) the distance travelled by the train in the first 30 seconds.

(6) Table 8.3 shows the depth of water in a harbour at various times during one particular morning.

Using a horizontal scale of 1 cm to represent 1 hour and a vertical scale of 1 cm to represent 1 metre, plot the graph.

A ship requires a depth of 6 metres of water in order to enter the harbour.

Table 8.3

Time in h	0.00	1.00	2.00	3.00	4.00	5.00	6.00
Depth in m	8.0	11.2	12.4	13.0	12.4	11.2	8.00

Time in h	7.00	8.00	9.00	10.00	11.00	12.00
Depth in m	4.8	3.6	3.0	3.6	4.8	8.0

Use your graph to estimate the time during which the ship will be unable to enter the port during the morning.

(7) A metal container was weighed with various volumes of water inside and the results are recorded in the table below:

Volume of water (cm^3)	112	188	256	312	417
Total mass (g)	527	603	671	727	832

(a) Plot these values on graph paper. Join the points to form a straight line graph.
(b) Using your graph write down
(i) the total mass in grams when the volume of water is 200 cm^3,
(ii) the volume of water in cubic centimetres when the total mass is 700 g.
(c) What is the mass of the container? (LEAG)

(8) A particle is moving with an initial speed of 4 m/s. In the next four seconds its speed increases uniformly to 10 m/s and then the speed decreases uniformly until the particle stops moving after a further eight seconds.
(a) Show this information on a speed–time graph.
(b) Find:
(i) the acceleration in the last eight seconds of the motion,
(ii) the total distance travelled by the particle.

(9) Figure 8.20 illustrates the performance of a runner in a 60 metre race.
(a) How far had the runner gone after 5.6 seconds?
(b) Describe, as fully as possible, what the graph tells you about the runner's speed during the race.
(c) On reaching the finish the runner slows down, until he stops 10 metres further on. Show on the diagram how the graph might continue. (MEG)

(10) At 10.00 a.m. Amrik set out from home to ride his cycle to a friend's home 12 km away. He rode at a steady speed of 18 km/h but, after 20 minutes,

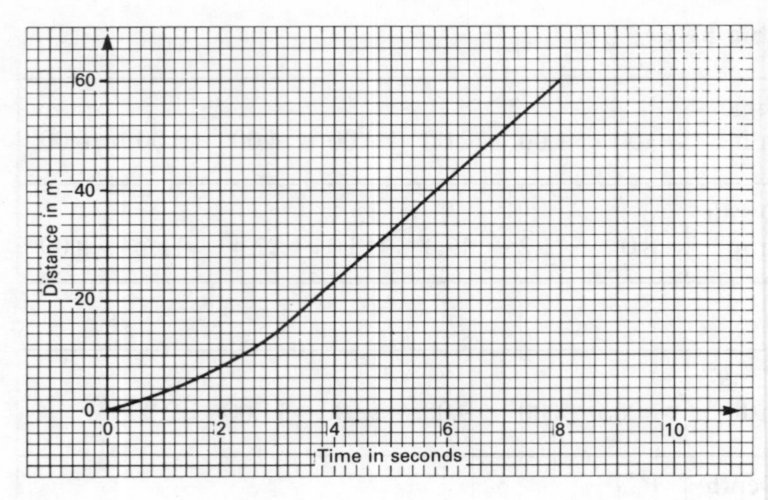

Figure 8.20

one of his tyres was punctured. After spending 5 minutes trying to repair it, Amrik walked the rest of the way at 6 km/h.

(a) Using a scale of 2 cm to represent 10 minutes on the time axis and 2 cm to represent 2 km on the distance axis, draw a distance–time graph for Amrik's journey.

(b) Find the time at which Amrik arrived at his friend's home.

(MEG)

8.9 Algebraic graphs

Graphs showing the relationship between two sets of data can often be expressed algebraically in the form of an *equation.* Each point on the graph gives a pair of values and these values will satisfy the equation of the graph. Look again at the graph for worked example 8.4 (Figure 8.9). The gradient was shown to be 20 pence per kilometre. So every additional kilometre travelled adds 20 pence or £1/5 to the total cost. If we therefore let the total cost per week in £s be C and the number of kilometres travelled in a week be k, then all co-ordinates on the graph satisfy the equation:

$$C = 5 + \frac{1}{5}k$$

By substituting $k = 100$ km into this equation we find that:

$$C = 5 + \frac{1}{5} \times 100$$

$$= £25.00$$

which agrees with the value read off the graph.

Using this equation we not only have an accurate way of determining total costs and kilometres travelled, but we also have an alternative way of labelling the axes, namely by using the letters C and k.

In most problems involving algebraic equations (but not all, as the previous example has shown), the letters x and y are used and by convention, x is used to label the horizontal axis and y is used for the vertical axis. Every co-ordinate on a graph can then be given in terms of an *x-co-ordinate* and a *y-co-ordinate.*

8.10 Straight line graphs

Figure 8.21 shows the graphs of lines parallel to the two axes, x and y. In the case of the vertical line, *all* the x-co-ordinates of the points on the line are 3, therefore the equation of the line is:

$$x = 3$$

In the case of the horizontal line, *all* the y-co-ordinates of the points on the line are 2, therefore the equation of the line is:

$$y = 2$$

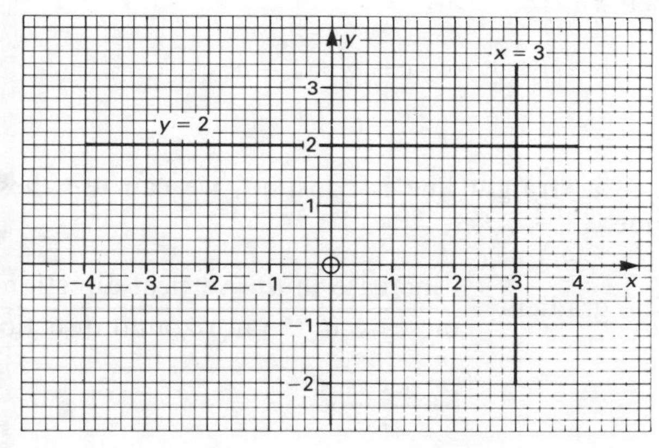

Figure 8.21

Worked example 8.4 is an example of a line which is not parallel to either axes. In terms of x and y, the general equation of the straight line is given by:

$$y = mx + c$$

When $x = 0$, this gives

$$y = m \times 0 + c$$

so $$y = c.$$

The co-ordinate $(0, c)$ therefore satisfies the equation and is the point where the graph crosses the y-axis. But what is m?

The equation

$$C = 5 + \frac{1}{2}k$$

may have already suggested to you that it has something to do with the gradient.

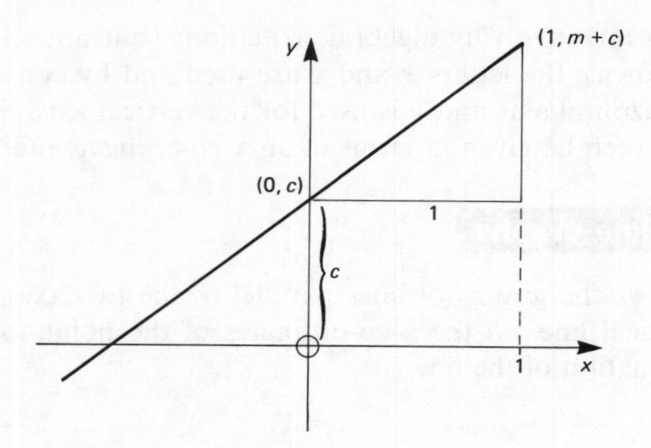

Figure 8.22

Now look at Figure 8.22. This shows the graph of '$y = mx + c$'. The co-ordinates of two points are marked where $x = 0$ and $x = 1$. We have already seen that when $x = 0$, $y = c$. When $x = 1$:

$$y = m \times 1 + c$$
$$= m + c$$

As the gradient of a straight line is constant, we can use these two points to determine its value.

$$\text{The gradient} = \frac{\text{increase in the vertical direction}}{\text{increase in the horizontal direction}}$$
$$= \frac{(m + c) - c}{1}$$
$$= m$$

So for the general equation of a straight line:

$y = mx$	$+$	c
↑		↑
Gradient of line		Where line crosses y-axis

As gradients are determined by measuring *increases* in both the horizontal and vertical directions, Figure 8.23 shows an example of a straight line with a *negative* gradient. The equation of this line is $y = -2x + 3$.

8.11 Solving simultaneous equations by graph

In order to plot a straight line graph, the co-ordinates of only *two* points are required. We shall now make use of this to solve two simultaneous equations by graphical means.

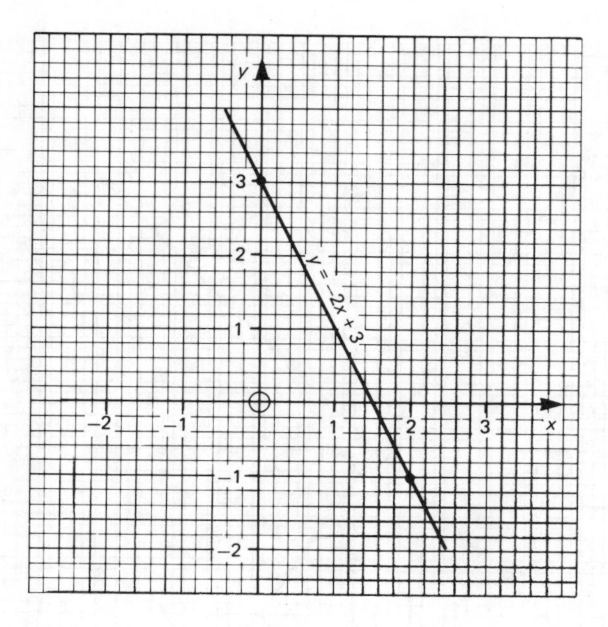

Figure 8.23

Worked example 8.8

Solve the simultaneous equations:

$$y - 2x = 1 \qquad \text{(i)}$$

$$y + 3x = 11 \qquad \text{(ii)}$$

In equation (i) when

$$x = 0,\ y = 1$$

and when

$$x = 1,\ y = 3.$$

In equation (ii) when

$$x = 0,\ y = 11$$

and when

$$x = 1,\ y = 8.$$

Plotting these four points and drawing the straight lines gives us Figure 8.24. The co-ordinates of the point of intersection of the two graphs, (2, 5) satisfies both relationships, therefore the solution to the two equations is:

$$x = 2 \qquad \text{and} \qquad y = 5$$

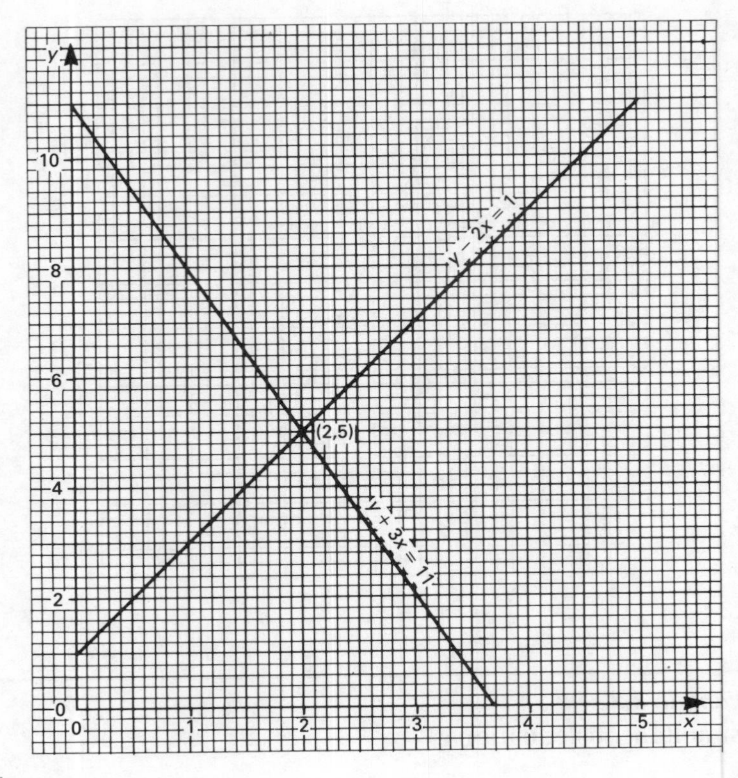

Figure 8.24

8.12 Quadratic graphs

Quadratic equations, involving terms in x^2, produce curved graphs so more than two points are needed for a graph to be drawn. A table can be drawn up which gives the values of y corresponding to the range of values of x. These points can then be plotted and a smooth curve drawn through them.

Worked example 8.9

(a) Complete the table below for the graph of $y = x^2 - 2x - 1$.

x	−2	−1	0	1	2	3	4
y	7		−1				7

(b) Draw the graph and write down the co-ordinates of the points where the graph crosses the x-axis.

(c) By drawing a suitable horizontal line on the graph, find the approximate solutions of

$$x^2 - 2x = 4$$

(a) To complete the table it is useful to break the right-hand side of the equation down into its various parts and then to add the parts. So:

x	-2	-1	0	1	2	3	4
x^2		1		1	4	9	
$-2x$		2		-2	-4	-6	
-1		-1		-1	-1	-1	
y	7	2	-1	-2	-1	2	7

Adding ↓

The graph is drawn in Figure 8.25.

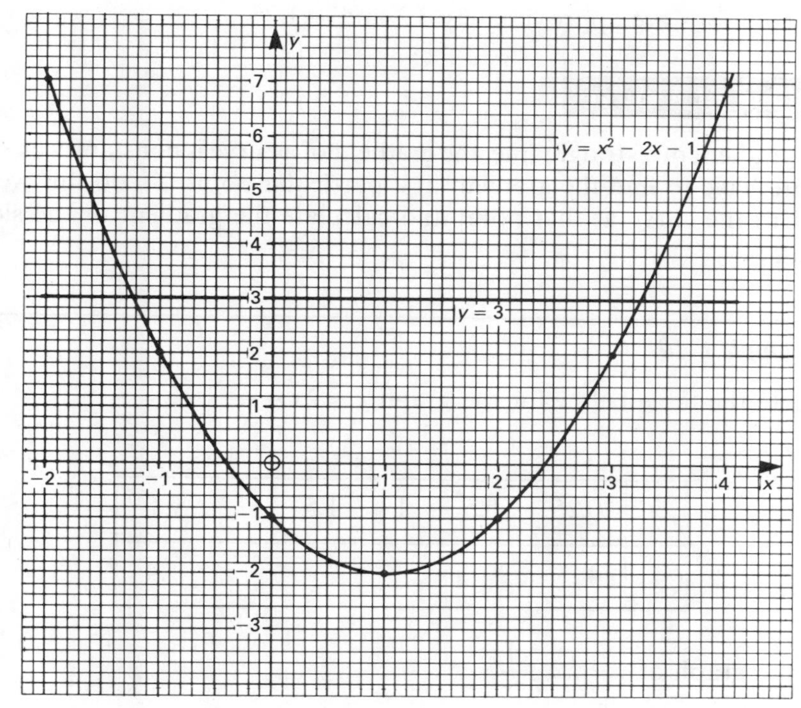

Figure 8.25

(b) The co-ordinates of the points where the graph crosses the x-axis are approximately

$$(2.4, 0) \qquad \text{and} \qquad (-0.4, 0)$$

Note that at these two points, $y = 0$ so the values of x are solutions to the equation

$$x^2 - 2x - 1 = 0$$

In Unit 11 we shall see that there are numerical methods which will give such solutions more accurately.

(c) To find the approximate solutions of the equation

$$x^2 - 2x = 4 \qquad \text{(i)}$$

we need to transform one side of the equation into the form

$$x^2 - 2x - 1$$

To do this we subtract 1 from both sides of equation (i):

So $$x^2 - 2x - 1 = 3$$

As we are given that y is equal to the left-hand side of this equation, so y must also be equal to the right-hand side.

Therefore the equation of the line we need to draw is $y = 3$ and the x-co-ordinates of the points of intersection give the solutions of equation (i).

The solutions are approximately 3.2 and -1.25.

8.13 Gradient of a curve

Unlike the gradient of a straight line, the gradient of a curve is constantly changing. We therefore have to find the gradient at a particular point. To do this we draw the *tangent* to the curve at that point and calculate the gradient or the slope of this tangent.

Worked example 8.10

(a) Complete the table below for the graph $y = 2x + \frac{1}{x}$.

x	0.25	0.5	1	1.5	2	2.5
y	4.5					5.4

(b) Draw the graph of $y = 2x + \frac{1}{x}$.

(c) Write down the minimum value of $2x + \frac{1}{x}$.

(d) Find the gradient of the curve at the point where $x = 0.5$.

(a)

x	0.25	0.5	1	1.5	2	2.5
$2x$		1	2	3	4	
$\frac{1}{x}$		2	1	0.7	0.5	
y	4.5	3	3	3.7	4.5	5.4

Adding ↓

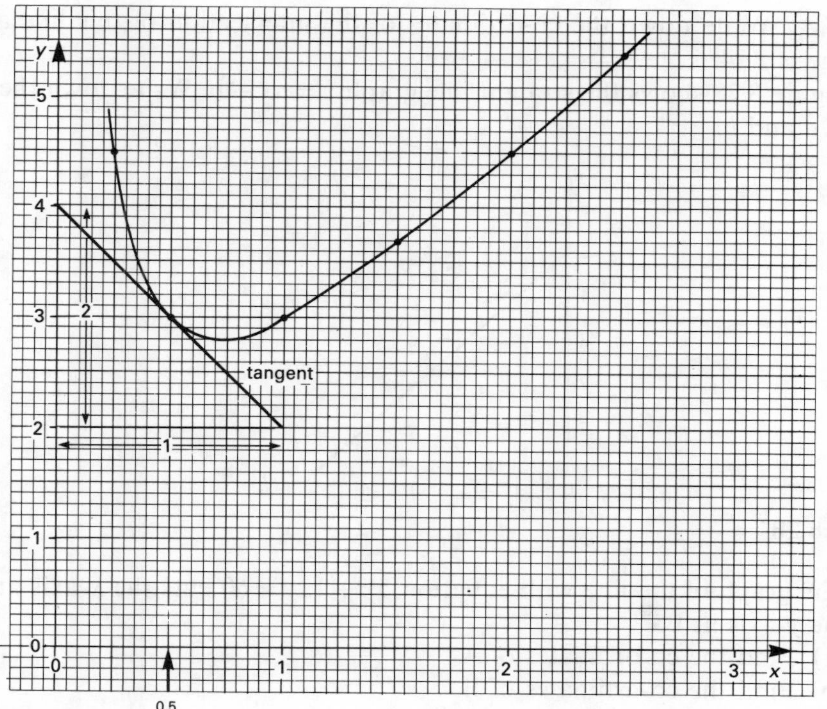

Figure 8.26

(b) The graph is drawn in Figure 8.26.
(c) From the graph, the minimum value is 2.8.
(d) Drawing a tangent at the point where $x = 0.5$, the gradient is found by constructing a right-angled triangle and reading off the lengths of the sides. Note that the slope of the line shows a *negative* gradient so the gradient is therefore:

$$-\frac{2}{1} = -2$$

Exercise 8c

(1) Figure 8.27 shows the graph of $y = mx + 2$. If $m = 3$, what are the co-ordinates

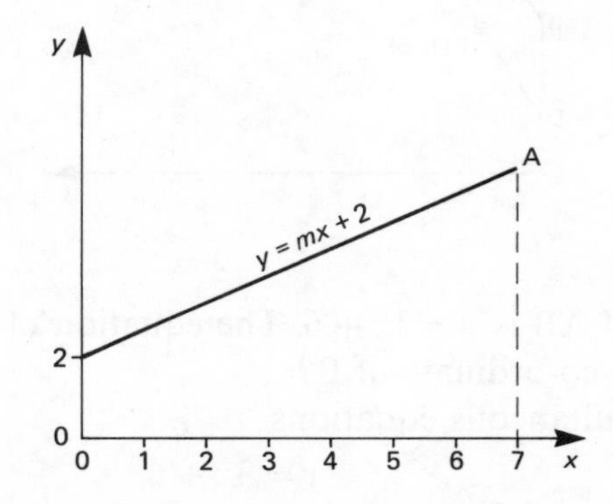

Figure 8.27

of A? Write down the co-ordinates of the intersection of this line and the line $y = 14$.

(2) Figure 8.28 shows the straight line graph $y = ax + b$. Write down the values of a and b.

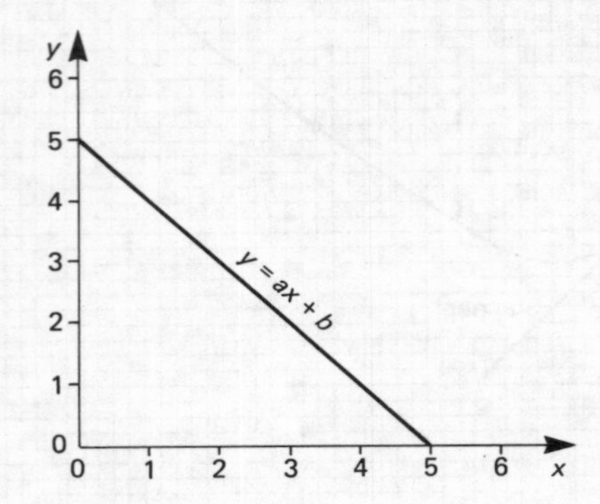

Figure 8.28

(3) The graph of the line with equation $5x + 12y = 60$ cuts the x-axis at A and the y-axis at B.
(a) Find the co-ordinates of A.
(b) Find the co-ordinates of B.
(c) Calculate the length of AB.

(4) Use a graphical method to find the solutions of the simultaneous equations:

(a) $x + y = 7$, $2x - y = 8$ (b) $2y - x = 7$, $3y + x = 8$ (c) $3x - 2y = 4$, $5x - 2y = 0$ (MEG)

(5) Figure 8.29 (which is not drawn accurately) illustrates a problem which occurs in Economics.

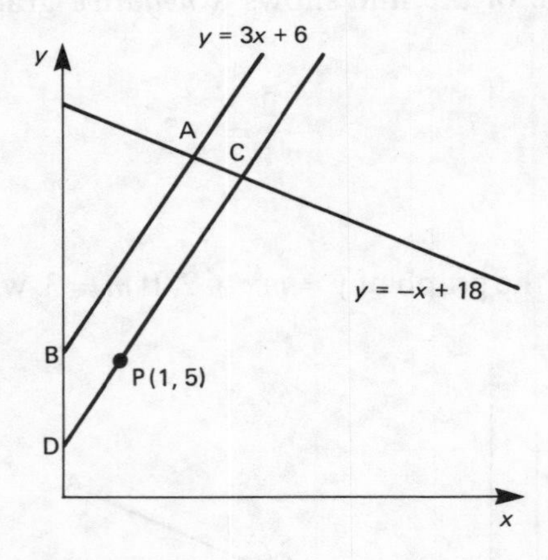

Figure 8.29

The equation of AB is $y = 3x + 6$. The equation of AC is $y = -x + 18$.

(a) What is the y-co-ordinate of B?

(b) Solve the simultaneous equations

$$y = 3x + 6$$
$$y = -x + 18$$

(c) What do your answers to part (b) represent?
(d) The line CD is parallel to AB and passes through the point P(1, 5). Find the equation of the line CD. (MEG)

(6) Figure 8.30 shows the relation $y = x^2$ for values of x from $x = 0$ to $x = 6$.

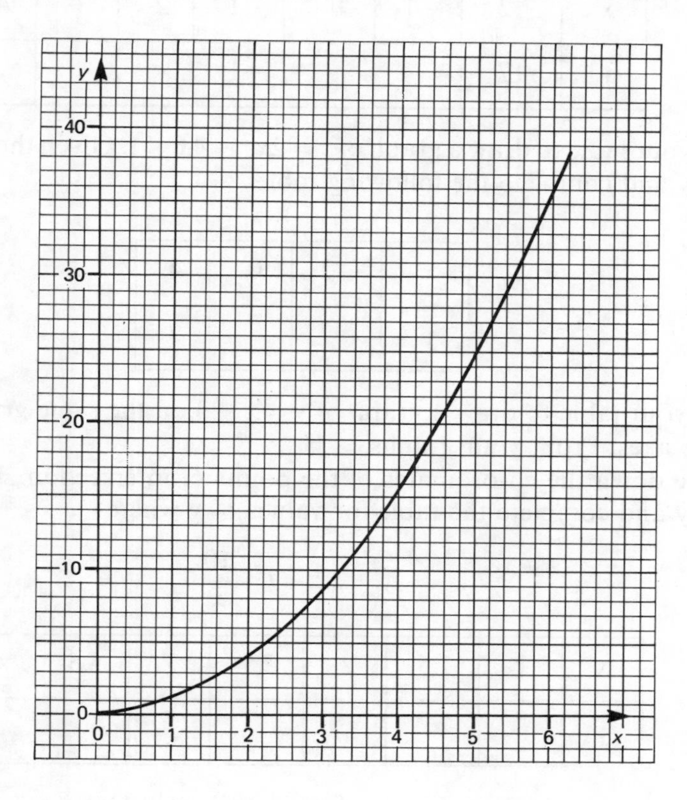

Figure 8.30

(a) Use the graph to find an approximate value for $(3.6)^2$.
(b) Use the graph to find an approximate value for $\sqrt{28}$.
(c) Use the graph to find an approximate value for $\sqrt{600}$.

(7) For the graph shown in Figure 8.25, find the gradient at the points (i) $x = 2$, (ii) $x = -1$.

(8) (a) Copy and complete the table below for the graph of

$$y = (x - 2)(x + 1)$$

x	-3	-2	-1	0	1	2	3	4
y	10		0		-2			10

(b) Draw the graph for values of x from -3 to $+4$ inclusive.
(c) From your graph, what is the minimum value of $y = (x - 2)(x + 1)$?
(d) By drawing a tangent at the point $(1, -2)$, estimate the gradient at that point.

(9) Given that $y = 2x^2 - 4x + 1$:

(a) Copy and complete the table below:

x	-2	-1	0	1	2	3	4
y	17			-1			

(b) Use your table to draw a graph of $y = 2x^2 - 4x + 1$. Given that $y = x + 3$,

(c) Copy and complete the following table:

x	-2	0	4
y			7

(d) Use your table to draw a graph of $y = x + 3$ on the same grid using the same axes. From your graphs,

(e) Write down the co-ordinates of the points of intersection. (LEAG)

(10) (a) Copy and complete the table for values of y where

$$y = 2x - 7 + \frac{10}{x}$$

x	1	1.5	2	2.5	3	3.5	4
y							

(b) Draw an x-axis, using a scale of 4 cm to 1 unit. Draw a y-axis using a scale of 2 cm to 1 unit.
Plot the points from your table and draw the graph of

$$y = 2x - 7 + \frac{10}{x}$$

(c) Use the graph to find the minimum value of $2x - 7 + \frac{10}{x}$ for values of x between 1 and 4.

(d) By drawing a suitable straight line on your graph parallel to the x-axis, find the approximate solutions of

$$2x + \frac{10}{x} = 10$$

(LEAG)

Suggestions for coursework 8

 1. Figure 8.31 shows a device for investigating the distance travelled in a horizontal

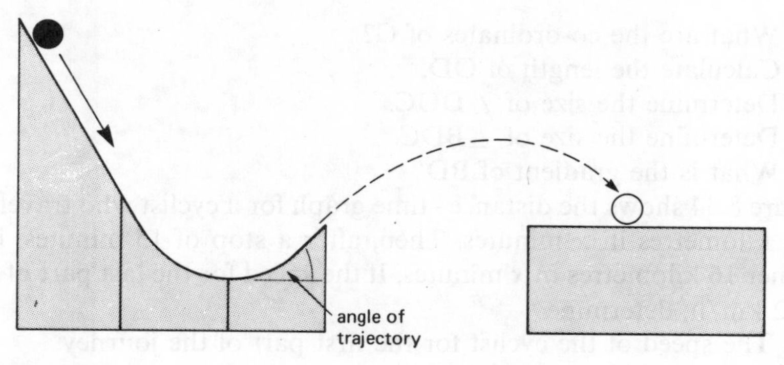

Figure 8.31

direction by a marble or ball for different angles of trajectory. Carry out your own investigation.

2. A rectangular framework is to be made as shown in Figure 8.32 from five lengths of wood. For a given area, investigate possible total lengths of wood required.

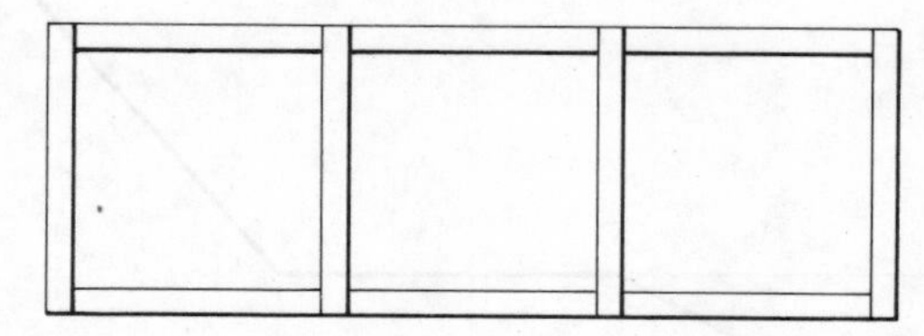

Figure 8.32

3. Investigate the height to which a ball will rebound to, when dropped from a given height.
4. Investigate a country's population growth for particular yearly percentage increases in population.

Miscellaneous exercise 8

1. Show graphically that the points $A(-2, 4)$, $B(1, 1)$ and $C(5, -3)$ lie in the same straight line. Find the gradient of this line.
2. Certain co-ordinates have been placed on Figure 8.33. Using this information:

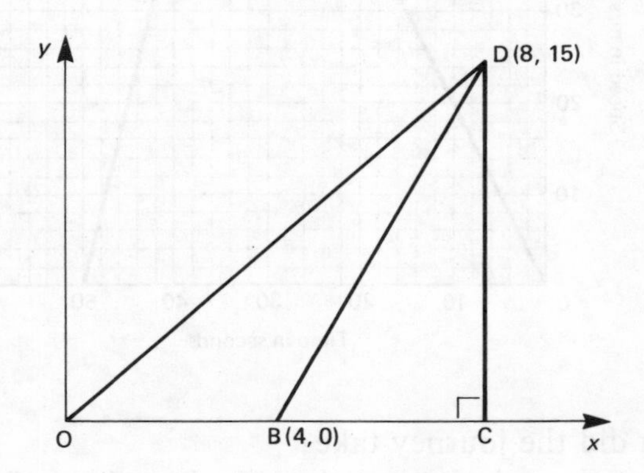

Figure 8.33

(a) What are the co-ordinates of C?
(b) Calculate the length of OD.
(c) Determine the size of ∠DOC.
(d) Determine the size of ∠BDC.
(e) What is the gradient of BD?

3. Figure 8.34 shows the distance–time graph for a cyclist who travels a distance of 8 kilometres in x minutes. Then, after a stop of 15 minutes, he travels a further 16 kilometres in x minutes. If the speed for the last part of the journey is 32 km/h, determine;
   (i) The speed of the cyclist for the first part of the journey.
   (ii) The value of x.
   (iii) The total time for the journey.
   (iv) The average speed of the cyclist for the complete journey.

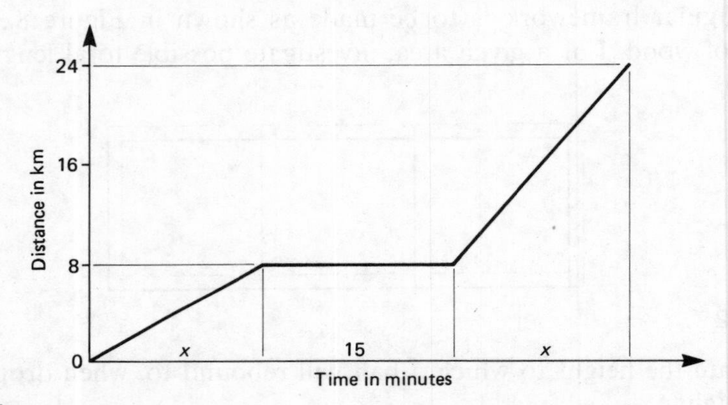

Figure 8.34

4. Figure 8.35 shows the speed–time graph of a car travelling between two sets of traffic lights.

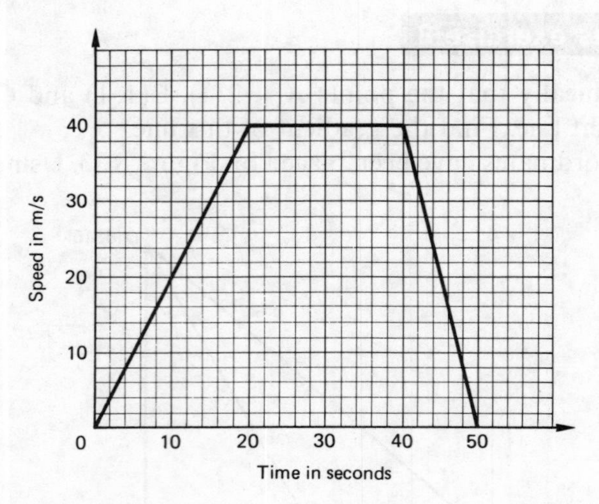

Figure 8.35

   (i) How long did the journey take?
   (ii) What was the maximum speed reached by the car?

(iii) What was the acceleration during the first 20 seconds?
(iv) What is the distance between the traffic lights?

5. A tour company representative records the temperature every three hours. The data she records between 6 a.m. on Thursday and noon on Friday is:

	Thursday							Friday			
Time	6 a.m.	9 a.m.	12 (noon)	3 p.m.	6 p.m.	9 p.m.	12 (midnight)	3 a.m.	6 a.m.	9 a.m.	12 (noon)
Temp. (°F)	59	73	77	79	77	71.5	65	62	61	76	80

(a) Plot these results on a graph, and join these points with a smooth curve.
(b) What was the minimum temperature at the resort on Thursday night?
(c) Anne says that she can only sunbathe when the temperature is over 75°F. Between what times could Anne sunbathe on Thursday?
(d) Julia wears a jacket outside when the temperature drops below 64°F. She did not wear a jacket that night. Between what times was Julia inside?
(SEG)

6. Midland Motors hire out lorries at a basic charge of £60, plus a further charge of 60p per km travelled.
(a) If £C is the total hire charge when the lorry travels x km, copy and complete the following table of values:

x (km)	0	50	100	150	200	250	300
C (£)	60		120	150			240

(b) Using a scale of 2 cm to represent 50 units on the x-axis and 2 cm to represent 20 units on the C-axis, draw a graph to show how C varies with x.
(c) Use your graph to find the distance travelled by a lorry for which the hire charge was £192.
(d) Write down a formula for C in terms of x. (MEG)

7. Figure 8.36 shows the line $y = 3x$ which meets the line $x = 2$ at the point A.
(i) What are the co-ordinates of A?
(ii) The x-axis is a line of symmetry of triangle OAB. What is the name we give to this type of triangle?
(iii) What are the co-ordinates of B?
(iv) What is the equation of the straight line which passes through O and B?

8. A straight line graph passes through (1, 3) and (3, 7). Plot these two points on graph paper and determine the gradient of this straight line. What is the equation of the line which passes through these two points?

9. When round steel bars are turned on a lathe the surface speed of the bar must be kept at a constant speed of 1 m/s.
The rotational speed varies with diameter.

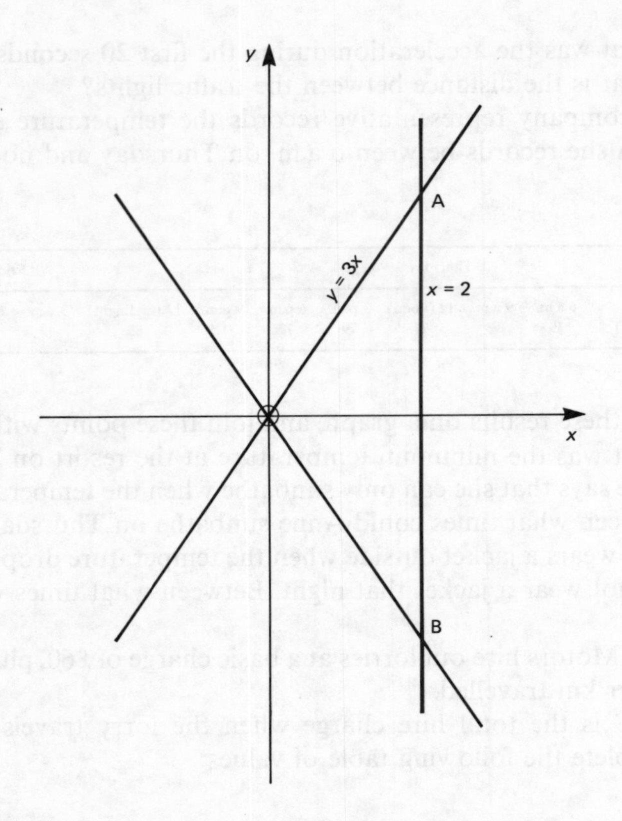

Figure 8.36

For example, for a bar of diameter 25 mm:

$$\text{Rotational speed} = \frac{1 \text{ m/s}}{\text{circumference of bar}} = \frac{1000 \text{ mm/s}}{\pi \times 25 \text{ mm}}$$

$$= 12.73 \text{ rev/s}$$

(a) Copy and complete the table below:

Diameter of bar (mm)	25	35	45	60	70
Rotational speed (revolutions/sec)	12.73				4.55

(b) Draw a graph to represent these figures.

(c) From the graph estimate the required rotational speed for turning a bar of 50 mm diameter. (LEAG)

10. (a) Given that $y = (x - 1)^2$, copy and complete the following table:

x	-1	-0.5	0	0.5	1	1.5	2	2.5	3
y		2.25	1	0.25		0.25	1		4

(b) (i) Using a scale of 4 cm to represent 1 unit on each axis, draw the graph of $y=(x-1)^2$ for values of x from -1 to 3 inclusive.
(ii) By drawing a tangent, estimate the gradient of the graph $y=(x-1)^2$ at the point (2, 1).

(c) Use your graph to find the values of x for which $(x-1)^2=3$. (MEG)

11. (a) Given that $y=4x^2-x^3$, copy and complete the following table:

x	0	0.5	1	1.5	2	2.5	3	3.5	4
y	0		3		8	9.375		6.125	0

Using a scale of 4 cm to represent 1 unit on the x-axis and 2 cm to represent 1 unit on the y-axis, draw the graph of $y=4x^2-x^3$, for values of x from 0 to 4 inclusive.

(b) By drawing appropriate straight lines on your graph:
(i) estimate the gradient of the curve $y=4x^2-x^3$ at the point (3.5, 6.125),
(ii) find two solutions of the equation

$$4x^2-x^3=x+2$$

(MEG)

12. The following table is based on the performance figures for a car as it accelerates from rest.

Time (seconds), t	0	2	4	6	8	10
Velocity (metres per second), v	0	10	18	23.5	27.5	31

(a) Draw the graph of v against t.
(b) By drawing the tangent to the curve at (4, 18), estimate the gradient of the curve at this point. State the significance of this value. (NEA)

13. (a) Complete the following table for the function $F=\dfrac{144}{d^2}$.

d	1	1.2	1.5	2	3	4	5	6	8	10	12
d^2		1.44			9						
$F=\dfrac{144}{d^2}$		100			16						

(b) Using a scale of 1 cm to represent 1 unit on the d axis, and 1 cm to represent 10 units on the F axis, draw the graph of $F=\dfrac{144}{d^2}$ for values of d from 1 to 12 inclusive.

(c) What happens to F as d increases?
(d) Where will the graph meet the axes?
(e) Use your graph to find the value of
 (i) F when $d = 4.6$,
 (ii) d when $F = 50$.
(f) By drawing a suitable tangent, estimate the gradient of the graph when $d = 2$. (MEG)

PROBABILITY

9

Not everything is certain in life, and this unit shows how probability is used to describe the likelihood of events happening.

9.1 Probability

Phrases such as 'million-to-one', 'fifty-fifty', '1 in 6' are familiar to us in everyday life as they represent numerical estimates, or values, of the chance that an *event* will or will not happen.

The headlines in the *Daily Blah* (Figure 9.1), suggest that it is extremely unlikely that a toddler would win a marathon. The toddler in question is more likely to throw a 6 with a fair die (1 in 6) or even to be able to throw a head with a fair coin (fifty-fifty).

Figure 9.1

Probability is the study of unpredictable events, which may or may not happen.

The numerical value which is assigned to the likelihood of an event happening is called the *probability* of the event, and is usually expressed as a fraction in the range 0 (impossible) to 1 (certainty)—see Figure 9.2.

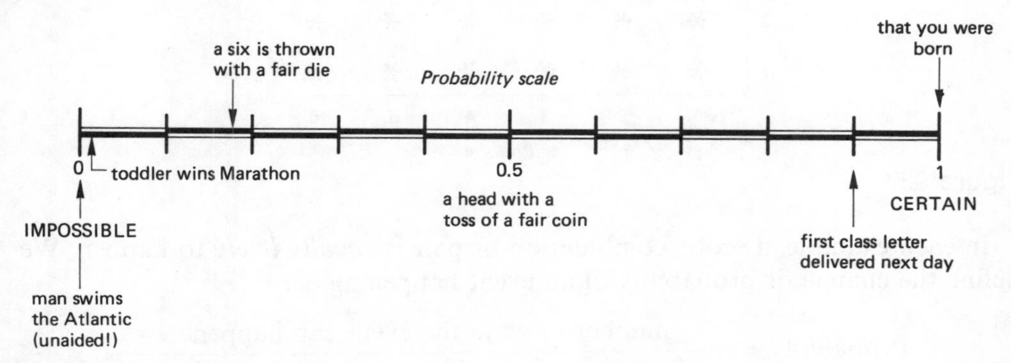

Figure 9.2

9.2 Possibility spaces

When a fair die is rolled, there are *six* possible scores or *outcomes* that can be obtained. These are shown in Figure 9.3.

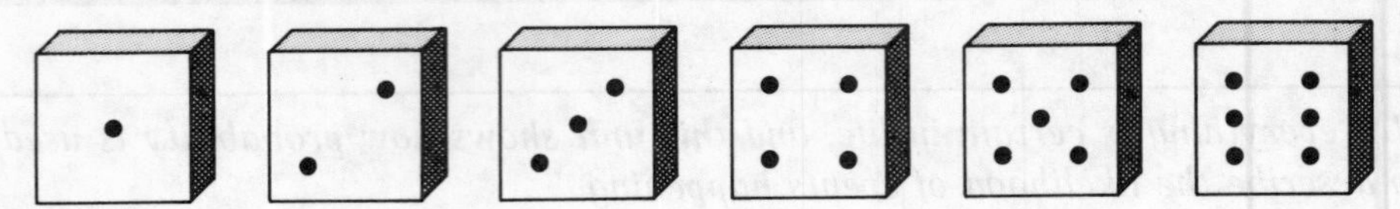

Figure 9.3

When two coins are thrown together, there are *four* possible outcomes—the combinations of heads and tails shown in Figure 9.4.

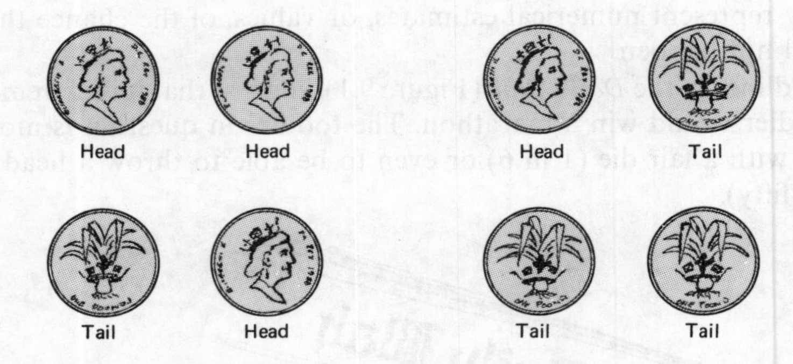

Figure 9.4

And when two dice are rolled, there are *36* possible pairings or 36 outcomes which can be represented in diagrammatical form—see Figure 9.5.

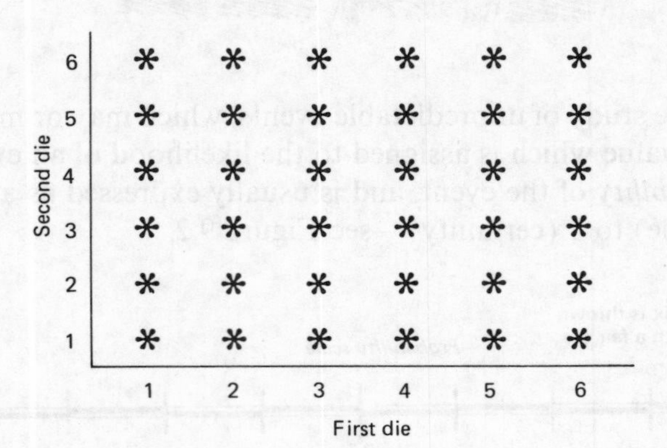

Figure 9.5

In each example, a score, combination or pair is *equally likely* to happen. We define the change or probability of an event happening as:

$$\text{Probability} = \frac{\text{number of ways the event can happen}}{\text{total number of possible } \textit{equally likely} \text{ outcomes}}$$

Worked example 9.1
A fair die is rolled. What is the probability of a score less than 5?

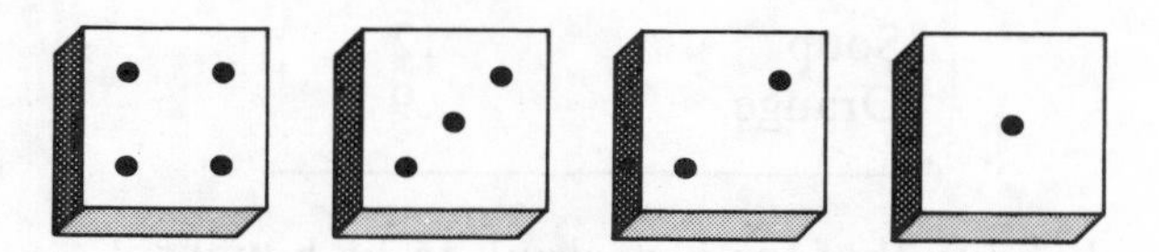

Figure 9.6

Number of ways the event can happen

$$=4$$

P (score less than 5) $= 4/6 = 2/3$.

9.3 Estimating probabilities

Consider an experiment of throwing a drawing pin and recording the possible outcomes. What is the probability of it landing point uppermost? Clearly there are two possible outcomes. See Figure 9.7.

Figure 9.7

We cannot say that both of these outcomes are equally likely and therefore we have to resort to a different method to determine a probability. We need to *estimate* the probability by carrying out an experiment in which the drawing pin is thrown a number of times and a tally kept of the results.

The estimated probability is defined as:

$$\text{Probability} = \frac{\text{number of successful outcomes in the experiment}}{\text{total number of trials in the experiment}}.$$

If the pin is thrown 400 times and 175 point uppermosts are recorded, then the probability:

$$P\ (\text{point uppermost}) = 175/400 = 7/16$$

Worked example 9.2
The drinks machine in the common room serves four types of drink and the sales for one coffee break were as follows:

Drink type	No. of drinks
Coffee	16
Cola	23
Soup	12
Orange	9

Estimate the probability that the next drink to be bought:

(i) will be Cola,
(ii) will not be Soup.

(i) $P(\text{Cola}) = \dfrac{23}{(16+23+12+9)} = \dfrac{23}{60}$

(ii) $P(\text{Not Soup}) = P(\text{Coffee, Cola or Orange is chosen})$

$$= \frac{16+23+9}{60} = \frac{48}{60} = \frac{4}{5} \text{ (in lowest terms)}$$

9.4 Complement

In the previous example, the probability that the next drink bought *is* Soup

$$= 12/60$$

So $P(\text{Soup}) + P(\text{Not Soup}) = 12/60 + 48/60 = 1$

The probability of *not* buying Soup is the *complement* of buying Soup. So:

$$P(\text{Event happening}) + P(\text{Event does not happen}) = 1$$

Worked example 9.3

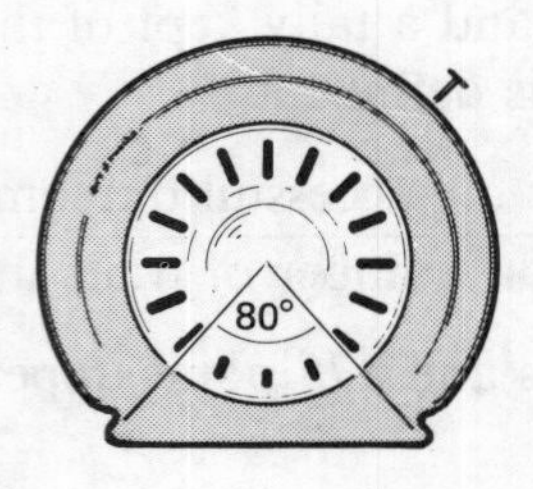

Figure 9.8

In my haste to reverse the car out of the garage, I inadvertently run over a large nail. I immediately stop to check the tyre. If the part of tyre touching the ground makes an angle of 80° at the centre of the wheel (see Figure 9.8), what is the probability that I am able to find the nail without moving the tyre?

I will not find the nail immediately if it is in that part of the tyre which is touching the ground.

Therefore, P(Not finding the nail)

$$= \frac{\text{part of circumference touching the ground}}{\text{total circumference}}$$

$$= \frac{80}{360} = \frac{2}{9}$$

so $P(\text{Finding the nail}) = 1 - \frac{2}{9} = \frac{7}{9}$

Exercise 9a

(1) Determine the probabilities of the following events:

(i) The person sitting closest to you has the same birthday as yourself.
(ii) Rolling a 7 with a fair die.
(iii) Guessing a two-digit number correctly.
(iv) An integer chosen at random between 1 and 30 which is divisible by 5.
(v) Drawing a 10 from a pack of playing cards.
(vi) Choosing a letter from the word SYMMETRY which has line symmetry.
(vii) A score of 8 with two dice.
(viii) A letter chosen at random from the alphabet is not a consonant.
(ix) Not picking a brazil nut from a bag containing 20 brazil nuts, 15 walnuts, 25 hazelnuts and 40 almonds.
(x) It will rain tomorrow, given that the probability that it will not rain is 3/11.
(xi) The sun rising in the morning.
(xii) Pigs flying.
(xiii) New Year's Day in the 20th Century chosen at random falling on a Monday.
(xiv) Drawing a red 6 from a pack of playing cards.
(xv) Moving from 'Mayfair' to 'Go' in successive moves at Monopoly.
(xvi) Guessing the 3-digit codes on the two locks of a briefcase.
(xvii) At least one head with the throw of two coins.
(xviii) Getting a head and a '3' when a coin and die are thrown together.
(xvix) Not drawing a blue ball from a bag which contains only red and blue balls, and twice as many red balls as blue balls.
(xx) A point chosen inside a circle of radius 5 cm is within 2 cm of the centre of the circle.

(2) Use the diagrammatical representation of the rolling of two dice (Figure 9.5) to find the following probabilities when two fair dice are thrown. You may find it helpful to draw the diagram and to ring the crosses which represent the favourable outcomes.

(i) A score greater than 9.
(ii) Two odd scores showing on the two dice.

(iii) A double score.
(iv) At least one 6 is thrown.
(v) The difference in the two scores is less than 3.

(3) Sandy Bunker, a well known golfer, keeps a record of the number of shots it has taken him at any hole on his local golf course. One hole always causes him problems, and on one particular day he looked up his record for this hole before beginning his round:

Number of shots taken	3	4	5	6	7
Number of times	7	15	27	36	15

Using the information given, what is the probability that on this round of golf, Sandy will
(i) take four shots;
(ii) take more than five shots;
(iii) not take three shots.

9.5 Addition of probabilities

The drinks machine in worked example 9.2 is susceptible to problems. In any given week (Monday to Friday) there is no coffee available on Monday, Tuesday and Wednesday, and for some unexplicable reason, although the machine accepts money, no drinks are served by the machine on Fridays as it is always broken down. What is the probability that in any given week on a day chosen at random, I choose a drink at random which is not available?

Let us examine the possibility space, shown in Figure 9.9.

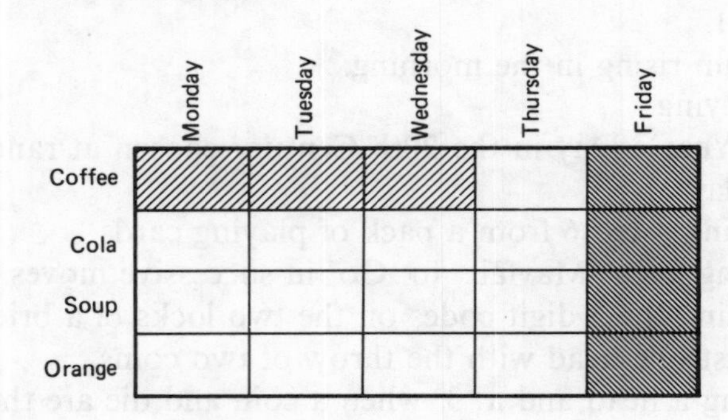

Figure 9.9

Assuming a random choice of drinks and random choice of day, each of the 'squares' in the possibility space are equally likely and therefore:

$$P(\text{A drink is chosen which is unavailable}) = \frac{3+4}{20} = \frac{7}{20}$$

Note that this is equivalent to:

$$P(\text{Coffee chosen on Monday, Tuesday or Wednesday})$$
$$+$$
$$P(\text{A drink is chosen on Friday})$$

It appears that we can add the probabilities of two alternative events to find the probability of *either* one *or* the other happening.

Worked example 9.4
Matters are now worse with the drinks machine. Coffee is now unavailable and the machine still breaks down every Friday. Determine the probability that, on a day chosen at random, a drink is chosen which is unavailable.

Again drawing the possibility space, we can see that

$$P(\text{Drink is unavailable}) = \frac{8}{20}$$

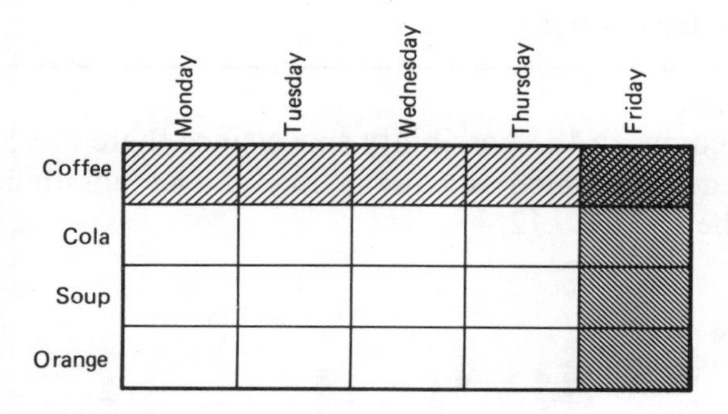

Figure 9.10

Note however that $P(\text{Coffee chosen}) = \frac{5}{20}$

and $P(\text{Drink chosen on Friday}) = \frac{4}{20}$

Adding these two probability fractions together we arrive at 9/20.

The reason for this is the overlap on Fridays when the unavailability of coffee is counted twice. *Direct adding of probabilities is only possible when there is no overlap.*

9.6 Independent events

Worked example 9.5
The bus services 27A, 27B and 27C into town from a college run very frequently with the same number of buses on each service. Unfortunately there is no indication on the destination board of any bus as to whether it is a 27A, 27B or 27C. Both the 27A and 27B terminate at the bus station while the 27C terminates on the estate. Michaela catches a bus at random and chooses one of the destinations

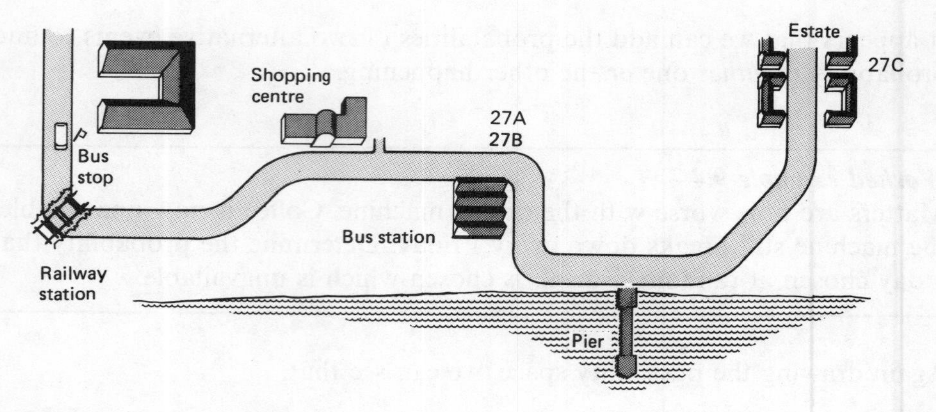

Figure 9.11

shown on the diagram at random. What is the probability that the bus she catches does not take her far enough?

There are 15 outcomes in the possibility space since there are 5 destinations and 3 bus services, each of which is equally likely. The possibility space may be represented by the Figure 9.12.

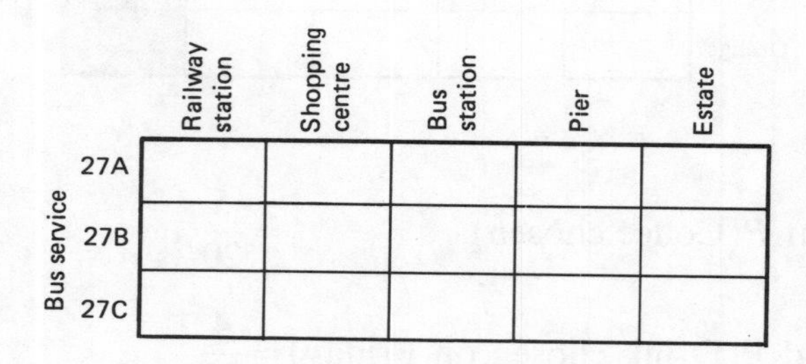

Figure 9.12

As all the services go as far as the bus station, it is only the destinations of the pier and the estate which should concern us. The services 27A and 27B do not go as far as these destinations, and this information is represented in the area shaded in Figure 9.13.

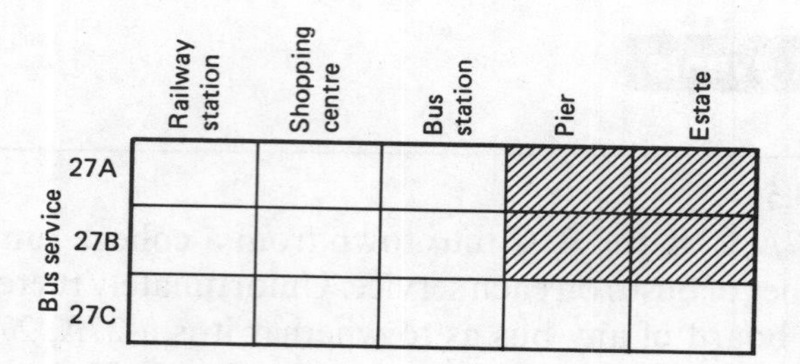

 Figure 9.13

The probability therefore that the bus does not take Michaela far enough is $\frac{4}{15}$. This could be written as $\frac{2}{3} \times \frac{2}{5}$. That is, the probability of both a 27A or 27B bus *and* a destination beyond the bus station

$$= P(\text{27A or 27B bus}) \times P(\text{Destination beyond bus station})$$

This is an example of two events which are *independent* of each other.

Worked example 9.6

Rizwan and Winston both work for a local service station on Saturdays. The probability that Rizwan arrives late for work is 1/10 and the probability that Winston arrives late is 1/8.

Assuming that their arrival times are independent, what is the probability that:

(i) they both arrive late?

(ii) they both arrive on time?

Deduce the probability that only one arrives late.

(i) $P(\text{Both arrive late}) = P(\text{Rizwan arrives late}) \times P(\text{Winston arrives late})$
$= 1/10 \times 1/8 = 1/80$

(ii) $P(\text{Both arrive on time}) = P(\text{Rizwan arrives on time}) \times P(\text{Winston arrives on time})$
$= (1 - 1/10) \times (1 - 1/8)$
$= 9/10 \times 7/8 = 63/80$

Now as we can be *certain* that both arrive late or both arrive on time or only one arrives late:

$1/80 + 63/80 + P(\text{One arrives late}) = 1$

So $P(\text{One arrives late}) = 1 - 64/80$
$= 16/80$

9.7 Probability trees

Worked example 9.7

Returning to worked example 9.5, we now find that service 27A terminates at the railway station. Given again a random choice of service and a random destination, what is the probability that Michaela catches a bus which does not take her far enough?

The solution can again be illustrated by a possibility space diagram, as can be seen in Figure 9.14.

The probability therefore that the bus does not take Michaela far enough

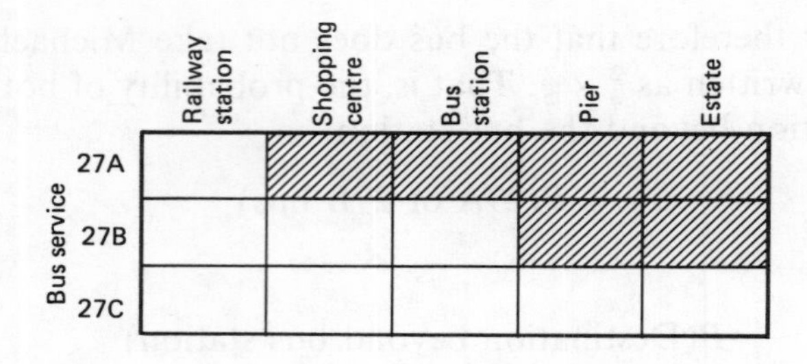

Figure 9.14

is 6/15. Although a possibility space diagram illustrates the answer here, this example can be tackled by drawing what is called a *probability tree*.

We begin the tree by showing three branches which represent the three possible choices of service bus (Figure 9.15).

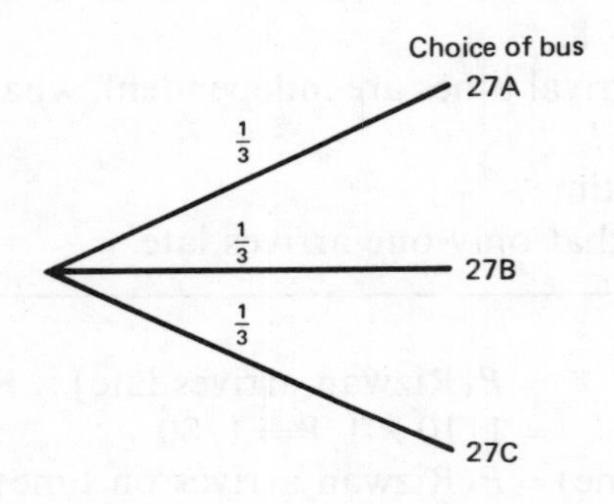

Figure 9.15

In Figure 9.16, the probability of the event at the end of the branch is marked on each branch. We now extend the tree by drawing in branches which represent the possible events 'taken far enough' and 'not taken far enough' (Figure 9.16).

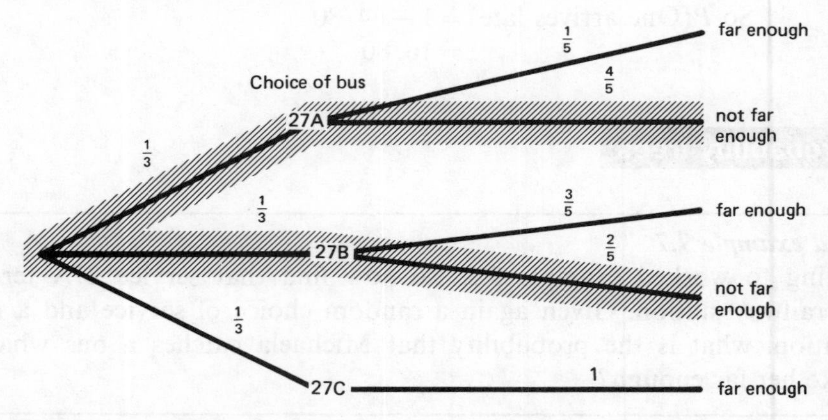

Figure 9.16

To determine the probability that Michaela is not taken far enough, we must trace the branches which agree with this condition, namely the two that are shaded in Figure 9.16. We then *multiply* the probabilities *along* the branches together and *add* the results.

So P(The bus does not take Michaela far enough)

$$= \underset{\text{service 27A}}{\frac{1}{3}} \times \underset{\text{any destination after the railway station}}{\frac{4}{5}} + \underset{\text{service 27B}}{\frac{1}{3}} \times \underset{\text{any destination after the bus station}}{\frac{2}{5}}$$

$$= \frac{4}{15} + \frac{2}{15}$$

$$= \frac{6}{15}$$

Worked example 9.8

From worked example 9.6, draw a probability tree to determine that only one arrives late.

The two branches which agree with the condition are shaded in Figure 9.17.

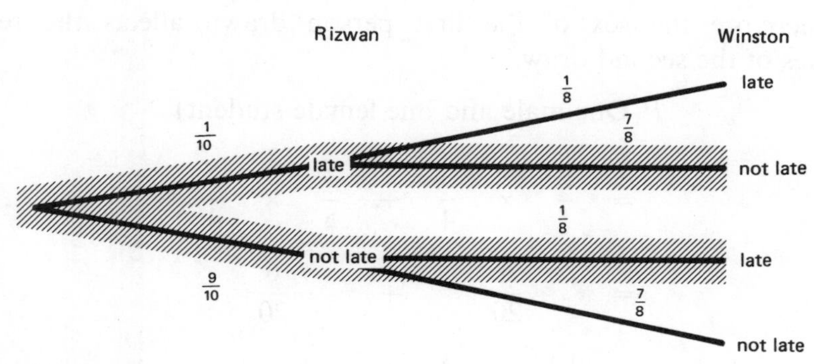

Figure 9.17

Therefore P(Only one arrives late)

$$= \frac{1}{10} \times \frac{7}{8} + \frac{9}{10} \times \frac{1}{8}$$

$$= \frac{7}{80} + \frac{9}{80}$$

$$= \frac{16}{80}$$

which agrees with the previous calculation.

Worked example 9.9

Arriving late for a concert, a group of five students find that there are only two seats left. They decide to put their names in a hat and the first two names drawn are to get the seats. If there are 3 females and 2 males in the group, what is the probability that the two seats go to 1 female and 1 male student?

Draw a probability tree—as shown in Figure 9.18.

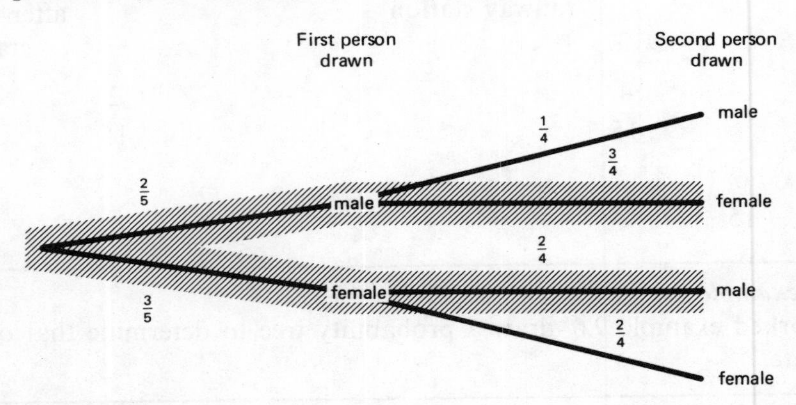

Figure 9.18

Note that the event of the second person drawn is *dependent* on the first person drawn. Therefore the sex of the first person drawn affects the respective probabilities of the second draw.

$$P(\text{One male and one female student})$$

$$= \frac{2}{5} \times \frac{3}{4} + \frac{3}{5} \times \frac{2}{4}$$

$$= \frac{6}{20} + \frac{6}{20}$$

$$= \frac{12}{20} = \frac{3}{5}$$

Exercise 9b

(1) Six students are to take part in a Psychology experiment on handwriting. Their names are put into a bag and drawn at random to decide the order in which they take part.

Two of the students, Lesley and Lionel, are left-handed; the others are right-handed.

Find the probability that:

(a) the first name to be drawn is Lionel;

(b) the first name to be drawn is that of a left-handed student;

(c) the first two names to be drawn are both of right-handed students.

(MEG)

(2) Corinne was asked, 'What is the probability that when a die is thrown the score is either odd or less than 4?'. She answered as follows:

$$P(\text{Odd}) = \frac{3}{6}$$

$$P(\text{Less than } 4) = \frac{3}{6}$$

$$\text{So } P(\text{Odd or less than } 4) = \frac{3}{6} + \frac{3}{6} = 1 \text{ or certainty!}$$

Where is Corinne's mistake?

(3) Four playing cards consisting of a 5, 6, 7 and 8 lie face downwards on the table. If one is chosen at random, what is the probability that it is a prime number? If two cards are chosen at random, what is the probability that the sum of their scores is exactly 14?

(4) Two students are to be chosen from Stenworth College to represent the College in a TV quiz, 'Mindbuster'. Reza, Lorraine, Nick and Liam all of equal ability, had their names placed in a hat and two names were drawn out.
   (a) Write down a list of all possible pairs which could be drawn.
   (b) What is the probability that Nick will be chosen?
   (c) What is the probability that either Lorraine or Liam will be chosen?

(5) At a local fete, Marcus decides to bet on two of the woodlice in a woodlice race. There are six woodlice in the race each with the same chance of winning. What is the probability that one of the woodlice chosen by Marcus wins? If he bets on two woodlice in two successive races, what is the probability that he has a winner in both races? What is the probability that he will have at least one winner in the two races?

(6) Nicky applies for a position with two credit card companies, American Suburban and Plasicard. She is selected for interview at both places. At American Suburban there are five candidates and at Plasicard there are four candidates. If each candidate is equally likely to be selected, what is the probability that Nicky is selected for:
   (i) a position at American Suburban;
   (ii) a position at both American Suburban and Plasicard;
   (iii) a position at either American Suburban or Plasicard, but not both.

(7) A coin is biased towards 'heads' such that the $P(\text{Head}) = 2/3$. If the coin is tossed twice, what is the probability of:
   (a) two heads;
   (b) at most one head.

(8) The North and South orbitals (Figure 9.19) are two parts of a motorway which surrounds a large city. Tom and Gerry are delivery drivers who travel daily on this motorway. On one particular day they decide to meet at one of the three service stations on the motorway for lunch. Unfortunately as Tom approached the motorway from the West and Gerry approached from the East, neither could remember at which service station they had agreed to meet. As they both approached the motorway, each threw a fair coin to

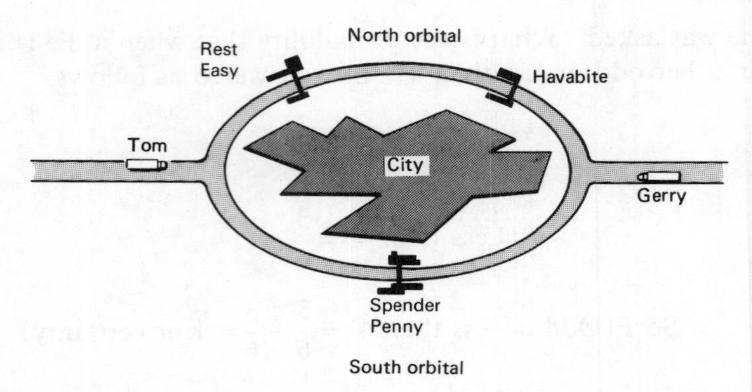

Figure 9.19

decide whether they would go on the North orbital or the South orbital. If either driver decides on the South orbital he will stop at Spender Penny, however if the North orbital is chosen the coin is thrown again to decide on which service station to stop at.

What is the probability that:

(a) Gerry stops at Spender Penny;
(b) Tom stops at the Havabite service station;
(c) both drivers choose the North orbital;
(d) both meet at the Rest Easy service station;
(e) they do not meet?

(9) Figure 9.20 shows a target for shooting; it is divided into three regions by circles which have the same centre and whose radii are 2 cm, 3 cm and 4 cm. You may assume that all shots hit the target and that the probability of any shot hitting any region is proportional to the area of that region.

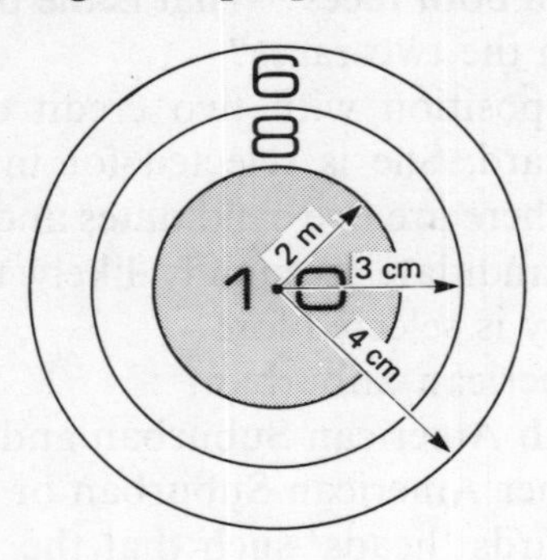

Figure 9.20

(a) If one shot is fired, calculate the probability of:
 (i) a score of 10;
 (ii) a score of more than 6.
(b) If two shots are fired, calculate the probability of a score of 12.

(10) A coin is tossed and a card taken at random from one red card and two black cards.

(a) Copy and complete the tree diagram in Figure 9.21 which shows the various probabilities when this is done.
(b) What is the probability of taking a black card?

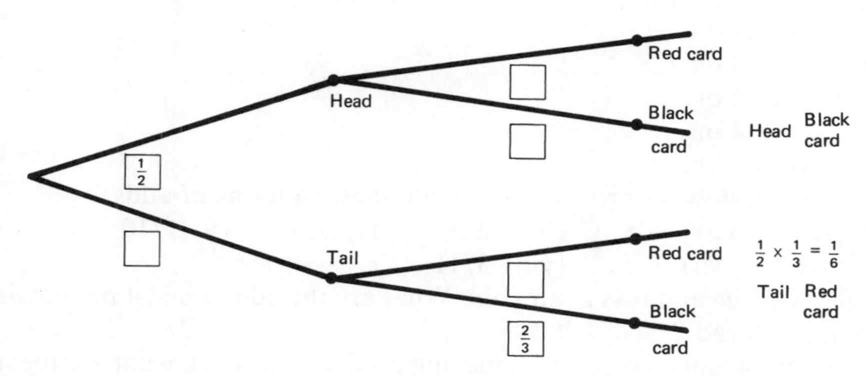

Figure 9.21

(c) What is the probability of:
 (i) Getting a tail and taking a red card;
 (ii) Getting a head and taking a black card? (SEG)

9.8 Odds

In the latest Cornhill Cricket test, England are quoted at 3 to 1 *against* winning the match. In terms of probability this means that the England team has 1 chance in $(3+1)$ of winning, i.e. the probability of success is $1/4$.

We arrive at this figure as follows:

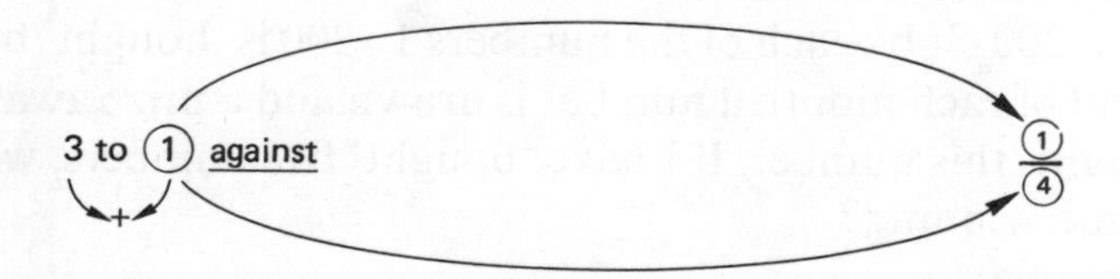

In the same match, the West Indies are quoted at 2 to 1 *on*. This time the probability fraction is arrived at as follows:

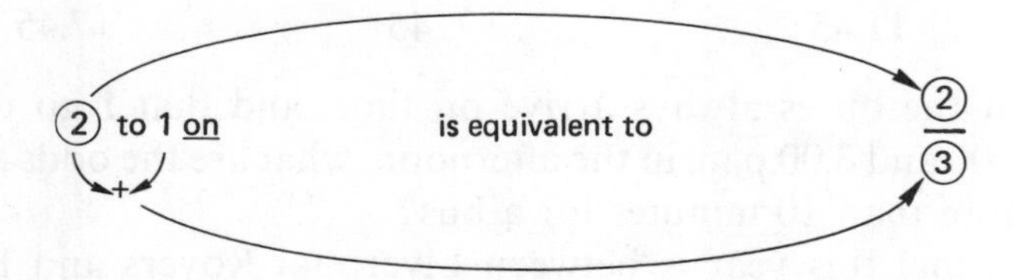

Odds of 'evens' is 1 to 1 and is equivalent to a probability of $1/2$.

Exercise 9c

(1) Given the following odds, write these in terms of a probability fraction:
 (i) 2 to 1 against.
 (ii) 9 to 1 against.
 (iii) 3 to 2 against.

(iv) 5 to 4 against.
(v) 13 to 2 against.
(vi) 3 to 1 on.
(vii) 3 to 2 on.
(viii) 5 to 4 on.
(ix) 6 to 4 on.
(x) 10 to 1 on.

(2) Given the following probabilities, write these in terms of odds:
(i) 1/4 (ii) 4/5 (iii) 2/3 (iv) 1/7 (v) 1/10
(vi) 5/6 (vii) 2/7 (viii) 3/11

(3) I roll a fair die and toss a fair coin. What are the odds against me obtaining a 6 and a 'head'?

(4) If the odds against an event happening are 2 to 1 against, what are the odds against the event not happening.

(5) The probability that it will rain tomorrow is 1/4. What are the odds against it not raining tomorrow?

(6) In the Cornhill Test, given that the only results possible are a win for either team or a draw, using the data of section 9.8 what are the fair odds on a draw?

(7) Haws Rider, a leading jockey, is to ride two horses at a race meeting this afternoon. In the first race he is to ride Nevercomeslast which has a probability of 1/10 of winning the race. In the second race he is to ride the favourite Cronzalea which has a probability of 1/3 of winning the race. Find the probability that both horses win their respective races. Determine the fair odds against both horses winning their races. What is the probability that Haws Rider has only one winner?

(8) In a local '200 club', each of the numbers 1–200 is 'bought' by club members. At the end of each month a number is drawn and a prize awarded to whoever has 'bought' this number. If I have 'bought' five numbers, what are the odds against me winning?

(9) Bus timetable:

11.00	then	00		17.00
11.15	after	15	until	17.15
11.30	each hour	30		17.30
11.45		45		17.45

Given that the buses always arrive on time, and that I go to catch a bus between 2.00 and 3.00 p.m. in the afternoon, what are the odds against having to wait more than 10 minutes for a bus?

(10) The Cup Final this year is between Liverpool Rovers and Bristol Forest. Liverpool are 2 to 1 against to win and the draw is evens. What are the fair odds on Bristol winning?

Suggestions for coursework 9

1. Figure 9.22 shows a network of roads in a 2 × 2 square. Starting in the bottom left corner (S) and finishing at the top right-hand corner (F), investigate how

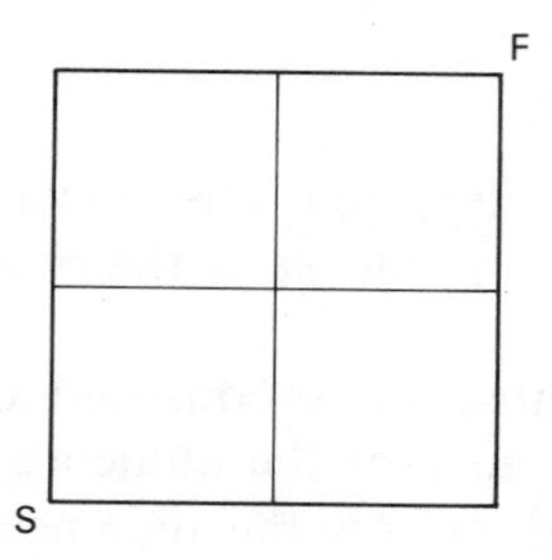

Figure 9.22

many different routes you can take without tracing the same path twice. Extend your investigations to 3×3, $4 \times 4, \ldots, n \times n$ squares. Further extensions could involve rectangular pattern of squares or even three dimensions.

2. Figure 9.23 shows a quadrant of a circle of radius 6 cm with a 6×6 grid superimposed. Each square on the grid can be given a co-ordinate from (1,1) to (6,6). By rolling two dice we can choose one of these squares at random, i.e. the cross marked in square (3,2) denotes a throw of a 3 with the first die and a 2 with the second die. By recording the number of crosses which are inside the quadrant and the total number of throws, we can find an approximation to π. Before you can carry out this investigation, however, you will need to devise a method of deciding what happens when your throw denotes squares such as (6,1), (6,2), (6,3), etc. You may wish to extend your ideas to other shapes.

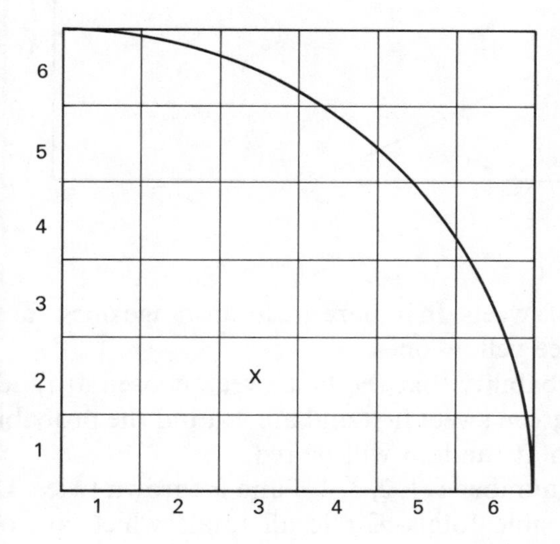

Figure 9.23

3. In this unit we gave as an example of estimating probabilities the experiment of throwing a drawing pin. Carry out your own investigations.
4. Choose a card game of which you are familiar. Investigate the probabilities of various distributions of cards in this game.
5. Investigate the odds offered by bookmakers for various events. Suggest reasons why you would never see a poor bookmaker!

Miscellaneous exercise 9

1. A fair coin is tossed three times and comes down 'heads' on each occasion. If the coin is tossed for a fourth time, state the probability that it comes down 'heads' again.
2. The audience at a TV game show consisted of a coachload of 30 from Stenworth while the remainder of the audience came from Brambing. If a contestant is chosen at random from the audience, the probability that he/she comes from Stenworth is 0.6. Find how many of the audience came from Brambing.
3. New computer equipment is ordered by phone and delivery is expected in a few days. The probability that it will be delivered by a courier service is 1/3, and the probability that it will be delivered by the computer company's own van is 1/4. What is the probability that the equipment will be delivered by one of these methods? What is the probability that it will be delivered by neither of these two methods?
4. In Figure 9.24, X and Y are the midpoints of adjacent sides of the rectangle ABCD. AB = 8 cm and BC = 4 cm. A point is selected at random in the rectangle. Calculate the probability that it lies in triangle XYD.

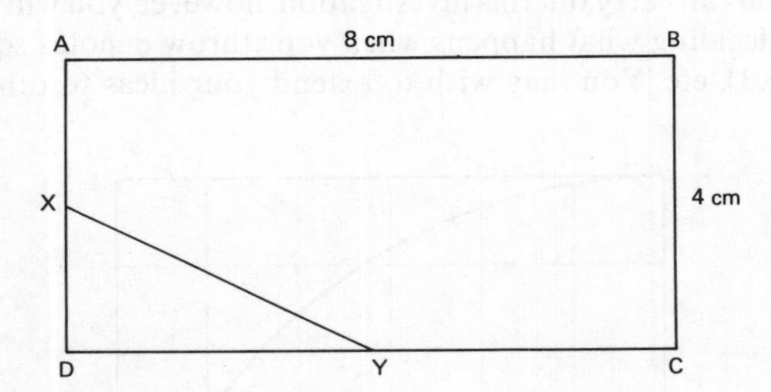

Figure 9.24

5. Latifa has a bag of sweets. In it there are four orange ones, three green ones, five red ones and three yellow ones.
   (a) Find the probability that the first sweet, chosen at random, is green.
   (b) She chose a green sweet first and ate it. Find the probability that the next sweet, chosen at random will be red. (LEAG)
6. A six-sided die is number 1, 1, 2, 3, 4, 4 and is thrown twice. Draw a possibility space for all possible totals. Circle all totals which are odd. What is the probability that the total is odd? What is the probability of a score greater than 6?
7. Using Figure 9.25, write down all the different routes from A to C which do not involve passing through the same point twice. Write your answers in the format A → B → C. If one of these routes is chosen at random, what is the probability that it passes through E?
8. One of the twenty whole numbers from 10 to 29 is chosen at random. Find the probability that the whole number is:

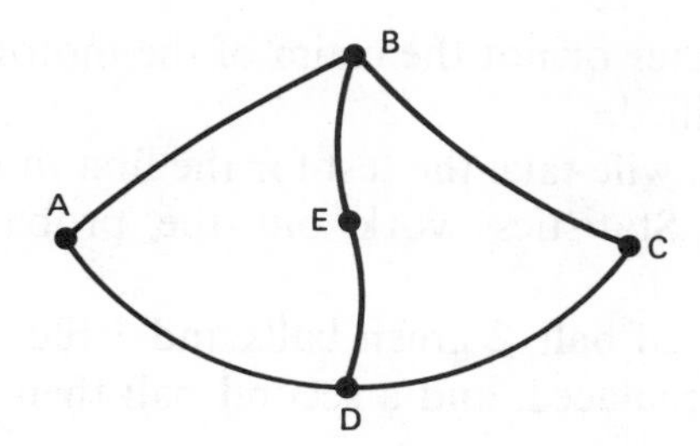

Figure 9.25

(a) divisible by 3,
(b) divisible by 3 *and* 2,
(c) divisible by 3 or by 2, but not by both.
(d) a prime number. (SEG)

9. Each morning Ann and Janet have to catch the same train, otherwise they are late for work. The probability that on any day Ann catches the train is 0.9 and the probability that Janet does is 0.8. (These probabilities are independent of each other.) Calculate the probability that on any given day
   (a) they both catch the train,
   (b) Ann catches the train but Janet does not,
   (c) just one of them catches the train,
   (d) at least one of them catches the train. (NEA)

10. A die is biased towards a 6 such that the probability of a six is 1/2. The probability of the remaining five numbers is the same.
    (a) What is the probability of:
        (i) not getting a 6,
        (ii) a score less than 3?
    (b) If the die is rolled twice what is the probability of:
        (i) a six followed by a 3,
        (ii) a score of 11?

11. During one month at a Driving Test Centre, 140 males took the Driving Test for the first time: 81 of them were successful. During the same month 110 females took the Driving Test for the first time: 62 were successful.
    In addition, of the 90 males repeating the Driving Test, 60 passed, and of the 70 females repeating, 45 passed.
    (a) Present all this information in the form of a table. Include under the main headings, 'First time Test' and 'Repeat Test', the number of passes and failures for both male and female candidates and the total number of passes and failures for all candidates.
    (b) A motoring organisation claims that this Centre is more severe than other Centres. Nationally, the probability of a male taking the test for the first time and passing is 0.7 and the probability of a female taking the test first time and passing is 0.6.
        (i) Using these probabilities, estimate the number of males and the number of females who could be expected to pass first time at this Centre.
        (ii) Compare these figures with the actual figures at the Centre and hence

state whether or not the claim of the motoring organisation appears to be justified.

(c) A man and his wife take the test for the first time on the same day. Using the National Statistics work out the probability that both will be successful. (LEAG)

12. A bag contains 1 red ball, 2 green balls and 3 blue balls. A ball is taken out at random, *NOT* replaced, and a second ball then removed.

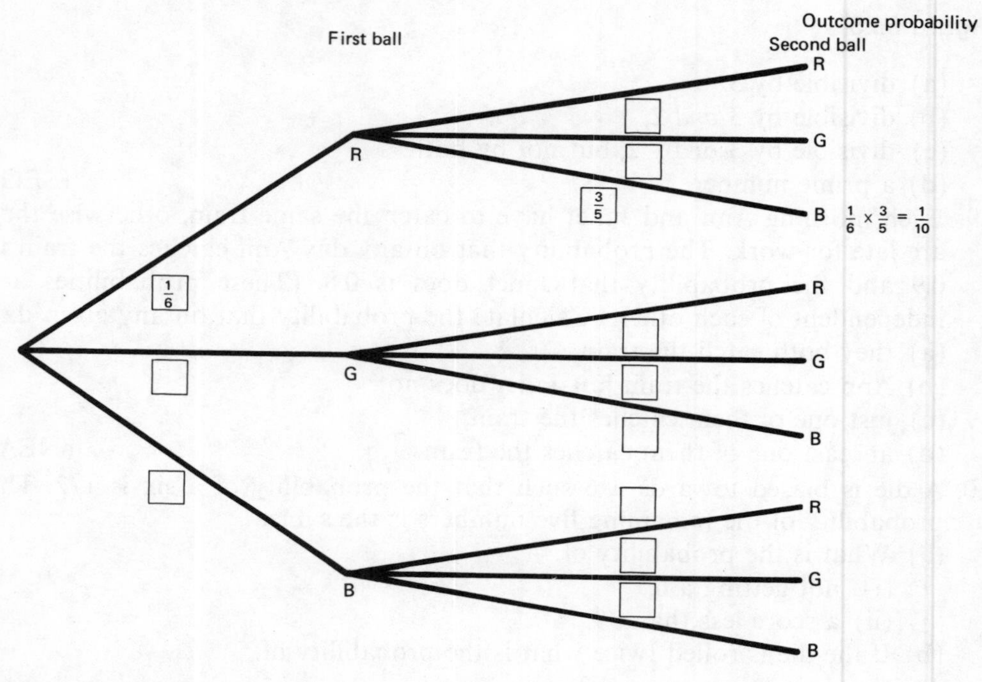

Figure 9.26

(a) Copy and complete the 'tree' diagram (Figure 9.26) which shows the various probabilities when this is done.

(b) Use your diagram to find the probabilities:

(i) of the 2 balls being the same colour,

(ii) of them being 1 green and 1 blue. (SEG)

STATISTICS 1

10

Statistics involves the collection, presentation and analysis of data. In this first unit of two, we will be looking at how data can be gathered and displayed.

10.1 What is statistics?

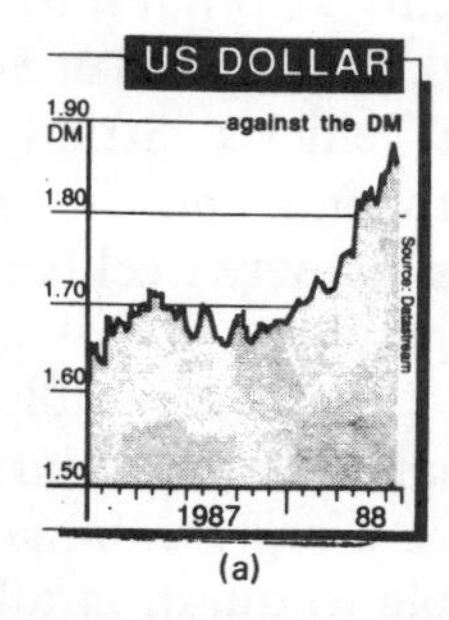

(a)

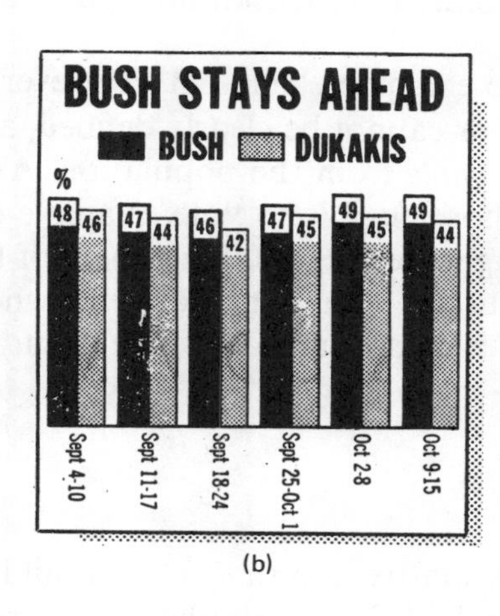

(b)

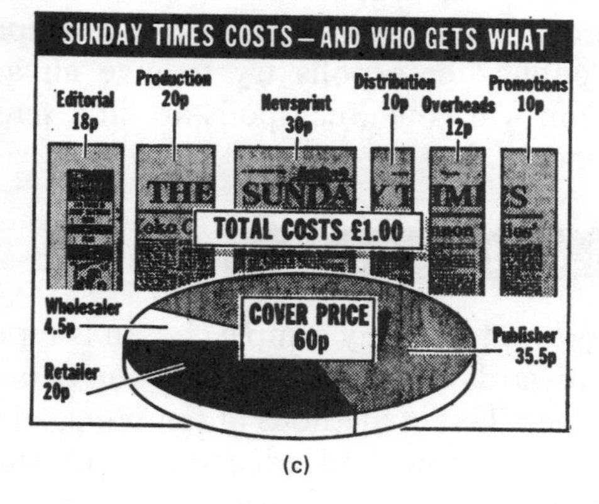

(c)

Figure 10.1

Figure 10.1 shows examples of diagrams which appear frequently in newspapers. Each diagram is meant to give a visual representation of *numerical facts* in such a way that the reader can interpret at a glance what information is being conveyed.

In all three cases, before the diagrams could be drawn, the numerical facts had to be collected. These facts are commonly referred to as *statistical data.* If meaningful conclusions are to be drawn from the diagrams, then the methods used to collect the data are crucial. Diagram (b) is a representation of opinion polls taken prior to the US presidential election in 1988. If polls are to be used as an accurate predictor of the outcome of such events, it is important that the data used is free from any *bias* which may lead to an incorrect representation.

Although they cannot be right everytime, opinion polls would lose credibility if they were wrong too often!

So what is statistics? Well it is concerned with drawing conclusions from data. In order to arrive at a conclusion however, it is necessary to *collect*, *organise* and *represent* the data in some way. The importance of correct procedures in collecting data cannot be over-emphasised, as conclusions drawn are only as good as the original figures on which they are based.

10.2 Sampling

In any statistical investigation, the data items to be collected will come from a *population*. For instance, to investigate the weekly income and expenditure of sixth formers in a particular school would involve collecting numerical data from some, if not all, of the students who fall into this category. An investigation into the reliability of a certain make and model of a car would perhaps involve collecting data from owners of such vehicles. Strictly speaking, the populations under consideration are not the students and car owners themselves, but the data that is provided by these persons, namely weekly income, and expenditure figures and reliability figures for the type of car.

If populations are small, it may be feasible to examine each item. However in the majority of cases, populations are too large or cannot be clearly defined, and it becomes necessary to take a *representative sample* from the population. As it would be virtually impossible to question all those eligible to vote, example (b) in Figure 10.1 uses such a method. By sampling as few as 1000 members of the voting public, these polls try to give an accurate picture of how the whole population will vote. Such polls are in a sense a compromise between accuracy and economy.

10.3 Methods of sampling

It is important that any sample chosen is representative and that there should be no bias towards any individual or group of individuals in the population. In order to examine different methods of sampling, let us imagine that we are investigating the weekly income and expenditure of students in a large sixth form. The admissions register for the year shows that the number of students in each year is as follows:

1st Year Sixth	2nd Year Sixth	3rd Year Sixth
320	140	40

It would seem an impractical task to obtain weekly income and expenditure figures from *all* students, therefore a representative sample will need to be chosen.

(i) Simple random sampling

All the students would be given a number in the range 001–500. Numbers in this range would then be written on pieces of paper and placed in a hat or a box and then drawn one by one until the required sample size is reached, or random number tables could be used to generate a random selection of the required sample size.

(ii) Systematic sampling

The above method could prove to be a little tedious, particularly for a large population. This second method involves setting up some system of selection, i.e. placing registers in any order and systematically working through them, choosing say every tenth student in each register. There is a danger here that all those chosen could share similar characteristics, i.e. they could all be female first year sixth students.

(iii) Stratified sampling

This method involves choosing random samples from subgroups defined on the population. In our example we have three clearly defined subgroups, namely the three year groups. Suppose we wish to take a sample of 50 students, then we would divide 50 in the ratio

$$320:140:40$$

which are the corresponding numbers for each year group. Now

$$320:140:40 = 32:14:4$$

so we would choose 3 random samples from the 3 year subgroups of sizes 32, 14 and 4.

To use this method, it is necessary to have a clearly defined population and clearly defined subgroups. However it does ensure that all sections of the population are adequately represented in any sample.

(iv) Quota sampling

This is an example of non-random sampling. *Quotas* for different sections of the population are set in a similar manner to stratified sampling. However it is then up to us whom we question or interview for the required information. Using the same figures generated from the stratified sampling technique, we could interview the first 32 1st year sixth students whom we happen to meet, the first 14 2nd year sixth students, etc. There is, of course, a great danger of bias here, however it does reduce considerably the amount of time required to carry out the survey. Because savings in time also mean savings in money, some regular opinion polls use this method of sampling.

10.4 Methods of collecting data

Once a method of sampling has been decided upon, the data can be collected by carrying out a *sample survey*. The two standard methods for this are either by

(a) interview, or (b) questionnaire

Both methods require questions to be asked, and we therefore must be clear what information is needed in order that the right questions can be asked:

The questions should be easy to understand and unambiguous, and there should be no leading or unreasonable questions.

If a questionnaire is to be filled in, the layout must be well designed with questions which are easily seen, and appropriate spaces left for the answers.

Table 10.1 Advantages and Disadvantages

	Interview	Questionnaire
Advantages	(1) High response rate (2) More detailed information can be collected	(1) Less time-consuming (2) Bias of interviewer removed
Disadvantages	(1) Expensive and time-consuming (2) Possible bias in the way questions are asked and data recorded	(1) Usually poor response rate (2) Bias may be introduced as only a particular group may reply

Worked example 10.1

Sam Pling, a student following a GCSE mathematics course, wishes to investigate sixth form student's weekly income and expenditure. In order to find relevant information, he requires to design a questionnaire which he will then distribute to a random sample of 50 sixth formers.

Design a possible questionnaire.

A possible questionnaire can be seen in Table 10.2.

Although questionnaires and interviews are suitable methods of collecting data for such problems, there are other methods for different situations. If, for instance, we require to log the arrival and departure times of trains at a station, we would carry out the task by *observation.* If, on the other hand, we were comparing the present year's rainfall to previous years' rainfall, we would not only need to observe the rainfall for the current year but also need to acquire the figures from previous years. Figures from previous years are known as *secondary data,* as undoubtedly someone else has collected this information. Many examples of secondary data can be found in the form of published tables.

Exercise 10a

(1) The *Daily Blah* newspaper wishes to determine the physical characteristics of its readers. 12 newsagents are selected and an observer is stationed at each. The observer then notes down the physical facts about each person

Table 10.2

MATHEMATICS GCSE PROJECT

A confidential survey investigating student's weekly income and expenditure.

Please tick appropriate boxes

1. Year Group

   1st Year Sixth ☐
   2nd Year Sixth ☐
   3rd Year Sixth ☐

2. Sex

   Male ☐
   Female ☐

3. Do you have a part-time job?

   Yes ☐
   No ☐

4. If your answer to the last question was YES, what is your approximate weekly income from your job?

   £
   0.00– 4.99 ☐
   5.00– 9.99 ☐
   10.00–14.99 ☐
   15.00–19.99 ☐
   20.00–24.99 ☐
   25.00 or more ☐

5. What is your approximate *total* weekly income including any jobs, any pocket money or money from any other source?

   £
   0.00– 4.99 ☐
   5.00– 9.99 ☐
   10.00–14.99 ☐
   15.00–19.99 ☐
   20.00–24.99 ☐
   25.00 or more ☐

6. How much do you spend approximately each week?

   £
   0.00– 4.99 ☐
   5.00– 9.99 ☐
   10.00–14.99 ☐
   15.00–19.99 ☐
   20.00–24.99 ☐
   25.00 or more ☐

7. From the list below, choose the item on which you spend most of your money each week.

Clothes	Food	Entertainment	Transport	Other (Please specify)
☐	☐	☐	☐	☐ ....

Thank you for your time and co-operation. Would you please return the completed form to Sam Pling (Form 6XX)

who bought a copy of the newspaper. Give two comments criticising this method of collecting data.

(2) Melissa wishes to carry out a survey on attitudes towards alcohol and driving among her Sixth Form colleagues. Rather than question them all, she decides

to take a stratified sample of size 60 from the Sixth Form population which can be divided into the following four subgroups:

	Drivers	Non-drivers
Male	54	63
Female	72	81

How many of each subgroup should she choose?

(3) It is intended to carry out an investigation into the consumption of alcohol by the adult population of England.

(a) Comment *briefly* on the bias that might be introduced if a sample were taken from:

(i) the customers in a wine bar,

(ii) the residents of an upper class area.

(b) Describe *briefly* a method of obtaining an unbiased sample from a town. (SEG)

(4) John and Mary want to carry out a survey in their school to find the most popular record in the top 20 during a certain week. There are 1500 pupils and they decide to question 150 of these as follows:

They will stand at the school gates one morning and stop every tenth pupil who passes them to ask his or her opinion.

(a) Write down two reasons why this may not give a good sample.

(b) Describe carefully how John and Mary could obtain a better sample. (MEG)

(5) Describe briefly a method of systematic sampling which could be used in order to obtain a sample of 20 from the 200 first-year students in a sixth-form college. (NEA)

(6) A questionnaire is to be sent to a random sample of 100 private (i.e. not a business or a firm) telephone subscribers in a certain town.

Criticise the following methods of obtaining the sample.

A. Using random number tables to select 100 names from the electoral roll for the town.

B. Using random number tables to select 100 names from the telephone directory for the town.

C. Choosing the first 100 private subscribers listed in the telephone directory for the town. (SEG)

(7) In order to determine the public's reaction to a particular issue in Parliament, a newspaper asks its readers for their views. Comment on any bias that may be introduced.

(8) List any advantages and any disadvantages of a postal questionnaire as opposed to a personal interview.

(9) An investigation into local public opinion into a by-pass which is to be built around a town is to be made by:

(i) interviewing people in the street,

(ii) asking questions by telephone,

(iii) sending a letter to each householder.

Give a disadvantage of each method.

(10) Comment on any bias that may arise in the response to the following questions to the general public.
(i) What is your age?
(ii) Have you a criminal record?

(11) Criticise the following question which was being asked of British Rail commuters.

'Do you agree that the cost of rail fares is too high?'

(12) Design a questionnaire for use in your school or college to investigate the television viewing habits of sixth-form students.

10.5 Tabulation of data

After the data has been collected, it is necessary to present the information in as concise a form as possible so that trends and patterns can be seen at a glance. This is the purpose of *tabulation*.

Worked example 10.2
In Sam Pling's investigation into the weekly income and expenditure of sixth-form students mentioned in the previous example, suppose he had also asked the question:

How many times a week do you go out in the evening for pleasure?

Suppose that the responses were as follows:

0	3	1	0	6	7	2	0	1	2
2	3	3	1	6	6	5	4	4	4
3	2	0	1	2	0	0	0	1	2
3	1	0	2	3	0	0	1	4	3
0	1	5	1	0	5	5	4	1	0

Is there a pattern in this data?

This is the *raw data* and it is difficult to see if there is any pattern. We therefore need to construct a table and count the number of times out in a week by means of *tally marks*. The distribution formed is called a *frequency distribution*.

The final column in Table 10.3 gives a much clearer picture than the raw data. This type of data is called *discrete* data because it can only take certain values, namely 0, 1, 2, 3,..., 7. Discrete data is not limited to whole numbers as, for example, shoe sizes which can take values of the form 6, $6\frac{1}{2}$, 7, $7\frac{1}{2}$, 8,..., etc.

Continuous data on the other hand has no restriction on specific values as it is measured on a continuous scale. Examples are people's height, weight and age. It is very difficult with such data to be precise with our measurements, and approximations are required.

Table 10.3

Number of times out	Tally	Number of students (frequency)
0	~~1111~~ ~~1111~~ 111	13
1	~~1111~~ ~~1111~~	10
2	~~1111~~ 11	7
3	~~1111~~ 11	7
4	~~1111~~	5
5	1111	4
6	111	3
7	1	1
	TOTAL	50

10.6 Grouping data together

Worked example 10.3

In order to see whether or not an extra secretary is required in a college office to answer the phone, the number of calls made and received per day were recorded over a 50 day period. The raw data was as follows:

19	7	24	17	8	4	32	21	18	22
23	22	36	2	11	16	31	24	27	19
8	3	37	31	29	28	26	18	17	20
21	25	19	12	14	33	17	16	7	12
26	19	18	20	15	30	11	19	24	14

Is there a pattern?

In the previous example the spread of data was small, however the range in the discrete data above is from 2 to 37. To count the number of days for each value in this range would not really be practical. We need to condense the data into groups called *classes* and to count the number of calls in each class. We can conveniently divide the calls into groups of 1–5, 6–10, ..., 36–40. The resulting distribution is known as a *grouped frequency distribution,* and can be seen in Table 10.4.

Note that the starting point and the finishing point of each class are quite clear when using discrete data. The next example illustrates the use of grouping for continuous data.

Table 10.4

Number of calls	Tally	Number of days
1–5	111	3
6–10	1111	4
11–15	~~1111~~ 11	7
16–20	~~1111~~ ~~1111~~ ~~1111~~	15
21–25	~~1111~~ 1111	9
26–30	~~1111~~ 1	6
31–35	1111	4
36–40	11	2
	TOTAL	50

Worked example 10.4

The miles covered, to the nearest mile, by the 30 mini-cab drivers who work for OZCABS on one particular day were as follows:

104	183	210	86	124	226	212	237	128	40
176	183	62	145	115	139	287	90	161	74
121	243	66	120	185	231	191	62	70	225

Draw up a tally chart for the classes 0–50 miles, 51–100 miles, 101–150 miles,..., 251–300 miles, and hence form a grouped frequency distribution.

Table 10.5 Number of miles covered by 30 mini-cab drivers

Distance	Tally	Number of cab drivers
0–50 miles	1	1
51–100 miles	~~1111~~ 11	7
101–150 miles	~~1111~~ 111	8
151–200 miles	~~1111~~ 1	6
201–250 miles	~~1111~~ 11	7
251–300 miles	1	1
	TOTAL	30

As the distances recorded are continuous data, there is no restriction on the value a distance can take in a given range. A distance of 100.4 miles would be rounded *down* to 100 miles and be recorded as an entry in the class 51–100 miles, whereas a distance of 100.6 miles would be rounded *up* to 101 miles and be included in the next class.

The *class boundaries* are the dividing lines between the classes, and the class boundary between the classes 51–100 and 101–150 is 100.5 miles. Following normal arithmetical practice, a value of 100.5 miles would be rounded *up* to 101 miles.

10.7 Presentation of data

(i) Pictograms
This form of presentation involves the use of pictures to represent data.

Worked example 10.5
Suppose Sam Pling's survey (Table 10.2) produced the following results for question 7:

Item	Frequency
Clothes	11
Food	5
Entertainment	17
Transport	15
Other	2

Represent this information in a pictogram, using the symbol

to represent 5 students.

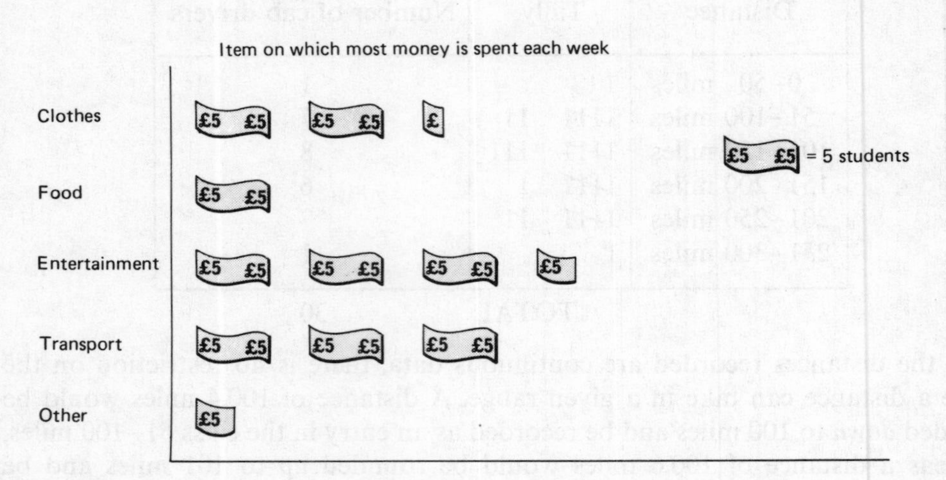

Figure 10.2

This method, shown in Figure 10.2, is useful to portray an 'instant' message as to what the diagram is representing and can often be quite eyecatching. However its accuracy is often limited by the scale used, as it is difficult to draw fractions of a symbol accurately.

(ii) Bar charts

Sometimes referred to as *bar graphs*, one axis represents the classes and the other axis represents the frequency. Each bar should be of the same width, and the length of a particular bar denotes the frequency for that class.

The data in worked example 10.5 could have been represented as in Figure 10.3.

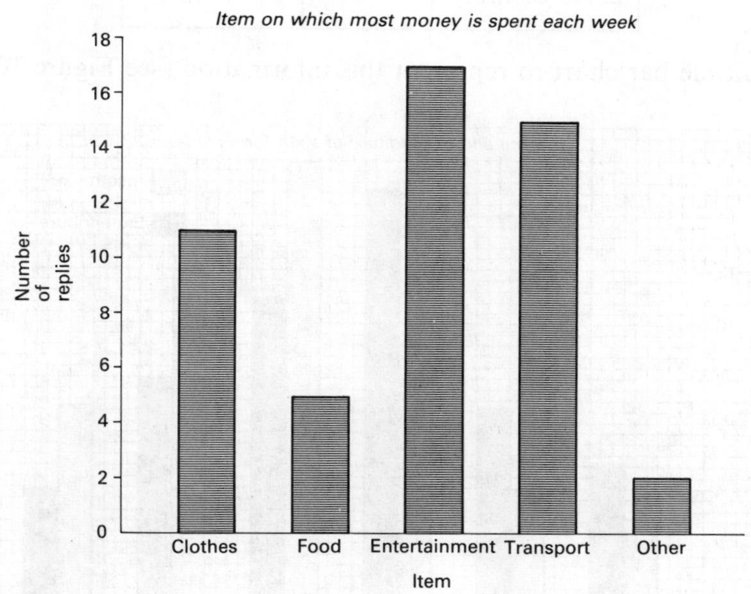

Figure 10.3

Although not as common, the bars can be drawn horizontally, with the classes marked on the vertical axis and the frequency along the horizontal axis.

(iii) Multiple bar charts

Two or more charts with the same classes can be compared side by side by drawing the bars for each class next to each other. Normally the bars would be shaded differently for each set of data. This technique is useful if we wish to compare *differences* between corresponding component values. Again the lengths of the bars represent the frequencies of the classes.

Worked example 10.6

Sam Pling's survey (Table 10.2) produced the following results for questions 5 and 6.

Table 10.6

£	Number of students	
	Income	Expenditure
0.00– 4.99	3	5
5.00– 9.99	4	7
10.00–14.99	8	15
15.00–19.99	16	11
20.00–24.99	12	8
25.00 or more	7	4

Draw a multiple bar chart to represent this information (see Figure 10.4).

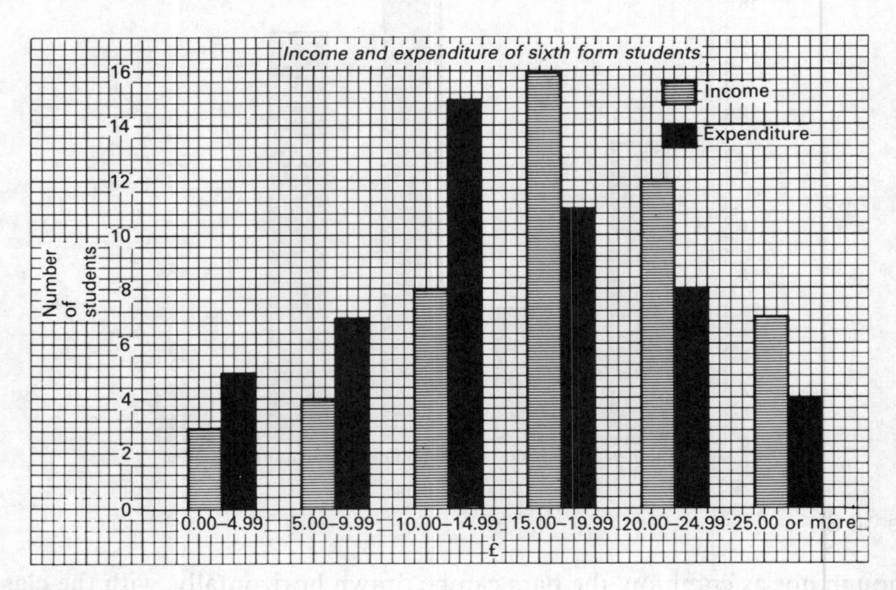

Figure 10.4

(iv) Component bar charts
Figure 10.5 shows an example of a *component bar chart*. The figure shows the number of students entering a Sixth form in a particular school for a period of four years. Each bar is broken down into three components, 1st Year Sixth, 2nd Year Sixth and 3rd Year Sixth.

The figure clearly shows not only changes in totals but also gives an indication of the size of each component.

Worked example 10.7
In Figure 10.5:

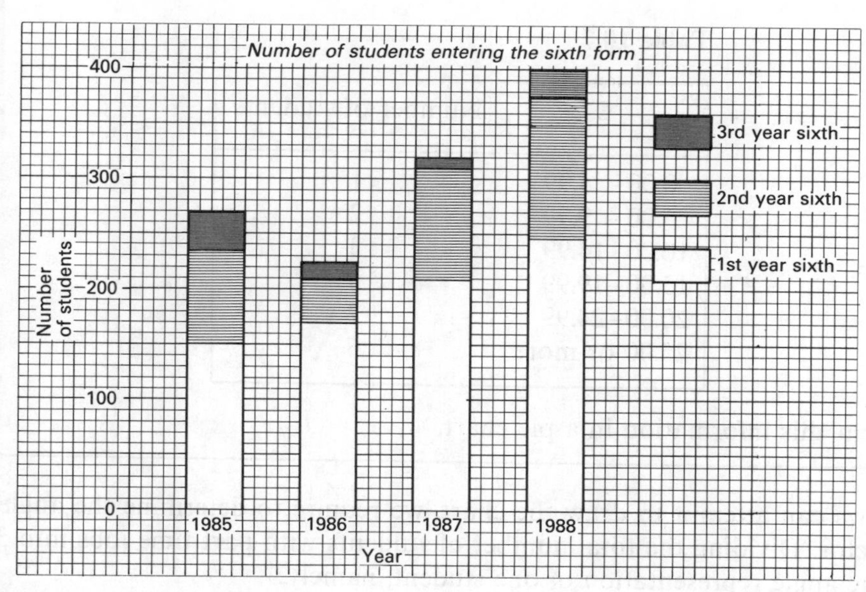

Figure 10.5

(i) What was the total number of 3rd Year Sixth students who entered the Sixth Form during this four year period?
(ii) What was the difference between the largest and the smallest entry of 2nd Year Sixth students during this period?

(i) The number of 3rd Year Sixth students during this period

$$= 35 + 15 + 10 + 25$$
$$= 85.$$

(ii) The year with the largest entry of 2nd Year Sixth students is 1988 with

$$375 - 245 \text{ students}$$
$$= 130 \text{ students.}$$

The year with the smallest entry of 2nd Year Sixth students is 1986 with

$$210 - 270 \text{ students}$$
$$= 40 \text{ students}$$

The difference is therefore $130 - 40$

$$= 90 \text{ students.}$$

(v) Pie charts

These are useful pictorial representations when we wish to show the *proportion* of each class to the whole. The chart is a circle divided into sectors and the area of each sector is proportional to the number in the corresponding class.

Worked example 10.8

Sam Pling's survey (Table 10.2) showed that 24 students in his sample had part-time jobs. The weekly income for these students was as follows:

Table 10.7

£	Number of students
0.00– 4.99	1
5.00– 9.99	2
10.00–14.99	3
15.00–19.99	5
20.00–24.99	7
25.00 or more	6

Represent this information in a pie chart.

Before we can attempt to draw the chart, we require to determine the angle for each sector. Dividing the total number of students with part-time jobs into 360° gives the angle representation of one student, namely:

$$360°/24 = 15°$$

Therefore, for 7 students the angle of the sector would be $7 \times 15° = 105°$.

Table 10.8 shows all the required angles and the chart can now be drawn (Figure 10.6).

Table 10.8

£	Number of students	Angle of sector
0.00– 4.99	1	15°
5.00– 9.99	2	30°
10.00–14.99	3	45°
15.00–19.99	5	75°
20.00–24.99	7	105°
25.00 or more	6	90°
Totals	24	360°

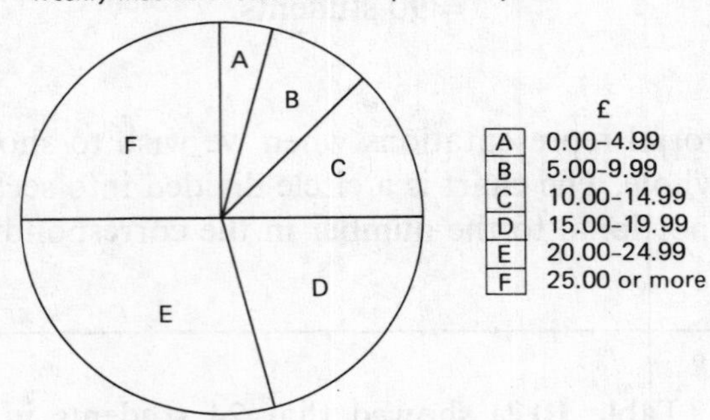

 Figure 10.6

Worked example 10.9
Trubshire police investigated the cause of 1800 accidents within the county and displayed their findings in the form of a pie chart (see Figure 10.7).

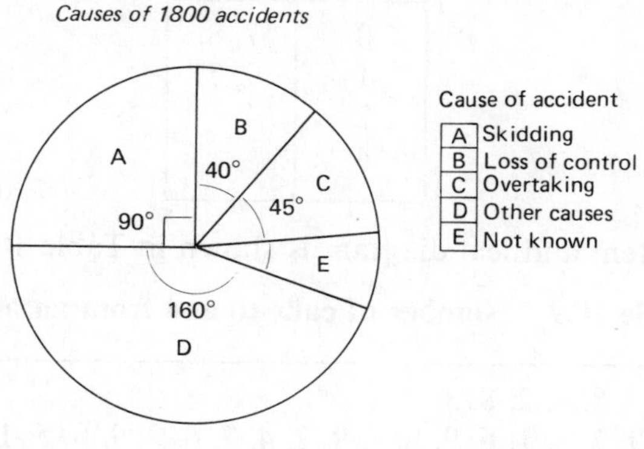

Causes of 1800 accidents

Figure 10.7

(i) How many accidents were due to loss of control?
(ii) In how many accidents was the cause unknown?

(i) The angle representing 'loss of control' = 40°
So the number of accidents due to this factor

$$= \frac{40^\circ}{360^\circ} \times 1800$$
$$= \frac{1}{9} \times 1800$$
$$= 200.$$

(ii) The angle representing 'unknown causes'

$$= 360^\circ - (90 + 40 + 45 + 160)^\circ$$
$$= 25^\circ$$

So the number of accidents $= \frac{25^\circ}{360^\circ} \times 1800$

$$= \frac{5}{72} \times 1800$$
$$= 125.$$

(vi) Stem and leaf diagrams

A stem and leaf diagram has the advantages of a bar chart, while at the same time it retains the detail of the original data. To give an example of a stem and leaf diagram, we will make use again of the data in worked example 10.3. Usually only 2 significant figures are shown, and in our example the stem of the diagram represents 'tens' and the leaves are the unit components of the data. So the first five data items

19 7 24 17 8

will be recorded as follows:

(tens)	(units)
0	7, 8
1	9, 7
2	4
4	

The completed, stem and leaf diagram is shown in Table 10.9.

Table 10.9 Number of calls to and from an office per day

0	7, 8, 4, 2, 8, 3, 7
1	9, 7, 8, 1, 6, 9, 8, 7, 9, 2, 4, 7, 6, 2, 9, 8, 5, 1, 9, 4
2	4, 1, 2, 3, 2, 4, 7, 9, 8, 6, 0, 1, 5, 6, 0, 4
3	2, 6, 1, 7, 1, 3, 0

(vii) Histograms

A *histogram* is the display of data in the form of a bar graph with no gaps between the classes. The bars in the histogram take the width of each class (called the class interval), and it is the *area* of each bar which is proportional to the number or frequency in that class. If the bars representing the frequencies for each class are the same width, then the heights of each bar in a histogram are proportional to the number or frequency in that class, and the histogram can be interpreted in the same way as a bar chart.

Worked example 10.10

Draw a histogram of the data represented in worked example 10.3 using the class intervals 1–5, 6–10,..., etc.

Each class is the same width so the bars in the histogram will be of equal width. Rather than use the starting and finishing points of each class as a label, the horizontal axis is labelled numerically on a continuous scale. We have chosen the *mid-class values* of 3, 8, 13,..., (see Figure 10.8).

By joining the mid-points of the tops of the bars of the histogram you can produce a *frequency polygon*, and this also has been drawn on Figure 10.8.

Worked example 10.11

Sam Pling requires to draw a histogram using the data obtained from the response to question 5 on the questionnaire (see Table 10.6).

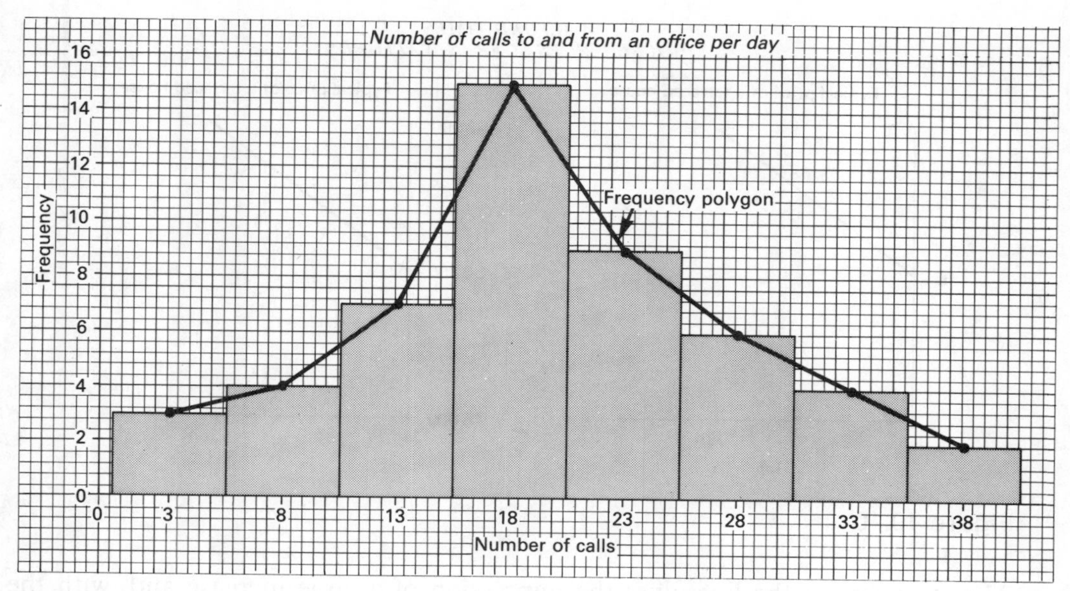

Figure 10.8

Before the histogram can be drawn, Sam has to decide upon a top limit to the last class. Without the actual figures, he must allocate a realistic value and he chooses £40.00. The last class is now 3 times wider than the others. The height of this last bar should therefore be 1/3rd of what it would be if a bar graph was to be drawn. The resulting histogram is shown in Figure 10.9.

Note that the vertical axis is labelled *frequency density*.

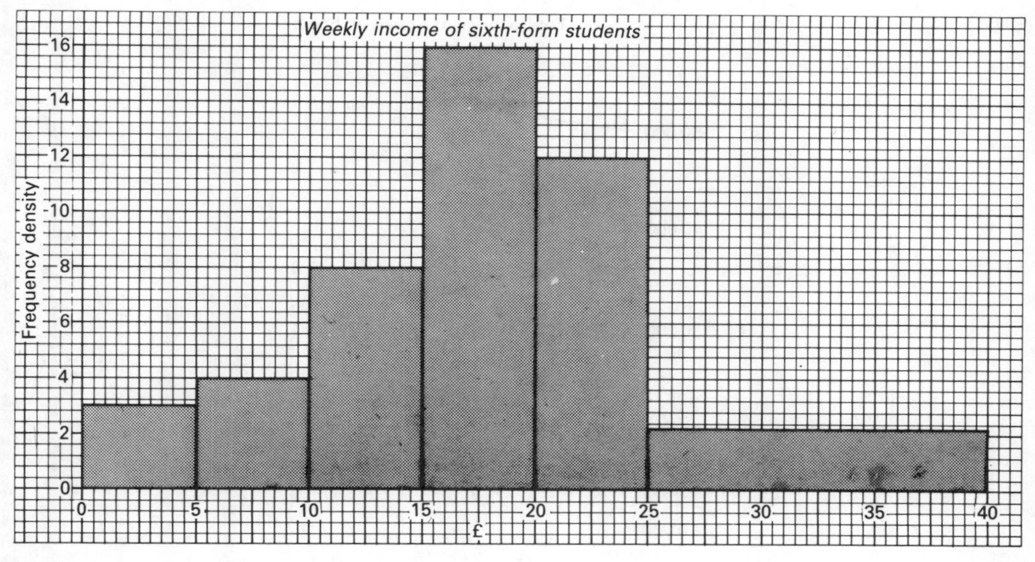

Figure 10.9

10.8 Misleading diagrams

Statistical diagrams are occasionally used to mislead people. Look at Figure 10.10(a).

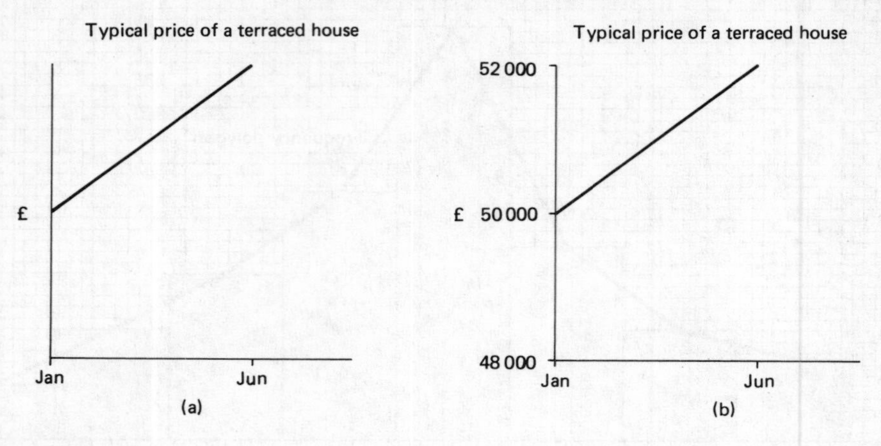

Figure 10.10

The steepness of the line gives the impression of a large increase and, with the absence of a vertical scale, it looks as if house prices have doubled in the six month period. Figure 10.10(b) is the same diagram but with a scale added to the vertical axis. This is only slightly better than the first diagram, as the altering of the base line (the starting value of the vertical axis) hardly detracts from the visual impression of the sloping line. The real story can be told in an accurate representation of the data as shown in Figure 10.11. This however is hardly a convincing justification of the *Walford Times* headline banner.

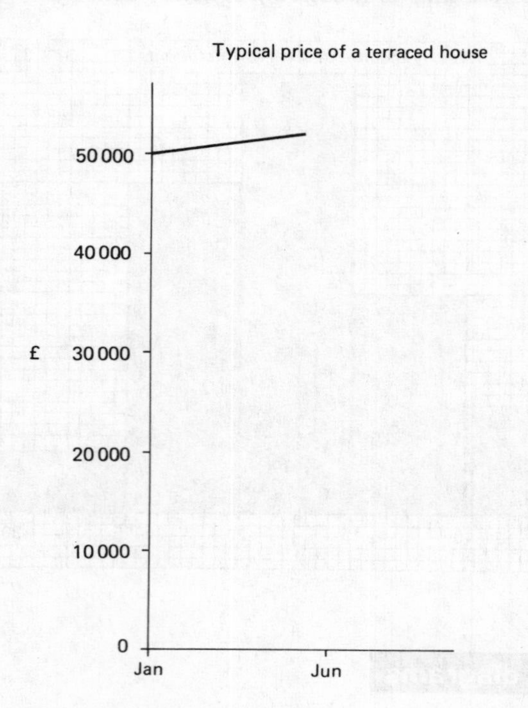

Figure 10.11

Another method that is sometimes used to misrepresent data is the drawing of three-dimensional figures to represent the information. Even though the correct figures may be quoted under the diagram, it is the diagram which catches the reader's eye.

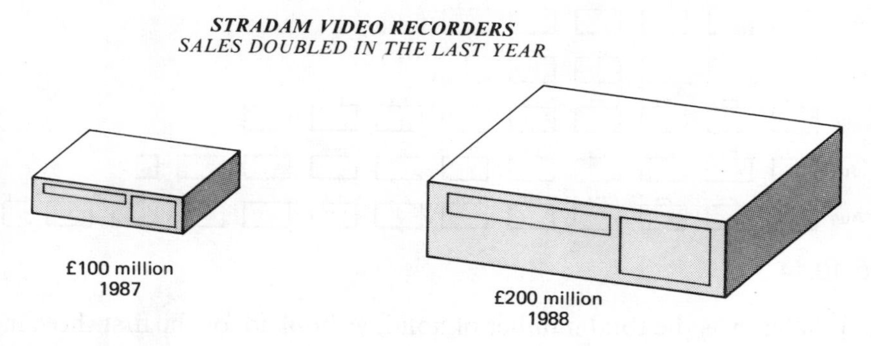

Figure 10.12

Figure 10.12 is such a misrepresentation. The sale of Stradam Videos may have doubled in the last year, but the *volume* of the video on the right is not twice the volume of the video on the left. You can see by measurement that all *lengths* on the larger video are twice the corresponding lengths of the smaller video. However the volume of the larger video is 8 times the volume of the smaller video. The numerical figures show a true 2-fold increase, whereas the diagrams reflect an 8-fold increase. Unfortunately it is often the diagrams that the reader is compelled to compare.

Exercise 10b

(1) Sheila did a survey at the local Sports centre. She displayed her results in the pictogram of Figure 10.13.

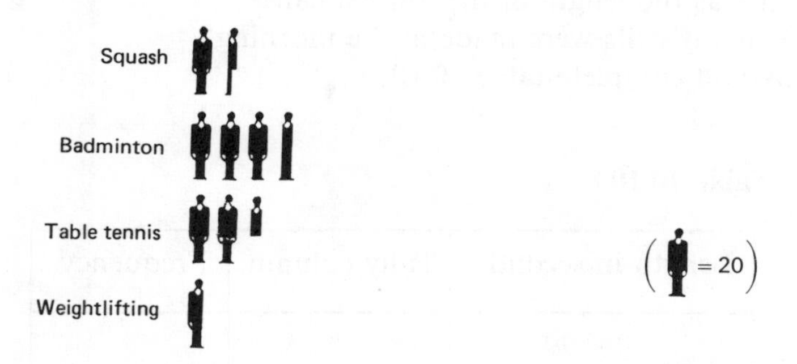

Figure 10.13

(a) Is this a useful pictogram? Give a reason for your answer.
(b) What other diagram could you use to represent this data? (LEAG)

(2) Figure 10.14 illustrates the numbers of holidays booked through a travel agency for each of the first 8 months of a year.
(a) How many holidays were booked for March?
(b) How many holidays were booked for April?

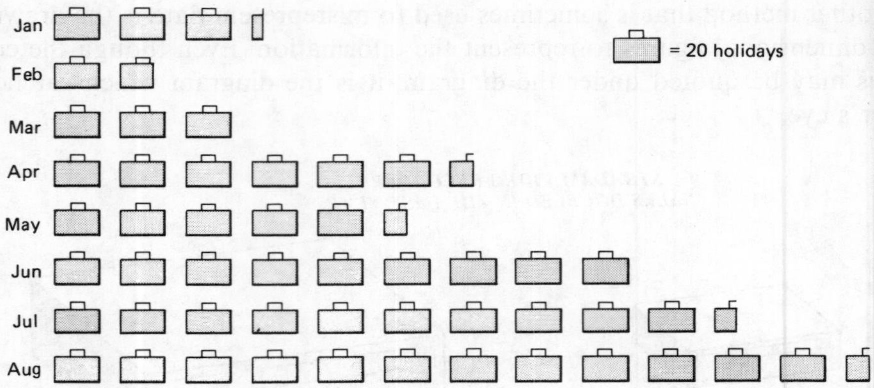

Jan
Feb
Mar
Apr
May
Jun
Jul
Aug
= 20 holidays

Figure 10.14

(c) What was the total number of holidays booked for the first three months of the year?

(d) 165 holidays were booked for September. On squared paper draw the line of symbols that should be added to the diagram for September.

(3) A company timed and noted down the length of telephone calls made in a morning.

Length of calls in seconds

342	46	427	202	449	221	468	538	275	182
356	520	185	430	140	375	225	278	170	180
321	268	303	369	282	263	254	93	402	307
87	193	146	273	385	227	410	120	311	291
342	362	153	402	318	197	135	339	251	179
85	223	349	356	382	314	385	401	463	510

(a) What was the length of the shortest call?

(b) What was the length of the longest call?

(c) How many calls were made in the morning?

(d) Copy and complete table 10.10.

Table 10.10

Length in seconds	Tally column	Frequency
0– 99		
100–199		
200–299		
300–399		
400–499		
500–599		
	Total	

(e) How many calls were longer than 399.5 seconds?
(f) What percentage of the calls were longer than 399.5 seconds?

(LEAG)

(4) Classify the following into either discrete or continuous data:
(i) number of days in a year,
(ii) daily temperatures for the month of August,
(iii) number of words in a sentence,
(iv) speed of motorists on a motorway,
(v) number of 'Smarties' in a tube,
(vi) gate receipts at all the concerts in one year at the Wembley Arena,
(vii) time taken by athletes to run 100 m in an Olympic final.

(5) An analysis of the frequency of the use of the vowels in the paragraph below was undertaken. The tally chart shows this analysis as far as the end of the first sentence. (Up to '... and character.')

Paragraph for analysis:

'Sam Peckinpah's second film *Guns in the Afternoon* is a gentle knockout, acutely sensitive to landscape and character. Pleasures include Lucien Ballard's beautiful photography.'

Table 10.11

Vowel	Number of times used	Totals
A	~~1111~~ ~~1111~~	
E	~~1111~~ ~~1111~~ 1	
I	~~1111~~ 1	
O	~~1111~~ 1	
U	111	

(a) Copy and complete the tally chart of Table 10.11 by inserting the data for the last sentence and filling the totals.
(b) How many E's are there in the paragraph?
(c) How many vowels are there in the paragraph?
(d) Draw a bar chart on graph paper to illustrate the frequency of the use of the various vowels in the paragraph. (SEG)

(6) Sales of saloon cars and hatchbacks by a certain firm for the years 1980 to 1983 are shown in Figure 10.15.
(a) How many cars did the firm sell altogether in 1981?
(b) (i) In which year were most hatchbacks sold?
(ii) How many were sold that year?

(7) When it opened for business in January, Tesburys hypermarket appointed 90 staff for part-time work, 79 of whom were male. By June, 17 of these part time-staff had left, 13 of whom were male. There were, however, 13 new part-time staff appointed of whom 3 were female. Draw a component bar chart for the months of January and June showing the number of part-time male and part-time female staff employed by Tesburys at these times.

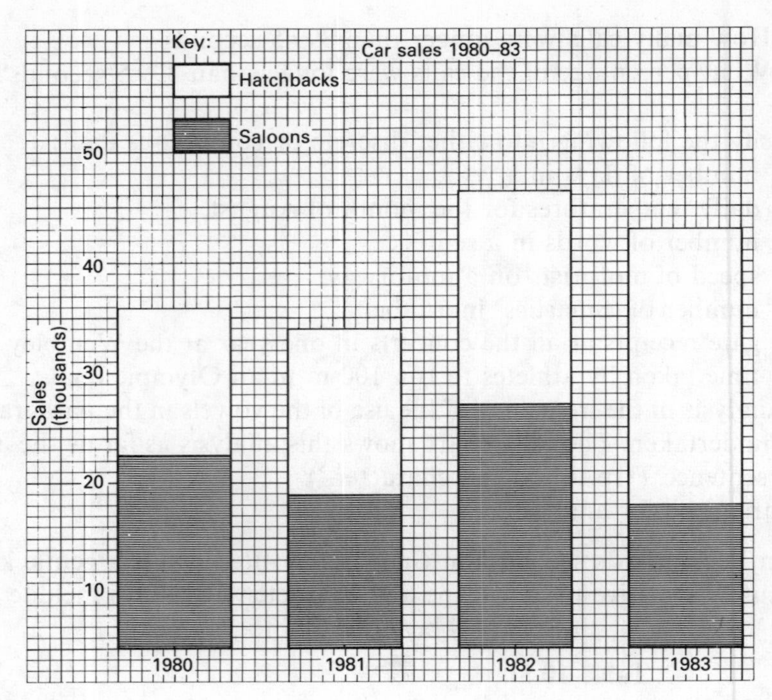

Figure 10.15

(8) The chart of Figure 10.16 gives information about road accidents in Belfast in 1982 and 1983.

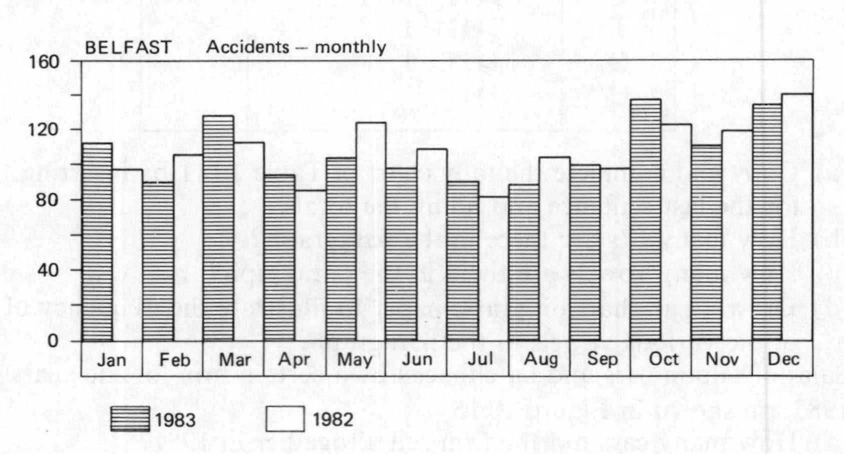

Figure 10.16

(a) To the nearest 10, how many accidents occurred in January 1983?
(b) In which month in 1982 did the least number of accidents occur?
(c) In which month in the 2 year period did the greatest number of accidents occur?
(d) In which months was there least difference between the 1982 and 1983 accident figures? (NI)

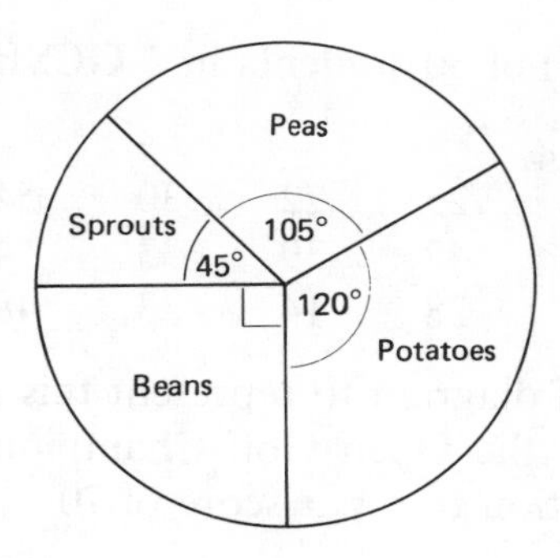

Figure 10.17

(9) A market gardener has 24 acres of land and uses it to grow peas, potatoes, beans and sprouts.

The pie chart (Figure 10.17) shows how the land is divided.

(a) How many acres are used to grow potatoes?

(b) What fraction of the land is used to grow sprouts? (MEG)

(10) The placement of students who left Varnworth Sixth-Form College in June 1988 is given by the table below:

University	24%
Polytechnic	38%
College of Further Education	12%

(i) If 60 students went to University, how many students left the College in June 1988?

(ii) If 55 students took up full-time employment, what percentage of student leavers did this represent?

(iii) If the remainder were unemployed at the time of the survey, how many students were in this category?

(iv) Draw a pie chart to represent this information.

(11) A pie chart is to be drawn to represent the information in Table 10.12.

Table 10.12 Shows where the fat we eat comes from

FOOD	Dairy foods	Meat and Meat products	Oils and Fats	Cake, Biscuits and Pastries	Other
% ANGLE	30	27	25	6	12

(a) Copy and complete the table to show the size of each angle, to the nearest degree, that will need to be drawn.

(b) Draw the pie chart. Label each sector. (LEAG)

(12) The marks (out of 60) of 30 students in a GCSE mathematics test were as follows:

35	22	49	25	22	30	54	25	28	42
45	27	29	42	39	33	14	46	29	40
25	35	11	28	14	3	48	7	27	30

Draw a stem and leaf diagram to represent this information.

(13) In the final round of the Open Golf Championship, the following scores were recorded in relation to a par score of 70.

Scores above or below par in the final round

less than −4	−4 to −2	−1 to +1	+2 to +4	+5 to +7	+8 to +10	more than 10
0	3	12	18	11	6	0

Draw a histogram of this information.

(14) The Sixth-Form students in a rural school were surveyed to determine how far they had to travel to school each day. The results of the survey were as follows:

Distance	*Number of students*
Under 1 mile	36
1–under 3 miles	27
3–under 5 miles	45
5–under 10 miles	12
10 miles or more	0

Construct a histogram to represent this information. Add a frequency polygon to your diagram.

(15) Study the graph shown in Figure 10.18.

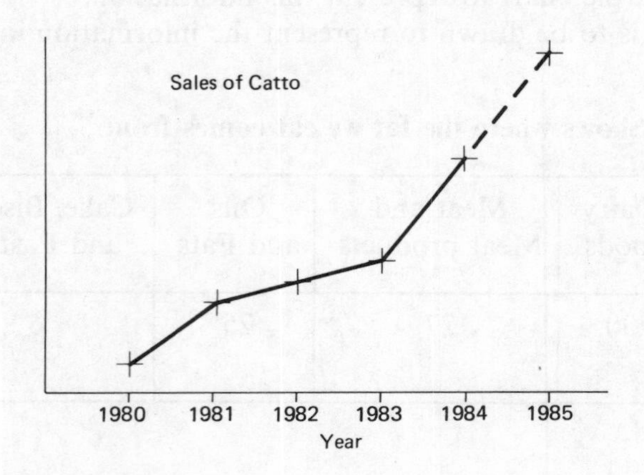

Figure 10.18

Write down two examples of bad practice in this graph. (MEG)

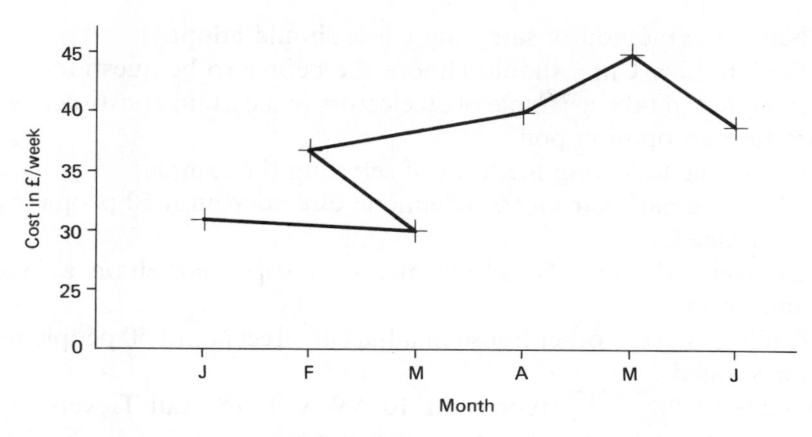

Figure 10.19

(16) Write down three deliberate mistakes, or examples of bad practice, used in Figure 10.19—a graph of rainfall for the first six months of a year.

(MEG)

Suggestions for coursework 10

1. Examine Radio and TV programme schedules and investigate the time devoted to various types of programme.
2. Investigate the way that data is presented visually in the media.
3. Carry out statistical investigations in your school or college to show different methods of choosing a random sample of students.
4. For a given field or plot of land, investigate a way of determining an estimate of the number of a particular variety of plant growing in the area.
5. Investigate the number of heads and tails obtained with the throwing of 2, 3, 4, 5,..., coins a number of times.

Miscellaneous exercise 10

1. A national survey on television viewing habits is to be conducted. Comment briefly on the bias in the sample:

   'People whose names are on the Electoral Register'
2. Give a reason why the following question is unsuitable for inclusion in a questionnaire:
   'Do you agree that stiffer prison sentences are the only cure for the present excess of football hooliganism?'
3. Chris has been employed by his Local Council to carry out a survey in his area. The survey involves obtaining responses to questions from the following age groups:

   18–26, 27–36, 37–46, 47–64, 65 and over

   Chris is provided with the ages of all the people in his area on the Electoral Roll.

(a) Name the method of sampling Chris should adopt.
(b) Explain how Chris should choose the people to be questioned. (LEAG)

4. It is intended to take a sample of 50 electors in a certain constituency in order to produce an opinion poll.
 Criticise the following methods of selecting the sample:
 (a) Choosing names from the telephone directory until 50 people have been questioned.
 (b) Choosing the first 50 adults met in a supermarket on a Wednesday afternoon.
 (c) Calling at every other house in adjacent streets until 50 people have been questioned. (SEG)

5. In Figure 10.20 [ticket symbol] represents 10 AWAYDAY Rail Tickets to London sold at Oldtown Station during a typical week.

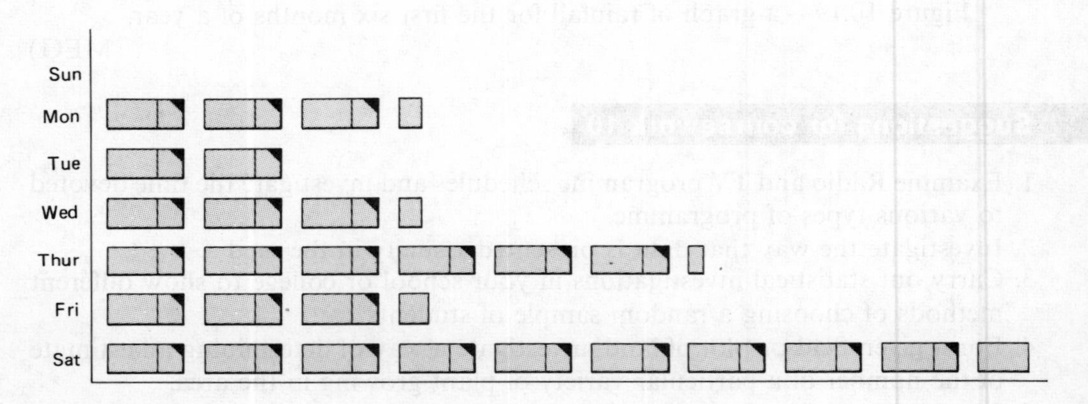

Figure 10.20

 (a) Calculate the total number of tickets sold.
 (b) If the AWAYDAY fare is £3.70 on Saturday and Sunday and £5.40 on all other days, calculate the total amount received by British Rail during the week for these tickets.
 (c) It is decided that the same fare will be charged on all days. What should that fare be if, with the same number of people travelling, British Rail wishes to receive the same total amount of money? (Give your answer to the nearest penny.)
 (d) Why do you think that no tickets were sold on Sunday?
 (e) Which day of the week do you think is early closing day in Oldtown? Give a reason for your answer.

6. The number of newspapers and magazines sold each day from Monday to Friday by a newsagent are shown in Figure 10.21.
 (a) How many newspapers were sold altogether?
 (b) On which two days were the number of magazines sold the same?
 (c) On which day was the least number of magazines sold? (SEG)

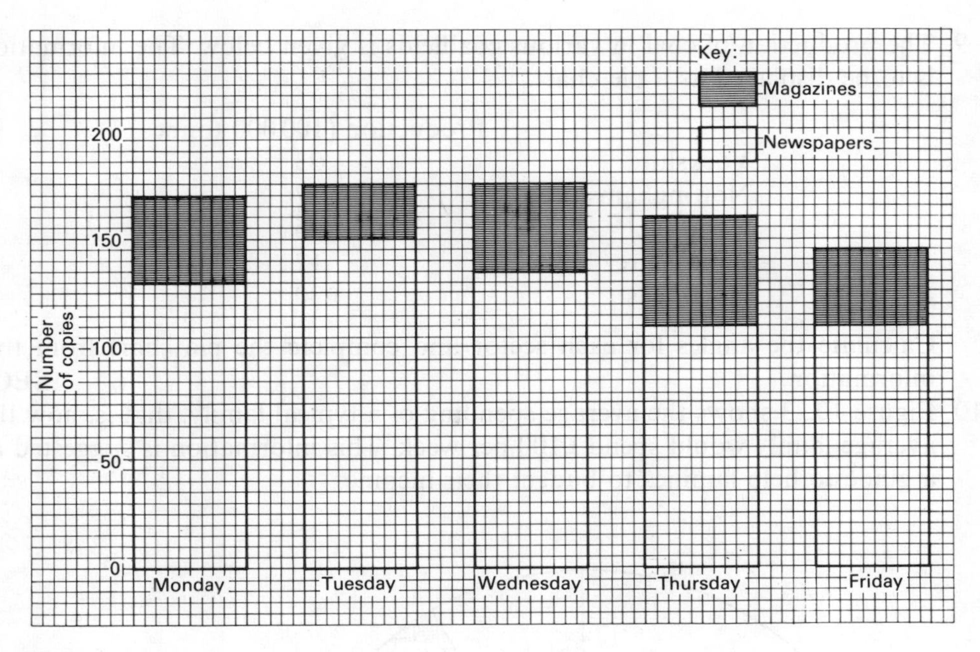

Figure 10.21

7. State which of the following are continuous, as opposed to discrete, variables:
   (a) the temperature in a room,
   (b) the number of people in a room,
   (c) the height of trees measured to the nearest metre,
   (d) the speed of a car. (SEG)
8. 72 students following a sixth-form course have to opt for one of four additional activities. Their choices are illustrated by the bar chart of Figure 10.22.

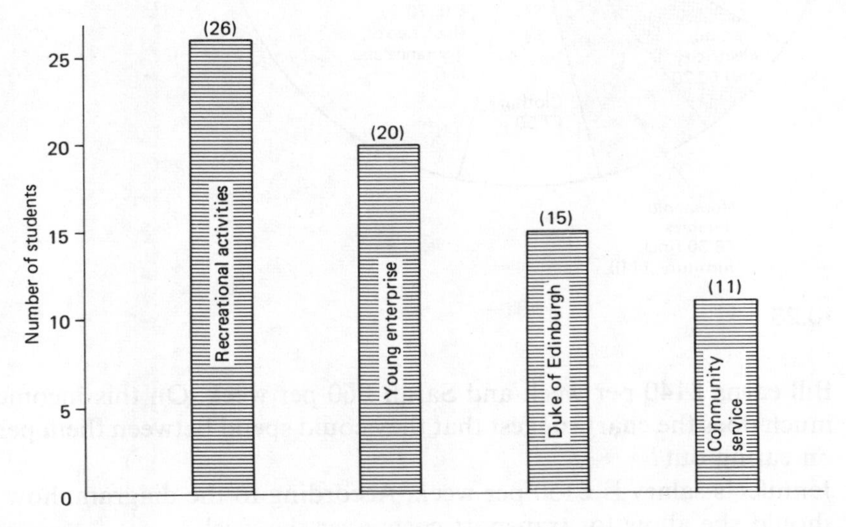

Figure 10.22

Illustrate this information on a clearly labelled pie chart.

9. The production of coal in various coalfields is given below. The information is to be displayed as a pie chart.

	Production (100 000 tonnes)
Yorkshire	33
Nottinghamshire	27
Kent	6
Lancashire	12
North-east	30

Calculate the angles for each sector and complete the pie chart from this information. (SEG)

10. Figure 10.23 shows the average spending of a typical family, that is, how the average family would spend £100 per week. This information is suggested as a guide to help families to budget their income.

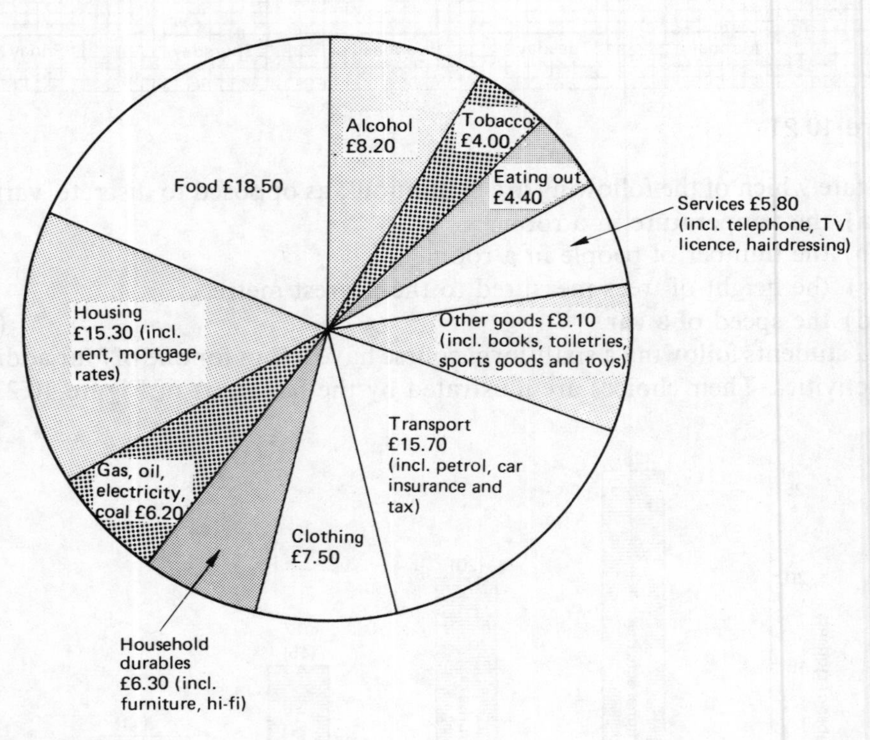

Figure 10.23

(a) Bill earns £140 per week and Sarah £60 per week. On this income, how much does the chart suggest that they could spend between them per week on eating out?

(b) Jennifer's salary is £150 per week. According to the diagram, how much should she allow for transport costs over the week?

(c) Derek receives only £60 per week. Find out how much the chart indicates that he might spend on food per week. (NEA)

11. The stem and leaf table (Table 10.13) shows the times, measured to the nearest second, taken by 20 students to complete an algebraic problem.

Table 10.13

Time (seconds)	
4	3, 4, 7
5	1, 1, 4, 4, 5
6	2, 6, 8, 8, 8, 9
7	0, 3, 8, 9
8	5, 5

(a) What was the shortest time taken to complete the problem?
(b) Another student took 5 seconds less than the longest time. Explain where this student's time would be entered on the diagram and in what form. (MEG)

12. The speeds of 200 cars passing a certain point subject to an 80 km/h speed limit were checked by the police. They are as shown in the table below:

Speeds (km/h)	0–55	55–60	60–65	65–70	70–75	75–80
Frequency	0	1	4	8	15	20

Speeds (km/h)	80–85	85–90	90–95	95–100	100–105	105–110
Frequency	35	50	40	21	5	1

(a) On graph paper, draw a histogram representing this information.
(b) State, with a reason, whether or not the majority of motorists broke the law. (SEG)

13. The histogram of Figure 10.24 represents the amount spent (in pounds) on Christmas presents by a group of families.
(a) Insert the scale on the frequency density axis given that there are 100 families in the group.
(b) From the histogram find:
(i) the number spending between £60 and £80,
(ii) the number spending less than £140,
(iii) the number spending more than £100. (SEG)

14. Comment critically on the advertisement for cheese, shown in Figure 10.25. (SEG)

15. The three diagrams of Figure 10.26 are intended to represent the sales of a product for each of the years 1980 to 1983.

State which diagram you think represents this information most fairly and list any criticisms you have of the other diagrams. (SEG)

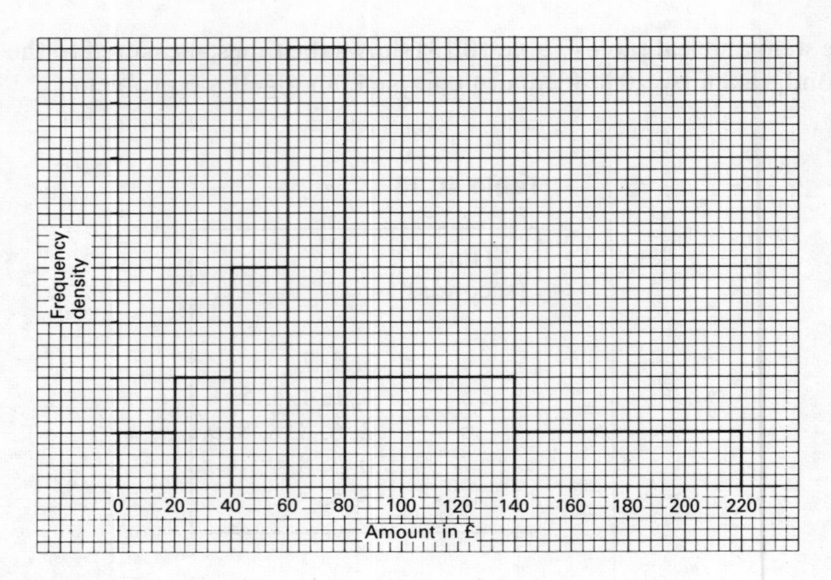

Figure 10.24

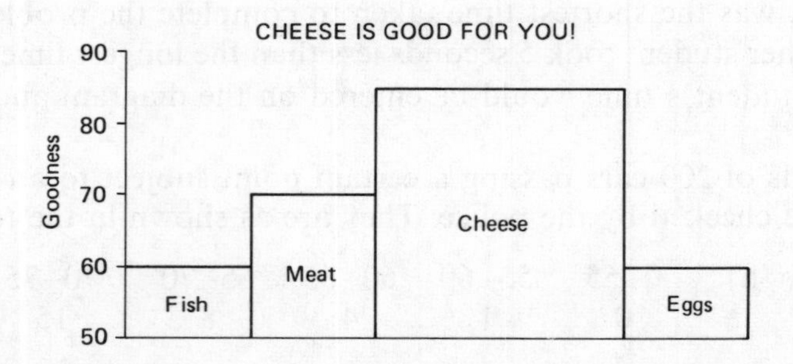

Figure 10.25

16. The receipts from the sale of four different types of fuel at a garage over a month are represented by the diagrams of Figure 10.27:
   1. A bar chart.
   2 and 3. Frequency polygons.

   Comment critically on each of these diagrams. (SEG)

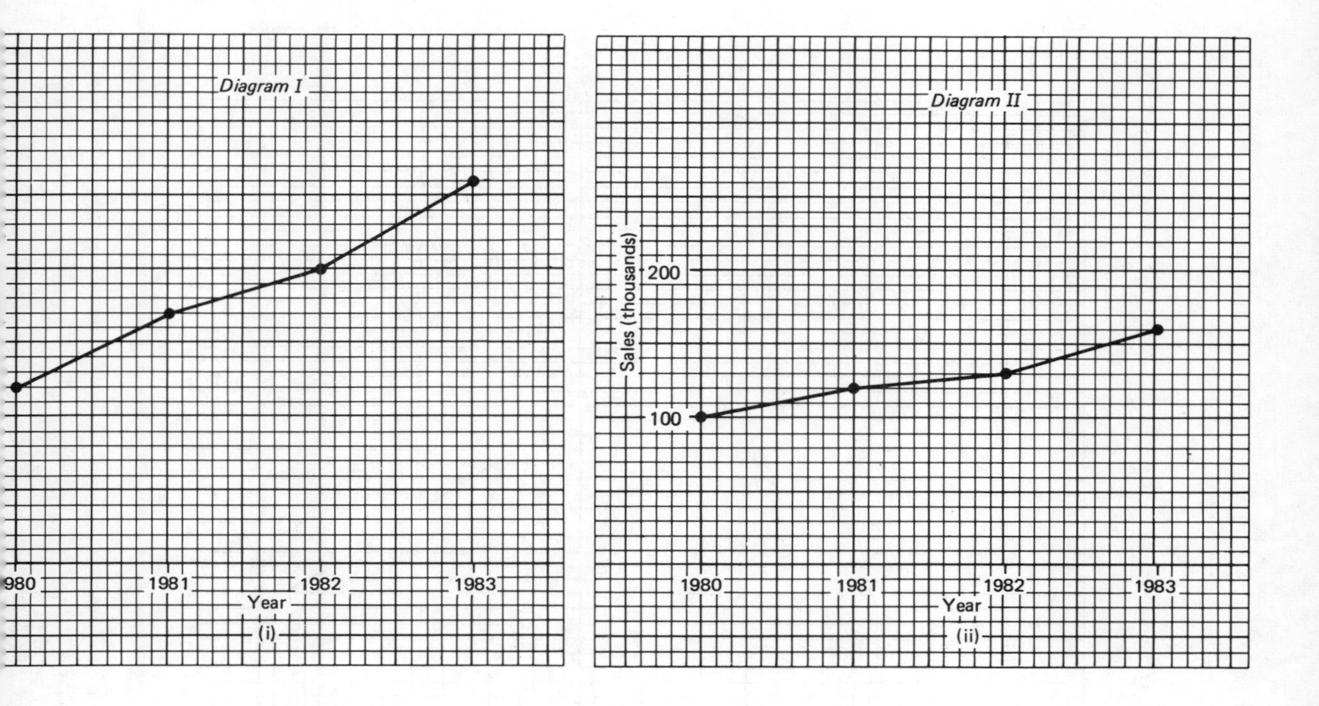
Diagram I
980
1981
1982
1983
Year
(i)
Diagram II
Sales (thousands)
200
100
1980
1981
1982
1983
Year
(ii)

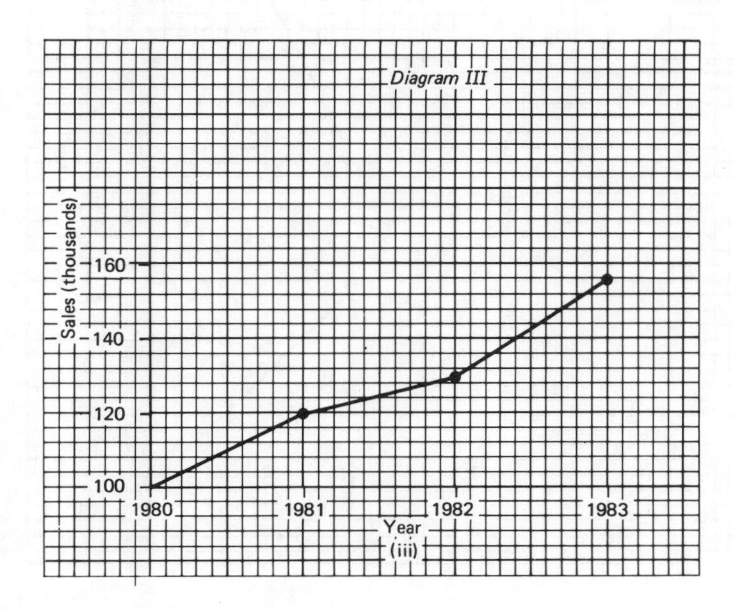
Diagram III
Sales (thousands)
160
140
120
100
1980
1981
1982
1983
Year
(iii)

Figure 10.26

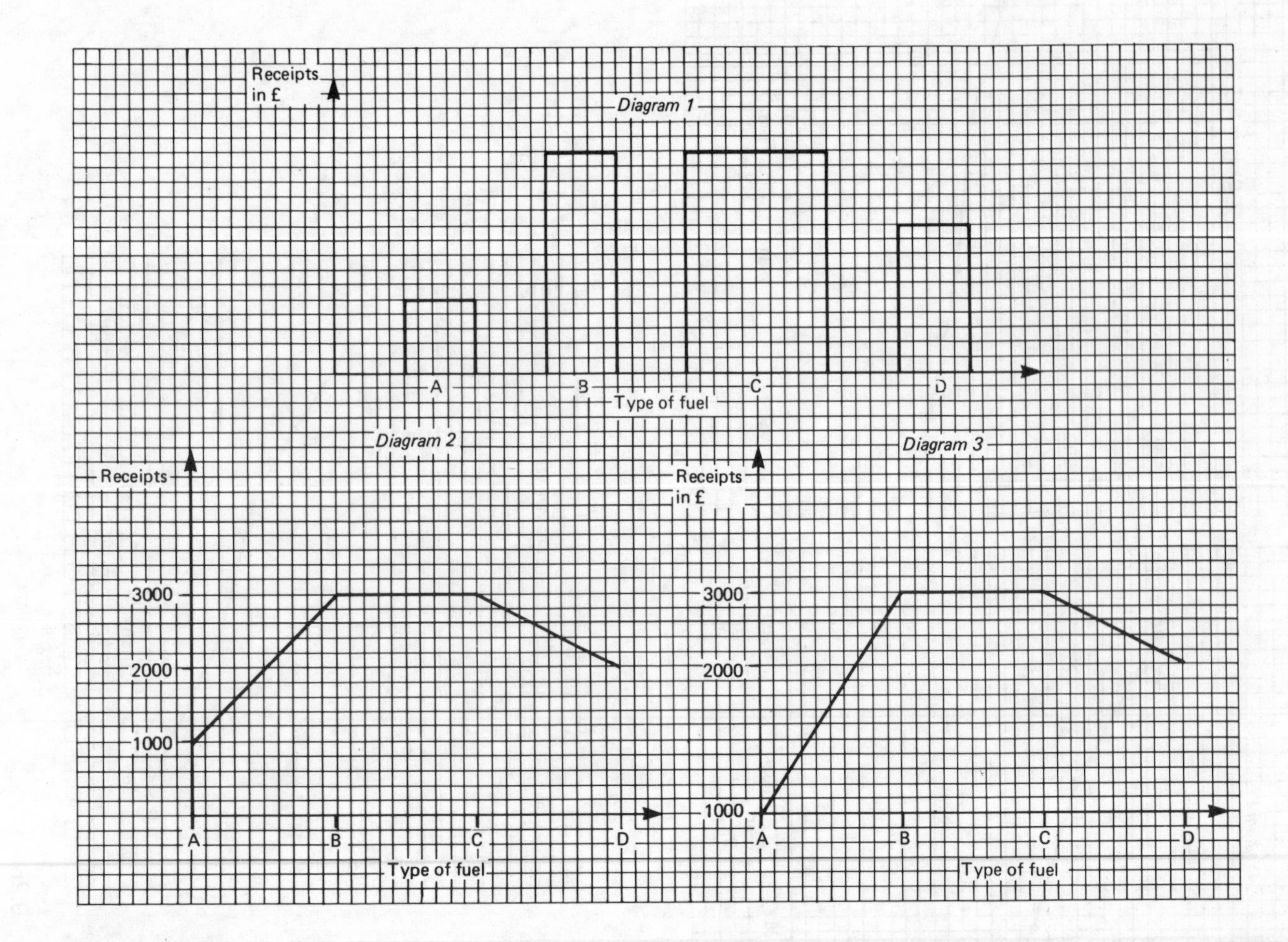
Diagram 1
Receipts
in £
A
B
C
D
Type of fuel
Diagram 2
Receipts
3000
2000
1000
A
B
C
D
Type of fuel
Diagram 3
Receipts
in £
3000
2000
1000
A
B
C
D
Type of fuel

Figure 10.27

11 ALGEBRA 2

If you are taking a higher level paper, you will need more algebra than was covered in Unit 2. The work in this unit should enable you to cope with any of the situations you might meet.

11.1 Harder substitution

If you are taking the higher level papers, it is likely that the formulae you will use become more complicated. Provided you do not invent your own rules, you should be able to master the problems that arise. The following example shows you what to look out for.

Worked example 11.1

In the following questions, take $a = -4$, $b = \frac{1}{2}$, $c = -2.4$, $d = 0$, $e = 3.6$, $f = -2\frac{1}{2}$.
In each example, use the given formula to find the value of x.

(i) $x = 4a^2 + 5$ (ii) $x = 3abc$

(iii) $x = 2f^2 + 1$ (iv) $x = 2ab + 3cd$

(v) $x = a^2 + b^2 + c^2$ (vi) $x = \dfrac{ab}{c^2}$

(vii) $x = \sqrt{\dfrac{eb}{a^2}}$ (viii) $x = 2\sqrt{\dfrac{b}{e}}$

(i) Remember $4a^2$ means $4 \times a^2$, not $(4a)^2$.
So $x = 4 \times (-4)^2 + 5$.
Negative $\times$ negative is positive, so $(-4)^2 = 16$.
Hence $x = 4 \times 16 + 5 = 64 + 5 = 69$.

(ii) $3abc$ means $3 \times a \times b \times c$.
So $x = 3 \times -4 \times \frac{1}{2} \times -2.4 = -12 \times \frac{1}{2} \times -2.4 = -6 \times -2.4 = 14.4$.
If the calculation is done on the calculator, proceed as follows:

3 [×] 4  [×] 0.5  2.4  [=] Display [14.4]

(0.5) or 1 [$a^{b/c}$] 2 if you have a fraction button

(iii) $x = 2 \times (-2\frac{1}{2})^2 + 1$.

Now $(-2\frac{1}{2})^2 = -2\frac{1}{2} \times -2\frac{1}{2} = -\frac{5}{2} \times -\frac{5}{2} = \frac{25}{4}$.

So $x = 2 \times \frac{25}{4} + 1 = \frac{25}{2} + 1 = 12\frac{1}{2} + 1 = 13\frac{1}{2}$.

On the calculator:

								Display
2	×	2.5	±	x^2	+	1	=	13.5

(iv) $x = 2 \times -4 \times \frac{1}{2} + 3 \times -2.4 \times 0 = 2 \times -4 \times \frac{1}{2} + 0 = -8 \times \frac{1}{2} = -4$.

(v) $x = (-4)^2 + (\frac{1}{2})^2 + (-2.4)^2 = 16 + 0.25 + 5.76[(\frac{1}{2})^2 = \frac{1}{2} \times \frac{1}{2} = \frac{1}{4} = 0.25]$
$= 22.01$.

(vi) $x = (-4 \times \frac{1}{2}) \div (-2.4)^2 = -2 \div 5.76 = -0.347$ (3 sig. figs.).

(vii) $x = \sqrt{\frac{3.6 \times \frac{1}{2}}{(-4)^2}} = \sqrt{\frac{1.8}{16}} = 0.335$ by calculator.

(viii) $x = 2\sqrt{\frac{\frac{1}{2}}{3.6}} = 2\sqrt{\frac{0.5}{3.6}} = 2\sqrt{0.139} = 0.745$.

Exercise 11a

In the following examples, take $a = 1.2$, $b = 0.4$, $c = -3$, $d = \frac{1}{4}$, $e = -3.5$. In each case, find the value of x. Use a calculator where necessary.

(1) $x = 2a + 4b$ (2) $x = 4abc$ (3) $x = e^2 + d^2$

(4) $x = \sqrt{a^2 - b^2}$ (5) $x = \frac{ab}{cd}$ (6) $x = e^2 - c^2$

(7) $x = \sqrt{\frac{ab}{d}}$ (8) $x = 2b^2 + 4$ (9) $x = 3e^2 - 2b$

(10) $x = a^2 + b^2 + c^2$ (11) $x = \sqrt{4ab}$ (12) $x = 3c^3$

(13) $x = b^2 d$ (14) $x = 4a^2 - 3b^2$ (15) $x = 5ab^2 c$

11.2 Further linear equations

In Unit 2, we dealt with simple linear equations. In this section, we extend the work to include equations containing fractions.

(i) $\frac{1}{2}x + \frac{2}{5}x = 4$.

The lowest common multiple of the bottom line is 10. Hence we multiply the equation term by term, by 10.

$$\cancel{10}^5 \times \frac{1}{\cancel{2}_1}x + \cancel{10}^2 \times \frac{2}{\cancel{5}_1}x = 10 \times 4$$

This simplifies to

$$5x + 4x = 40, \text{ i.e. } 9x = 40$$

Hence $x = \frac{40}{9} = 4\frac{4}{9}$

(ii) $\frac{2}{3}+\frac{(x+1)}{2}=1\frac{1}{4}$.

The lowest common multiple of the bottom line is 12. Hence we multiply the equation term by term by 12.

$$\cancel{12}^{4}\times\frac{2}{\cancel{3}_{1}}+\cancel{12}^{6}\times\frac{(x+1)}{\cancel{2}_{1}}=12\times 1\tfrac{1}{4}$$

Hence $8+6(x+1)=15$.
Remove the bracket:

$$8+6x+6=15$$

Simplify:
$$6x+14=15.$$

Hence
$$6x=15-14=1 \qquad \text{So } x=\tfrac{1}{6}.$$

(iii) $\frac{2}{x+4}=\frac{3}{x}$.

This example is harder, but the principles are the same. The lowest common multiple of the bottom line is $x(x+4)$.
Hence

$$x\overset{1}{\cancel{(x+4)}}\times\frac{2}{\underset{1}{\cancel{(x+4)}}}=\overset{1}{\cancel{x}}(x+4)\times\frac{3}{\underset{1}{\cancel{x}}}$$

so

$$2x=3(x+4)=3x+12$$

$$2x-3x=12$$

hence
$$x=-12.$$

Exercise 11b

Solve the following equations:

(1) $\frac{1}{2}x+\frac{1}{3}x=5$ (2) $\frac{1}{4}x+\frac{2}{3}x=11$

(3) $\frac{2}{3}x+\frac{(x+1)}{2}=11$ (4) $5x+\frac{2}{3}(x-1)=22$

(5) $\frac{1}{5}x-\frac{1}{2}=\frac{2}{3}x-14\frac{1}{2}$ (6) $\frac{x+1}{2}+\frac{x-1}{2}=6\frac{3}{4}$

(7) $\frac{3}{5}x+4(x+1)=6\frac{3}{10}$ (8) $\frac{2}{5}x+\frac{3}{5}(x+2)=4\frac{1}{5}$

(9) $\frac{2}{x}=\frac{3}{(x+4)}$ (10) $\frac{2}{x}=\frac{1}{(x-3)}$

(11) $\frac{4}{(x-5)}=\frac{5}{(x+1)}$ (12) $\frac{3}{(x-5)}=\frac{1}{(4x+1)}$

11.3 Further rearrangement of formulae

As with the previous section on equations, we will not extend the work on changing the subject of a formula to include fractions and the type where the new subject appears in more than one place.

Worked example 11.2

Make x the new subject of the following equations:

(i) $M = \dfrac{bc}{x}$ (ii) $t = \dfrac{a}{x+y}$

(iii) $q = \dfrac{4x}{t} + y$ (iv) $x = 4q + tx$ [Note: x appears twice.]

(i) $\times x$: $Mx = \dfrac{bc}{\cancel{x}1} \times \cancel{x}^1$ i.e. $Mx = bc$ $\div M$: $x = \dfrac{bc}{M}$

(ii) $\times (x+y)$: $t(x+y) = \dfrac{a}{\cancel{(x+y)}1} \times \cancel{(x+y)}^1 = a$

So $t(x+y) = a$

Remove brackets: $tx + ty = a$

$-ty$ from each side: $tx = a - ty$

$\div t$: $x = \dfrac{a - ty}{t}$

(iii) $\times t$: $qt = \dfrac{4x}{\cancel{t}1} \times \cancel{t}^1 + yt$

So $qt = 4x + yt$

$-yt$ from each side: $qt - yt = 4x$ i.e. $4x = qt - yt$

$\div 4$: $x = \dfrac{qt - yt}{4}$

(iv) $x = 4q + tx$

$-tx$ from each side: $x - tx = 4q$

factorise: $x(1-t) = 4q$ This always happens if the new subject appears twice.

$\div(1-t)$: $x = \dfrac{4q}{(1-t)}$

Exercise 11c

Make x the new subject of the following formulae:

(1) $t = \dfrac{4x}{y}$ (2) $N = \dfrac{tx}{2y}$ (3) $p = \dfrac{3}{x+1}$

(4) $r = \dfrac{3x}{y} + z$ (5) $y = ax + t - qx$

(6) $y = \dfrac{a+x}{b+x}$ (7) $N = \dfrac{x}{2} + \dfrac{y}{4}$

(8) $z = \frac{t^2 x}{5}$ (9) $T = \frac{2}{x} + y$

(10) $q = \frac{ax}{b} + c$

11.4 Simultaneous equations

In section 8.11 we saw how to solve simultaneous equations using graphs. The following method can be easilly learned if graph paper is not available.

Worked example 11.3

Solve the equations $y = 3x + 4$ (1)

$y = 2x + 9$ (2)

Since '$y = y$', we can write:

$$3x + 4 = 2x + 9$$

$$3x - 2x = 9 - 4 \qquad \text{so } x = 5$$

Substitute for x into (1) gives $y = 3 \times 5 + 4 = 19$.

The solution is $x = 5$, $y = 19$.

Worked example 11.4

Solve the equations $y + 2x = 6$ (1)

$y - 3x = 1$ (2)

From (1): $y = 6 - 2x$ (3)

From (2): $y = 1 + 3x$ (4)

So $6 - 2x = 1 + 3x$

$6 - 1 = 3x + 2x$, $5 = 5x$. Hence $x = 1$

Substitute for x into (4): $y = 1 + 3 \times 1 = 4$

The solution is $x = 1$, $y = 4$.

Worked example 11.5

Solve the equations $2y + 3x = 7$ (1)

$3y - 4x = 2$ (2)

From (1): $2y = 7 - 3x \div 2$: $y = \frac{7 - 3x}{2} = \frac{7}{2} - \frac{3x}{2}$ (3)

From (2): $3y = 2 + 4x \div 3$: $y = \frac{2 + 4x}{3} = \frac{2}{3} + \frac{4x}{3}$ (4)

Hence $\frac{7}{2} - \frac{3}{2}x = \frac{2}{3} + \frac{4}{3}x$

$\times 6$: $21 - 9x = 4 + 8x$, $21 - 4 = 8x + 9x$

i.e. $17 = 17x$, so $x = 1$

Substitute into (1) for x: $2y + 3 \times 1 = 7$

Hence $2y = 7 - 3 = 4$, $y = 2$

The solution is $x = 1$, $y = 2$.

Exercise 11d

Solve the following pairs of simultaneous equations:

(1) $y = 2x + 3$, $y = 3x - 1$ (2) $y = x + 2$, $y = 4x - 4$ (3) $2y = x + 6$, $y = 2x - 3$

(4) $y + 2x = 9$, $y = 3x - 8$ (5) $2y + 3x = 1$, $3y + x = 3$ (6) $2y - x = 11$, $y - 3x = 3$

(7) $2y + 3x = 2$, $5y + 6x = 4$ (8) $5y = 7x + 1$, $2y - 3x = 3$

11.5 Simplifying products and quotients

The word *product* means *multiply* in mathematics.

(i) $2p \times 2q$ means $2 \times p \times 2 \times q = 4 \times p \times q = 4pq$

(ii) $2x \times 3x$ means $2 \times x \times 3 \times x = 6 \times x \times x = 6x^2$

The word *quotient* means *divide* in mathematics.

(iii) $12x \div 4x = \frac{\not{12}^3 \times \not{x}^1}{\not{4}_1 \times \not{x}_1} = 3$

(iv) $4t^2 \div 3t = \frac{4 \times t \times \not{t}^1}{3 \times \not{t}_1} = \frac{4t}{3}$

Exercise 11e

Simplify the following as much as possible:

(1) $4x \times 2x$ (2) $3t \times 5t$ (3) $6q \times 2q$

(4) $4t \times 3q$ (5) $t^2 \times 2t$ (6) $at \times 4$

(7) $5p \times 3p^2$ (8) $(2p)^2$ (9) $(4x)^2 \times 2x$

(10) $4u \times 3v$ (11) $12p \div 2p$ (12) $6x^2 \div 2x$

(13) $2t^2 \div 3t$ (14) $4pq \div 2p$ (15) $8t^2 \div 4p$

(16) $(3x)^2 \div x$ (17) $5p \div p^2$ (18) $2x \div (3x)^2$

(19) $4ab \div 8ab$ (20) $(4x \times 3y) \div 2xy$

11.6 The product of two brackets

In this section, the ideas developed in section 2.3 will be extended. The rectangle PQRS is this time divided into 4 rectangles by the lines TU and VW, labelled A, B, C and D, see Figure 11.1.

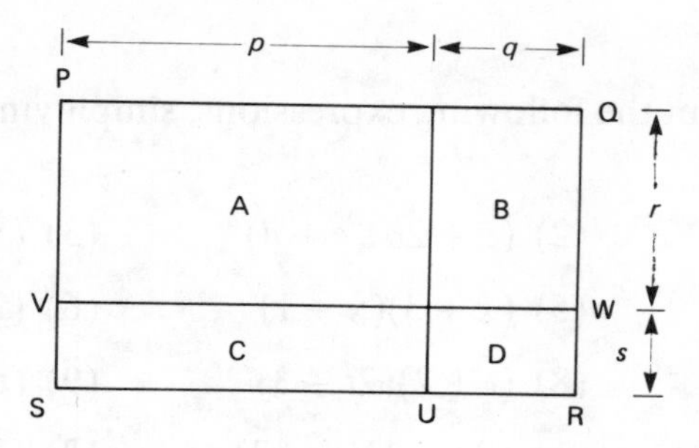

Figure 11.1

The area of PQRS $= (r+s) \times (p+q)$ or $(r+s)(p+q)$.
The area of A $= r \times p$.
The area of B $= r \times q$.
The area of C $= s \times p$.
The area of D $= s \times q$.
Hence $(r+s) \times (p+q) = r \times p + r \times q + s \times p + s \times q$.
Without the $\times$ sign, this becomes:

$$(r+s)(p+q) = rp + rq + sp + sq$$

It is important to see that you get four terms, although it may be possible to simplify them. Se also worked example 11.6(iii) and (iv) for examples with negative numbers.

Worked example 11.6
Remove the brackets from the following expressions, simplifying your answers if possible.

(i) $(x+2)(x+3)$ (ii) $(2u+v)(u+3v)$
(iii) $(x-3)(2x+1)$ (iv) $(2x-1)(3x-2)$

(i) $(x+2)(x+3) = x \times x + x \times 3 + 2 \times x + 2 \times 3$
$= x^2 + 3x + 2x + 6$
$= x^2 + 5x + 6$

(ii) $(2u+v)(u+3v) = 2u \times u + 2u \times 3v + v \times u + v \times 3v$
$= 2u^2 + 6uv + vu + 3v^2$
$= 2u^2 + 7uv + 3v^2$ [*Note*: uv is the same as vu.]

(iii) $(x-3)(2x+1) = x \times 2x + x \times 1 - 3 \times 2x - 3 \times 1$
$= 2x^2 + x - 6x - 3$
$= 2x^2 - 5x - 3$

(iv) $(2x-1)(3x-2) = 2x \times 3x + 2x \times -2 - 1 \times 3x - 1 \times -2$
$= 6x^2 - 4x - 3x + 2$ [*Note*: $-1 \times -2 = +2$.]
$= 6x^2 - 7x + 2$

Exercise 11f
Remove brackets from the following expressions, simplifying your answers where possible:

(1) $(x+y)(p+q)$ (2) $(a+2b)(c+d)$ (3) $(x+3y)(3x+2y)$
(4) $(2x+1)(x+2)$ (5) $(x+3)(x-1)$ (6) $(2x+1)(x+3)$
(7) $(x+3)(x+5)$ (8) $(t+2)(2t+3)$ (9) $(t-2)(t-3)$
(10) $(x-5)(x+6)$ (11) $(2x-1)(x-2)$ (12) $(3x+5)(2x+1)$
(13) $(u+3v)(u+2v)$ (14) $(3u+4v)(2u-v)$ (15) $(3x-5)(x+7)$
(16) $(t-2)(t+2)$ (17) $(3x-5)(3x+5)$ (18) $(7x-2)^2$
(19) $(x+4)^2$ (20) $(3x-1)^2$

11.7 Factorisation

In section 2.3, we saw how to remove brackets from simple expressions containing 2 terms. In this section, we look at how to replace brackets, or *factorise*. The important thing when factorising is to try and factorise completely.

(a) Two terms
(i) $8x+16y=4(2x+4y)$. Check from the right-hand side.
However, it can be taken further, and 8 can be put outside the bracket.
So $8x+16y=8(x+2y)$.
(ii) $4p^2-8pq=4(p^2-2pq)$.
In this case, however, p can also be put outside the bracket.
So $4p^2-8pq=4p(p-2q)$.

(b) Three terms (quadratic)
$(2x-1)(3x+1)=6x^2-x-1$ using the method of section 11.6.
If we are given the right-hand side, however, can we work out the left-hand side?
The expression $6x^2-x-1$ is called a *quadratic expression.* The process of factorising a quadratic is not easy and will be considered in stages.

(i) $$\begin{aligned} x^2+5x+6 &= x^2+2x+3x+6 \\ &= x(x+2)+3(x+2) \\ &= (x+3)(x+2) \end{aligned}$$

(ii) $$\begin{aligned} x^2+x-6 &= x^2+3x-2x-6 \\ &= x(x+3)-2(x+3) \\ &= (x-2)(x+3) \end{aligned}$$

(iii) $$\begin{aligned} x^2-7x+10 &= x^2-5x-2x+10 \\ &= x(x-5)-2(x-5) \\ &= (x-2)(x-5) \end{aligned}$$

 The process is really trial and error, and practice makes perfect.

(iv) $2x^2 - x - 6 = 2x^2 + 3x - 4x - 6$
$= x(2x+3) - 2(2x+3)$
$= (x-2)(2x+3)$

(v) $18x^2 + 45x - 8 = 18x^2 + 48x - 3x - 8$
$= 6x(3x+8) - (3x+8)$
$= (6x-1)(3x+8)$

When splitting up into four terms, notice that the product of the two middle terms equals the product of the two outside terms, e.g.:

(i) $x^2 \times 6 = 2x \times 3x = 6x^2$.
(iii) $x^2 \times 10 = -5x \times -2x = 10x^2$.
(v) $18x^2 \times -8 = 48x \times -3x = -144x^2$.

It is not always possible to put a quadratic expression into brackets; the following rule can save you a lot of time.

For the expression $ax^2 + bx + c$, work out $b^2 - 4ac$. If it is a perfect square it factorises, if not it does not factorise.

For example:

(i) $x^2 + 6x + 8 \qquad a = 1, b = 6, c = 8;$
$b^2 - 4ac = 36 - 32 = 4$, which *is* a perfect square.
In fact $x^2 + 6x + 8 = (x+4)(x+2)$.

(ii) $x^2 - 4x - 10 \qquad a = 1, b = -4, c = -10;$
$b^2 - 4ac = 16 + 40 = 56$, which *is not* a perfect square, hence it does not factorise.

The alternative method which also involves trial and error is illustrated as follows:

Factorise $6x^2 + 19x - 20$.

↑ Consider this number first (the $6x^2$) ↑ Consider this number second (the -20)

$6x^2$ suggests $3x \times 2x$, or $6x \times 1x$
-20 suggests 5×-4 or 2×-10, etc.

Try $(3x+5)(2x-4) = 6x^2 + 10x - 12x - 20$
$= 6x^2 - 2x - 20.$ ×

Try $(2x-10)(3x+2) = 6x^2 + 4x - 30x - 20$
$= 6x^2 - 26x - 20.$ ×

There are many possible combinations.

In fact $(6x-5)(x+4) = 6x^2 - 5x + 24x - 20$
$= 6x^2 + 19x - 20.$ √

(c) $a^2 - b^2$

This expression is known as a *difference of two squares.*

$$a^2 - b^2 = (a-b)(a+b)$$

Example: $16x^2 - 25y^2 = (4x)^2 - (5y)^2 = (4x - 5y)(4x + 5y)$.

(d) $a^2 \pm 2ab + b^2$

The expression is a *perfect square.*

$$a^2 + 2ab + b^2 = (a+b)(a+b) = (a+b)^2$$
$$a^2 - 2ab + b^2 = (a-b)^2$$

Exercise 11g

Factorise if possible the following expressions:

(1) $4x + 16$ (2) $3x - 9$ (3) $t^2 + t$

(4) $ab + 2a^2$ (5) $6pq + 2p$ (6) $ac + 4a^2c^2$

(7) $9t^2 - 3t$ (8) $pq + ab$ (9) $6b + 5b^2$

(10) $2p^3 + p^2$ (11) $x^2 + 3x + 2$ (12) $x^2 + 10x + 24$

(13) $x^2 + 5x + 4$ (14) $x^2 + 14x + 48$ (15) $x^2 + 9x + 18$

(16) $x^2 + 2x + x$ (17) $x^2 - 6x + 5$ (18) $x^2 - 6x + 8$

(19) $x^2 + 2x - 15$ (20) $2x^2 + 5x + 2$ (21) $2x^2 + 3x - 2$

(22) $2x^2 - x - 3$ (23) $x^2 + 6x + 9$ (24) $4x^2 + 4x + 1$

(25) $x^2 - 16$ (26) $4x^2 - 9$ (27) $2x^2 - 50$

(28) $x^2 + x + 4$ (29) $x^3 - x$ (30) $x^2 + 1$

11.8 Quadratic equations

(a) By factors

Consider $x^2 - 5x + 6 = 0$. In section 11.7 we factorised quadratic expressions.

Hence $(x-3)(x-2) = 0$

either $x - 3 = 0$, or $x - 2 = 0$.

We get two solutions: $x = 3$ or $x = 2$.

A further example: $6x^2 - x - 2 = 0$

factorising: $(2x+1)(3x-2) = 0$

Hence $2x + 1 = 0$ or $3x - 2 = 0$

i.e. $2x = -1$ or $3x = 2$

therefore $x = -\frac{1}{2}$ or $x = \frac{2}{3}$.

(c) By formula

The equation $x^2 + x - 1 = 0$ cannot be solved by factors, because it does not factorise. The following formula can be used.

For the equation $ax^2 + bx + c = 0$

$$x = \frac{-b \pm \sqrt{b^2 - 4ac}}{2a}$$

If $b^2 - 4ac$ is negative, the equation has no solution. In the equation $x^2 + x - 1 = 0$, $a = 1$, $b = 1$, $c = -1$.

$$\text{So } x = \frac{-1 \pm \sqrt{1+4}}{2} = \frac{-1 \pm \sqrt{5}}{2}$$

 Hence: $x = 0.618$ or -1.618.

An equation may not always be obviously a quadratic. This is illustrated in the following worked example.

Worked example 11.7

Solve the equation:

$$\frac{1}{2x}+\frac{3}{(x+1)}=2$$

Multiply the equation by $2x(x+1)$.

Hence $\cancel{2x}(x+1)\times\frac{1}{\cancel{2x}}+2x\cancel{(x+1)}\times\frac{3}{\cancel{(x+1)}}=2\times 2x(x+1)$

therefore $(x+1)+6x=4x^2+4x$

So $0=4x^2-3x-1$

i.e. $0=(4x+1)(x-1)$

The solutions are $x=1,\ -\frac{1}{4}$.

Exercise 11h

Solve where possible the following quadratic equations:

(1) $x^2-2x-3=0$

(2) $2x^2+x-1=0$

(3) $x^2-2x+1=0$

(4) $x^2+3x-9=0$

(5) $x^2+8x=3$

(6) $x+\frac{1}{3x}=2$

(7) $4x^2+x+7=0$

(8) $\frac{1}{x}+x=2$

(9) $x^2-10x+16=0$

(10) $3x^2-10x+3=0$

(11) $x^2-9x=0$

(12) $x^2+2x-1=0$

(13) $4x^2=9x+1$

(14) $\frac{x+2}{x+5}=\frac{2x+3}{x-1}$

(15) $\frac{1}{x-1}-\frac{1}{x+1}=6$

11.9 Forming quadratic equations

The same techniques that were used in section 2.6 will often give rise to a quadratic equation. This is illustrated in the following worked example.

Worked example 11.8

A machine produces two types of bolt. Bolt A is produced at the rate of x per minute, and bolt B is produced at the rate of $(x-6)$ per minute.

(a) Write down an expression in seconds for the time taken to produce each bolt.

(b) If it takes $\frac{5}{6}$ second longer to produce bolt B than bolt A, write down an equation in x and solve it.

(a) If x of bolt A are produced in 60 seconds, then one is produced in $\frac{60}{x}$ seconds.

Similarly, one of B is produced in $\frac{60}{x-6}$ seconds.

(b) Seeing that the difference is $\frac{5}{6}$ second, and $\frac{60}{x-6}$ is larger than $\frac{60}{x}$ (because fewer bolts are produced per minute), it follows that:

$$\frac{60}{x-6} = \frac{60}{x} + \frac{5}{6}$$

Multiply by common denominator $6x(x-1)$.

Hence: $360x = 360(x-6) + 5x(x-6)$

i.e. $\quad 360x = 360x - 2160 + 5x^2 - 30x$

Rearranging: $5x^2 - 30x - 2160 = 0 \qquad \div 5: \qquad x^2 - 6x - 432 = 0$

Factorising: $(x+180)(x-24) = 0$

x cannot be -18, hence $x = 24$.

11.10 Algebraic fractions

We consider here examples illustrating the four rules of addition, subtraction, multiplication and division.

(a) Addition and subtraction

(i) $\frac{2}{x} + \frac{4}{3x} = \frac{6}{3x} + \frac{4}{3x} = \frac{10}{3x}$

(ii) $\frac{2}{a} + \frac{1}{a^2} = \frac{2a}{a^2} + \frac{1}{a^2} = \frac{2a+1}{a^2}$

(iii) $\frac{4}{x} - \frac{3}{y} = \frac{4y}{xy} - \frac{3x}{xy} = \frac{4y-3x}{xy}$

(b) Multiplication

(i) $\frac{a}{b} \times \frac{c}{d} = \frac{ac}{bd}$

(ii) $\frac{4}{x} \times \frac{x}{2} = \frac{4x}{2x} = \frac{4}{2} = 2$

(iii) $\frac{x}{a^2} \times \frac{a}{b} = \frac{x\not{a}^1}{a^{\not{2}}b} = \frac{x}{ab}$

(c) Division

(i) $\frac{a}{b} \div \frac{c}{d} = \frac{a}{b} \times \frac{d}{c} = \frac{ad}{bc}$

(ii) $\frac{2}{x} \div \frac{3}{x} = \frac{2}{\cancel{x}_1} \times \frac{\cancel{x}^1}{3} = \frac{2}{3}$

(iii) $\frac{t}{x^2} \div \frac{t^2}{x} = \frac{\cancel{t}^1}{x^{\cancel{2}}} \times \frac{\cancel{x}^1}{t^{\cancel{2}}} = \frac{1}{xt}$

Exercise 11i

Simplify, as much as possible, the following:

(1) $\frac{x}{3} + \frac{x}{4}$ (2) $\frac{x}{2} - \frac{x}{3}$ (3) $\frac{2x}{3} + \frac{x}{4}$ (4) $\frac{2y}{5} + \frac{y}{2}$

(5) $\frac{x}{4} + 1$ (6) $\frac{1}{y} + \frac{1}{2y}$ (7) $\frac{1}{3m} + \frac{1}{2n}$ (8) $\frac{1}{5q} - \frac{1}{10q}$

(9) $\frac{1}{4t} + \frac{2}{3t}$ (10) $\frac{x}{y} + 2$ (11) $\frac{a}{b} \times \frac{2}{3}$ (12) $\frac{1}{x} \times \frac{x}{2}$

(13) $\frac{1}{a} \times a$ (14) $\frac{a}{c} \times \frac{c}{d}$ (15) $\frac{x^2}{y} \times \frac{2y}{x}$ (16) $\frac{t}{y} \times \frac{y}{t}$

(17) $\frac{4}{3} \times \frac{3x}{y}$ (18) $2t \times \frac{4}{t}$ (19) $\frac{a}{q} \times \frac{2q}{t}$ (20) $\frac{3t}{m} \times \frac{m}{t^2}$

(21) $\frac{x}{2} \div \frac{x}{3}$ (22) $\frac{a}{b} \div \frac{2a}{b}$ (23) $\frac{1}{x^2} \div \frac{1}{x}$ (24) $\frac{x}{y} \div \frac{1}{2y}$

(25) $pq^2 \div \frac{p}{q}$

11.11 Further Venn diagrams

Sometimes, when using Venn diagrams the following symbols are used to save writing long sentences.

The symbol $\cap$ means where 2 sets overlap, i.e. the intersection.

The symbol $\cup$ means the 2 sets combined together, i.e. the union.

A' means outside set A, i.e. the complement.

The use of these symbols is illustrated in Figure 11.2.

Exercise 11j

Draw suitable diagrams and shade the following sets:

(1) $P \cup Q$ (2) $P \cup Q'$ (3) $P' \cup Q'$
(4) $A \cap (B \cup C)$ (5) $(A' \cup B) \cap C$ (6) $(P \cap Q')'$
(7) $(X \cup Y) \cap X'$ (8) $(A \cup B) \cup C'$

11.12 Numerical methods

In worked example 8.9 we solved the equation $x^2 - 2x - 1 = 0$ by inspection from the graph. The answer was only an approximation and we shall now look at a

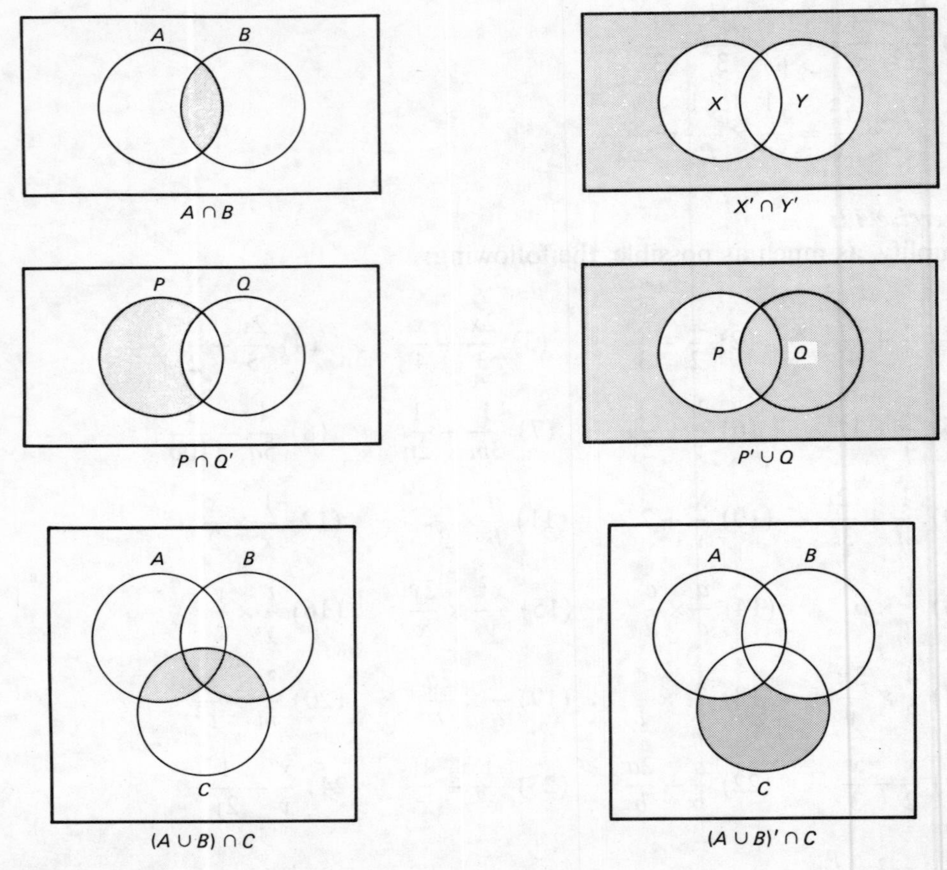

Figure 11.2

numerical method which will help us to solve such equations to any degree of accuracy. Although we could use the formula solution for a quadratic equation (see section 11.8), numerical methods can be extended to virtually *any* equation.

The technique requires us to rearrange the original quadratic equation into a form with a single term of x on one side of the equation and the rest of the terms on the other side of the equation.

So for $x^2 - 2x - 1 = 0$

$$x^2 = 2x + 1$$

therefore $$x = \frac{2x+1}{x}$$

is one possible arrangement.

We now make an initial estimate of a possible answer and substitute this value of x into the right-hand side of this new equation. For example, we could estimate the value $x = 2$ which is a first approximation to the positive solution of the equation. By substitution this gives a new value of

$$x = \frac{4+1}{2}$$

$$= 2.5$$

In turn, substituting this value into the right-hand side of the equation:

$$x = \frac{5+1}{2.5}$$

$$= 2.4$$

Repeating this process gives further successive values of x of 2.417, 2.414, 2.414, 2.414... (to 3 dec. places). So the positive root of the equation correct to 3 decimal places is 2.414.

This is an example of an *iterative method* where, by starting at an initial approximation, we obtain a sequence of approximations to *a* solution. The equation used is known as an *iterative equation.* For the technique to be successful, the approximations must get better and better until the degree of accuracy required is reached.

The initial approximation is usually denoted by x_0 from which we obtain a sequence of successive approximations $x_1, x_2, x_3, \ldots$, and so on. Using our example this can be generalised in the form:

$$x_{n+1} = \frac{2x_n + 1}{x_n}$$

i.e. $x_0 = 2$ generates $x_1 = 2.5$, $x_2 = 2.4, \ldots$

We can present successive approximations to the solution in tabular form:

n	x_n	$\frac{2x_n+1}{x_n}$
0	2	2.5
1	2.5	2.4
2	2.4	2.417
3	2.417	2.414
4	2.414	2.414
.	.	.
.	.	.
.	.	.

The entries in the third column are the generated values of x read from the calculator and correct to 3 decimal places.

Now, as an exercise, try other initial starting values, i.e. $x_0 = 3, 4, 5, -100$. Now try $x_0 = 0$. What do you notice? No matter what values you try, the iteration will not converge to the negative solution.

Worked example 11.9

The iteration formula for a sequence is

$$u_{n+1} = \frac{u_n^2 - 1}{2}$$

By writing x for u_{n+1} and u_n, show that this formula gives a solution to the quadratic equation $x^2 - 2x - 1 = 0$; and by taking a starting value of $u_0 = 0$, find this solution correct to 2 decimal places.

Substituting x for u_n and u_{n+1}:

$$x = \frac{x^2 - 1}{2}$$

so $$2x = x^2 - 1$$

therefore $$x^2 - 2x - 1 = 0$$

So the iterative formula will provide a solution to the quadratic. Presenting the successive approximations in tabular form:

n	u_n	$\dfrac{u_n^2 - 1}{2}$
0	0	−0.5
1	−0.5	−0.375
2	−0.375	−0.430
3	−0.430	−0.408
4	−0.408	−0.417
5	−0.417	−0.413
6	−0.413	−0.415
7	−0.415	−0.414
8	−0.414	−0.414
9	−0.414	−0.414
10	.	.
11	.	.

So correct to 3 decimal places, the solution found is −0.414.

We have now found the negative solution for Worked example 11.8.

11.13 Area under a curve

In section 8.7 we have seen that the area under a speed–time graph is equivalent to the total distance travelled. Where the speed–time graph is made up of a number of straight line segments, the area can be divided into a number of triangles and rectangles, and it is then a straightforward task to calculate the area. If however the graph is a smooth curve, we need to adopt an *estimation* technique by dividing the area under the curve into trapezia. By then working out the area of each trapezium, an approximation to the area can be found. This method will be an under- or over-estimation, depending on the shape of the curve, and is commonly referred to as the *trapezium rule*.

Worked example 11.10

Figure 11.3 shows part of the graph of $y = \dfrac{10}{x}$ for values of x between 1 and 4. Use the trapezium rule with 3 strips of equal width to find the approximate area between $x = 1$ and $x = 4$. Is the area found an under- or over-estimate of the true area? Justify your answer.

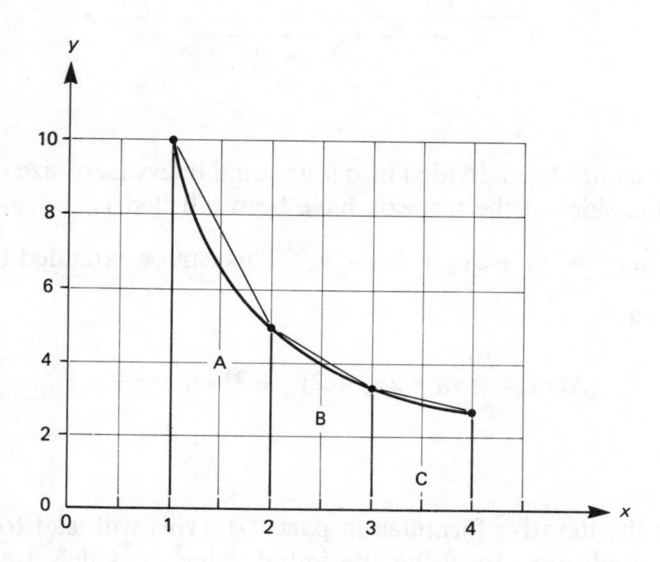

Figure 11.3

In order to find the area, we need to determine the lengths of the parallel sides of the trapezia. These lengths are the y-co-ordinates corresponding to $x = 1$, $x = 2$, $x = 3$ and $x = 4$, i.e.

x	1	2	3	4
y	10	5	3.3	2.5

↑ an approximation to 1 decimal place

The approximate area therefore
= area of trapezium A + area of trapezium B + area of trapezium C
$= \frac{1}{2}(10 + 5) \times 1 + \frac{1}{2}(5 + 3.3) \times 1 + \frac{1}{2}(3.3 + 2.5) \times 1$
$= 7.5 + 4.15 + 2.9$
$= 14.55 = 14.6$ units^2.

Because of the shape of the curve, the trapezia used are larger in area than the corresponding areas under the curve, so the area found is an over-estimate of the true area. (The actual area, correct to 1 decimal place is 13.9 units^2.)

There is a formula that can be used for this rule. In Figure 14.4, we have a

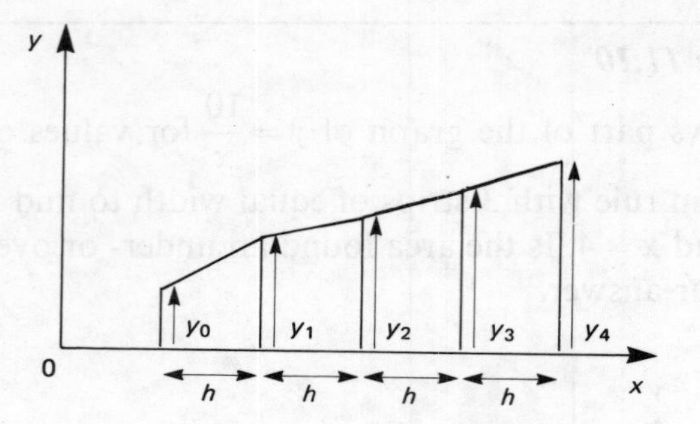

Figure 11.4

graph whose area has been divided into four equal intervals of size h. The y values of the top of the sides of the trapezia have been labelled $y_0, \ldots, y_4$.

The area $= \frac{h}{2}(y_0 + 2y_1 + 2y_2 + 2y_3 + y_4)$. This can be extended to any number n in the formula:

$$\text{Area} = \frac{h}{2}(y_0 + 2y_1 + 2y_2 + 2y_3 + \cdots + y_n)$$

Exercise 11k

(1) Show that the iterative formulae in parts (i)–(viii) will lead to a solution of the given quadratics. By using the initial values x which are given, find a solution to each equation correct to 3 decimal places.

(i) $x^2 - 7x - 4 = 0;\ x_{n+1} = \dfrac{x_n^2 - 4}{7};\ x_0 = 0$

(ii) $x^2 - 7x - 4 = 0;\ x_{n+1} = 7 + \dfrac{4}{x_n};\ x_0 = 7$

(iii) $x^2 - 5x - 6 = 0;\ x_{n+1} = 5 + \dfrac{6}{x_n};\ x_0 = 4$

(iv) $x^2 - 5x - 6 = 0;\ x_{n+1} = \dfrac{6}{x_n - 5};\ x_0 = 0$

(v) $x^2 - 7x - 7 = 0;\ x_{n+1} = \dfrac{1}{7}x_n^2 - 1;\ x_0 = 0$

(vi) $x^2 - 7x - 7 = 0;\ x_{n+1} = 7\left(1 + \dfrac{1}{x_n}\right);\ x_0 = 10$

(vii) $x^2 + 11x - 5 = 0;\ x_{n+1} = \dfrac{5 - x_n^2}{11};\ x_0 = 1$

(viii) $x^2 - 15 = 0;\ x_{n+1} = \dfrac{x_n^2 + 15}{2x_n};\ x_0 = 4$

(This last example is a very useful iterative formula as it enables us to find the square root of any number very quickly. All that is necessary is to replace '15' by the number we wish to square root.)

(2) A sequence of numbers $x_1, x_2, x_3, \ldots,$ is related by the iterative formula:

$$x_{n+1} = \frac{1}{2}\left(x_n + \frac{5}{x_n}\right)$$

For example, when $x_1 = 2$, $x_2 = \frac{1}{2}\left(2 + \frac{5}{2}\right) = 2.25$.

(a) Use this value of x_2 to calculate the value of x_3.
(b) This formula is used to find the square root of a number. Which number?
(c) Use this formula to find this square root correct to four places of decimals.
(d) How did you use the formula to make sure that your answer was correct to four places of decimals?
(e) Write down an iterative formula which could be used to find $\sqrt{11}$.

(NEA)

(3) In Figure 11.5, rectangle ABCD is divided into two smaller rectangles by the line LM.

AD = 1 cm, DM = 2 cm, AB = x cm

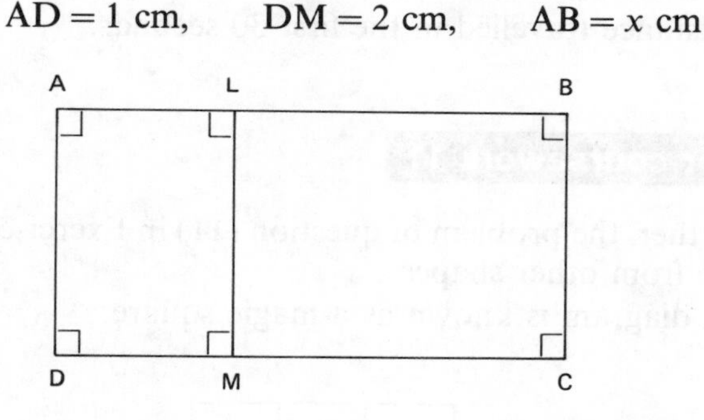

Figure 11.5

(a) Write down the length of MC in terms of x.
Rectangles ABCD and BCML are similar.
(b) Use this fact to write down two expressions in x which must be equal and show that

$$x^2 - 2x - 1 = 0$$

(c) Show that $x^2 - 3x - 1 = 0$ can be written in the form:

$$x = 2 + \frac{1}{x}$$

(d) Taking $x_0 = 3$, use

$$x_{n+1} = 2 + \frac{1}{x_n}$$

to calculate x_1, x_2 and x_3. Write down ALL the figures shown on your calculator.

(e) Continue the iteration and find the solution of the equation

$$x^2 - 2x - 1 = 0$$

correct to 4 significant figures. (LEAG)

(4) Using Figure 8.30 (page 217), determine the area under the curve $y = x^2$ between $x = 0$ and $x = 6$, using 6 equal class intervals and the trapezium rule. Give your answer to 1 decimal place.

(5) A car travels along a road and at various intervals its speed was measured. The results are shown in the table below:

Time(s)	0	5	10	15	20	25	30
Speed (m/s)	0	7	18	25	22	11	4

(i) Sketch the speed–time graph representing the data.

(ii) Determine the gradient of the graph at 10 seconds. What does this represent?

(iii) Use the trapezium rule with 6 equal intervals to determine an estimate of the distance travelled in the first 30 seconds.

Suggestions for coursework 11

1. Investigate further, the problem of question (11) in Exercise 2i. Look at similar patterns made from other shapes.
2. The following diagram is known as a magic square:

4	9	2
3	5	7
8	1	6

The numbers in each row, column and diagonal add up to 15. Investigate other magic squares.

3. How many matches are necessary in a knock-out tournament to produce a winner from n entries. Investigate similar problems.
4. A room contains 30 people. If everybody shakes hands with everybody else, how many handshakes is this? Can you find the result for a room containing n people. Think of similar problems.
5. Investigate more fully, the sequence

$$1, 5, 14, 30, 55, 91, \ldots$$

6. Write a programme for finding the HCF and LCM of any two numbers.

Miscellaneous exercise 11

1. (a) Write down, in terms of x, an expression for the area of the shape of Figure 11.6.

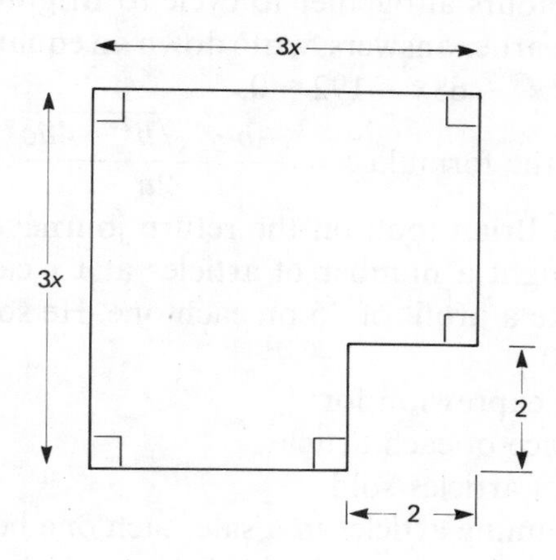

Figure 11.6

(b) Multiply out $(3x+2)(3x-2)$, giving your answer in its simplest form.
(c) By making one cut and reassembling, the above shape can be made into a rectangle. Using your answers to (a) and (b), draw a diagram to show how this can be done. Mark the dimensions of the rectangle on your diagram. (NEA)

2. The total distance d metres travelled by a bouncing ball is given by the formula $d = \dfrac{2h}{1-f}$ where h metres is the height to which the ball is thrown initially, and the height of each bounce is a fraction, f, of the height of the previous bounce.
(a) Find d if $h = 8$ and $f = \frac{3}{4}$.
(b) Rearrange the formula to give f in terms of d and h.
(c) Find f if $d = 45$ and $h = 15$. (SEG)

3. A firm manufactures n identical toys per hour. The cost of manufacturing *each toy* is C pence, this cost being represented by the formula $C = A + \dfrac{k}{n}$, where A and k are constants. Given that $C = 60$ when $n = 5$, and $C = 45$ when $n = 20$, form two equations and thus find the values of A and k.
Hence calculate:
(a) the cost of each toy when 100 are produced in 4 hours,
(b) the number of toys being produced per hour when the cost of manufacturing each toy is 42 pence.
Find a formula for n in terms of C.

4. Brian, who lives in London, decided to have a day out cycling to Brighton and back. He cycled the 48 miles there at an average speed of x m.p.h. On the way back he was in a hurry to reach home and so he increased his average speed by 4 m.p.h.

(a) Write down, in terms of x:

(i) The time, in hours, Brian took to cycle to Brighton.

(ii) The average speed, in m.p.h., at which he returned to London.

(iii) The time, in hours, he took to return to London.

(b) Brian took 7 hours altogether to cycle to Brighton and back. Using this fact and your earlier answers, write down an equation in x and show that it simplifies to $7x^2 - 68x - 192 = 0$.

(c) Find x using the formula $x = \dfrac{-b \pm \sqrt{b^2 - 4ac}}{2a}$, or otherwise.

(d) Find the time Brian took on the return journey. (SEG)

5. A shopkeeper bought a number of articles at £x each and fixed the selling price so as to make a profit of £5 on each one. He sold some of these articles for a total of £300

Write down an expression for:

(i) the selling price of each article,

(ii) the number of articles sold.

He sold all remaining articles in a sale, each one being sold for £1 less than the cost price of £x. In the sale he sold these articles for a total of £120.

Write down an expression for:

(iii) the number of articles he sold in the sale,

(iv) the total number of articles he bought.

The total number of articles he bought was 15.

(v) Write down the equation for x and solve it to find the value of x.

6. $p = \dfrac{5}{x+3}$ and $q = \dfrac{2}{2x-1}$.

(a) Obtain an expression in x, in its simplest form, for $p + q$.

(b) Find the value of x which satisfies the equation $p + q = pq$. (MEG)

7. Mrs Jones has 18 m of fencing, all of which she intends to use to make a rectangular border for a flower bed which she intends to dig in her garden.

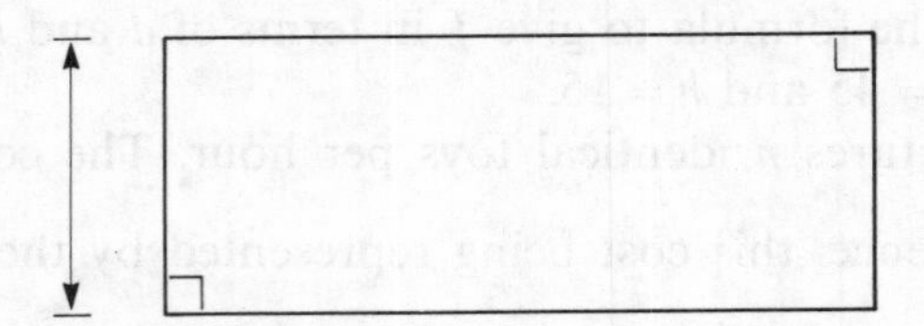

Figure 11.7

(a) Given that one side of the rectangle will be of length x m, write down the length of an adjacent side, in m.

(b) Hence write down an expression for the area of the proposed flower bed, in m^2.

Mrs Jones thinks that 20 m^2 of flower bed will be as much as she can manage to weed.

(c) Calculate the values of x which will give this area.

8. Figure 11.8 shows a cylindrical tube standing on a horizontal table. The tube is cut along the vertical line AB and opened out.

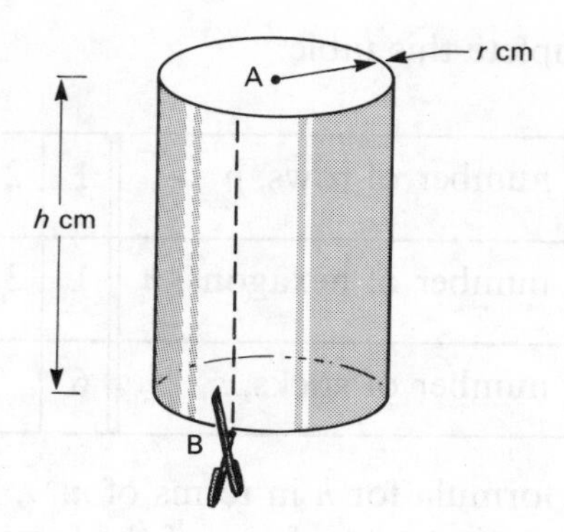

Figure 11.8

(a) Show that the area of the resulting rectangle is $2\pi rh$ cm^2. A closed metal can of end radius r cm and height h cm is made of tin plate.

(b) Show that the area, A cm^2, of the tin plate required is given by

$$A = 2\pi rh + 2\pi r^2$$

(c) Given that $A = 132$ and $r = 3$, find h. Give your answer to the nearest whole number. (LEAG)

9. A satellite is orbiting the earth at a height of H km. If the radius of the earth is R km, then the time T hours that it takes the satellite to complete one orbit of the earth is given by the formula:

$$T = 2\pi\sqrt{\frac{(R+H)^3}{10R^2}}, \text{ where } R = 6400 \text{ km}$$

(i) Find how long it takes for a satellite to orbit the earth at a height of 20 km.

(ii) If the satellite takes 200 hours to orbit the earth, at what height is it?

10. Bridget is doing an investigation which involves making patterns of hexagons with sticks.

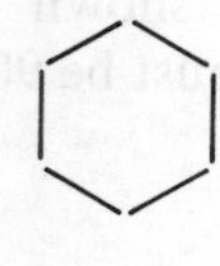
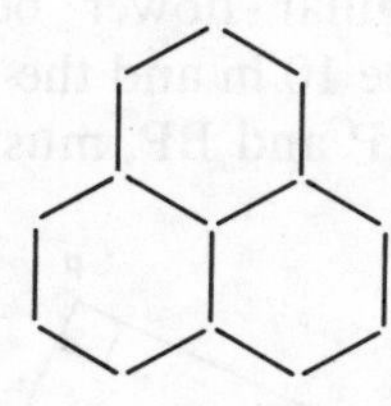
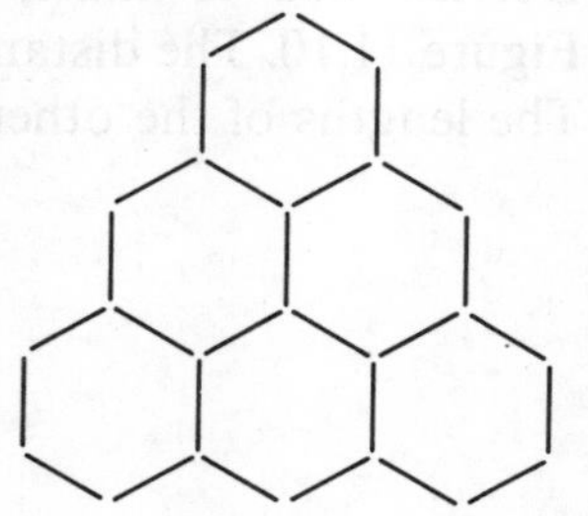

Figure 11.9

The first three of these patterns are shown above.

(a) Copy and complete this table.

number of rows, n	1	2	3	4	5
number of hexagons, h	1	3			
number of sticks, s	6				

(b) Write down a formula for h in terms of n.
(c) The formula for s in terms of n is of the form

$$s = an^2 + bn$$

where a and b are constants.
Find the value of a and the value of b. (MEG)

11. The time, T minutes, taken by the Moon to eclipse the Sun totally is given by the formula:

$$T = \frac{1}{v}\left(\frac{rD}{R} - d\right)$$

d and D are the diameters, in kilometres, of the Moon and Sun respectively; r and R are the distances, in kilometres, of the Moon and Sun respectively from the earth;
v is the speed of the Moon in kilometres per minute.
Given that

$$d = 3.48 \times 10^3$$
$$D = 1.41 \times 10^6$$
$$r = 3.82 \times 10^5$$
$$R = 1.48 \times 10^8$$
$$v = 59.5$$

calculate the time taken for a total eclipse, giving your answer in minutes, correct to 2 significant figures. (MEG)

12. Derrick has to mark out a triangular flower bed ABC, as shown in Figure 11.10. The distance AB must be 10 m and the angle APB must be 90°. The lengths of the other two sides, AP and BP, must total 13 m.

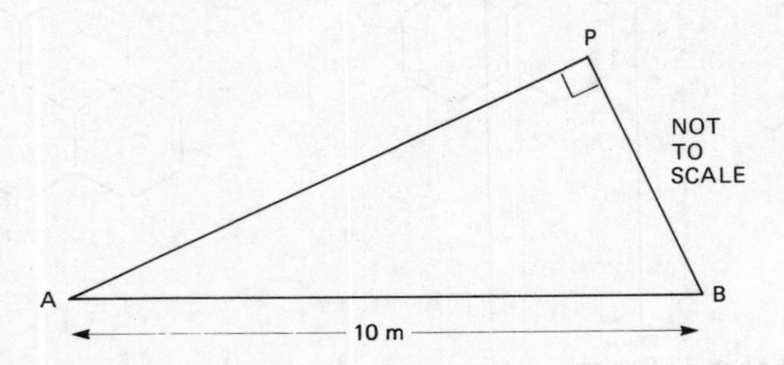

 Figure 11.10

(a) Taking the length of AP as x metres, form an equation in x, and show that it simplifies to $2x^2 - 26x + 69 = 0$.

(b) Solve this equation to find the two possible values of x, correct to 2 decimal places. (MEG)

13. The following table is based on the performance figures for a car as it accelerates from rest:

Time (seconds), t	0	2	4	6	8	10
Velocity (metres per second), v	0	10	18	23.5	27.5	31

(a) Draw the graph of v against t.

(b) By drawing the tangent to the curve at (4, 18), estimate the gradient of the curve at this point. State the significance of this value.

(c) Estimate the area of the region bounded by the curve, the t-axis and the line $t = 10$ by approximating this area to a triangle and four trapezia. State the significance of this value. (NEA)

14. In question 9 of exercise 8c, the graph of $y = 2x^2 - 4x + 1$ was drawn. The curve crosses the x-axis twice: once between 0 and 1, and once between 1 and 2. The x-co-ordinates of these two points satisfy the equation:

$$2x^2 - 4x + 1 = 0$$

(a) Show that $2x^2 - 4x + 1 = 0$ can be written in the form $x = \dfrac{4x - 1}{2x}$.

(b) Take $x_0 = 1$, and use $x_{n+1} = \dfrac{4x - 1}{2x_n}$ to calculate x_1, x_2, x_3.

(c) Continue the iteration and find one solution of the equation

$$2x^2 - 4x + 1 = 0$$

correct to 4 significant figures.

15. The iteration formula for a sequence is

$$x_{n+1} = \sqrt{5 + \frac{12}{x_n}}$$

(a) Starting with $x_1 = 2.5$, calculate x_2, x_3, x_4, x_5, x_6 and x_7. For each of these values, write down all the figures shown on your calculator.

(b) Suggest the limit to which the sequence appears to approach.

(c) (i) By writing x for x_n and for x_{n+1}, show that the iteration formula gives a solution to the equations

$$x^3 - 5x - 12 = 0$$

(ii) Show that the limit suggested in (b) is a solution of this equation.

(d) Find an iterative formula which can give you a solution of

$$x^3 + 7x^2 - 11 = 0$$

(SEG)

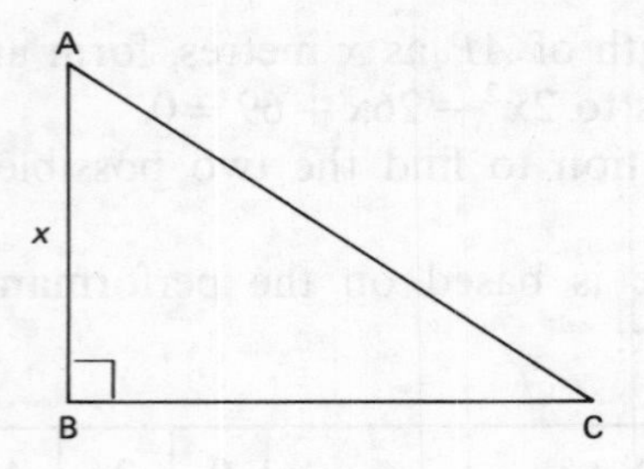

Figure 11.11

16. Figure 11.11 shows triangle ABC with angle ABC = 90°. The side AB has length x cm. Side BC is 2 cm longer than AB and side AC is 3 cm longer than AB.
    (a) Write down, in terms of x, the lengths of AC and BC.
    (b) Form an equation in x and show that it can be written as

$$x^2 - 2x - 5 = 0$$

    (c) Show that this can be rewritten as

$$x = 2 + \frac{5}{x}$$

    (d) Use the iterative formula

$$x_{n+1} = 2 + \frac{5}{x_n}$$

    with a starting value of $x_1 = 3$, to calculate x_2 and x_3 correct to 5 significant figures.
    (e) Continue the iteration to estimate a root of the equation correct to 2 decimal places, showing that your answer is correct to that accuracy. (LEAG)

17. Figure 11.12 shows a sketch of the graph of the curve with equation $y = x^3$ for values of x between 0 and 4.

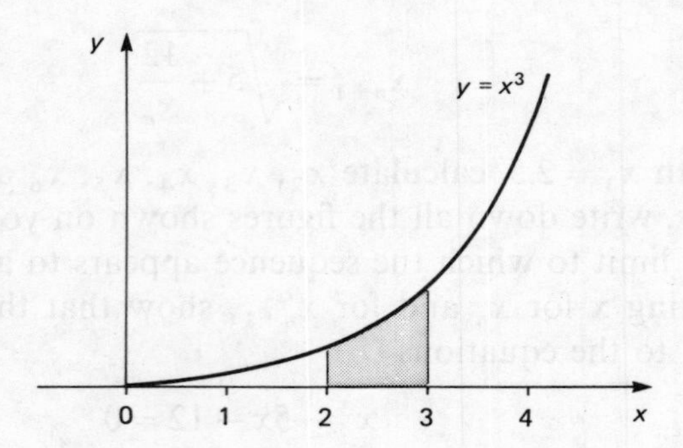

Figure 11.12

    (a) Use the trapezium rule with 5 strips of equal width to find the approximate area of the shaded region, working throughout to 3 decimal places.

(b) The percentage error in the approximate area is given by

$$\frac{\text{approximate area} - \text{exact area}}{\text{exact area}} \times 100.$$

Given that the exact area of the shaded region is 16.25, find the percentage error, correct to 3 decimal places. (LEAG)

18. The speed of a train is given by the following table:

Time (seconds)	0	10	20	30	40
Speed (m/s)	10	13	18	28	40

Draw a sketch of the speed-time graph of the train.

Use the trapezium rule with four equal intervals to find the area under the curve. What does this value represent?

19. (a) Copy and complete this pattern:

$3^2 - 1 = \ldots\ldots\ldots\ldots\ldots = 8\ldots\ldots\ldots\ldots\ldots$

$5^2 - 1 = \ldots\ldots\ldots\ldots\ldots = 8\ldots\ldots\ldots\ldots\ldots$

$7^2 - 1 = \ldots\ldots\ldots\ldots\ldots = 8\ldots\ldots\ldots\ldots\ldots$

$9^2 - 1 = \ldots\ldots\ldots\ldots\ldots = 8\ldots\ldots\ldots\ldots\ldots$

(b) Write the next three lines of the pattern.

(c) (i) What name is given to the sequence of numbers in the final column?

(ii) What is the tenth number in this sequence?

(d) Simplify $\frac{(2n+1)^2 - 1}{8}$, factorising your answer.

(e) Use part (d) to show that the pattern in parts (a) and (b) is true for all positive integers, n.

20. The iteration formula for a sequence is

$$x_{n+1} = \sqrt{4 - \frac{3}{x_n}}$$

(a) Starting with $x_1 = 1.1$, calculate x_2, x_3, x_4, x_5, x_6 and x_7. For each of these values, write down all the figures shown on your calculator.

(b) Suggest the limit to which the sequence appears to approach.

(c) (i) By writing x for x_n and for x_{n+1}, show that the iteration formula gives a solution to the equation $x^3 - 4x + 3 = 0$.

(ii) Show that the limit suggested in (b) is a solution of this equation.

21. John wants to see his brother. He walks x km at 6 km h^{-1} to his sister. When he arrives at his sister's, she drives him the rest of the way by car at an average speed of 60 km h^{-1}. The total journey is 90 km long and takes 3 hours. How far did he walk?

VECTORS, MATRICES AND TRANSFORMATIONS

12

Vectors and matrices will probably only appear in the higher level papers of the GCSE. The treatment here assumes a small amount of knowledge about matrices.

12.1 Routes and vectors

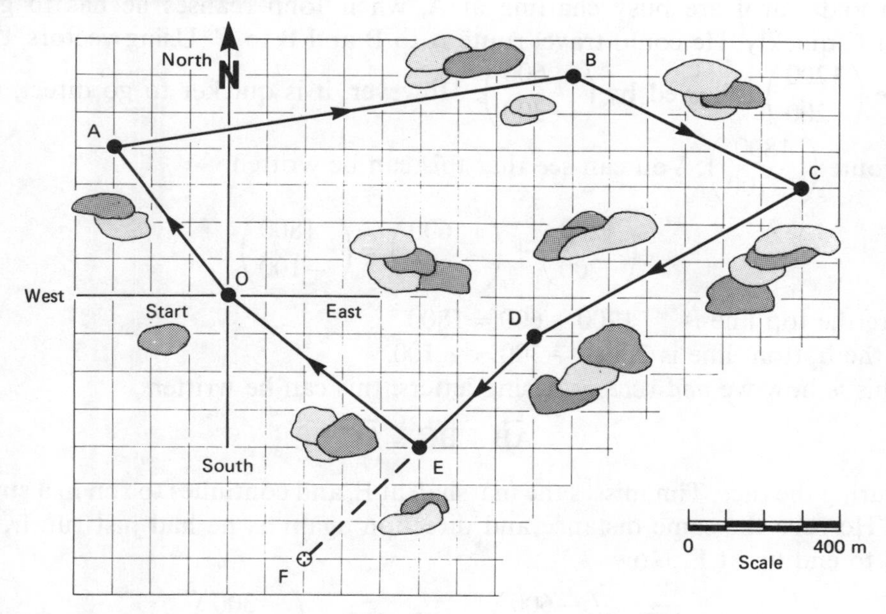

Figure 12.1

Figure 12.1 shows the cross-country course for Pete's school's championships. The start is at O. In order to mark out point A, a marshall is asked to walk 300 m West, and then 400 m North. A convenient way of showing this is

$$\begin{pmatrix} 300 \text{ m West} \\ 400 \text{ m North} \end{pmatrix}$$

If we use the convention used in co-ordinates, then West is negative and South is negative.

The route is then written

$$\overrightarrow{OA} = \begin{pmatrix} -300 \\ 400 \end{pmatrix}$$

This is called a *vector*, or *column vector*. Note that the arrow over OA points from O to A. $\overleftarrow{OA}$ would mean A to O. The route from A to B is

$$\overrightarrow{AB} = \begin{pmatrix} 1200 \\ 200 \end{pmatrix}$$

A vector quantity is any quantity which has both *length* and *direction*. The length of $\overrightarrow{AB}$ is written $|\overrightarrow{AB}|$ and is sometimes called its *modulus*. By Pythagoras, $|\overrightarrow{AB}| = \sqrt{(1200)^2 + (200)^2} = 1217$ m.

Exercise 12a

Write down as a vector, the following routes:

(1) B to C (2) C to D (3) D to E (4) E to O
(5) A to C (6) C to A (7) A to D

John and Carol are busy chatting at A, when John realises he has to get to point C quickly. He could travel route A to B and B to C. Using vectors, this is route $\begin{pmatrix} 1200 \\ 200 \end{pmatrix}$ followed by $\begin{pmatrix} 600 \\ -300 \end{pmatrix}$. However, it is quicker to go direct, using the route $\begin{pmatrix} 1800 \\ -100 \end{pmatrix}$. You can see that this can be written

$$\begin{pmatrix} 1200 \\ 200 \end{pmatrix} + \begin{pmatrix} 600 \\ -300 \end{pmatrix} = \begin{pmatrix} 1800 \\ -100 \end{pmatrix}$$

where the top line is $1200 + 600 = 1800$
and the bottom line is $200 + -300 = -100$.

This is how we *add vectors*. Using letters, this can be written

$$\overrightarrow{AB} + \overrightarrow{BC} = \overrightarrow{AC}$$

During the race, Tim misses the marshall at E, and continues to run in a straight line. He runs the same distance and direction again as he had just run from D to E to end up at F. Now

$$\overrightarrow{DF} = \begin{pmatrix} -600 \\ -600 \end{pmatrix} \quad \text{and} \quad DE = \begin{pmatrix} -300 \\ -300 \end{pmatrix}$$

Since DF is twice DE, then

$$\begin{pmatrix} -600 \\ -600 \end{pmatrix} = 2 \times \begin{pmatrix} -300 \\ -300 \end{pmatrix} = 2\begin{pmatrix} -300 \\ -300 \end{pmatrix} \quad \text{[the } \times \text{ sign can be omitted]}$$

This is how we multiply a vector by an ordinary number. (An ordinary number is sometimes called a *scalar*.)

We can also *subtract vectors* as follows:

If $\overrightarrow{PQ} = \begin{pmatrix} 40 \\ 80 \end{pmatrix}$, and $\overrightarrow{RS} = \begin{pmatrix} 20 \\ 90 \end{pmatrix}$, then

$$\overrightarrow{PQ} - \overrightarrow{RS} = \begin{pmatrix} 40 \\ 80 \end{pmatrix} - \begin{pmatrix} 20 \\ 90 \end{pmatrix} = \begin{pmatrix} 20 \\ -10 \end{pmatrix}$$

Here the top line is $40 - 20 = 20$
and the bottom line is $80 - 90 = -10$.

Exercise 12b

In the following examples, take $\overrightarrow{AB} = \begin{pmatrix} 20 \\ 40 \end{pmatrix}$, $\overrightarrow{BC} = \begin{pmatrix} 30 \\ -10 \end{pmatrix}$ and $\overrightarrow{CD} = \begin{pmatrix} -20 \\ 20 \end{pmatrix}$.

Write as a column vector:

(1) $\overrightarrow{AB} + \overrightarrow{BC}$ (2) $2\overrightarrow{CD} + \overrightarrow{BC}$ (3) $\overrightarrow{BA}$

(4) $\overrightarrow{AB} + \overrightarrow{BC} + \overrightarrow{CD}$ (5) $\overrightarrow{AB} - \overrightarrow{BC}$ (6) $2\overrightarrow{AB} - \overrightarrow{BC}$

(7) $3\overrightarrow{AB} + 2\overrightarrow{BC} - \overrightarrow{CD}$ (8) $2\overrightarrow{AB} + \overrightarrow{BC}$

(9) $\overrightarrow{AD}$ (10) $\overrightarrow{CA}$

12.2 Vectors in geometry

Mathematicians have found that vectors can be used to solve problems in geometry. The techniques are surprisingly similar to those used in the cross-country race situation except that a slightly more precise notation is used, where a single letter ***a***, ***b*** will often be used instead of the notation $\overrightarrow{XY}$ etc.

Worked example 12.1

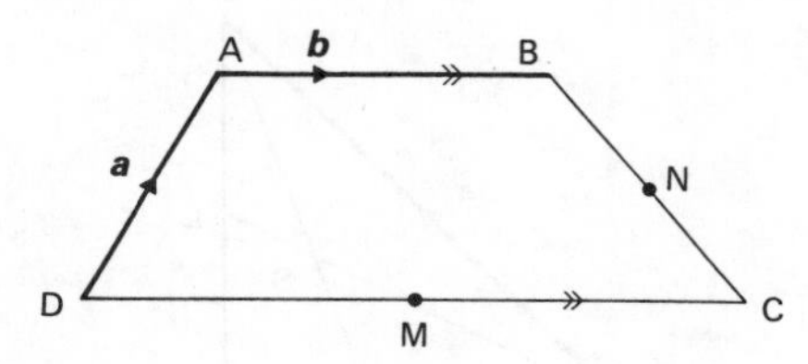

Figure 12.2

In Figure 12.2, ABCD is a trapezium. AB is parallel to DC, and DC = 2AB. M is the mid-point of DC, and N is the mid-point of BC. If $\overrightarrow{AB} = \boldsymbol{b}$ and $\overrightarrow{DA} = \boldsymbol{a}$, express in terms of ***a*** and ***b*** only:

(i) $\overrightarrow{DB}$ (ii) $\overrightarrow{DC}$ (iii) $\overrightarrow{BC}$ (iv) $\overrightarrow{BN}$

(v) $\overrightarrow{AM}$ (vi) $\overrightarrow{NM}$

The secret of this type of problem is to still think of a route around the diagram as we did on the cross-country course.

(i) $\overrightarrow{DB}$ means D to B which is the same as D to A to B $= \boldsymbol{a} + \boldsymbol{b}$.

(ii) $\overrightarrow{DC}$ is in the same direction as $\overrightarrow{AB}$ and is twice the length. So $\overrightarrow{DC} = 2\boldsymbol{b}$.

(iii) BC is a little more involved.
B to C is B to A to D to C $= -\boldsymbol{b} + -\boldsymbol{a} + 2\boldsymbol{b} = \boldsymbol{b} - \boldsymbol{a}$. [Note that $-\boldsymbol{b}$ is in the opposite direction to $\boldsymbol{b}$.]

(iv) $\overrightarrow{BN}$ is half the route from B to C
so $\overrightarrow{BN} = \frac{1}{2}(\boldsymbol{b} - \boldsymbol{a})$.

(v) $\overrightarrow{AM}$ is A to D to M, and D to M is half of D to C, which is the same as A to B.
So $\overrightarrow{AM} = -\boldsymbol{a} + \boldsymbol{b} = \boldsymbol{b} - \boldsymbol{a}$.

(vi) $\overrightarrow{NM}$ can be travelled in several ways. The quickest is N to C to M.
Now $\overrightarrow{NC} = \overrightarrow{BN} = \frac{1}{2}(\boldsymbol{b} - \boldsymbol{a})$
$\overrightarrow{CM} = -\boldsymbol{b}$
So $\overrightarrow{NM} = \frac{1}{2}(\boldsymbol{b} - \boldsymbol{a}) - \boldsymbol{b} = \frac{1}{2}\boldsymbol{b} - \frac{1}{2}\boldsymbol{a} - \boldsymbol{b} = -\frac{1}{2}\boldsymbol{b} - \frac{1}{2}\boldsymbol{a}$.

Worked example 12.2

In the triangle OAB, $\overrightarrow{OA} = \boldsymbol{a}$ and $\overrightarrow{OB} = \boldsymbol{b}$. L is a point on the side AB. M is a point on the side OB, and OL and AM meet at S. It is given that AS = SM and OS/OL $= \frac{3}{4}$; also that OM/OB $= h$ and AL/AB $= k$.

(a) express the vectors $\overrightarrow{AM}$ and $\overrightarrow{OS}$ in terms of a, b and h;

(b) express the vectors $\overrightarrow{OL}$ and $\overrightarrow{OS}$ in terms of a, b and k.

Find h and k, and hence find the values of the ratios OM/MB and AL/LB.

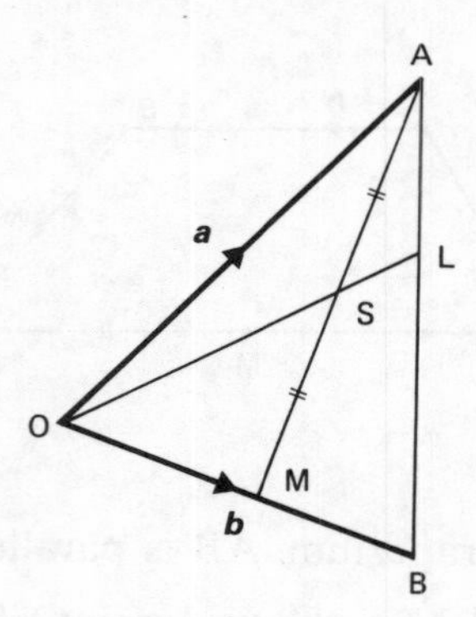

Figure 12.3

Referring to Figure 12.3, since

$$\frac{OS}{OL} = \frac{3}{4}, \ \overrightarrow{OS} = \frac{3}{4}\overrightarrow{OL}$$

$$\frac{OM}{OB} = h, \text{ so } OM = hOB$$

Using vectors, therefore:

$$\overrightarrow{OM} = h\boldsymbol{b}$$

also $\quad \overrightarrow{AL} = k\overrightarrow{AB}$

Now $\quad \overrightarrow{AB} = -\boldsymbol{a} + \boldsymbol{b} = \boldsymbol{b} - \boldsymbol{a}$

Hence $\quad \overrightarrow{AL} = k(\boldsymbol{b} - \boldsymbol{a})$

(a) $\overrightarrow{AM} = \overrightarrow{AO} + \overrightarrow{OM} = -\boldsymbol{a} + h\boldsymbol{b}$

$$\begin{aligned}\overrightarrow{OS} &= \overrightarrow{OA} + \overrightarrow{AS}\\ &= \overrightarrow{OA} + \tfrac{1}{2}\overrightarrow{AM}\\ &= \boldsymbol{a} + \tfrac{1}{2}(-\boldsymbol{a} + h\boldsymbol{b})\\ \overrightarrow{OS} &= \tfrac{1}{2}\boldsymbol{a} + \tfrac{1}{2}h\boldsymbol{b}\end{aligned}$$

(b) $\overrightarrow{OL} = \overrightarrow{OA} + \overrightarrow{AL} = \boldsymbol{a} + k(\boldsymbol{b} - a) = (1-k)\boldsymbol{a} + k\boldsymbol{b}$

$\overrightarrow{OS} = \tfrac{3}{4}\overrightarrow{OL} = \tfrac{3}{4}(1-k)\boldsymbol{a} + \tfrac{3}{4}k\boldsymbol{b}$

Since both expressions for $\overrightarrow{OS}$ must be the same:

$$\tfrac{1}{2}\boldsymbol{a} + \tfrac{1}{2}h\boldsymbol{b} = \tfrac{3}{4}(1-k)\boldsymbol{a} + \tfrac{3}{4}k\boldsymbol{b}$$

Hence $\quad [\tfrac{1}{2} - \tfrac{3}{4}(1-k)]\boldsymbol{a} = (\tfrac{3}{4}k - \tfrac{1}{2}h)\boldsymbol{b}$

Since $\boldsymbol{a}$ and $\boldsymbol{b}$ are not parallel:

$$\tfrac{1}{2} - \tfrac{3}{4}(1-k) = 0$$

and $\quad \tfrac{3}{4}k - \tfrac{1}{2}h = 0$

The solutions of these are $k = \tfrac{1}{3}$, $h = \tfrac{1}{2}$.

Hence the ratios $\dfrac{OM}{MB} = 1$, $\dfrac{AL}{LB} = \tfrac{1}{2}$.

12.3 Unit vectors *i*, *j*

A vector which has length 1, is called a unit vector. The unit vectors $\boldsymbol{i}$ and $\boldsymbol{j}$ are defined by $\boldsymbol{i} = \begin{pmatrix}1\\0\end{pmatrix}$ and $\boldsymbol{j} = \begin{pmatrix}0\\1\end{pmatrix}$. These unit vectors can also be used to represent vectors. Consider the vector $\overrightarrow{AB} = \begin{pmatrix}4\\-2\end{pmatrix}$. We can write $\begin{pmatrix}4\\-2\end{pmatrix} = \begin{pmatrix}4\\0\end{pmatrix} - \begin{pmatrix}0\\2\end{pmatrix} =$ $4\begin{pmatrix}1\\0\end{pmatrix} - 2\begin{pmatrix}0\\1\end{pmatrix} = 4\boldsymbol{i} - 2\boldsymbol{j}$ (see Figure 12.4 overleaf)

Exercise 12c

(1) If $\boldsymbol{p} = 2\boldsymbol{i} - \boldsymbol{j}$, $\boldsymbol{q} = 3\boldsymbol{i} + 2\boldsymbol{j}$, $\boldsymbol{r} = 4\boldsymbol{i} - 2\boldsymbol{j}$:
Find (i) $\boldsymbol{p} + 3\boldsymbol{q}$, (ii) $2\boldsymbol{p} - 3\boldsymbol{q}$, (iii) $\boldsymbol{r} - 2\boldsymbol{q}$.

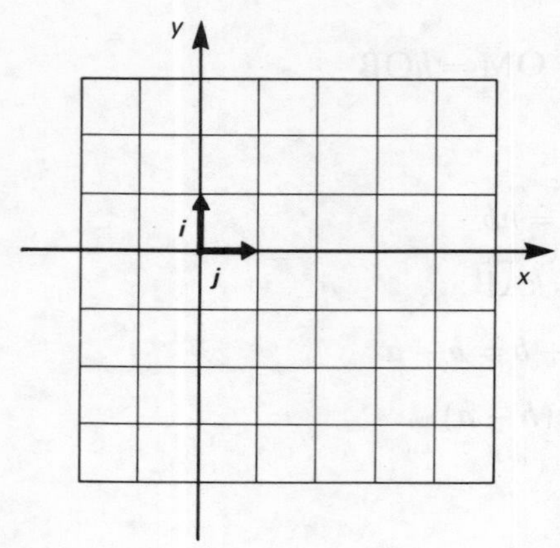

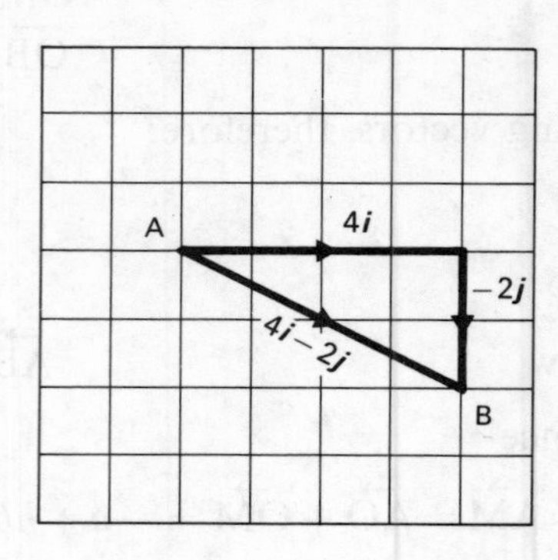

Figure 12.4

(2) OABC is a quadrilateral $\overrightarrow{OA} = \boldsymbol{a}$, $\overrightarrow{AB} = \boldsymbol{b}$, and $\overrightarrow{BC} = \boldsymbol{c}$. Write down the following vectors in terms of $\boldsymbol{a}$, $\boldsymbol{b}$ and $\boldsymbol{c}$:

(a) $\overrightarrow{AC}$ (b) $\overrightarrow{OC}$ (c) $\overrightarrow{OB}$

(3) RSTU is a rectangle. The diagonals of the rectangle meet in M. If $\overrightarrow{MR} = \boldsymbol{a}$ and $\overrightarrow{MS} = \boldsymbol{b}$, write down in terms of $\boldsymbol{a}$ and $\boldsymbol{b}$:

(a) $\overrightarrow{RS}$ (b) $\overrightarrow{ST}$ (c) $\overrightarrow{US}$

(4) X is the point (4, 1), Y is the point (−2, −1) and Z is the point (5, −3). Express, as a 2×1 column vector, the following:

(a) $\overrightarrow{XY}$ (b) $\overrightarrow{YZ}$ (c) $\overrightarrow{XZ}$

(5) O (0, 0), A (3, 1) and B (2, −1) are three vertices of a quadrilateral OABC. If $\overrightarrow{BC} = \begin{pmatrix} 3 \\ -2 \end{pmatrix}$, what are the co-ordinates of C? Find $\overrightarrow{AC}$.

(6) ABCD is quadrilateral, and A, B, D have co-ordinates (0, 2), (2, 5), (8, 0) respectively. If $\overrightarrow{AD} = 2\overrightarrow{BC}$, find the co-ordinates of C.

AC is extended to P so that $\overrightarrow{CP} = 2\overrightarrow{AC}$. Find the co-ordinates of P.

12.4 Matrix multiplication

It is assumed in the following work that the reader has a knowledge of simple matrices, and a reminder of how to multiply them is given here.

The numbers in the following example have been carefully chosen to show the method clearly:

$$\begin{pmatrix} 2 & 5 \\ 6 & 7 \\ 9 & 8 \end{pmatrix} \begin{pmatrix} 10 & 0 & 4 \\ 3 & 16 & 1 \end{pmatrix}$$

3×2 order ↗ rows ↖ columns $\qquad 2 \times 3$ order

$$= \begin{pmatrix} 2\times 10+5\times 3 & 2\times 0+5\times 16 & 2\times 4+5\times 1 \\ 6\times 10+7\times 3 & 6\times 0+7\times 16 & 6\times 4+7\times 1 \\ 9\times 10+8\times 3 & 9\times 0+8\times 16 & 9\times 4+8\times 1 \end{pmatrix}$$

$$= \begin{pmatrix} 35 & 80 & 13 \\ 81 & 112 & 31 \\ 114 & 128 & 44 \end{pmatrix}$$

3×3 order

Two matrices can only be multiplied if the number of columns in the first matrix equals the number of rows in the second. This means that

$$\underset{2\times 3}{\begin{pmatrix} 10 & 0 & 4 \\ 3 & 16 & 1 \end{pmatrix}} \quad \underset{3\times 2}{\begin{pmatrix} 2 & 5 \\ 6 & 7 \\ 9 & 8 \end{pmatrix}} = \underset{2\times 2}{\begin{pmatrix} 56 & 82 \\ 111 & 135 \end{pmatrix}}$$

12.5 Matrix inverse

If $A = \begin{pmatrix} 4 & 3 \\ 5 & 6 \end{pmatrix}$ and $B = \begin{pmatrix} \frac{6}{9} & -\frac{3}{9} \\ -\frac{5}{9} & \frac{4}{9} \end{pmatrix}$,

then $AB = \begin{pmatrix} 4 & 3 \\ 5 & 6 \end{pmatrix}\begin{pmatrix} \frac{6}{9} & -\frac{3}{9} \\ -\frac{5}{9} & \frac{4}{9} \end{pmatrix} = \begin{pmatrix} 1 & 0 \\ 0 & 1 \end{pmatrix}$, ← this is called the *identity matrix*

and $BA = \begin{pmatrix} \frac{6}{9} & -\frac{3}{9} \\ -\frac{5}{9} & \frac{4}{9} \end{pmatrix}\begin{pmatrix} 4 & 3 \\ 5 & 6 \end{pmatrix} = \begin{pmatrix} 1 & 0 \\ 0 & 1 \end{pmatrix}$. ←

We say that A is the *inverse* of B, or B is the inverse of A.

In general, if $A = \begin{pmatrix} a & b \\ c & d \end{pmatrix}$ then the inverse of $A = \begin{pmatrix} \frac{d}{\Delta} & \frac{-b}{\Delta} \\ \frac{-c}{\Delta} & \frac{a}{\Delta} \end{pmatrix}$ where $\Delta = ad - bc$.

Δ is called the *determinant of the matrix*.

If $\Delta = 0$, then the matrix will have no inverse, and is called a *singular* matrix. A matrix which has an inverse is called *non-singular*.

Exercise 12d

Find the inverse of the following matrices where possible:

(1) $\begin{pmatrix} 4 & 2 \\ 5 & 3 \end{pmatrix}$ (2) $\begin{pmatrix} 6 & -7 \\ -5 & 6 \end{pmatrix}$ (3) $\begin{pmatrix} 4 & 2 \\ -8 & -4 \end{pmatrix}$

(4) $\begin{pmatrix} 2 & 3 \\ -6 & 9 \end{pmatrix}$ (5) $\begin{pmatrix} 1 & 0 \\ 0 & -1 \end{pmatrix}$ (6) $\begin{pmatrix} 6 & -4 \\ 3 & 11 \end{pmatrix}$

12.6 Moving points with matrices

We can move a point on a co-ordinate grid using a 2×2 matrix, which we shall call a *transformation matrix*, using simple matrix multiplication.

Consider the matrix $M = \begin{pmatrix} 2 & 3 \\ 1 & 1 \end{pmatrix}$. If we want to move the point A$(1, -1)$, then write the co-ordinates of A as a column vector $\begin{pmatrix} 1 \\ -1 \end{pmatrix}$, and then multiply this vector by the matrix M.

That is:

$$\begin{pmatrix} 2 & 3 \\ 1 & 1 \end{pmatrix}\begin{pmatrix} 1 \\ -1 \end{pmatrix} = \begin{pmatrix} -1 \\ 0 \end{pmatrix}.$$

Since M is in front of the vector, it is called premultiplying. The new position of A, usually denoted by A′, is called the *image* of A under the transformation M. If we were trying to find the image of more than one point, say A$(1, -1)$, B$(2, 3)$, C$(3, 0)$ and D$(2, -3)$, this can be done in one procedure.

$$\begin{pmatrix} 2 & 3 \\ 0 & 1 \end{pmatrix}\overset{\begin{matrix}\text{A} & \text{B} & \text{C} & \text{D}\end{matrix}}{\begin{pmatrix} 1 & 2 & 3 & 2 \\ -1 & 3 & 0 & -3 \end{pmatrix}} = \overset{\begin{matrix}\text{A}' & \text{B}' & \text{C}' & \text{D}'\end{matrix}}{\begin{pmatrix} -1 & 13 & 6 & -5 \\ -1 & 3 & 0 & -3 \end{pmatrix}}$$

If the points A, B, C and D are plotted to form a shape, the image shape A′B′C′D′ can then be drawn as shown in Figure 12.5.

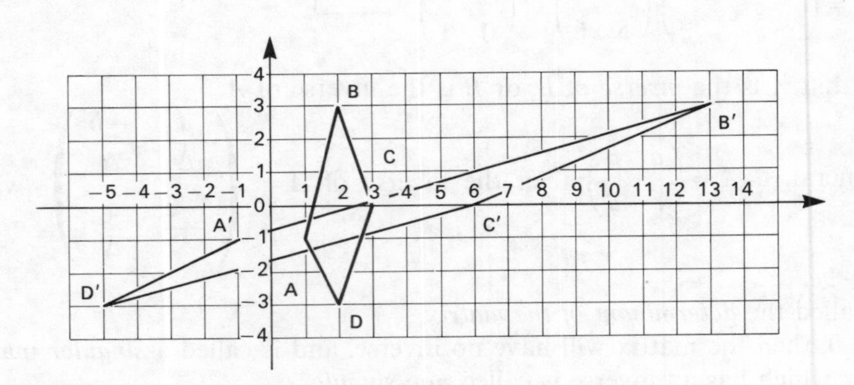

Figure 12.5

If you divide the shapes up into suitable triangles, then you find:

(i) the area of ABCD = 6,
(ii) the area of A′B′C′D′ = 12.

[Try and check these answers.]

Now the determinant of $\begin{pmatrix} 2 & 3 \\ 0 & 1 \end{pmatrix} = 2 \times 1 - 3 \times 0 = 2.$

The area of A′B′C′D′ = 2 × the area of ABCD.

The determinant of a matrix gives the multiplication factor for area in a transformation.

12.7 The unit square

The unit square is the name usually given to the set of points O(0, 0), A(1, 0) B(1, 1) and C(0, 1). This set of points simply forms a square with side of length 1 unit. The unit square is very helpful, if we are trying to interpret what effect a transformation matrix produces. The following section illustrates how we can use the unit square.

12.8 Transformation using vectors and matrices

(i) Translation

In Figure 12.6, each point of P has been mapped onto Q, by a simple *displacement* or *translation.* A translation can be represented by a vector.

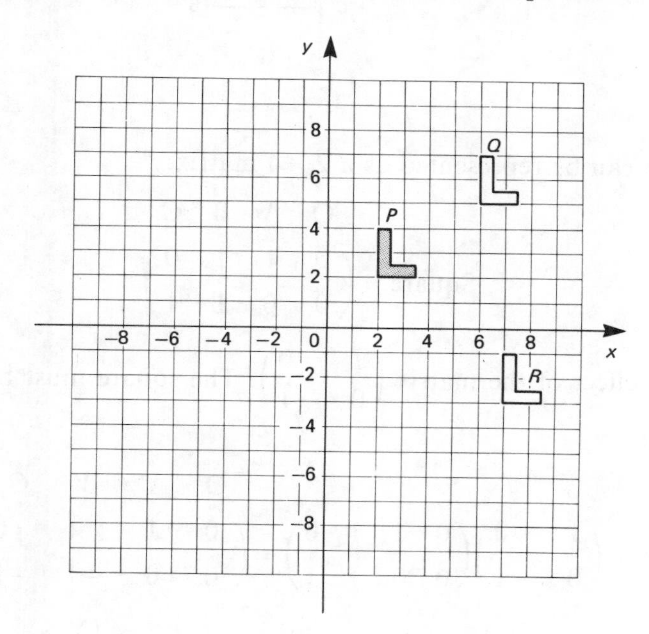

Figure 12.6

We write $P \to Q$ under the translation $\boldsymbol{T}_1 = \begin{pmatrix} 4 \\ 3 \end{pmatrix}$, or P *maps onto* Q. Similarly, $Q \to R$ under the translation $\boldsymbol{T}_2 = \begin{pmatrix} 1 \\ -8 \end{pmatrix}$.

If we wish to carry out one transformation followed by another, i.e. $\boldsymbol{T}_1$ followed by $\boldsymbol{T}_2$, it is denoted by $\boldsymbol{T}_2\boldsymbol{T}_1$. *Note the order of letters.*

Now $P \to R$ under the translation $\boldsymbol{T}_3 = \begin{pmatrix} 5 \\ -5 \end{pmatrix}$ but $\boldsymbol{T}_3$ is the same as $\boldsymbol{T}_2\boldsymbol{T}_1$. It can be seen that combining translations is the same as adding vectors.

Hence $$\boldsymbol{T}_2\boldsymbol{T}_1 = \begin{pmatrix} 1 \\ -8 \end{pmatrix} + \begin{pmatrix} 4 \\ 3 \end{pmatrix} = \begin{pmatrix} 5 \\ -5 \end{pmatrix} = \boldsymbol{T}_3$$

(ii) Reflection

The unit square will be used here.

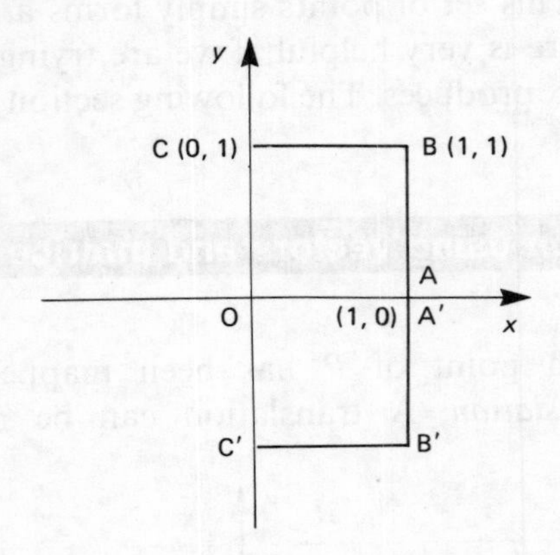

Figure 12.7

The square can be represented as a 2×4 matrix:

$$\begin{matrix} & \text{O} & \text{A} & \text{B} & \text{C} \end{matrix}$$
$$\text{Square} = \begin{pmatrix} 0 & 1 & 1 & 0 \\ 0 & 0 & 1 & 1 \end{pmatrix}$$

Consider the effect of the matrix $\begin{pmatrix} 1 & 0 \\ 0 & -1 \end{pmatrix}$. The square must be *pre-multiplied* by the matrix.

$$\begin{matrix} \text{O}' & \text{A}' & \text{B}' & \text{C}' \end{matrix}$$
So $$\begin{pmatrix} 1 & 0 \\ 0 & -1 \end{pmatrix}\begin{pmatrix} 0 & 1 & 1 & 0 \\ 0 & 0 & 1 & 1 \end{pmatrix} = \begin{pmatrix} 0 & 1 & 1 & 0 \\ 0 & 0 & -1 & -1 \end{pmatrix}$$

Note: the first column of the matrix is the image of $\begin{pmatrix} 1 \\ 0 \end{pmatrix}$, *the second of* $\begin{pmatrix} 0 \\ 1 \end{pmatrix}$.

We can say that the square ABCD has been reflected in the *x*-axis.

Similarly, $\begin{pmatrix} -1 & 0 \\ 0 & 1 \end{pmatrix}$ represents reflection in the *y*-axis

$\begin{pmatrix} 0 & 1 \\ 1 & 0 \end{pmatrix}$ represents reflection in the line $y = x$

$\begin{pmatrix} 0 & -1 \\ -1 & 0 \end{pmatrix}$ represents reflection in the line $y = -x$.

(iii) Rotation

Consider the matrix $\begin{pmatrix} 0 & 1 \\ -1 & 0 \end{pmatrix}$:

$$\begin{matrix} & & \text{A} & \text{B} & \text{C} & & \text{A}' & \text{B}' & \text{C}' \end{matrix}$$
$$\begin{pmatrix} 0 & 1 \\ -1 & 0 \end{pmatrix}\begin{pmatrix} 1 & 1 & 0 \\ 0 & 1 & 1 \end{pmatrix} = \begin{pmatrix} 0 & 1 & 1 \\ -1 & -1 & 0 \end{pmatrix}$$

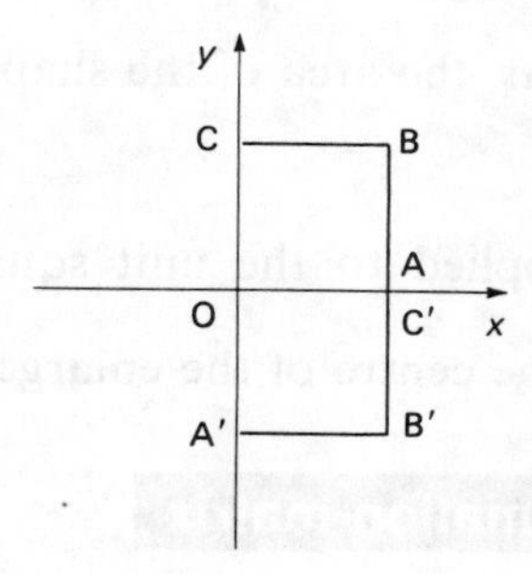

Figure 12.8

The square appears to be in the same place as a reflection (Figure 12.8), but on looking at the labelling of the points, it can be seen that the square has been rotated *clockwise* by 90° about O, i.e. −90°. Clockwise angles are designated as being negative. Similarly, $\begin{pmatrix} 0 & -1 \\ 1 & 0 \end{pmatrix}$ is a rotation of +90° about O and $\begin{pmatrix} -1 & 0 \\ 0 & -1 \end{pmatrix}$ is a half-turn about O.

(iv) Shear

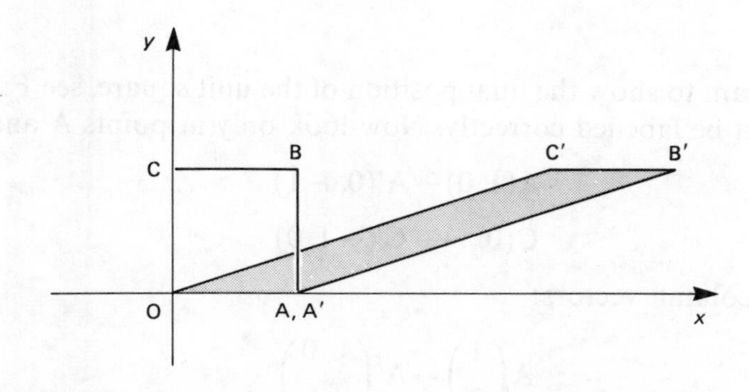

Figure 12.9

Consider the effect of $\begin{pmatrix} 1 & 3 \\ 0 & 1 \end{pmatrix}$ on the unit square

$$\begin{pmatrix} 1 & 3 \\ 0 & 1 \end{pmatrix}\begin{pmatrix} 1 & 1 & 0 \\ 0 & 1 & 1 \end{pmatrix} = \begin{pmatrix} 1 & 4 & 3 \\ 0 & 1 & 1 \end{pmatrix}$$

The square has been sheared parallel to the x-axis, the x-axis is the *invariant line* and $(1, 1) \rightarrow (4, 1)$. [Invariant means it stays the same.]

In general, $\begin{pmatrix} 1 & k \\ 0 & 1 \end{pmatrix}$ is a shear of shear factor k parallel to the x-axis

$\begin{pmatrix} 1 & 0 \\ k & 1 \end{pmatrix}$ is a shear of shear factor k parallel to the y-axis.

You can see that in a shear, the area of the shape stays the same.

(v) Enlargement

If the matrix $\begin{pmatrix} k & 0 \\ 0 & k \end{pmatrix}$ is applied to the unit square, it simply enlarges it by a scale factor k, with O as the centre of the enlargement.

12.9 Finding the transformation matrix

If you are asked, what is the 2×2 matrix that will reflect a shape in the line $y = -x$, without remembering the answer, how can it be found? The unit square will come to our aid.

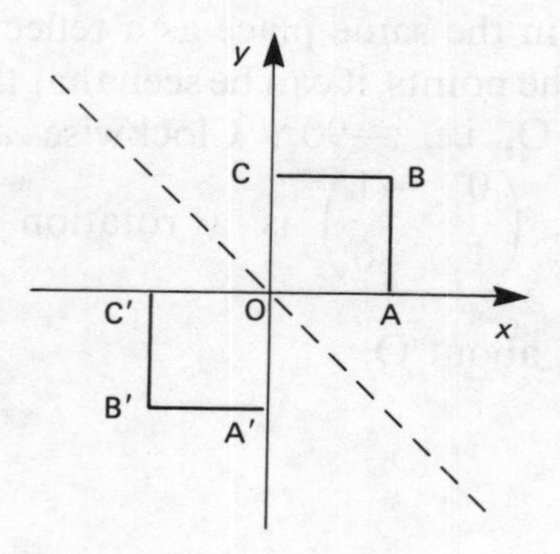

Figure 12.10

Draw a diagram to show the final position of the unit square, see Figure 12.10. The points must be labelled correctly. Now look only at points A and C.

$$A(1, 0) \rightarrow A'(0, -1)$$

$$C(0, 1) \rightarrow C'(-1, 0)$$

Write these as column vectors:

$$A\begin{pmatrix} 1 \\ 0 \end{pmatrix} \rightarrow A'\begin{pmatrix} 0 \\ -1 \end{pmatrix}$$

$$C\begin{pmatrix}0\\1\end{pmatrix}\rightarrow C'\begin{pmatrix}-1\\0\end{pmatrix}$$

Write down column A′ followed by column C′ (it must be in this order):

$$A'C'\begin{pmatrix}0 & -1\\-1 & 0\end{pmatrix}$$

The 2×2 matrix you require is $\begin{pmatrix}0 & -1\\-1 & 0\end{pmatrix}$.

12.10 Combined transformation

If a shape X is transformed by a matrix A into shape Y, and then shape Y is transformed into shape Z by a matrix B, then to find a single matrix to represent A followed by B, we evaluate BA. (The order is important.)

Worked example 12.3

A triangle ABC has co-ordinates A(2, 2), B(2, 4) and C(0, 4). Plot the points ABC, and also find the co-ordinates of the triangle A′B′C′, obtained from ABC by the transformation matrix $M=\begin{pmatrix}1 & 0\\0 & -1\end{pmatrix}$. A second triangle A″B″C″ is obtained from A′B′C′ by the transformation $N=\begin{pmatrix}0 & 1\\-1 & 0\end{pmatrix}$. Plot the triangle A″B″C″ and find the single matrix that will map the triangle ABC onto the triangle A″B″C″. Describe the inverse transformation that maps A″B″C″ back onto ABC, and find the matrix of this transformation.

$$\begin{pmatrix}1 & 0\\0 & -1\end{pmatrix}\overset{\text{A}\quad\text{B}\quad\text{C}}{\begin{pmatrix}2 & 2 & 0\\2 & 4 & 4\end{pmatrix}}=\overset{\text{A}'\quad\text{B}'\quad\text{C}'}{\begin{pmatrix}2 & 2 & 0\\-2 & -4 & -4\end{pmatrix}}$$

$$\begin{pmatrix}0 & 1\\-1 & 0\end{pmatrix}\overset{\text{A}'\quad\text{B}'\quad\text{C}'}{\begin{pmatrix}2 & 2 & 0\\-2 & -4 & -4\end{pmatrix}}=\overset{\text{A}''\quad\text{B}''\quad\text{C}''}{\begin{pmatrix}-2 & -4 & -4\\-2 & -2 & 0\end{pmatrix}}$$

The triangles ABC, A′B′C′ and A″B″C″ are shown in Figure 12.11.

The single transformation for M followed by N is NM, i.e.

$$NM=\begin{pmatrix}0 & 1\\-1 & 0\end{pmatrix}\begin{pmatrix}1 & 0\\0 & -1\end{pmatrix}=\begin{pmatrix}0 & -1\\-1 & 0\end{pmatrix}$$

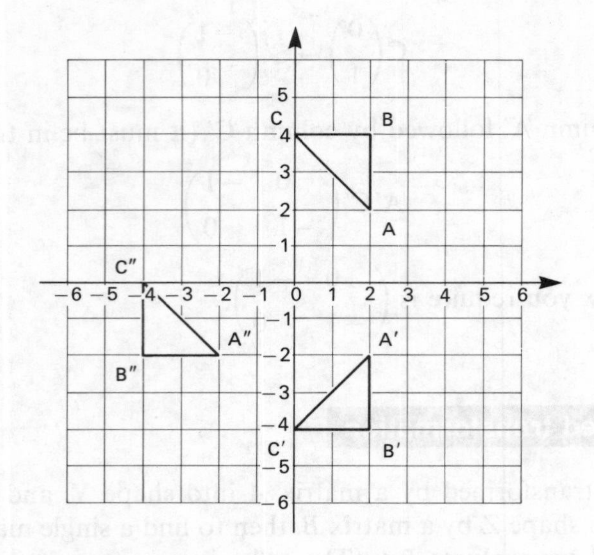

Figure 12.11

The inverse transformation is reflection in the line $y = -x$. The matrix of this is $\begin{pmatrix} 0 & -1 \\ -1 & 0 \end{pmatrix}$.

Suggestions for coursework 12

1. Investigate the transformation effects that can be produced by a matrix $\begin{pmatrix} a & b \\ c & d \end{pmatrix}$ when a, b, c and d can only be 0 or 1.
2. In 'gridland', all places are represented by points on a 2-dimensional grid. The grid lines are at right angles, and places are represented by co-ordinates (x, y), where $1 \leqslant x \leqslant 10$, and $1 \leqslant y \leqslant 10$. If you lived in gridland at (1, 4) and wanted to visit a friend at (5, 7), which way would you go? Would your answer be different if the grid lines are not at right angles? How many different ways are there of travelling from (1, 4) to (5, 7)? Consider similar problems.
3. Investigate the uses of vectors in a subject area of your choice.
4. Write a computer program that will carry out rotations, by the use of matrices.
5. Design a wallpaper pattern.
6. Investigate matrices of the type $\begin{pmatrix} a & b \\ c & d \end{pmatrix}$, where $ad - bc = 1$.

Miscellaneous exercise 12

1. On the given isometric grid in Figure 12.12 $\overrightarrow{OA}$ and $\overrightarrow{OB}$ represent $\boldsymbol{a}$ and $\boldsymbol{b}$ respectively.

   Express, in terms of $\boldsymbol{a}$ and $\boldsymbol{b}$, the translation which maps
   (a) shape X onto shape Y,
   (b) Shape X onto shape Z. (NEA)

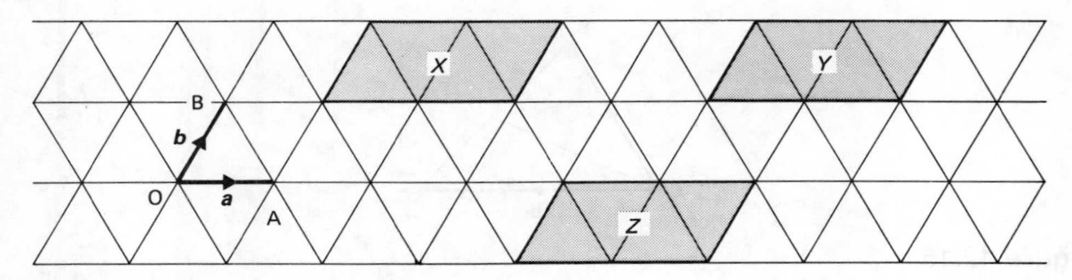

Figure 12.12

2. Figure 12.13, not drawn to scale, shows a parallelogram OABC, with diagonal OB. D is on OB so that $OD = \frac{2}{3}OB$. CD extended meets AB at E. $\overrightarrow{OA} = \boldsymbol{a}$ and $\overrightarrow{OC} = \boldsymbol{c}$.

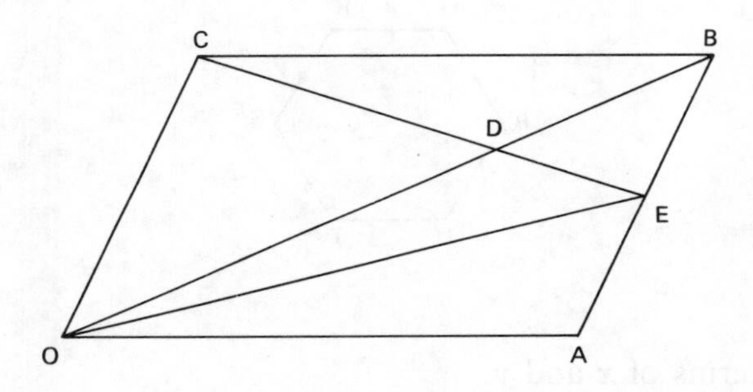

Figure 12.13

(a) Express $\overrightarrow{OB}$, $\overrightarrow{OD}$ and $\overrightarrow{CD}$ in terms of $\boldsymbol{a}$ and $\boldsymbol{c}$.

(b) (i) Write down $\overrightarrow{OA} + \frac{1}{2}\overrightarrow{AB}$ in terms of $\boldsymbol{a}$ and $\boldsymbol{c}$.

(ii) Write down, and simplify, $\overrightarrow{OC} + \frac{3}{2}\overrightarrow{CD}$ in terms of $\boldsymbol{a}$ and $\boldsymbol{c}$.

(iii) Hence write down $\overrightarrow{OE}$ in terms of $\boldsymbol{a}$ and $\boldsymbol{c}$.

(iv) What is the value of $\frac{AE}{EB}$? (SEG)

3. In Figure 12.14, ABCD is a parallelogram. M is the mid-point of BC, and N is the mid-point of CD. $\overrightarrow{AB} = \boldsymbol{a}$ and $\overrightarrow{AD} = \boldsymbol{b}$. Express in terms of $\boldsymbol{a}$ and $\boldsymbol{b}$:

(i) $\overrightarrow{BM}$, (ii) $\overrightarrow{CD}$, (iii) $\overrightarrow{MD}$, (iv) $\overrightarrow{AC}$.

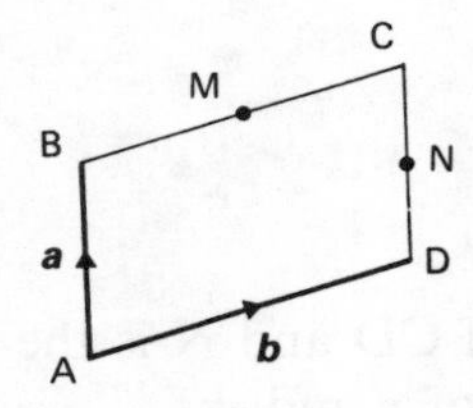

Figure 12.14

4. In Figure 12.15, D is the mid-point of BC in triangle ABC. $\overrightarrow{BA} = \boldsymbol{a}$ and $\overrightarrow{AC} = \boldsymbol{b}$.

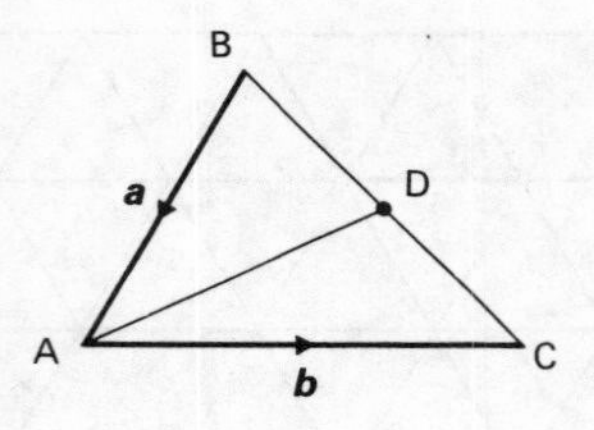

Figure 12.15

Express in terms of $\boldsymbol{a}$ and $\boldsymbol{b}$:

(i) $\overrightarrow{BC}$, (ii) $\overrightarrow{BD}$, (iii) $\overrightarrow{AD}$.

5. In this question, you may find it helpful to draw an accurate regular hexagon PQRSTU as shown in Figure 12.16. $\overrightarrow{QR} = \boldsymbol{x}$ and $\overrightarrow{RS} = \boldsymbol{y}$.

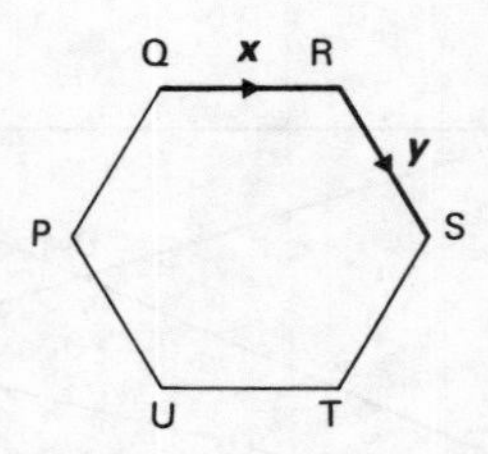

Figure 12.16

Express in terms of $\boldsymbol{x}$ and $\boldsymbol{y}$:

(i) $\overrightarrow{UP}$, (ii) $\overrightarrow{QT}$, (iii) $\overrightarrow{ST}$.

6. The point O is the centre of the regular hexagon ABCDEF. $\overrightarrow{FA} = \boldsymbol{x}$ and $\overrightarrow{FE} = \boldsymbol{y}$.

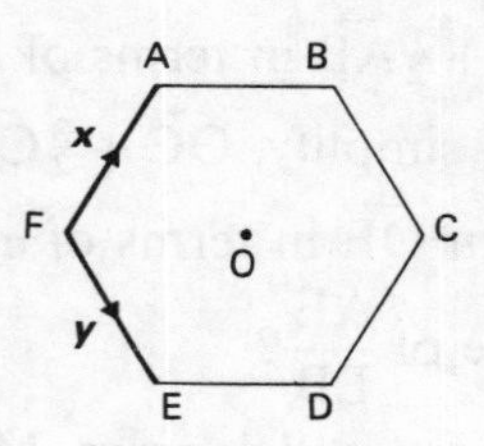

Figure 12.17

(a) Express in terms of $\boldsymbol{x}$ and $\boldsymbol{y}$:

(i) $\overrightarrow{FO}$

(ii) $\overrightarrow{FC}$

(iii) $\overrightarrow{FB}$

(iv) $\overrightarrow{FD}$.

(b) If M is the midpoint of CD and N is the point which divides BE in the ratio 5:3, express in terms of $\boldsymbol{x}$ and $\boldsymbol{y}$:

(i) $\overrightarrow{FM}$

(ii) $\overrightarrow{FN}$.

(c) What conclusions can you draw about the points M and N? (MEG)

7. A, B and C are three points. B is 6 km East and 3 km North of A. C is 2 km West and 1 km North of B. The unit base vectors $\boldsymbol{i}$ and $\boldsymbol{j}$ have lengths 1 km and point East and North respectively.

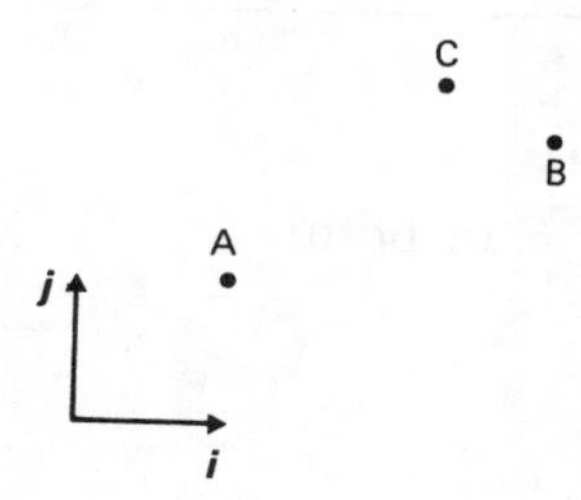

Figure 12.18

(a) Find, in terms of $\boldsymbol{i}$ and $\boldsymbol{j}$, expressions for the vectors:
(i) $\overrightarrow{AB}$, (ii) $\overrightarrow{BC}$, (iii) $\overrightarrow{AC}$.

(b) Calculate the distance in kilometres from A to C, giving your answer correct to 2 decimal places. (SEG)

8. The point X lies on the side OA of the triangle OAB and $OX = \frac{1}{3}OA$. The vectors $\boldsymbol{a}$ and $\boldsymbol{b}$ are such that $\overrightarrow{OA} = \boldsymbol{a}$ and $\overrightarrow{OB} = \boldsymbol{b}$. The point P is such that $\overrightarrow{OP} = \frac{1}{5}\boldsymbol{a} + \frac{2}{5}\boldsymbol{b}$.

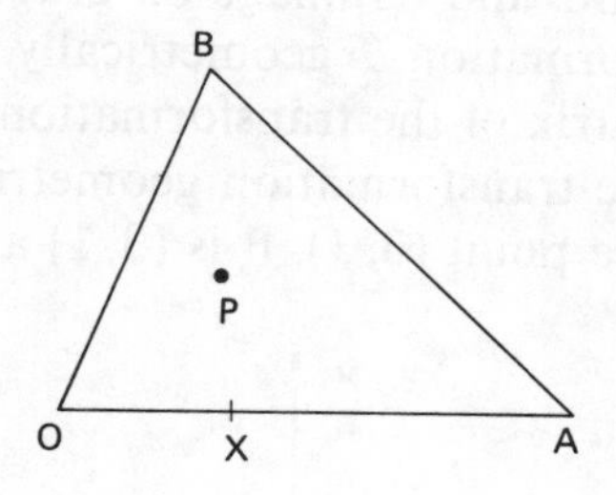

Figure 12.19

(a) Express, in terms of $\boldsymbol{a}$ and $\boldsymbol{b}$, the vectors:
(i) $\overrightarrow{BX}$
(ii) $\overrightarrow{BP}$.

(b) By considering the results of part (a), or otherwise, show that the line BX passes through P and write down the numerical value of $\frac{BP}{BX}$.

(c) Given that the area of triangle OAB is 30 cm^2, calculate:
(i) the area of triangle ABX,
(ii) the area of triangle ABP,
(iii) the area of triangle OPA. (MEG)

9. In the $\triangle OAB$ $\overrightarrow{OA} = \boldsymbol{a}$, $\overrightarrow{OB} = \boldsymbol{b}$. $\overrightarrow{OX} = \frac{3}{4}\overrightarrow{OA}$ and Y divides AB in the ratio 1:3.

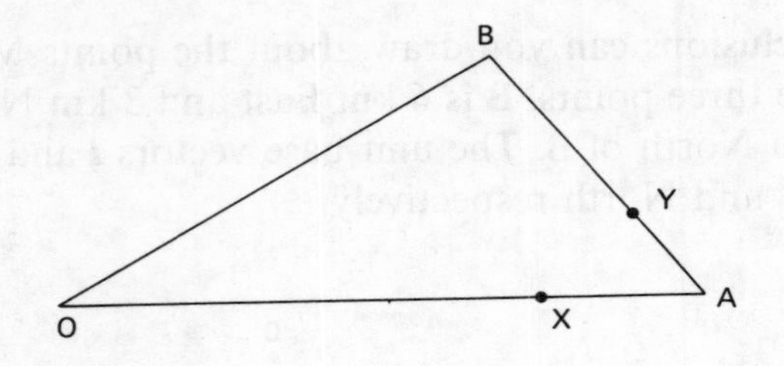

Figure 12.20

(a) Find in terms of $\boldsymbol{a}$, or $\boldsymbol{b}$, or both:

(i) $\overrightarrow{OX}$

(ii) $\overrightarrow{AB}$

(iii) $\overrightarrow{AY}$

(iv) $\overrightarrow{OY}$

(v) $\overrightarrow{XY}$. (SEG)

10. Under a transformation $\boldsymbol{T}$, the point (x, y) is mapped to the point (x', y') where

$$\begin{pmatrix} x' \\ y' \end{pmatrix} = \begin{pmatrix} 0 & 1 \\ -1 & 0 \end{pmatrix} \begin{pmatrix} x \\ y \end{pmatrix}$$

The triangle ABC has vertices A(1, 0), B(2, 1), C(4, 0).

(a) Find the co-ordinates of the images of A, B and C under the transformation $\boldsymbol{T}$.

(b) Plot the triangle ABC and its image on graph paper.

(c) Describe the transformation $\boldsymbol{T}$ geometrically.

(d) Write down the matrix of the transformation which is the inverse of $\boldsymbol{T}$.

(e) Describe the inverse transformation geometrically. (SEG)

11. In Figure 12.21, A is the point (3, 3), B is (3, 1) and C is (4, 1).

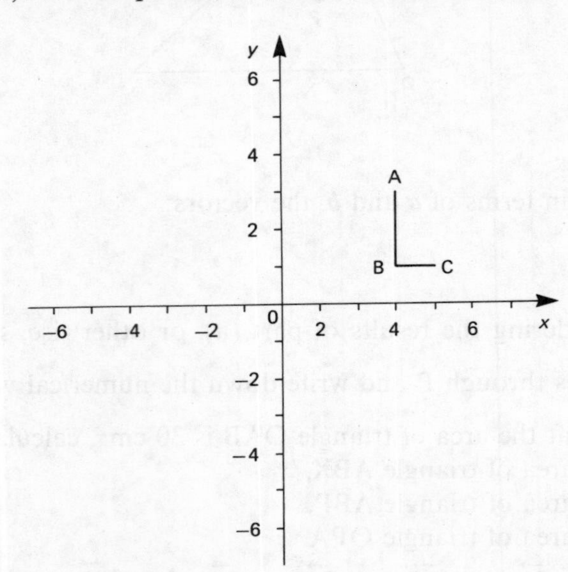

 Figure 12.21

(a) Using a scale of 1 centimetre to represent 1 unit on each axis, draw the diagram on graph paper.

(b) Draw, on your diagram, the image of the shape ABC under the transformation represented by the matrix $\begin{pmatrix} 1 & 0 \\ 0 & -1 \end{pmatrix}$.

Label the image $A_1B_1C_1$.

(c) Draw, on your diagram, the image of the shape $A_1B_1C_1$ under the transformation represented by the matrix $\begin{pmatrix} 0 & 1 \\ -1 & 0 \end{pmatrix}$.

Label the image $A_2B_2C_2$.

(d) (i) Describe fully the single transformation which would map ABC onto $A_2B_2C_2$.

(ii) Find the matrix which represents this transformation. (MEG)

12. Transformation V of the plane has matrix $\begin{pmatrix} 2 & 2 \\ -1 & 1 \end{pmatrix}$.

(a) A finite region R is mapped by V onto the region R′. Calculate the ratio of the area of R′ to the area of R.

(b) Write down the matrix of the inverse transformation V^{-1} and find the co-ordinates of the point which is mapped onto the point $(1, \frac{1}{2})$ by V.

(c) A second transformation U has matrix $\begin{pmatrix} 1 & 2 \\ 3 & 4 \end{pmatrix}$. Obtain the matrix of the transformation UV. (LEAG)

13. The vertices of an isosceles triangle E have co-ordinates A(0, 5), B(−1, 8), C(4, 8).

(a) Using a scale of 1 cm to 1 unit, draw x and y axes, taking values of x from −2 to 14 and values of y from 0 to 16.

Draw and label the icosceles triangle E and the point A.

(b) Transformation S is defined by

$$S: \begin{pmatrix} x \\ y \end{pmatrix} \mapsto \begin{pmatrix} 2 & 0 \\ 0 & 2 \end{pmatrix} \begin{pmatrix} x \\ y \end{pmatrix}$$

(i) Draw and label the isosceles triangle S(E) and the point S(A).

(ii) Describe fully the single transformation S.

(c) Transformation U is defined by

$$U: \begin{pmatrix} x \\ y \end{pmatrix} \mapsto \begin{pmatrix} -1.2 & 1.6 \\ 1.6 & 1.2 \end{pmatrix} \begin{pmatrix} x \\ y \end{pmatrix}$$

(i) Find the co-ordinates of the vertices of the isosceles triangle U(E).

(ii) Draw and label U(E) and the point U(A).

(d) Describe fully, in geometrical terms, a single transformation T that maps the triangle S(E) onto the triangle U(E) so that S(A) maps onto U(A). (LEAG)

14. The vertices of a rectangle OABC are O(0, 0), A(5, 0), B(5, 2) and C(0, 2).

(a) Taking 1 cm to represent 1 unit on each axis and marking each axis from -6 to 6, draw and label the rectangle OABC.

(b) The rectangle OABC is mapped onto rectangle $OA_1B_1C_1$ by the transformation represented by the matrix $\boldsymbol{P}$ where

$$\boldsymbol{P} = \begin{pmatrix} 0 & 1 \\ 1 & 0 \end{pmatrix}$$

Draw and label rectangle $OA_1B_1C_1$ on your diagram, and describe the transformation fully in geometrical terms.

(c) The original rectangle OABC is mapped onto another rectangle $OA_2B_2C_2$ by reflection in the x-axis. Draw and label the rectangle $OA_2B_2C_2$ on your diagram. Write down the matrix $\boldsymbol{Q}$ which represents this transformation.

(d) The rectangle $OA_1B_1C_1$ can be mapped onto the rectangle $OA_2B_2C_2$ by a single transformation represented by matrix $\boldsymbol{R}$. Describe this transformation fully in geometrical terms and state the relationship between the matrices $\boldsymbol{P}$, $\boldsymbol{Q}$, $\boldsymbol{R}$.

(e) Find the smallest positive integer n for which $\boldsymbol{R}^n = \boldsymbol{I}$, where $\boldsymbol{I}$ is the identity matrix.

(MEG)

15. The transformation of a point P with co-ordinates (x, y) into a point P′ with co-ordinates (x', y') is defined by the matrix multiplication:

$$\begin{pmatrix} x' \\ y' \end{pmatrix} = \begin{pmatrix} 3 & 0 \\ 0 & 3 \end{pmatrix} \begin{pmatrix} x \\ y \end{pmatrix}$$

A triangle ABC has co-ordinates P(1, 2), Q(2, 1), R($-2\frac{1}{2}$, 3).

(a) Find the co-ordinates of P′, Q′, R′ after transformation.

(b) Plot both $\triangle$PQR and $\triangle$P′Q′R′ on graph paper.

(c) Describe the transformation geometrically.

(d) Write down the matrix of the transformation which is the inverse of that described in (c).

16. $\boldsymbol{S}$ is the translation represented by the column vector $\begin{pmatrix} 4 \\ -2 \end{pmatrix}$.

$\boldsymbol{T}$ is the translation which maps the point A(3, 2) to the point A′(1, 4).

(a) Find the column vector which represents $\boldsymbol{T}$.

(b) Find the co-ordinates of the image of A under the transformation $\boldsymbol{ST}$.

17. $\boldsymbol{M}$ is the transformation represented by the matrix $\begin{pmatrix} 2 & 0 \\ 4 & 1 \end{pmatrix}$. X is the point (p, q) and $\boldsymbol{M}$ maps X to the point (4, -1). Find the values of p and q.

18. $\boldsymbol{M}$ is the matrix $\begin{pmatrix} 2 & 3 \\ 1 & -2 \end{pmatrix}$.

(a) Find the matrix $\boldsymbol{M}^2$.

(b) Hence or otherwise, find a matrix $\boldsymbol{N}$ such that $\boldsymbol{MN} = \boldsymbol{I}$, where $\boldsymbol{I}$ is the identity matrix.

(c) What is the relationship between $\boldsymbol{M}$ and $\boldsymbol{N}$?

19. (a) Draw an x-axis from 0 to 18 and a y-axis from -2 to 14. Use a scale of 1 cm to 1 unit on each axis.

(b) Plot the points A(1, 2), B(4, 2), C(4, 4) and D(3, 4). Join them to produce the trapezium ABCD.

The transformation $\boldsymbol{M}$ is defined by:

$$\boldsymbol{M}: \begin{pmatrix} x \\ y \end{pmatrix} \mapsto \begin{pmatrix} 1.5 & 2 \\ 2 & -1.5 \end{pmatrix} \begin{pmatrix} x \\ y \end{pmatrix}$$

$\boldsymbol{M}$ maps ABCD to $A_1B_1C_1D_1$.

(c) (i) Calculate the co-ordinates of $A_1B_1C_1D_1$.

(ii) Draw and label the shape $A_1B_1C_1D_1$.

(d) (i) Calculate the area of ABCD. Give your answer in cm^2.

(ii) Calculate the area $A_1B_1C_1D_1$.

The transformation $\boldsymbol{M}$ consists of an enlargement centre (0, 0) followed by a reflection in a line passing through the origin.

(e) (i) Calculate the scale factor of enlargement.

(ii) Draw and label the line of reflection. (LEAG)

20. *Graph paper must be used for this question.*

Using a scale of 1 cm to represent 1 unit, draw x and y axes, both from -6 to $+6$.

(a) Draw and label the triangle PQR in which P is $(4, -1)$, Q is $(2, 2)$ and R is $(2, 4)$.

(b) The matrix $\boldsymbol{A}$ is $\begin{pmatrix} 0 & 1 \\ -1 & 0 \end{pmatrix}$, and represents the transformation which maps triangle PQR onto triangle $P_1Q_1R_1$.

Draw and label triangle $P_1Q_1R_1$.

(c) (i) Find the inverse, $\boldsymbol{A}^{-1}$, of matrix $\boldsymbol{A}$.

(ii) The matrix $\boldsymbol{A}^{-1}$ represents the transformation which maps the triangle PQR onto triangle $P_2Q_2R_2$.

Draw and label triangle $P_2Q_2R_2$.

(d) (i) Find $\boldsymbol{A}^2$.

(ii) The matrix $\boldsymbol{A}^2$ represents the transformation which maps the triangle PQR onto triangle $P_3Q_3R_3$.

Draw and label triangle $P_3Q_3R_3$.

(e) (i) Find $\boldsymbol{A}^3$.

(ii) What do you notice?

(f) Describe the symmetry of the set of four triangles that you have drawn. (MEG)

21. Under a transformation $\boldsymbol{M}$, the point (p, q) is mapped onto the point (r, s) by a matrix such that

$$\begin{pmatrix} r \\ s \end{pmatrix} = \begin{pmatrix} -1 & 0 \\ 0 & 1 \end{pmatrix} \begin{pmatrix} p \\ q \end{pmatrix}.$$

The triangle ABC has vertices A(4, 2), B(3, 0) and C(2, 2).

(i) Find the images of the points A, B, C under the transformation $\boldsymbol{M}$.

(ii) Describe the transformation represented by $\boldsymbol{M}$.

(iii) Write down the matrix for the inverse transformation of $\boldsymbol{M}$.

(iv) If $r = 3$ and $s = 4$, find p and q.

STATISTICS 2

13

In this second of two units on Statistics we look at different methods of analysing the data we have collected.

13.1 Averages

A DALEY & Co
Reputable Second-hand Car Dealers
require
Sales Executive
Nice little earner – High salary – Good prospects

Figure 13.1

Terry saw an advert (Figure 13.1) in the local press and decided to apply. On his interview he was told by the managing director, Mr Daley, that the average earnings of the 7 sales executives was at present £15 000 per year. Following a successful interview and appointment, Terry was then informed by the junior sales executive, Mr Chisholm, the 'real' average salary was £9000 per year. Who was right—Mr Daley or the sales executive?

Well both are right in that they have used different interpretations of the word 'average'. Mr Daley's interpretation was probably a deciding factor in Terry choosing the job. Below are the salaries of the 7 present sales executives:

£8000 £8000 £8000 £9000
£10 000 £11 000 £51 000 (Mr Daley's son)

In order to justify his figure of £15 000, Mr Daley used the *arithmetic mean* as his definition of the average. This is obtained by adding up all seven salaries and dividing by the number of salaries which in this case is 7. This form of average is commonly referred to as the *mean*. So the mean salary

$$= \frac{£(8000 + 8000 + 8000 + 9000 + 10\,000 + 11\,000 + 51\,000)}{7}$$

$$= \frac{£105\,000}{7}$$

$$= £15\,000$$

Mr Chisholm's representation of the average is the *median*. This is the salary which has the same number of salaries above it as below it. In order to determine this value the data *must* first be placed in order so that the middle term can be found:

£8000 £8000 £8000 £9000 £10 000 £11 000 £51 000

↑ The middle term

The median therefore is £9000.

An average should be a typical example of the data it is representing and clearly the figure arrived at by Mr Daley, although valid, is much too high. The calculation of a mean is a common process when determining an average, but this example shows that it can produce an unrepresentative result if there are any extreme values within the data. As an exercise, imagine that Mr Daley's son has left the business. Determine the mean salary of the remaining six sales executives.

A third measure of average could also have been used on this data, namely the *mode*. This is the most frequent or most popular data value. In this instance there are three sales executives with the same salary of £8000, and this value is taken as the mode. Although this value is not quite as good a representation of the 'average' as perhaps the median is, it is certainly better than the mean for this set of data.

Worked example 13.1

On the TV quiz show, 'Guess the Price', 6 contestants were asked to give their estimate of the retail price (in £s) of an electric hedge trimmer. The values estimated were:

Contestant no.	1	2	3	4	5	6
Estimate	£47	£51	£42	£42	£49	£48

Determine (i) the mode, (ii) the median, and (iii) the mean of these six estimated values.

(i) By inspection, the mode can be seen to be £42 (2 values).

(ii) To determine the median we need to place the estimates in order:

£42 £42 £47 £48 £49 £51

↑ The middle term

There is no actual middle term and so we have to find the half way value between the two data items £47 and £48. So the median

$$= \frac{£(47 + 48)}{2}$$

$$= £47.50$$

(iii) The mean of the estimates

$$= \frac{£(47 + 51 + 42 + 42 + 49 + 48)}{6}$$

$$= \frac{£279}{6}$$

$$= £46.50.$$

With a calculator the mean value can be determined without too much trouble. If the figures are a little more complicated it may make the arithmetic easier if we *subtract* a fixed value from each number first. This value is known as a *working mean.* Once this has been done we can then work out the mean of the numbers that are left, and finally add the value that was initially subtracted: i.e. from each of the data items above subtract 40. The values left are 7, 11, 2, 2, 9 and 8.
The mean of these numbers

$$= \frac{7 + 11 + 2 + 2 + 9 + 8}{6}$$

$$= \frac{39}{6}$$

$$= 6.50.$$

Adding the working mean value 40 to this mean gives a mean of the original data as £(40 + 6.50) = £46.50.

We can set such problems out in tabular form (see Table 13.1).

Table 13.1

Estimate	Difference from working mean
47	7
51	11
42	2
42	2
49	9
48	8
TOTAL	39

The mean of the estimates $= 40 + \frac{39}{6} = 46.50.$

As an exercise, find out what would happen if we subtract (i) 42, (ii) 50, (iii) 25, from each number and repeat the above method.

13.2 Averages of a frequency distribution

Worked example 13.2
Using Sam Pling's data collected in Unit 10 (Table 10.3) for the number of times a student goes out in the evening each week, determine (i) the mode, (ii) the median, and (iii) the mean of the data.

To remind ourselves of the data, it is shown again in Table 13.2

Table 13.2

Number of times out in a week	0	1	2	3	4	5	6	7
Frequency	13	10	7	7	5	4	3	1

(i) Again, by observation, we can see that the greatest frequency is 13. Therefore the mode is 0.
(ii) As there are 50 data items, there must be 25 below and 25 above the median. The median is therefore half-way between the 25th and 26th data items which are both 2. So the median is 2.
(iii) To calculate the mean number of times out in a week we need to add up Sam's 50 data items and divide by 50, i.e. we need to add:

	13	0s	=	0
	10	1s	=	10
	7	2s	=	14
	7	3s	=	21
	5	4s	=	20
	4	5s	=	20
	3	6s	=	18
and	1	7	=	7
	50			110

So the mean $= \dfrac{110}{50}$

$= 2.2$

We can present the working for this in tabular form (Table 13.3) using Table 10.3 as a basis.

Worked example 13.3
The maximum daily temperature at a particular seaside resort for the month of June was recorded and the results displayed as a bar chart (Figure 13.2).

Table 13.3

Number of times out	Number of students (frequency)	Number of times out × frequency
0	13	0
1	10	10
2	7	14
3	7	21
4	5	20
5	4	20
6	3	18
7	1	7
TOTALS	50	110

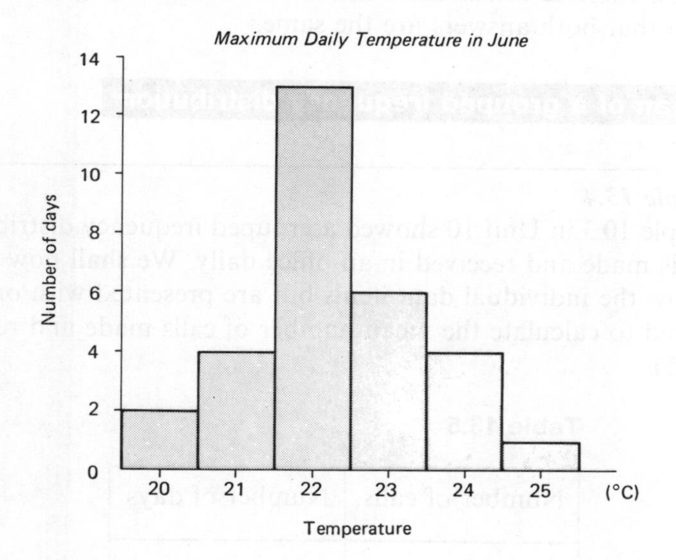

Figure 13.2

Calculate the mean daily maximum temperature during the month of June.

We can use a working mean for a frequency distribution in a similar way to worked example 13.1. After subtracting the working mean from the values representing the maximum daily temperatures, we then need to multiply these differences by the corresponding frequencies. Table 13.4 shows the necessary calculations using a working mean of 22°C.
So the mean maximum daily temperature

$$= \left(22 + \frac{9}{30}\right)^{\circ}\mathrm{C}$$

$$= 22.3^{\circ}\mathrm{C}$$

Table 13.4

Maximum daily temperature (°C)	Number of days (frequency)	Difference from working mean	Difference × frequency
20	2	−2	−4
21	4	−1	−4
22	13	0	0
23	6	1	6
24	4	2	8
25	1	3	3
TOTALS	30		9

It is left as an exercise to calculate the mean without using a working mean, and to confirm that both answers are the same.

13.3 The mean of a grouped frequency distribution

Worked example 13.4
Worked example 10.3 in Unit 10 showed a grouped frequency distribution of the number of calls made and received in an office daily. We shall now assume that we do not know the individual data items but are presented with only the table and are required to calculate the mean number of calls made and received daily (see Table 13.5).

Table 13.5

Number of calls	Number of days
1– 5	3
6–10	4
11–15	7
16–20	15
21–25	9
26–30	6
31–35	4
36–40	2

In this type of problem where the individual data items are unknown we have to make an assumption about the values within each class. We make the assumption that the values in each class are equal to the *mid-class value* for that class. So the three values in the class 1–5 are assumed to be 3, 3 and 3; the four values in the class 6–10 are assumed to be 8, 8, 8 and 8. This can only lead to an estimate of the mean but without the original data, it is the best we can do. The solution can now be arrived at using the method of worked example 13.2 (see Table 13.6).

Table 13.6

Number of calls	Mid-class value	Number of days (frequency)	Mid-class value × frequency
1– 5	3	3	9
6–10	8	4	32
11–15	13	7	91
16–20	18	15	270
21–25	23	9	207
26–30	28	6	168
31–35	33	4	132
36–40	38	2	76
TOTALS		50	985

So the mean number of calls made and received per day

$$= \frac{985}{50}$$

$$= 19.7.$$

This result, although only an approximation, compares favourably with the true value of 19.44 which can be determined from the original data.

The class 16–20 contains the most data items and is called the *modal class.*

Worked example 13.5
Determine the mean weekly expenditure of the 50 students sampled by Sam Pling in Unit 10.

Note that we have made the assumption that the top limit to the last class is £40.00 (see also worked example 10.11). Therefore the mean weekly expenditure

$$= \frac{£755}{50}$$

$$= £15.10.$$

Table 13.7

Weekly expenditure (£)	Mid-class value	Number of students (frequency)	Mid-class × frequency
0.00– 4.99	2.5	5	12.5
5.00– 9.99	7.5	7	52.5
10.00–14.99	12.5	15	187.5
15.00–19.99	17.5	11	192.5
20.00–24.99	22.5	8	180.0
25.00 or more	32.5	4	130.0
TOTALS		50	755.0

Note that the arithmetic could have been made easier in both worked examples 13.5 and 13.6 by the use of a working mean. In both cases a working mean would be chosen as one of the mid-class values towards the middle of the distribution, and a column of differences would then need to be produced. The working is then similar to worked example 13.3.

13.4 Weighted averages

Worked example 13.6

Maurice Traveller, a sales representative, completed a journey by travelling at an average speed of 90 km/h for 3 hours and then at an average speed of 50 km/h for a further 2 hours. What was Maurice's average speed per hour?

$$\text{Average speed} = \frac{\text{total distance}}{\text{total time taken}}$$

So the distance travelled in the first 3 hours $= 3 \times 90$ km.
The distance travelled in the next 2 hours $= 2 \times 50$ km.

$$\text{Total distance} = 3 \times 90 + 2 \times 50 \text{ km}$$

$$\text{Total time} = 2 + 3 \text{ hours}$$

So the average speed

$$= \frac{3 \times 90 + 2 \times 50}{2 + 3} \text{ km/h}$$

$$= \frac{270 + 100}{5} = \frac{370}{5}$$

$$= 74 \text{ km/h}.$$

This combined average or mean is often referred to as a *weighted average* because each individual average is 'weighted' by multiplying it by the number of hours travelled at that average speed. The two 'weights' are 3 and 2 and the final average is found by dividing by $(3+2)$.

Worked example 13.7

In order to save money, an Education Authority decide to join two schools together. Burnham School has 10 classes with an average of 18 pupils per class. Houghton High School has 6 classes with an average of 30 pupils per class. What is the average number of pupils per class in the combined school if there are still 16 classes?

The combined mean is determined using weighted averages, i.e. the average (mean) number of pupils per class

$$= \frac{10 \times 18 + 6 \times 30}{10 + 6}$$

$$= \frac{180 + 180}{16} = 22.5$$

Exercise 13a

(1) Calculate (i) the mode, (ii) the median, and (iii) the mean of the following numbers:
(a) 3, 4, 4, 4, 5, 8, 8, 9, 9;
(b) 10, 19, 17, 14, 12, 19, 18, 14, 16, 17, 19, 11.

(2) Find the mean value of 13, 17, 18, 21, 24. Hence write down the mean value of the numbers 513, 517, 518, 521, 524.

(3) Find the mean of the figures -4, 5, -9, -5, 7, -1. Hence write down the mean of the figures 216, 225, 211, 215, 227, 219.

(4) The mean weight of 8 occupants of a lift is 72 kg. By how much does their total weight exceed the maximum safe load of 550 kg?

(5) Find the median of the following sets of figures:
(i) 26, 41, 20, 85, 41, 27, 72, 65, 64;
(ii) 57, 71, 72, 99, 44, 32;
(iii) 17, 15, 13, 20, 17, 17, 14, 11, 14, 15.

(6) The median of a set of 8 numbers is $4\frac{1}{2}$. Given that 7 of the numbers are 3, 8, 1, 11, 2, 12 and 2, find the eighth number.

(7) The number of kittens in each of 50 litters was noted and the results appear below. Construct a frequency table and hence work out the mean number of kittens per litter.

1 2 1 4 3 3 2 4 1 5 3 4 2 1 3 3 1 2 2 2
2 3 3 5 4 2 1 3 2 4 1 5 1 3 3 4 2 4 2 1
3 3 4 2 1 3 2 5 1 2

What is the mode and the median of this distribution? (SEG)

(8) The manager of a shoe shop wants to know which size of shoe he should stock in the largest quantity. From a recent survey he knows the mean, the mode and the median. Which of these would he use to help him to decide his stock level? Give a reason for your answer. (LEAG)

(9) The numbers of aircraft landing at a small airport each week in 1987 were as follows:

Number of aircraft	36	37	38	39	40	41	42	43
Frequency	1	5	11	10	14	6	4	1

(a) Represent this information in the form of a bar chart.

(b) Write down the median number and the modal number of aircraft landing per week.

Based on the information in the table:

(c) Calculate the probability that the number of aircraft landing in one week will be more than 40.

(d) Calculate the probability that the number of aircraft landing in two consecutive weeks will be more than 40 in each week. (LEAG)

(10) 40 students were asked how many brothers and sisters they have. The results are shown in the table below:

Number of brothers and sisters	0	1	2	3	4	5	6 or more
Number of students	11	10	5	7	4	3	0

Find (i) the mode, (ii) the median, and (iii) the mean number of brothers and sisters.

(11) Amjad carried out a survey of 50 households in his district to find the number of occupants per house. The results are displayed in Figure 13.3.

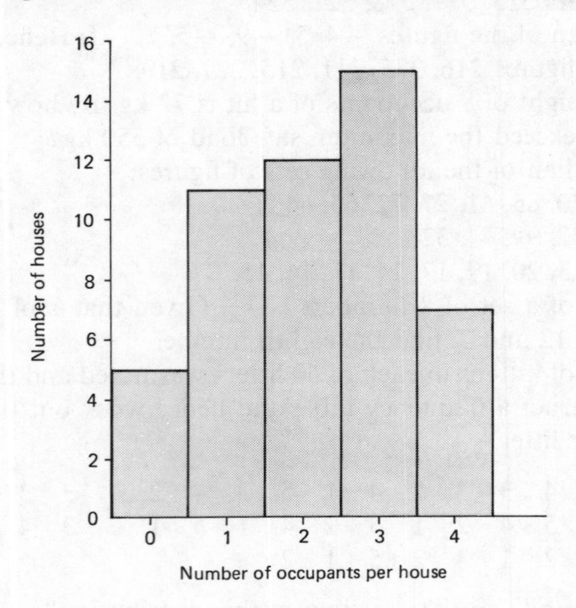

 Figure 13.3

(i) State the mode and median number of occupants per house.
(ii) How many occupants in total live in the 50 houses surveyed?
(iii) Calculate the mean number of occupants per house.
(iv) What is the probability that a house chosen at random from the houses surveyed will have at least one occupant?

(12) Shona starts work as a sales assistant for Askeys on a Saturday six weeks before Christmas and besides her wages she gets 2% commission on all sales that she makes. For example, on the first Saturday she sells £625 worth of goods. Her commission will therefore be 2% of £625 $= \frac{2}{100} \times £625 = £12.50$.
Copy and complete the table below for the six Saturdays before Christmas to show her commission.

Saturday	Week 1	Week 2	Week 3	Week 4	Week 5	Week 6
Sales	£625	£700	£800	£725	£800	£1000
Commission	£12.50					

(a) Calculate the mode, median and mean commission received for the Saturdays shown.
(b) Which of the averages do you consider to be the most useful for Shona to know and why?
(c) The commission that Shona receives for the seventh Saturday is £17.00. Determine the value of the goods that Shona has sold on this Saturday. Which of the three averages in part (a) will change? Find this new value.

(13) A student asked 30 people arriving at a football ground how long, to the nearest minute, it had taken them to reach the ground. The times they gave (in minutes) are listed below.

35	41	22	15	31	19	12	12	23	30
30	38	36	24	14	20	20	16	15	22
34	28	25	13	19	9	27	17	21	25

(a) (i) Copy and complete the following frequency table.
(ii) Draw a histogram to represent the information in the frequency table.

Time taken in minutes (to nearest minute)	8–12	13–17	18–22	23–27	28–32	33–37	38–42
Number of people	3	6	7	5	4		

(b) Of the 30 people questioned:

6 paid £2 each to see the football match
8 paid £3 each

4 paid £4 each
10 paid £5 each and
2 paid £6 each

(i) Calculate the total amount paid by these 30 people.
(ii) Calculate the mean amount paid by these 30 people. (MEG)

(14) For the group frequency distribution Table 10.5 (page 253), what is the modal class? Determine an estimate of the mean to the nearest 10 miles of the distribution from the table.

(15) Before installing a controlled crossing in place of a zebra crossing, a survey was taken to determine how long it took people to cross the road. The times measured to the nearest second, taken by 55 pedestrians, are as shown below:

11	16	22	26	28	20	22	31	32	33	11
21	12	16	22	27	26	18	21	29	30	26
21	19	12	24	24	23	28	19	23	25	17
27	21	17	13	33	24	34	24	17	29	22
25	30	17	25	15	22	35	33	34	20	33

(a) Copy and complete the grouped frequency table below.

	Frequency (f)	Mid Value (x)	fx
11–15			
16–20			
21–25			
26–30			
31–35			
		Total	

The green light on the controlled crossing is to show for 5 seconds longer than the mean time taken to cross the road.

(b) Using your table, estimate, to the nearest half second, how long the green light should show. (LEAG)

(16) The following table gives the weight of 100 newly born babies.

Weight (kg)	0–0.5	0.5–1.0	1.0–1.5	1.5–2.0	2.0–2.5	2.5–3.0	3.0–3.5	3.5–4.0	4.0–4.5
Frequency	0	3	8	15	21	25	18	6	4

Find the mean weight of the babies. (SEG)

(17) In a factory the mean weekly wage of all employees is £180 and the median weekly wage is £160.

The Management claims that the 10% pay offer means an average increase of £18 per week. The Unions claim that the majority of employees will not get a rise of £18.

Explain why both claims could be justified. (LEAG)

(18) The mean of a set of nine numbers is 8.

(i) Write down the sum of the nine numbers.

(ii) What number needs to be added to this set of numbers in order to make a new mean of 9?

(19) The average (mean) age of 6 students in a GCSE Mathematics set is $16\frac{1}{2}$. If 4 more students, whose average age is 17, join the set, what is the new average age of the set?

13.5 Cumulative frequencies

The median of a grouped frequency distribution can be estimated if we first find the *cumulative frequencies* of the distribution. This can be done by adding an extra column to the group frequency distribution table. The values placed in this column are found by adding each frequency in turn to the total of those above. Let us look again at the distribution of weekly expenditures found from Sam Pling's survey. Table 13.8 shows the frequencies and cumulative frequencies.

Table 13.8

Weekly expenditure (£)	Number of students (frequency)	Cumulative frequency
0.00– 4.99	5	5
5.00– 9.99	7	12 (7 + 5)
10.00–14.99	15	27 (15 + 12)
15.00–19.99	11	38 (11 + 27)
20.00–24.99	8	46 (8 + 38)
25.00 or more	4	50 (4 + 46)

The median, being the middle data value, lies in the class £10.00–£14.99. Strictly speaking, it should be half-way between the 25th and 26th data values. However for an even number of data values, as we can only *estimate* the median, its value is taken from the term which corresponds to

$$\frac{\text{number of data values}}{2}$$

which in this case is the 25th data value.

Figure 13.4 shows the position of the median. There are 12 data values below the class that we are interested in. The median therefore divides the class £10.00–£14.99 in the ratio 13:2. With a class width of £5.00, this gives a value of

$$\frac{13}{15} \text{ of } £5.00 = £4.33$$

As the lowest value of this class is £10.00, the approximation of the median is therefore £10.00 + £4.33 = £14.33.

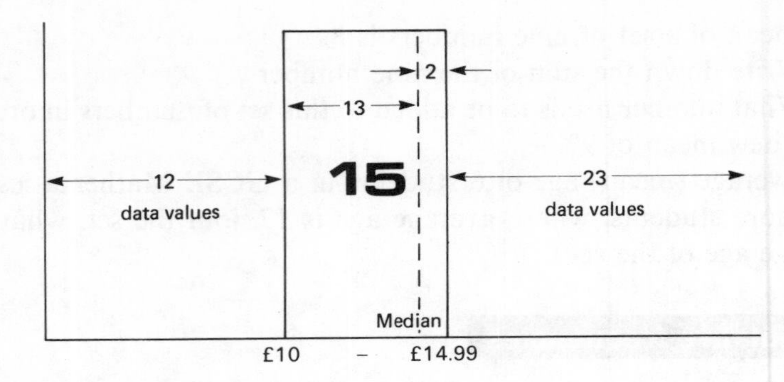

Figure 13.4

Perhaps a slightly easier method of determining an estimate of the median is by drawing a *cumulative frequency curve* (or graph). To draw this we require to plot the cumulative frequency values at the *upper class boundaries* of the appropriate classes. So for our example, the upper class boundary for the class £0.00–£4.99 would be a 'theoretical' £4.995. We should therefore try to plot this point very slightly to the left of (5, 5). Similarly, points should be plotted slightly to the left of (10, 12), (15, 27), (20, 38), (25, 46) and (40, 50). The last point is again assuming that the top limit for the last class is 'under £40.00'. The completed cumulative frequency curve is shown in Figure 13.5.

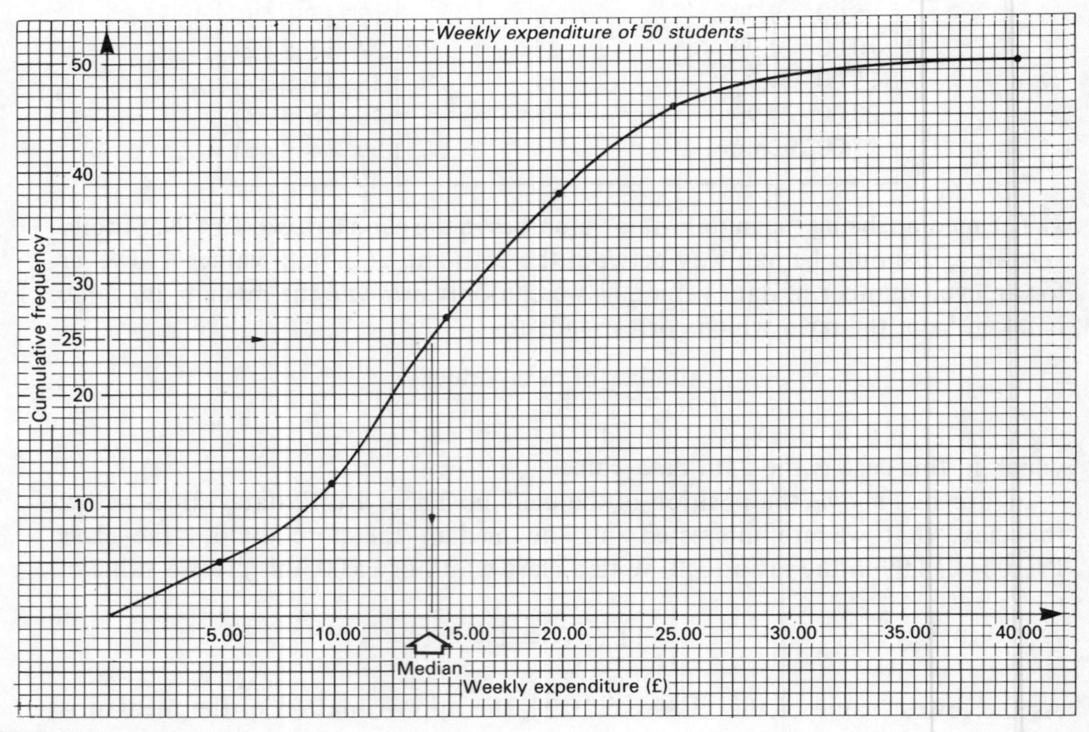

 Figure 13.5

To find the median we now draw a horizontal line half-way up the vertical scale and where this line meets the curve, we drop a perpendicular. The median is then read directly off the horizontal axis. This value is approximately £14.25.

Worked example 13.8

The No. 27 bus to Varnworth College is notorious for arriving late at the College gates (or so the students say). It is scheduled to arrive at 8.52 a.m. each day with lessons starting at 9.00 a.m. A record of its arrival time for 60 days was recorded with the following results:

Number of minutes late	$\leqslant 2$	$\leqslant 4$	$\leqslant 6$	$\leqslant 8$	$\leqslant 10$	$\leqslant 12$
Number of days	9	24	48	55	58	60

Draw a cumulative frequency curve and use it to estimate the median time of lateness of the No. 27 bus.

If it takes a student 3 minutes to walk to class from the College gates, on how many of the 60 days would you have expected a student arriving on the No. 27 to be late to class?

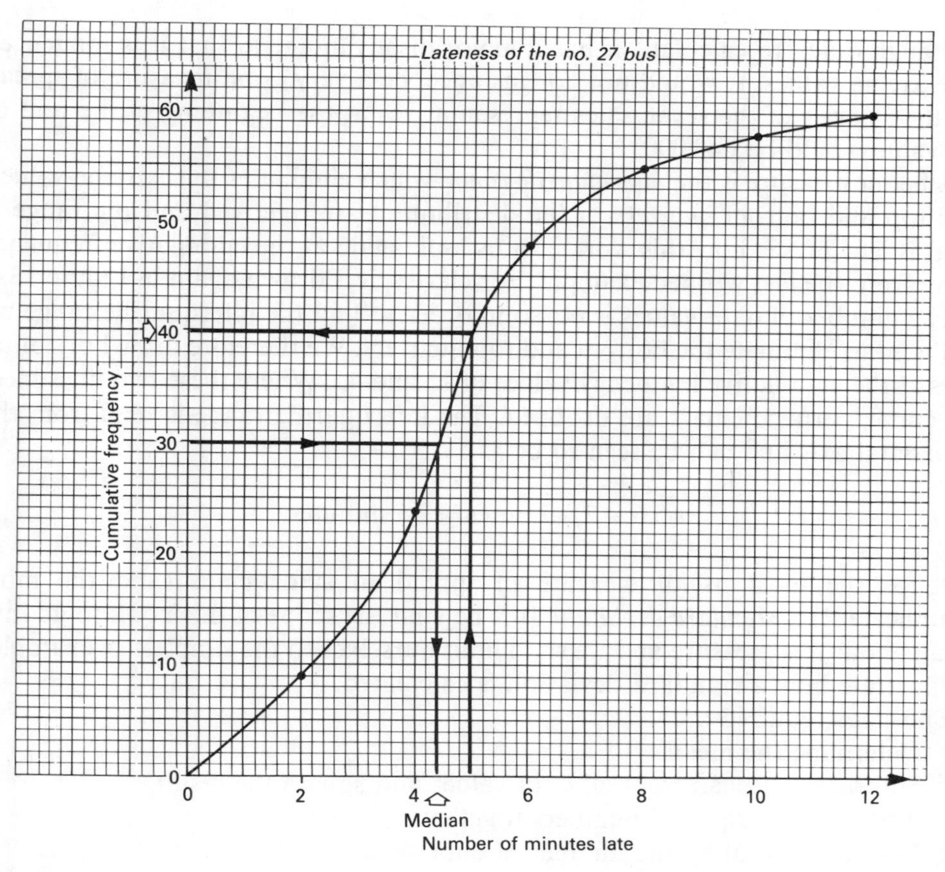

Figure 13.6

The table of results is already a cumulative frequency distribution and the values given for the number of minutes late are the class boundaries that are required for the plotting of the curve. The resultant curve is drawn in Figure 13.6 with the median marked. From the scale, the value of the median is approximately 4.4 minutes late.

For a student to be late to class, the bus would have to arrive after 8.57, i.e. more than 5 minutes late. Reading from the graph, the number of days on which the bus arrived *after* this time is 20 (60 – 40 from the graph). So we would have expected the student to be late a third of the time.

13.6 Measures of spread

Averages give a useful indication as to the position of the middle of a *distribution*. What they fail to tell, however, is how close to the average the values are. We therefore require to define some *measure of spread*.

The most obvious measure is the *range* which is the difference between the smallest and largest data values, i.e.

$$\text{range} = \text{highest value} - \text{lowest value}$$

Although quick to calculate, it has a major disadvantage in that it is influenced by extreme values. It would not, for instance, be a very good measure of spread for the salaries of the seven sales representatives at the beginning of the unit as the value would be £43 000.

A better measure is the *interquartile range* which eliminates the extreme values. The top and bottom *quarters* of the distribution are removed and the range is what is left. To see how this works, let us look again at the cumulative frequency curve from the last worked example. For clarity, it is drawn again in Figure 13.7.

The number of data values is 60, so in order to find the interquartile range we will need to eliminate the bottom and top quarters. We therefore draw horizontal lines at the cumulative frequency values of 15 and 45. Where these two lines meet the curve, perpendiculars are dropped. The interquartile range is read from the horizontal axis and can be seen to be

$$(5.6 - 3) \text{ minutes} = 2.6 \text{ minutes}$$

Perhaps the most useful measure of spread to statisticians (and the most complicated!) is *standard deviation*. It is beyond the scope of this book to go into detail about this measure and we shall limit ourselves to the simplest of examples to illustrate how a standard deviation is calculated.

The steps required are:

(i) Determine the mean.
(ii) Subtract the mean from all data values and square the result.
(iii) Add all these squared numbers together.
(iv) Divide the total by the number of data values.
(v) Square root your answer.

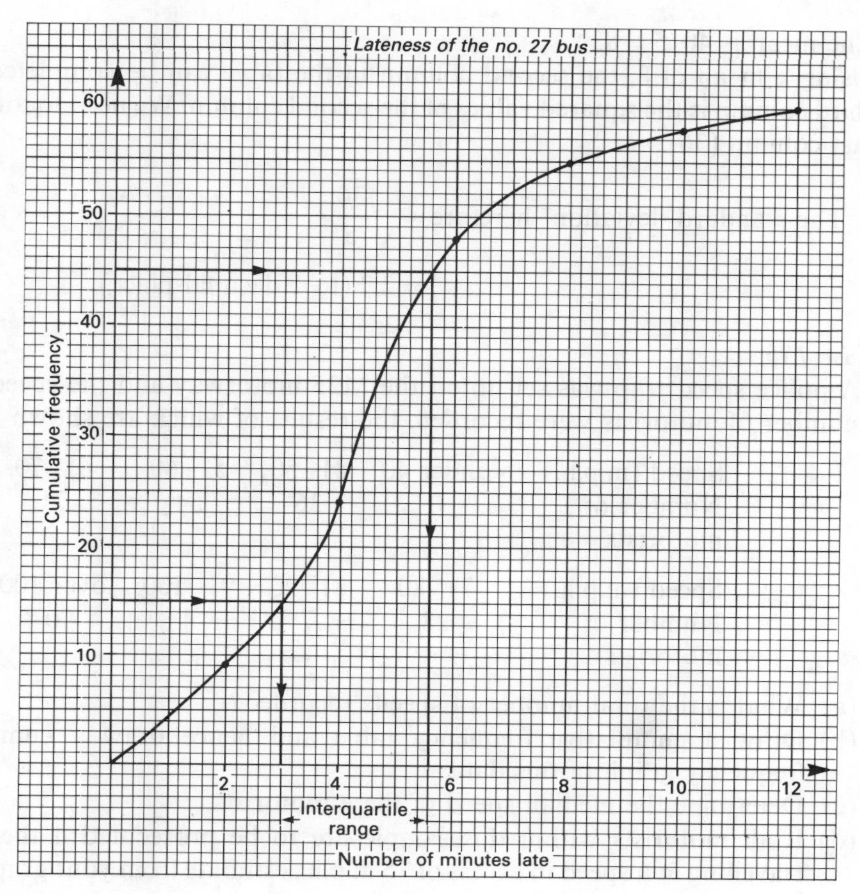

Figure 13.7

Worked example 13.9
For the figures 3, 8, 9, 9, 11 determine (i) the mean, and (ii) the standard deviation, correct to 1 decimal place.

Setting the solution out in tabular form:

Data values	values − mean	(values − mean)^2
3	−5	25
8	0	0
9	1	1
9	1	1
11	3	9
40		36

(i) The mean is $40/5=8$.

(ii) Using a mean of 8, the second column in the table can be completed. The third column is the squared values of the second column. We need to total this last column (36).

$$\text{The standard deviation therefore} = \sqrt{\frac{36}{5}}$$

$$= 2.7 \text{ (to 1 decimal place).}$$

Exercise 13b

(1) A police speed trap was set up on the M25 motorway and the speeds of a number of motorists were recorded, the results of which are shown below:

Speed (m.p.h.)	under 40	41–50	51–60	61–70
Number of motorists	0	36	54	71

Speed (m.p.h.)	71–80	81–90	91–100	over 100
Number of motorists	28	7	4	0

(a) What is the total number of speeds recorded?

(b) Draw a cumulative frequency table and hence draw a cumulative frequency graph of this data.

(c) Determine the median speed of these motorists.

(d) If all motorists within this sample are to be prosecuted if they were travelling at a speed in excess of 75 m.p.h., estimate from your graph how many motorists will be prosecuted.

(2) Miles Perower recorded his travelling time to work over a period of 100 days, with the results shown below:

Travelling time	less than 30 min	less than 35 min	less than 40 min	less than 45 min
Number of days	0	40	70	86

Travelling time	less than 50 min	less than 55 min	less than 60 min
Number of days	92	98	100

Draw a cumulative frequency graph to represent this information and determine the median time to travel to work.

To arrive at work by 9.00 a.m., he leaves home everyday at 8.22 a.m. From your graph, estimate the number of days that he will be late to work in the 100 day period.

(3) Calculate to 1 decimal place the standard deviation of the numbers:

1, 1, 3, 5, 6, 7, 8, 9

(4) Calculate to 2 decimal places the standard deviation of the contestant's estimates in worked example 13.1.

(5) Calculate to 1 decimal place the arithmetic mean and the standard deviation of the numbers:

4, 5, 7, 9, 10

A sixth number is added to this set of five numbers. Choose a sixth value from the numbers:

3, 6, 10

which will (a) increase the mean,
(b) decrease the mean,
(c) increase the range,
(d) decrease the standard deviation.

(6) Apples are graded into 5 categories according to their mass. The distribution of the masses of a sample of 100 apples from an orchard is shown in the following table:

Mass	Grade E less than 115	Grade D 115 but less than 125	Grade C 125 but less than 130	Grade B 130 but less than 145	Grade A 145 but less than 165
Number of apples	15	27	16	33	9

Copy and complete the cumulative frequency table below.

Mass (grams)	Number of apples
less than 115	15
less than 125	42
less than 130	
less than 145	
less than 165	

(b) Using your completed cumulative frequency table draw a cumulative frequency diagram.

New EEC regulations come into force. The new Grade 1 apple is one which has a mass of at least 135 grams.

(c) From your diagram estimate the number of Grade 1 apples in the sample.

(LEAG)

(7) The weights of 100 new-born babies are given, to the nearest 10 g, in the following table.

Weight (kg)	Frequency
1.00–1.49	3
1.50–1.99	8
2.00–2.49	12
2.50–2.99	22
3.00–3.49	27
3.50–3.99	18
4.00–4.49	9
4.50–4.99	1

Weight	Cumulative frequency
Not more than 0.995 kg	0
Not more than 1.495 kg	3

(a) Copy and complete the cumulative frequency table above.
(b) Draw a cumulative frequency diagram to illustrate the data.
(c) From your diagram, and making your methods clear, estimate
(i) the median weight,
(ii) the probability that a new-born baby chosen at random from this sample weighs between 2.4 kg and 3.2 kg. (NEA)

(8) The number of telephone calls each day from an office over a period of 100 days were as follows:

No. of calls	1–30	31–60	61–90	91–120	121–150	151–180
No. of days	8	18	30	24	16	4

Draw up a table showing the *cumulative* frequency totals of the number of calls less than or equal to 30, less than or equal to 60, etc., and hence plot the cumulative frequency graph.

From your graph estimate
(a) the median number of calls,
(b) the interquartile range of calls,
(c) the number of days with more than 140 calls. (SEG)

(9) The table below shows the number of students gaining marks in a test in each of the ranges 1–10, 11–20, ..., 91–100 (no-one scores 0!).

1–10	11–20	21–30	31–40	41–50	51–60	61–70	71–80	81–90	91–100
2	3	7	14	22	28	15	6	2	1

Draw up a table showing the *cumulative* frequency totals of students scoring 10 or less, 20 or less, etc., and hence plot the cumulative frequency graph.

Use your graph to estimate:

(a) the median mark,
(b) the interquartile range of marks,
(c) the number of marks needed to pass if 15% of the students failed the exam. (SEG)

(10) Mrs Tyson, an agent for a firm, kept a record of the time she spent (including travelling time) on each customer she saw.

During one particular 5 day week, she saw 80 customers and the record of the times spent on them is summarised in the following table.

Time (t minutes)	$20 < t \leqslant 25$	$25 < t \leqslant 30$	$30 < t \leqslant 35$	$35 < t \leqslant 40$	$40 < t \leqslant 45$	$45 < t \leqslant 50$
Number of customers	8	10	10	30	18	4

(a) Find the mean number of customers Mrs Tyson saw per day during this week.
(b) Mrs Tyson's normal working week is 40 hours. Calculate an estimate of the number of hours overtime which she worked during this week.
(c) Calculate an estimate of the mean length of time Mrs Tyson spent per customer.
(d) On graph paper, draw a cumulative frequency diagram for this distribution. (Use a scale of 2 cm to represent 5 minutes on the time axis and 2 cm to represent 10 customers on the cumulative frequency axis.)
(e) Use your diagram to estimate:
(i) the interquartile range for this distribution,
(ii) the number of customers on each of whom Mrs Tyson spent more than the mean length of time found in part (c). (MEG)

Worked example 13.10
Megan Watts quarterly electricity bills in £s for the last three years are as follows:

	Quarter 1	Quarter 2	Quarter 3	Quarter 4
1986	120	85	66	95
1987	130	82	79	102
1988	134	94	76	110

Is there reason to believe that she is now spending more on electricity than she has in the past?

The data can be illustrated in the form of a line graph (see Figure 13.8).

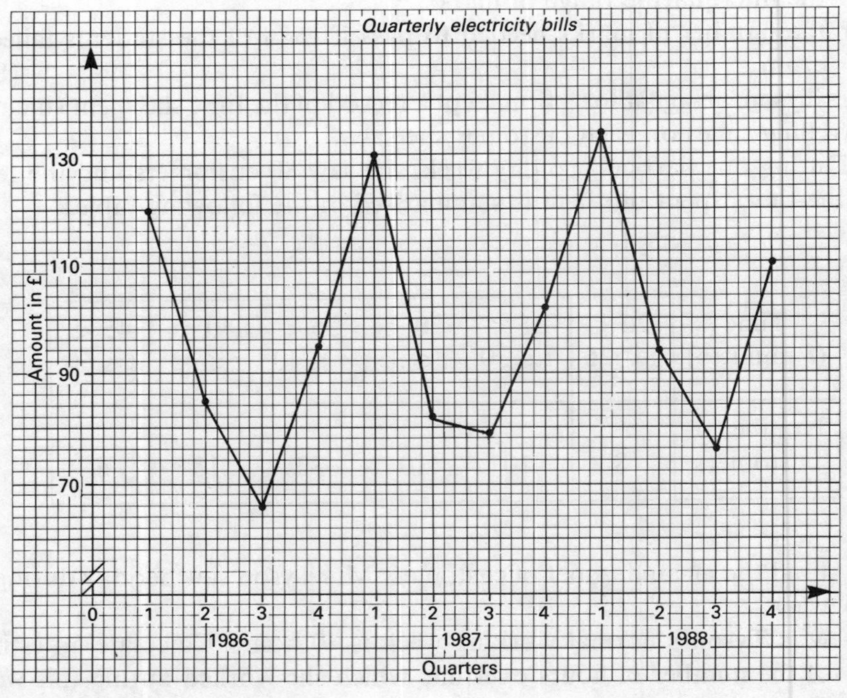

Figure 13.8

The fluctuations in the graph are due in the main to the colder weather at certain times of the year which leads to a greater consumption of electricity. These fluctuations, or *seasonal variations*, make it difficult to see whether there is an overall increase in the cost of electricity. To smooth out these variations we need to average the bills over a period of time and as the variation seems to repeat itself every 4th quarter, we calculate a series of *four-point* averages. The first of these is the result of adding the first four quarters and dividing by four, i.e.

$$(120 + 85 + 66 + 95)/4 = 91.5$$

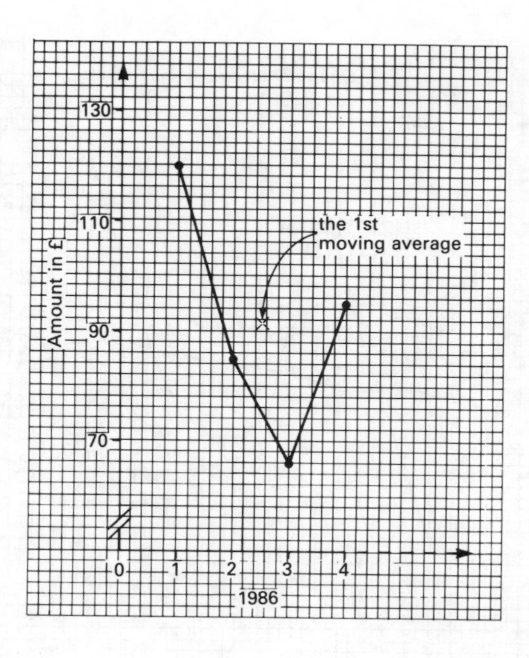

Figure 13.9

This point can then be plotted on the line graph as shown in Figure 13.9. Note that the point is positioned *half-way* between the 2nd and 3rd quarters.

The second average is calculated by using the 2nd, 3rd and 4th quarters of 1986 and the 1st quarter of 1987 and is then plotted half-way between the 3rd and 4th quarters of 1986. Continuing in this way we can produce a series of averages which we call *moving averages.*

Table 13.9

	Quarters	Electricity bill (£)	Four-point moving average
1986	1	120	
	2	85	91.5
	3	66	94
	4	95	93.25
1987	1	130	96.5
	2	82	98.25
	3	79	99.25
	4	102	102.25
1988	1	134	101.5
	2	94	103.5
	3	76	
	4	110	

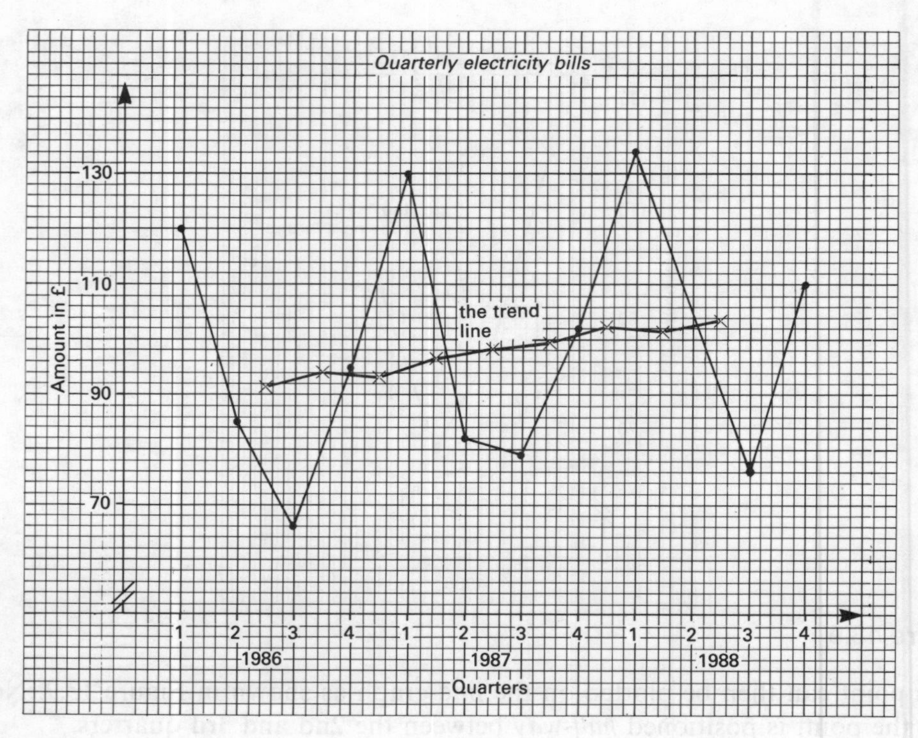

Figure 13.10

The completed calculations are shown in Table 13.9, and the line graph showing the trend line which has eliminated the seasonal variation is shown in Figure 13.10.

The *trend* line shows that there is an increase in the cost of electricity.

13.8 Cause and effect

It is often useful to compare two sets of data to see whether or not there is any *relationship* between them. A comparison of ice cream sales from a shop at a seaside resort with maximum daily temperatures at the resort would probably show that as temperatures rise, then so does the sale of ice cream. Does that then mean that, in general, temperature rises *cause* an increase in the sales of ice cream? Perhaps, but there could be other contributory factors such as an increase in visitors when the weather is warmer which in turn leads to greater sales of ice cream. Sometimes a clear relationship can be shown between two sets of data which have no connection whatsoever, i.e. as ice cream sales in the UK increase, so do the number of road accidents in France. It would certainly be wrong to conclude that the sale of ice creams in the UK causes road accidents in France.

13.9 Scatter diagrams

The relationship between two sets of data can be visually compared using a *scatter diagram.* The corresponding values from the two sets are plotted on a graph to see whether or not a straight line can be drawn which will pass very close to every point. Such a line is called a *line of best fit*. If the points lie very close to a line then there is a high degree of *correlation* between the two sets of data. If both sets of data are increasing, then the correlation is said to be *positive*. If one set of data decreases while the other is increasing, the correlation is said to be *negative*. A high degree of correlation allows us to use the line of best fit to *predict* values. If the points are so scattered that it would seem impossible to fit any line, the data is said to have *zero* or *no correlation*. Figure 13.11 shows examples of correlation.

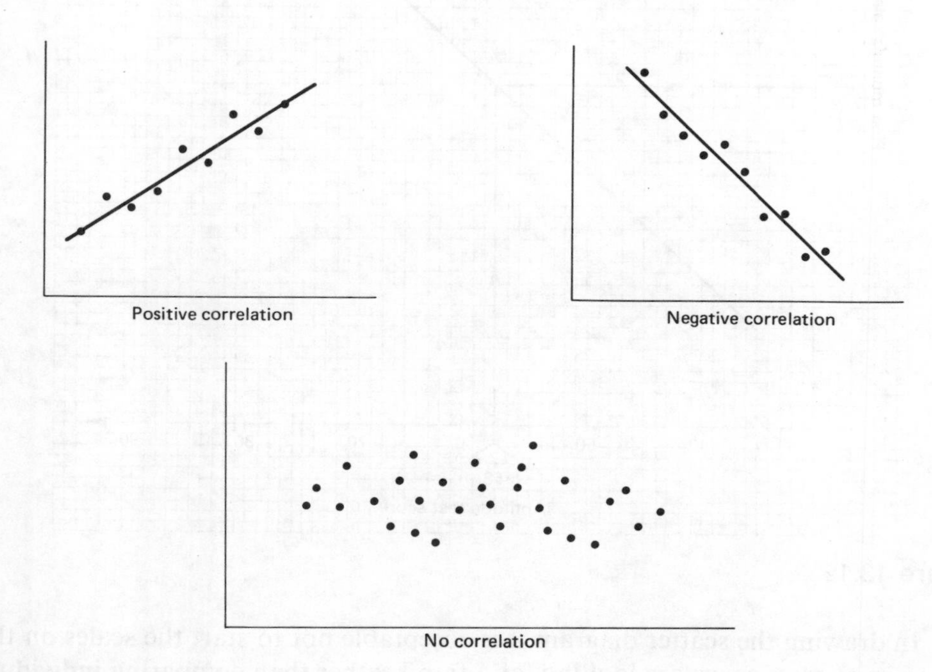

Figure 13.11

Worked example 13.11

Harrass the Motors give an aptitude test to all new salespersons. The data below shows the scores achieved by 10 salespersons and their respective first month's sales (£ thousands).

Salesperson	A	B	C	D	E	F	G	H	I	J
Aptitude test	65	73	42	52	84	60	70	79	60	83
First month's sales (£1000s)	78	88	54	68	92	77	84	89	70	99

(i) Draw a scatter diagram to illustrate this information.
(ii) Draw a line of best fit.
(iii) A salesperson joins the company and scores 58 on the aptitude test. Use your line of best fit to estimate this salesperson's first month's sales.

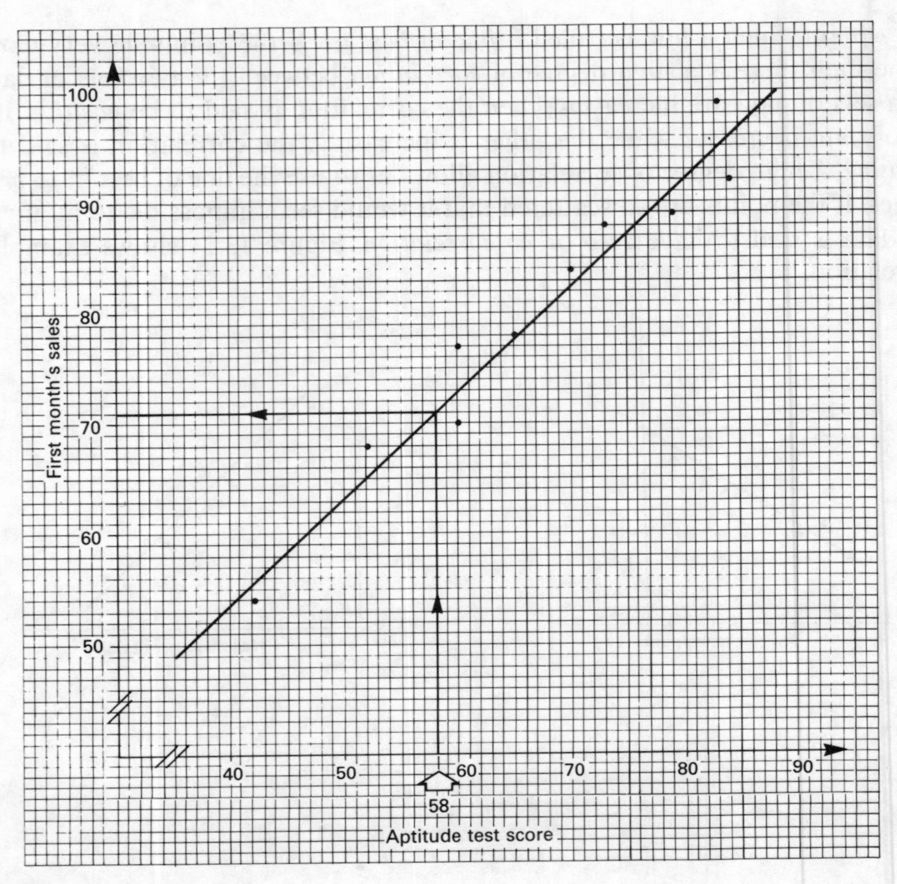

Figure 13.12

(i) In drawing the scatter diagram it is acceptable not to start the scales on the axes at zero, as we are looking for a trend rather than comparing individual values. The plotted points can be seen in Figure 13.12 and they clearly indicate a positive correlation.
(ii) In drawing the line of best fit, we should try to draw it passing centrally through the plotted points.
(iii) Using the line we can estimate the salesperson's first month's sales as £71 000.

Exercise 13c

(1) The table below shows the number of plane departures from a city airport over a two week period.

	Mon	Tues	Wed	Thurs	Fri	Sat	Sun
Week 1	41	27	25	31	39	52	57
Week 2	38	24	20	26	34	44	46

(a) Calculate, correct to 1 decimal place, the 7-point moving averages for this data.
(b) Draw a line graph to show the data and the moving averages. Comment on any trend.
(c) Why is a 7-point average used in this case?

(2) The quarterly telephone bills (in £s) of a household over 8 consecutive quarters were as follows:

Quarter	1	2	3	4	5	6	7	8
Bill	42	37	44	50	43	40	51	59

Calculate the first three *four*-point moving averages for the data, giving your answers correct to the nearest penny. (SEG)

(3) The total number of lunches served in a school canteen each term for six consecutive terms is shown in the following table:

Winter	Spring	Summer	Winter	Spring	Summer
700	720	600	720	800	450

(a) Calculate the first two 3-point moving averages for the data.
(b) Why is a 3-point moving average used in this case? (SEG)

(4) For each of the following sets of data draw a scatter diagram:

(i)

X	1	2	2	3	4	5	7	9
Y	8	6	5	4	5	2	3	1

(ii)

X	1	3	4	4	5	7	8	9
Y	2	1	3	5	7	6	9	7

(iii)

X	2	3	4	4	6	6	7	8
Y	4	7	6	8	5	8	3	6

Comment on any correlation.

(5) The District Football League consists of ten teams and at the end of the season the number of points scored and goals conceded for nine of the teams were as follows:

Team	A	B	C	E	F	G	H	I	J
Number of points scored	34	30	27	18	16	12	10	7	4
Number of goals conceded	49	50	42	48	55	57	71	60	67

(i) Draw a scatter diagram to represent this information and comment on any correlation.

(ii) Draw a line of best fit for the data.

(iii) Team D has ended the season with 22 points. Use your line of best fit to estimate the number of goals that team D conceded.

(6)

1980 Ford Escort	£1600
1979 Ford Escort	£1295
1981 Ford Escort	£1795
1979 Morris Marina	£1325
1979 Ford Escort	£1595
1980 Austin Allegro	£1600
1982 Ford Escort	£2100
1983 Ford Escort	£2400

This advertisement for used cars appeared in the local newspaper last week.

(a) Draw a scatter diagram for 'year' against 'price'.

(b) What does your diagram suggest about the correlation between the cars' age and price? (LEAG)

(7) (a) The stem and leaf diagram below shows the times, measured to the nearest second, taken by 11 workers to complete a job in a factory.

Time in seconds	
4	3, 7
5	1, 4, 4
6	2, 8, 8, 8
7	0, 3

(i) What was the shortest time taken to complete the job?

(ii) What was the median for the distribution?

(b) The table below shows, over a period of 8 years, the number of days with more than 4 hours of sunshine during April and the number of therms of gas to heat a house during those months.

Number of days	15	21	20	4	10	9	26	16
Number of therms	115	90	146	213	138	189	82	174

(i) Draw a scatter diagram to show this information. (Use a horizontal scale of 2 cm to represent 5 days and a vertical scale of 2 cm to represent 50 therms.)

(ii) Does this diagram suggest that there is any correlation between the two quantities? If so, describe it *very* briefly. (MEG)

Suggestions for coursework 13

1. Investigate whether there is any correlation between the age in years of a vehicle and the month in which the vehicle's tax disc expires.
2. Investigate whether there is any correlation between a person's height and their shoe size.
3. Investigate how many tube trains London Regional Transport will need to run per hour in the rush hour in 5 years' time. [If the data for this investigation is difficult to obtain, choose an appropriate travel situation near to where you live.]
4. Examine Football League statistics and investigate any correlation between number of points achieved, number of drawn games, goals scored and goals conceded.
5. Investigate which is more economical—a petrol driven car or a diesel driven car.
6. Compare the writing styles of two different authors. From your analysis, could you determine which of the two authors wrote a particular passage chosen at random? Ask a friend to put your theory to the test.

Miscellaneous exercise 13

1. A set of 7 positive integers has a mode of 4, a median of 6 and a mean of 8. Write down a set of seven integers which satisfies these properties.
2. During a given period the number of people in each car passing a point was recorded and the results are shown in the following table:

Number of people	1	2	3	4	5 or more
Number of cars	x	26	9	5	0

   If the mode is 1, state the minimum value of x.
   If the median is 2, state the maximum value of x.
3. The median of a set of six numbers is $4\frac{1}{2}$. If five of the numbers are 8, 3, 5, 1, 9, what is the mean of the six numbers?
4. Pandahire, a small car hire firm, have 6 cars to hire on a daily basis. During the month of November (30 days) the number of cars hired per day is shown in the table below:

Number of cars on hire	0	1	2	3	4	5	6
Number of days	3	2	9	6	5	3	2

   Draw a bar chart to represent this information. Calculate the mean number of cars hired per day.

5. The table below shows the quarterly rainfall (in cm) for a particular town for the period 1985–87.

	Quarter 1	Quarter 2	Quarter 3	Quarter 4
1985	170	224	148	190
1986	184	212	135	182
1987	190	226	152	186

Calculate a 4-point moving average for the data.

6. Name the three commonly used measures of average.

   Find each average for the following set of values, distinguishing clearly between them:

   3, 3, 4, 4, 5, 5, 5, 6, 6, 6, 6. (SEG)

7. Figure 13.13 shows the marks scored out of 10 in a recent test.

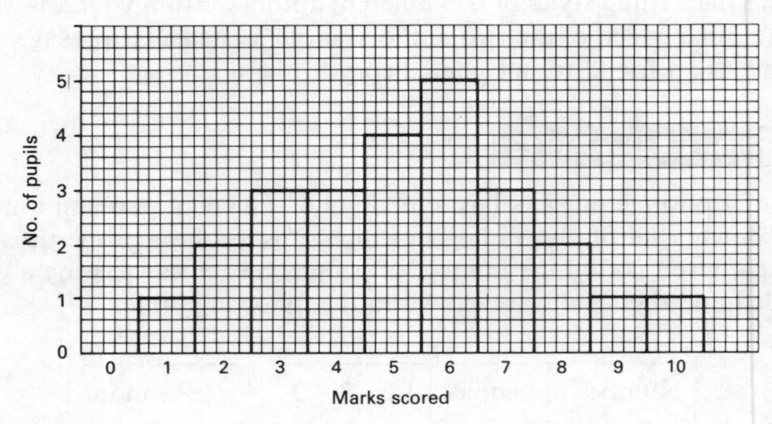

Figure 13.13

   (a) Use the graph to find the following:
      (i) Number of pupils taking the test.
      (ii) Modal score.
      (iii) Median score.
      (iv) Mean score.

   (b) After the bar graph had been drawn, two pupils who had scored 6 marks were found to have been marked incorrectly and so were given one extra mark each. State the mode, median and mean values after these two changes had been made.

8. A magazine invited a sample of 500 readers to keep records of the number of times repairs were needed to their washing machine during the first 12 months from new. The table below gives the results for one make of washing machine.

Number of repairs	0	1	2	3	4 or more
Number of machines	240	100	65	55	40

(a) Represent this information in the form of a bar chart.
(b) Write down:
  (i) the median and
  (ii) the modal number of repairs.
(c) Write down, on the basis of this sample, the probability that a washing machine of this make will not require any repairs during the first 12 months from new.
(d) Mrs Cockcroft buys this make of washing machine. She also buys a freezer which has a probability of 0.8 that it will not require any repairs during the first 12 months from new.

Work out the probability that neither appliance will require repairing during the first 12 months. (LEAG)

9. The graphs in Figure 13.14 show the monthly rainfall in mm for four towns over a twelve-month period.

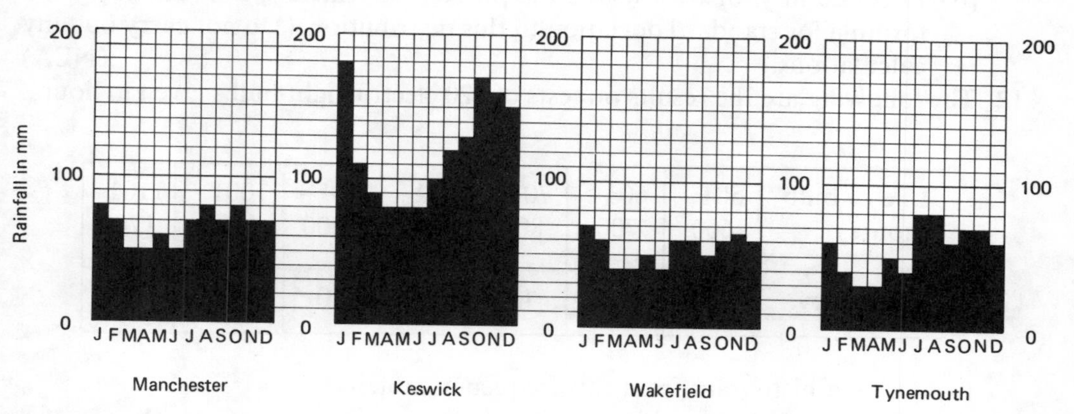

Figure 13.14

(a) Which town had the driest January?
(b) Which town had the wettest summer?
(c) Calculate the mean monthly rainfall for Manchester, giving your answer to the nearest mm.
(d) What is the range in monthly figures for the year in Keswick?
(e) During the month of July, Keswick had rain on 12 days. What is the probability that during a one-day visit to Keswick in July, it will not rain? (NEA)

10. A health food manufacturer offered prizes to groups of 'weight-watchers' who reduced their mean (average) weight by at least 3 kg in a fortnight. The 10 members of the MEGSLIM group decided to try to achieve this.

At their meeting at the end of the fortnight the average weight loss for 9 of the members was 2.9 kg. The tenth member, Mr Lean, had been delayed but when he arrived it was found that he had lost 5.9 kg.

What was the mean (average) weight loss for the whole group? (MEG)

11. The number of full-time female students in the United Kingdom in 1981 in various age groups is shown in the following table. (Frequencies are given to the nearest thousand.)

Age (years)	16–	20–	25–	35–	45–54	Total
Number (in thousands)	672	139	34	13	4	862

(*Source*: *Annual Abstract of Statistics* 1984, HMSO)

(a) This information is to be represented in a histogram using a scale of 2 cm to represent 5 years on the age axis and 4 cm^2 to represent 100 thousand students. Given that the width of the first rectangle is 1.6 cm, calculate its height.
(b) On graph paper, draw the histogram.
(c) By using mid-interval values, estimate the mean age of female students, giving your answer correct to the nearest tenth of a year.
(d) Describe in your own words the process by which you would apply the formula for standard deviation to this distribution. (Do not carry out any calculations.) (NEA)

12. The following are the results on tests of 120 electric light bulbs given in hours:

Life of bulb (hours)	501–600	601–700	701–800	801–900	901–1000	1001–1100	1101–1200
Frequency	7	12	19	36	40	4	2

(a) Draw a histogram from this frequency table.
(b) Use the frequency table to:
(i) estimate the mean life of a bulb;
(ii) calculate the median life of a bulb;
(c) When testing a batch of electric light bulbs, why is it quicker to find the median than the mean? (LEAG)

13. (a) For the values

2, 5, 5, 7, 8, 9

(i) calculate the mean,
(ii) calculate the standard deviation correct to one decimal place, showing clearly your working.

(b) The weight of individual luggage taken by 100 passengers on a cruise is given in the following table:

Weight (kg)	20–24	25–29	30–34	35–39	40–44
Frequency	15	23	30	19	13

(i) Calculate an estimate of the mean weight.
(ii) Why is this an estimate? (SEG)

14. In an examination there were 5000 candidates and the marks they obtained

are summarised in the table below:

Mark obtained	20 or less	30 or less	40 or less	50 or less	60 or less	70 or less	80 or less	100 or less
Number of candidates	250	500	1000	2000	3200	4000	4500	5000

(a) Draw a cumulative frequency diagram to represent these results. (Use a scale of 2 cm to represent 20 marks on the x-axis and 2 cm to represent 500 candidates on the cumulative frequency axis.)
(b) Using the cumulative frequency diagram, or otherwise, find:
(i) the number of candidates who scored 82 marks or less,
(ii) the median mark,
(iii) the interquartile range,
(iv) the percentage of candidates who scored more than 70 marks,
(v) the minimum mark required for a pass if 80% of the candidates passed the examination. (MEG)

15.

Month	Sales	3 Monthly moving average
January	4	—
February	2	3
March	3	3
April	4	
May	5	
June	9	
July	13	
August	20	
September	15	
October	10	10
November	5	6
December	3	—

The above table shows the ice-cream sold (thousands of litres) in a supermarket throughout the year.
(a) Draw the graph.
(b) Comment on your graph.
(c) Complete the table.
(d) Plot the averages on the graph. (LEAG)

16. The table overleaf shows the height in centimetres of 11 tomato plants in a greenhouse together with the number of tomatoes on each plant.

Height (cm)	146	154	166	145	139	148	147	150	161	150	159
Number of tomatoes	10	12	14	9	8	10	12	13	13	11	13

(a) Find the median number of tomatoes per plant.
(b) Calculate the mean plant height, giving your answer correct to the nearest centimetre.
(c) If a tomato plant is chosen at random from the greenhouse, what is the probability that it is both over 152 cm tall and carries more than 12 tomatoes? Give your answer as a fraction.
(d) Draw a scatter diagram to illustrate this information. Use a horizontal scale of 2 cm to represent plant height 5 cm, starting at height 130 cm. Use a vertical scale of 2 cm to represent 1 tomato, starting with 7 tomatoes.
(e) Describe briefly any correlation suggested by this diagram. (MEG)

17. An intelligence test was given to 15 girls of different ages (in months) and the following results were obtained:

Girl	A	B	C	D	E	F	G	H	I	J	K	L	M	N	O
Age (months)	134	141	154	144	163	164	175	185	175	188	185	202	205	202	200
Mark	62	63	64	62	72	69	80	76	72	82	80	82	82	85	81

(a) Using these figures, draw a scatter diagram of marks and age.
(b) Does your diagram show evidence of correlation between age and intelligence?
(c) The mean age of the girls is 174.5 months. Calculate the mean mark. Use this information to draw the line of best fit for the data.
(d) A sixteenth girl, P, whose age was 180 months, was absent for the test. Estimate, from your graph, the mark she might have obtained.

(MEG)

14 FURTHER TOPICS, INCLUDING SURVEYING

14.1 Further circle topics

In Unit 5 we defined the component parts of a circle and deduced the property that the angle in a semicircle is 90°. We shall now demonstrate further properties of angles in a circle.

Worked example 14.1

In Figure 14.1, O is the centre of the circle, $\angle$ BAO is $x°$ and $\angle$ CAO $= y°$. Find $\angle$ BOD and $\angle$ COD. What is the relationship between $\angle$ COB and $\angle$ CAB?

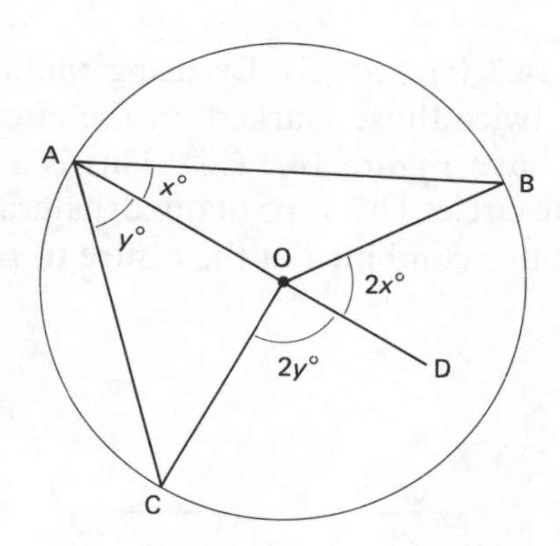

Figure 14.1

As triangles ABO and AOC are isosceles:

$$\angle \text{BOA} = \angle \text{BAO} = x°$$

and

$$\angle \text{ACO} = \angle \text{CAO} = y°$$

So by the property of exterior angles of a triangle:

$$\angle \text{BOD} = 2x°$$

and $$\angle COD = 2y°$$

Now $\angle COB = \angle BOD + \angle COD = 2x° + 2y°$ and $\angle CAB = \angle BAO + \angle CAO = x° + y°$.

So $$\angle COB \text{ is } \textit{double } \angle CAB.$$

Property:	The angle at the centre of a circle is twice the angle at the circumference standing on the same arc.

Figure 14.2 shows examples of this property.

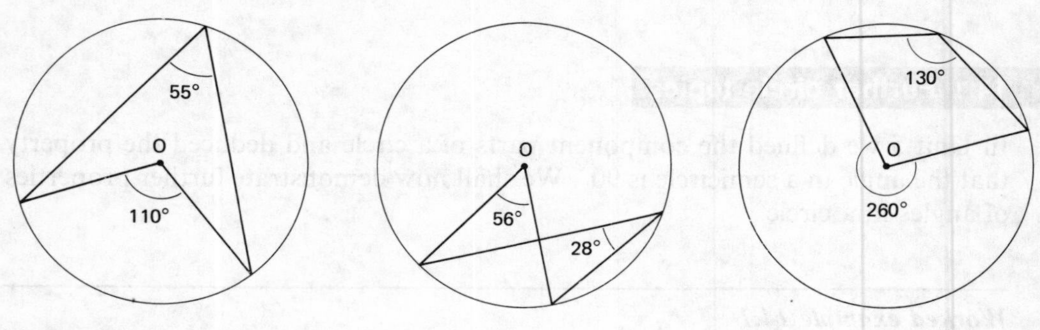

Figure 14.2

Note: the angle in a semicircle, shown in section 5.14, is a special case of this property.

Now look at Figure 14.3 (i) and (ii). By using the last property, the shaded angles at the centre are twice those marked on the circumference. Now putting (i) and (ii) together we have Figure 14.3 (iii). This is a quadrilateral which has four vertices touching the circle. This type of quadrilateral is a *cyclic quadrilateral.* The two shaded angles have combined at the centre to produce an angle of 360°.

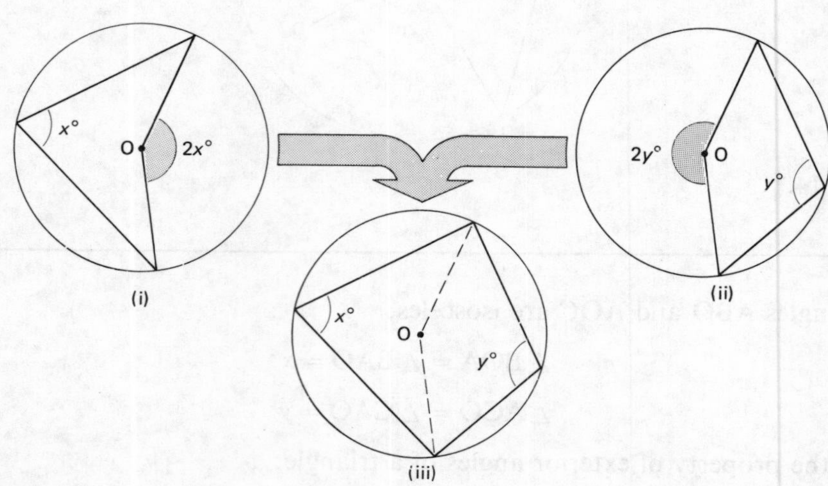

 Figure 14.3

So $$x^\circ + y^\circ = 180^\circ$$

Property: The opposite angles of a cyclic quadrilateral add up to 180°.

The third property relates to angles in the same segment. To remind ourselves, Figure 14.4 shows the chord AB which divides the circle into two segments. ∠ACB and ∠ADB are said to be angles in the same segment. If we use the first property above, we can show that these two angles are equal.

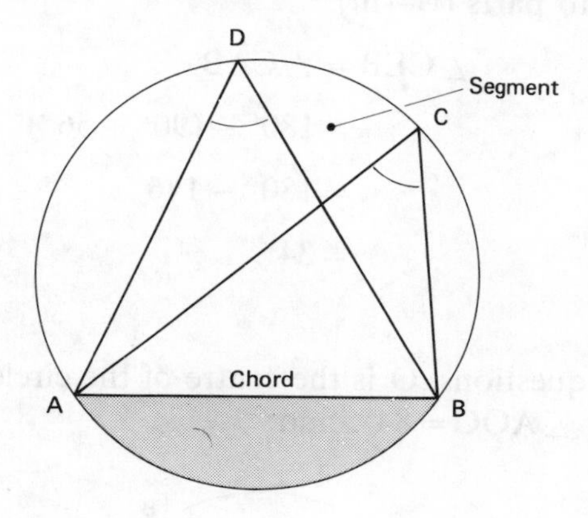

Figure 14.4

Property: Angles in the same segment of a circle are equal.

Worked example 14.2

In Figure 14.5, O is the centre of the circle with AB a diameter and ∠ADC = 124°.

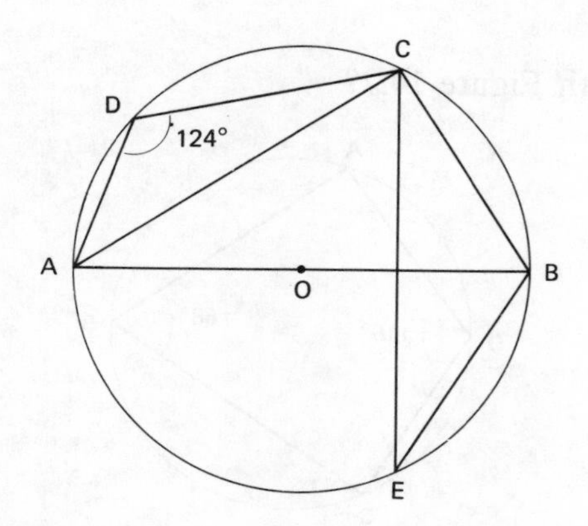

Figure 14.5

(i) Calculate the value of $\angle ABC$.
(ii) Name a right angle in the diagram.
(iii) Name an angle equal to $\angle CEB$.
(iv) Calculate the value of $\angle CEB$.

(i) $\angle ABC = 180° - 124° = 56°$ (opposite angles of a cyclic quadrilateral).
(ii) $\angle ACB$ is a right angle (angle in a semicircle).
(iii) The chord CB divides the circle into two segments.
So $\angle CEB = \angle CAB$ (angles in the same segment).
(iv) Using answers to parts (i)–(iii):

$$\begin{aligned}\angle CEB &= \angle CAB\\ &= 180° - (90° + 56°)\\ &= 180° - 146°\\ &= 34°.\end{aligned}$$

Exercise 14a

In all the following questions, O is the centre of the circle.

(1) In Figure 14.6, $\angle AOC = 84°$, find:

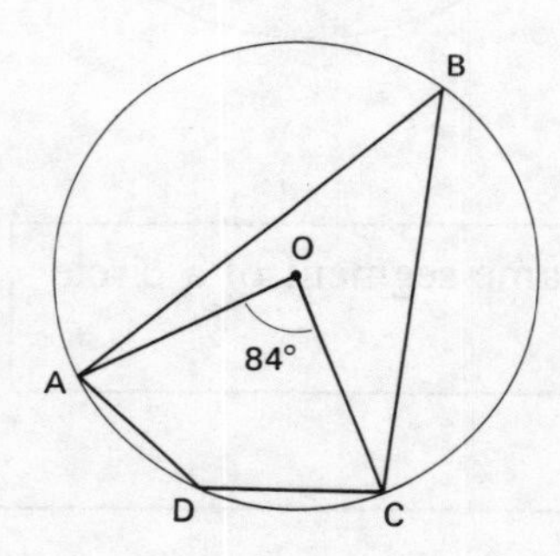

Figure 14.6

(i) $\angle ABC$,
(ii) $\angle ADC$.

(2) What is wrong with Figure 14.7?

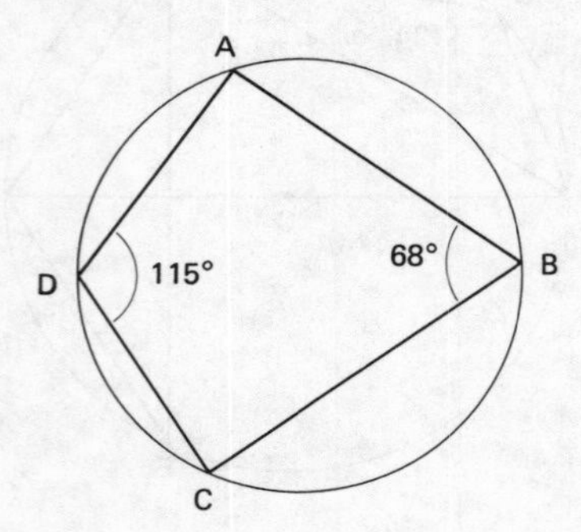

 Figure 14.7

(3) Figure 14.8 shows a cyclic quadrilateral ABCD with AB parallel to DC. If $\angle BAD = 110°$, find:

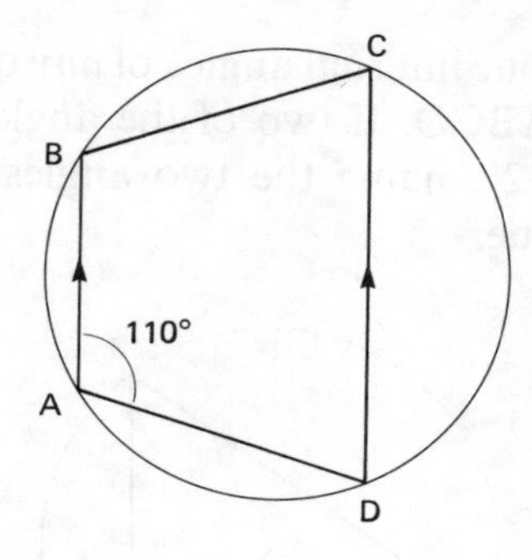

Figure 14.8

(i) $\angle ADC$,
(ii) $\angle ABC$.

(4) In Figure 14.9, AO is parallel to BC and $\angle AOB = 30°$.

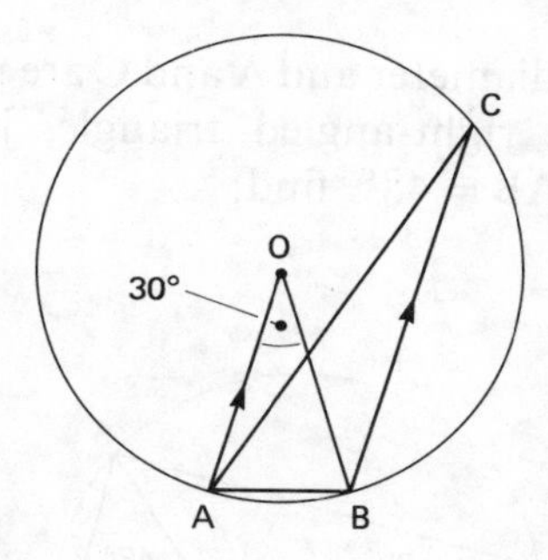

Figure 14.9

(i) Find $\angle ACB$.
(ii) Name another angle equal to $\angle ACB$.
(iii) What type of triangle is $\triangle OAB$?
(iv) Find $\angle OBA$.
(v) Find $\angle CAB$.

(5) Figure 14.10 shows a cyclic quadrilateral ABCD. If $\angle ACB = 43°$ and $\angle BDC = 37°$, find:

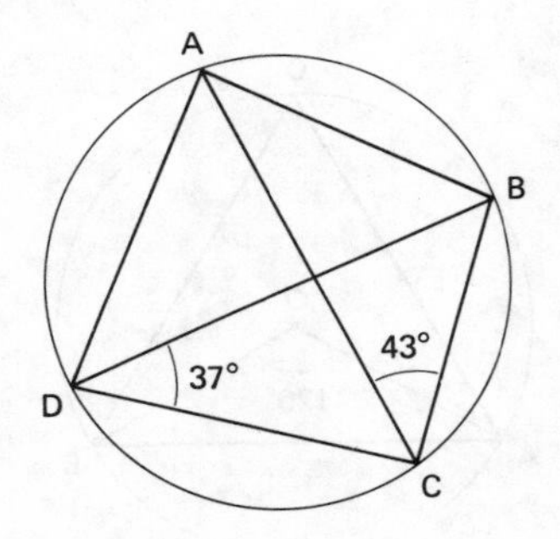

Figure 14.10

(i) $\angle$ ADB,
(ii) $\angle$ ABC.
Justify your answers.

(6) What is the sum of the four internal angles of *any* quadrilateral? Figure 14.11 shows a quadrilateral ABCO. If two of the angles of the quadrilateral are equal and $\angle \text{AOC} = 112°$, name the two angles which are equal to each other and find their value.

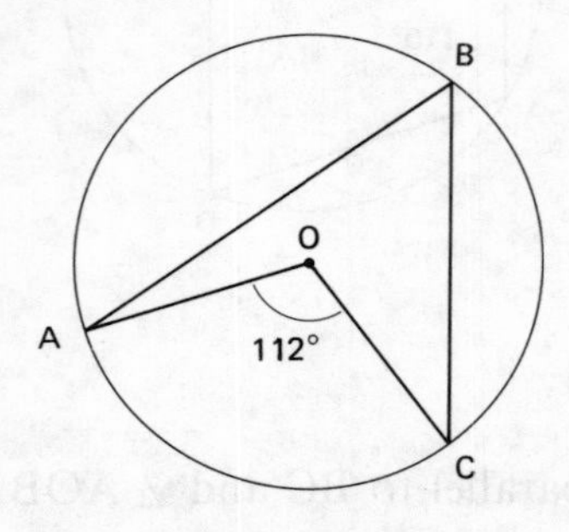

Figure 14.11

(7) In Figure 14.12, DB is a diameter and A and C are points on the circumference of the circle. Name a right-angled triangle, justifying your answer. If $\angle \text{ADB} = 25°$ and $\angle \text{CAB} = 48°$, find:
(i) $\angle$ ACB,
(ii) $\angle$ AEB.

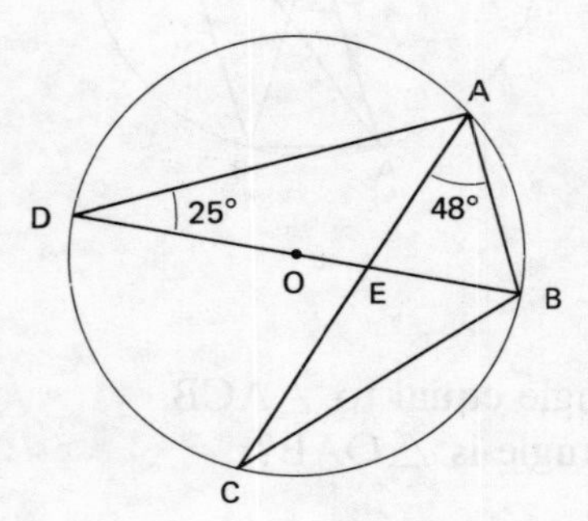

Figure 14.12

(8) Figure 14.13 shows a triangle ABC inscribed within a circle, and with AC = BC. $\angle \text{AOB} = 120°$. Show that triangle ABC is an equilateral triangle. Find $\angle$ CAO.

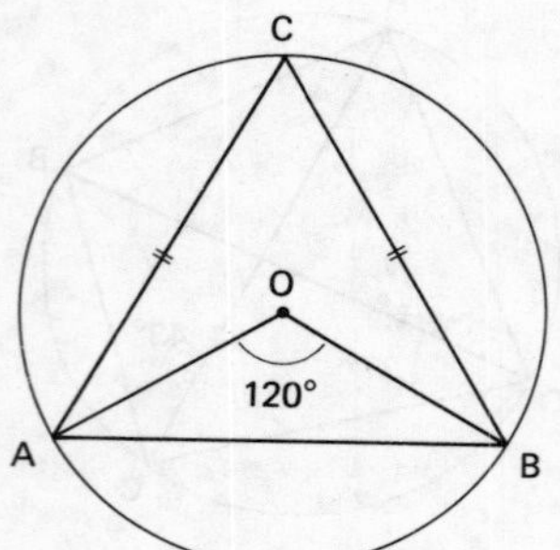

 Figure 14.13

(9) Figure 14.14 shows a quadrilateral ABCO with A, B and C on the circumference of the circle; $\angle BCO = 70^\circ$ and AB is parallel to OC.

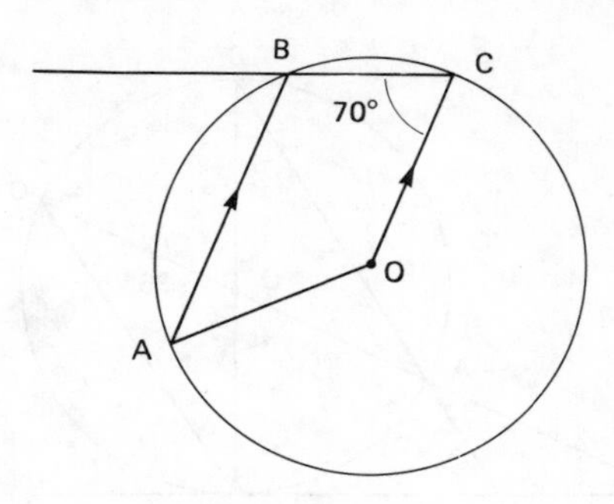

Figure 14.14

(i) Write down the value of $\angle ABC$.
(ii) Determine the value of the obtuse angle, $\angle AOC$.
(iii) Is the quadrilateral ABCO a cyclic quadrilateral? Justify your answer.

(10) Figure 14.15 shows a cyclic quadrilateral ABCD with AB and DC extended to meet at E. If $\angle BEC = 35^\circ$ and $\angle ADC = 75^\circ$, find the value of the other two angles of $\triangle BEC$.

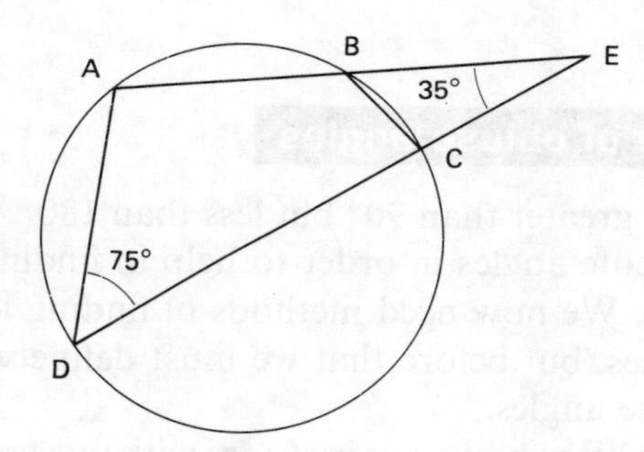

Figure 14.15

(11) In Figure 14.16, ABCDE is a pentagon inscribed in a circle, and AB = AE. $\angle ABE = 28^\circ$ and $\angle BCD = 110^\circ$. Write down the values of:
(i) $\angle BAE$,
(ii) $\angle AED$.

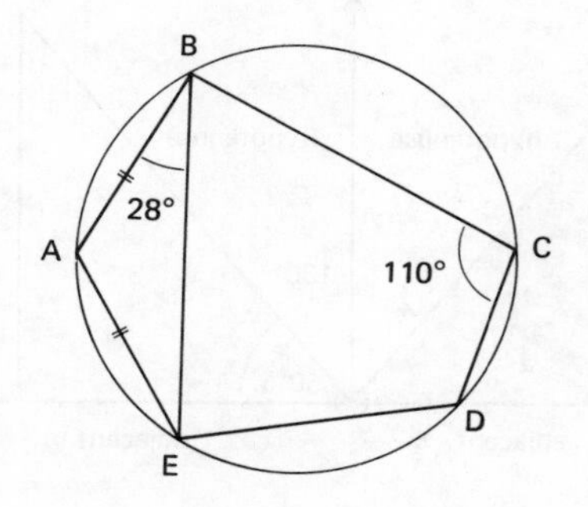

Figure 14.16

(12) Figure 14.17 shows a tangent FD drawn to a circle and touching at D. DB is a diameter and ∠OCD = 36°.

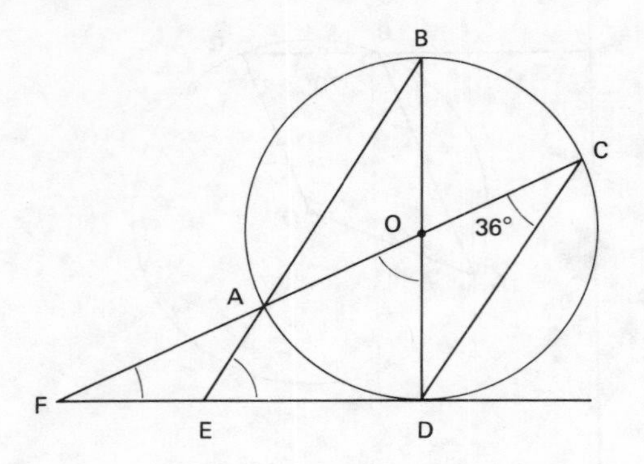

Figure 14.17

(a) Write down the name of a right-angled triangle in the diagram.
(b) Name *all* other angles in the diagram which are equal to ∠OCD.
(c) Determine the values of:
(i) ∠AOD,
(ii) ∠OFD,
(iii) ∠BED.

14.2 Sine and cosine of obtuse angles

Obtuse angles are angles greater than 90° but less than 180°. In Unit 7 we defined the sine and cosine of acute angles in order to help in finding lengths and angles of right-angled triangles. We now need methods of finding lengths and angles of non-right-angled triangles, but before that we must define what is meant by the sine and cosine of obtuse angles.

Look at Figure 14.18 which shows a pair of axes with two congruent right-angled triangles drawn. The common point of both triangles is taken as the origin. Both triangles are equal in all respects and the triangle on the right can be used to determine the values of sin 50° and cos 50° by measuring the adjacent, opposite and hypotenuse. The shaded triangle on the left can be used to determine the sine

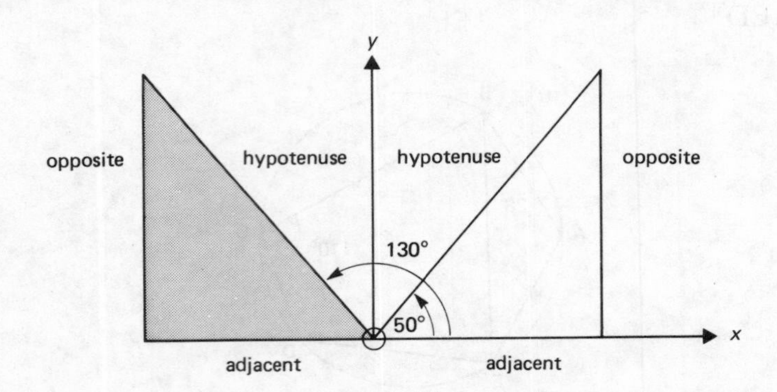

 Figure 14.18

and cosine of 130°. As all corresponding sides of the two triangles are equal in length, it can be seen that:

$$\sin 130° = \sin 50°$$

In a similar way, *numerically* $\cos 50° = \cos 130°$. However, the adjacent side of the shaded triangle is measured in the negative x direction. So:

$$\cos 130° = -\cos 50°$$

In general, therefore, for an obtuse angle $x°$:

$$\sin x° = \sin(180 - x)°$$

and

$$\cos x° = -\cos(180 - x)°$$

Exercise 14b

(1) Given the following information:

$\sin 20° = 0.342$	$\sin 30° = 0.500$	$\sin 70° = 0.940$	$\sin 80° = 0.985$
$\cos 20° = 0.940$	$\cos 30° = 0.866$	$\cos 70° = 0.342$	$\cos 80° = 0.174$

Write down, without using a calculator, the value of:

(i) $\cos 150°$ (ii) $\sin 160°$ (iii) $\cos 100°$ (iv) $\sin 150°$
(v) $\sin 100°$ (vi) $\cos 160°$ (vii) $\sin 110°$ (viii) $\cos 110°$

(2) What do you think is the value of (i) $\sin 180°$, (ii) $\cos 180°$?

(3) If $\sin x° = 0.75$, write down two possible values of x to the nearest 1/10th of a degree.

(4) If $\cos x° = -0.8$, write down a value of x to the nearest 1/10th of a degree.

We will now look at three formulae for non-right-angled triangles which will help us to find lengths, angles and areas. Each formula can be arrived at algebraically, however we will confine ourselves to stating the formulae and applying them to problems.

14.3 Sine rule

Figure 14.19 shows a triangle ABC. We shall denote the sides opposite the vertices A, B and C with the letters a, b and c respectively.

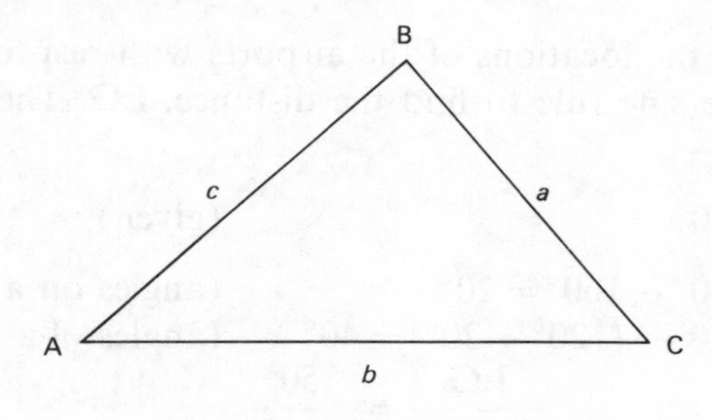

Figure 14.19

The *sine rule* states:

$$\frac{a}{\sin A} = \frac{b}{\sin B} = \frac{c}{\sin C}$$

It is sometimes useful to use the inversion of the rule:

$$\frac{\sin A}{a} = \frac{\sin B}{b} = \frac{\sin C}{c}$$

Worked example 14.3

'Shuttleair' fly a helicopter service between three airports, Lutwick, Heaton and Gatrow. Lutwick is 50 km due North of Heaton and Gatrow is on a bearing of 160° from Lutwick and 120° degrees from Heaton. Find the distance from Lutwick to Gatrow. How long to the nearest minute would a flight take from Lutwick to Gatrow if the average speed of the helicopter is 140 km/h?

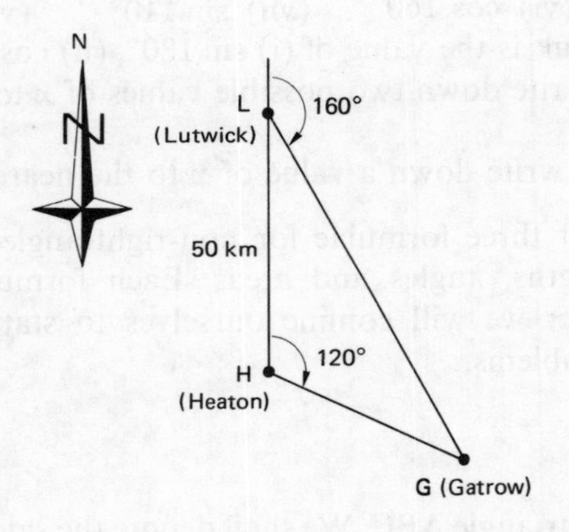

Figure 14.20

Figure 14.20 shows the locations of the airports with respect to each other. We will need to use the sine rule to find the distance, LG. The three angles of the triangle are:

$\angle \text{LHG} = 120°$ (given)

$\angle \text{HLG} = 180° - 160° = 20°$ (angles on a straight line)

and $\angle \text{HGL} = 180° - (120° + 20°) = 40°$ (angles of a triangle)

So

$$\frac{\text{LG}}{\sin 120°} = \frac{50}{\sin 40°}$$

$$\frac{LG}{0.8660} = \frac{50}{0.6428}$$

$$\frac{LG}{0.8660} = 77.78$$

Therefore $LG = 77.78 \times 0.8660$ km

$= 67.4$ km

If the helicopter flies at 140 km/h, the time taken to travel 67.4 km

$$= \frac{67.4}{140} \times 60 \text{ minutes}$$

$= 29$ minutes (to nearest minute).

The sine rule is used:

To find	when
A side	Two angles and one side are given
An angle	Two sides and one angle are given The angle is not between the two sides

14.4 Cosine rule

The sine rule does not help us to solve all problems related to non-right-angled triangles and we have to introduce a further rule, the *cosine rule*.

Using the lettering of Figure 14.19, the cosine rule can be stated as follows:

$$a^2 = b^2 + c^2 - (2bc \cos A)$$

This will allow us to find the length of a side of the triangle. The letters can be interchanged so that the sides b and c can be found from the formulae:

$$b^2 = a^2 + c^2 - (2ac \cos B)$$

and

$$c^2 = a^2 + b^2 - (2ab \cos C)$$

If we want to find the angle instead, the rearranged formulae become:

$$\cos A = \frac{b^2 + c^2 - a^2}{2bc}$$

$$\cos B = \frac{a^2 + c^2 - b^2}{2ac}$$

and

$$\cos C = \frac{a^2 + b^2 - c^2}{2ab}$$

Worked example 14.4

Eve N Parr, a keen local golfer, was about to play from the 17th tee. The distance from the tee to the hole is 350 m. Unfortunately she slices her shot at an angle of 20° from the direction of the hole and the ball travels a distance of 180 m. How far will her golf ball be from the hole?

Eve's partner, Sandy Bunker, hits his shot a distance of 220 m but is still 175 m from the hole. At what angle, to the nearest 1/10th of a degree, did he slice his shot?

Figure 14.21 shows the path taken by Eve's sliced ball and the distance to the flag is marked with the letter *d*.

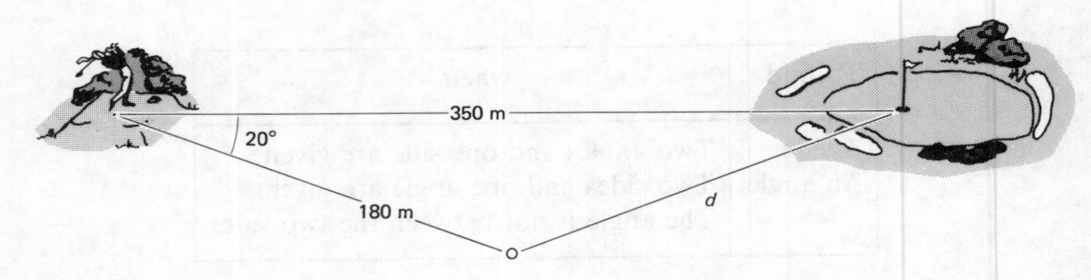

Figure 14.21

Using the cosine rule:

$$d^2 = 350^2 + 180^2 - (2 \times 350 \times 180 \cos 20°)$$
$$= 122\,500 + 32\,400 - (700 \times 180 \times 0.9397)$$
$$= 154\,900 - 118\,400$$
$$= 36\,500$$

So the distance to the hole $= \sqrt{36\,500} = 191$ metres.

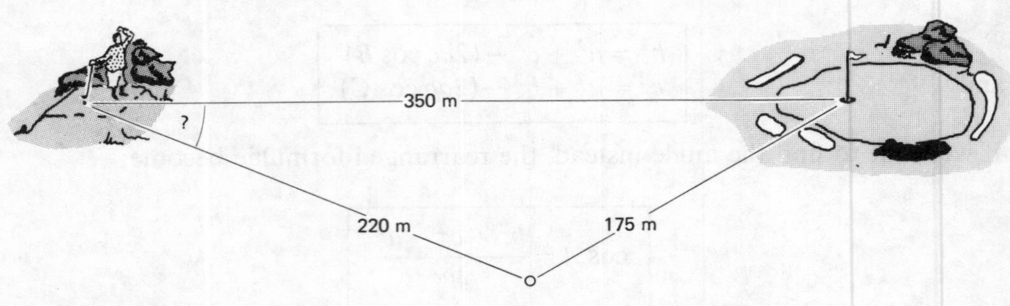

Figure 14.22

Figure 14.22 shows the path of Sandy's sliced shot. This time we have the measurements of three sides of a non-right-angled triangle and we require to work out an angle. Calling the angle *T*, we use the alternative form of the cosine rule:

$$\cos T = \frac{350^2 + 220^2 - 175^2}{2 \times 350 \times 220}$$

$$= \frac{122\,500 \times 48\,400 - 30\,625}{154\,000}$$

$$= \frac{140\,275}{154\,000}$$

$$= 0.9109$$

So $T = 24.4°$.

Note that the cosine of the largest angle of this triangle is given by:

$$\frac{220^2 + 175^2 - 350^2}{2 \times 220 \times 175}$$

$$= \frac{48\,400 + 30\,625 - 122\,500}{77\,000}$$

$$= \frac{-43\,475}{77\,000}$$

$$= -0.5646$$

Giving an angle of 124.4°.

The cosine rule is used:

To find	when
A side An angle	Two sides and the included angle are given Three sides are given

14.5 Area of a triangle

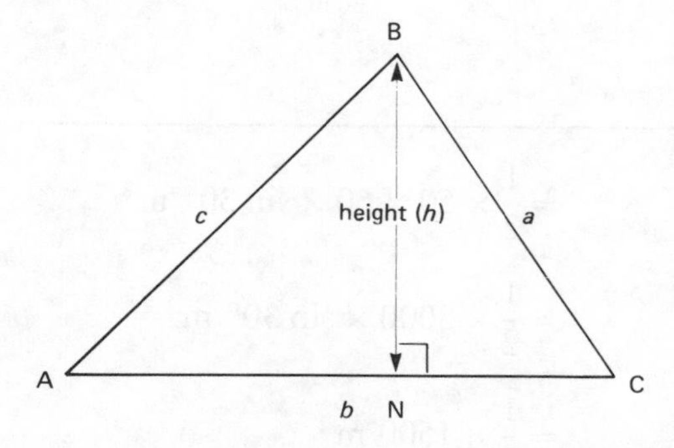

Figure 14.23

The area of the triangle shown in Figure 14.23 is given by the standard formula:

half of the base length × height

$$= \frac{1}{2} \times b \times h$$

But using the sine ratio for the right-angled triangle ABN:

$$h = c \times \sin A$$

So if the height is unknown, we can determine the area of a triangle by using the formula:

$$\frac{1}{2} \times (c \times \sin A) \times b$$

or

$$\boxed{\text{Area of triangle ABC} = \frac{1}{2} bc \sin A}$$

Again, like the cosine rule, the letters can be interchanged. One way of remembering how to use the formula is to:

(i) multiply two sides together,
(ii) multiply the result by the sine of the angle between the two sides.
(iii) half this result.

Worked example 14.5

Figure 14.24 shows a triangular village green. If the length of two of the sides are 50 m and 60 m and the included angle is 30°, find the area of the green.

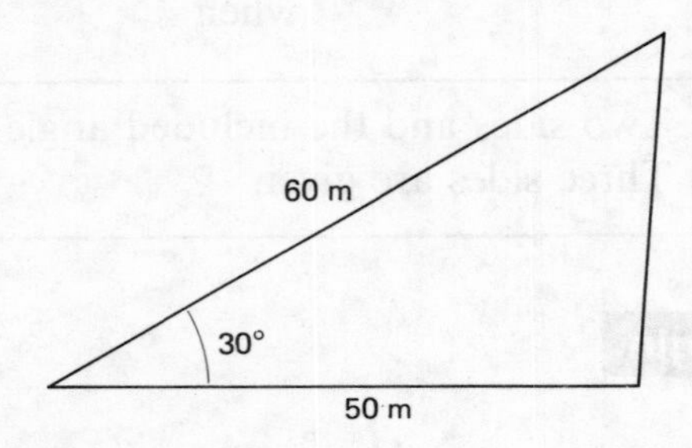

Figure 14.24

The area $= \frac{1}{2} \times 50 \times 60 \times \sin 30° \text{ m}^2$

$$= \frac{1}{2} \times 3000 \times \sin 30° \text{ m}^2$$

$$= \frac{1}{2} \times 1500 \text{ m}^2$$

$$= 750 \text{ m}^2.$$

Exercise 14c

(1) In Figure 14.19, use the sine rule to find the length of side BC, correct to 3 significant figures, if:

(i) $c = 12$ cm, $\angle$BAC $= 30°$, $\angle$ACB $= 45°$;
(ii) $b = 8$ cm, $\angle$BAC $= 50°$, $\angle$ACB $= 70°$;
(iii) $b = 9$ cm, $\angle$ABC $= 25°$, $\angle$ACB $= 35°$;
(iv) $c = 5$ cm, $\angle$BAC $= 100°$, $\angle$ACB $= 35°$.

(2) In Figure 14.19, use the sine rule to find $\angle$ACB, correct to the nearest 1/10th of a degree, if:

(i) $c = 7$ cm, $a = 8$ cm, $\angle$BAC $= 40°$;
(ii) $c = 15$ cm, $b = 11$ cm, $\angle$ABC $= 30°$;
(iii) $c = 13$ cm, $a = 20$ cm, $\angle$BAC $= 110°$.

(3) In Figure 14.19, use the cosine rule to find the length of side AB, correct to 3 significant figures, if:

(i) $a = 3$ cm, $b = 4$ cm, $\angle$ACB $= 75°$;
(ii) $a = 20$ cm, $b = 14$ cm, $\angle$ACB $= 115°$;
(iii) $a = 7$ cm, $b = 12$ cm, $\angle$ACB $= 60°$;
(iv) $a = 7$ cm, $b = 12$ cm, $\angle$ACB $= 120°$.

(4) In Figure 14.19, use the cosine rule to find $\angle$BAC, correct to the nearest 1/10th of a degree, if:

(i) $a = 4$ cm, $b = 3$ cm, $c = 5$ cm;
(ii) $a = 12$ cm, $b = 5$ cm, $c = 10$ cm;
(iii) $a = 11$ cm, $b = 7$ cm, $c = 8$ cm.

(5) In Figure 14.19, find the area of the triangle, correct to 3 significant figures, if:

(i) $a = 14$ cm, $b = 4$ cm, $\angle$ACB $= 45°$;
(ii) $b = 25$ cm, $c = 20$ cm, $\angle$BAC $= 42°$;
(iii) $a = 12$ cm, $c = 17$ cm, $\angle$ABC $= 138°$.

(6) If the area of Figure 14.19 is 25 cm^2, the length of side $a = 12$ cm and the length of side $b = 10$ cm, write down the value of sin $\angle$ACB as a fraction in its lowest terms. If the triangle is an acute angled triangle, use your result to find $\angle$ACB.

(7) A balloon is situated immediately above an arena. To highlight the balloon, two spotlights on level ground and which are 500 m apart are trained on the cabin hanging below the balloon (see Figure 14.25). One spotlight is at an angle of 70°, the other at an angle of 50°. Calculate the distances that

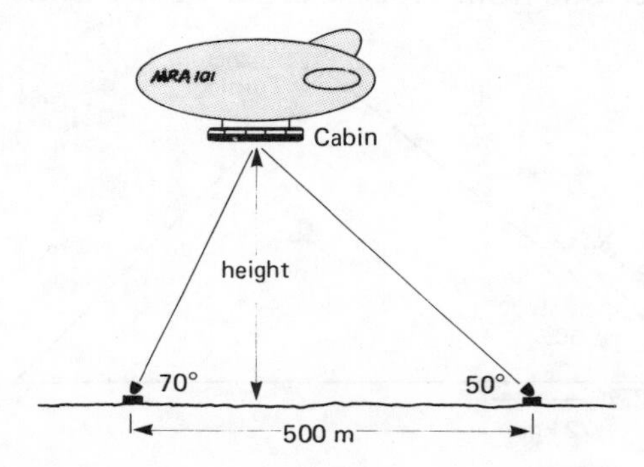

Figure 14.25

both beams of light have to travel to reach the cabin to the nearest 10 m. Using either of your answers, calculate the height of the cabin above ground to the nearest 10 m.

(8) To measure the height of a lighthouse, Winston measures the angle of elevation from two points at ground level which are 100 m apart. Both points are in a horizontal straight line with the base of the lighthouse. The angles of elevation were noted as 14° and 25°, and Winston was able to draw the diagram shown in Figure 14.26.

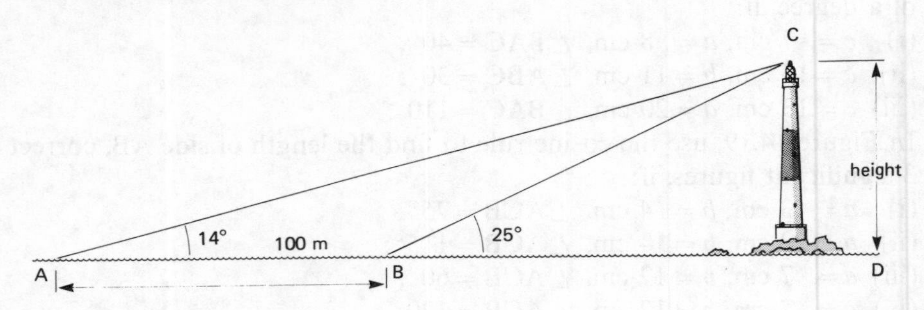

Figure 14.26

(i) Write down the values of $\angle ABC$ and $\angle ACB$.
(ii) Use the sine rule to find the length of BC.
(iii) Find the height, CD, of the lighthouse.

(9) The *Flying Kipper*, a small ship, is tied to a capstan on the quayside by two horizontal cables. One cable of length 20 m is attached to the front of the ship and the other of length 25 m to the rear (see Figure 14.27). The angle between the two cables is 140°. Find the length of the ship.

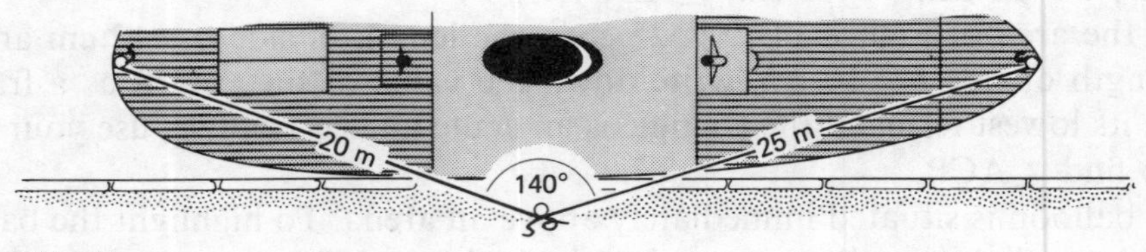

Figure 14.27

(10) Figure 14.28 shows two roads which meet at Apex Corner at an angle of 50°. Two cars set out from Apex Corner at the same time, one travelling

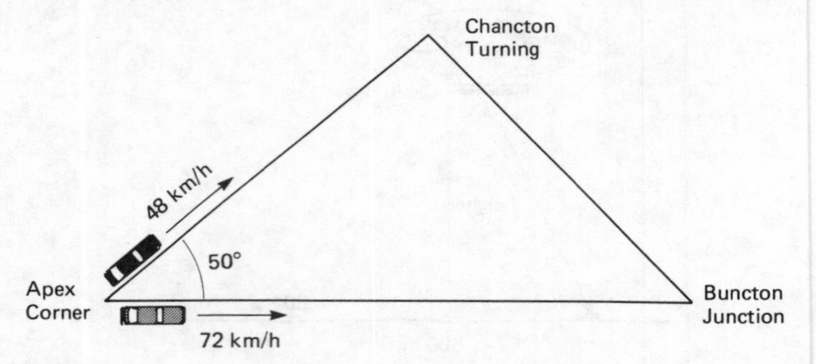

 Figure 14.28

towards Buncton Junction at a speed of 72 km/h and the other towards Chancton Turning at a speed of 48 km/h. After 5 minutes, the cars arrive at Buncton and Chancton respectively.

(i) How far are Buncton Junction and Chancton Turning from Apex Corner?

(ii) Find the length of the straight road joining Buncton Junction and Chancton Turning to the nearest 1/10th of a kilometre.

(11) Figure 14.29 shows a ladder resting against a vertical wall and the base is in contact with the ground which slopes at an angle of 10° to the horizontal. The height of the ladder up the wall is 8 m and the distance of the ladder from the base of the wall is 3 m.

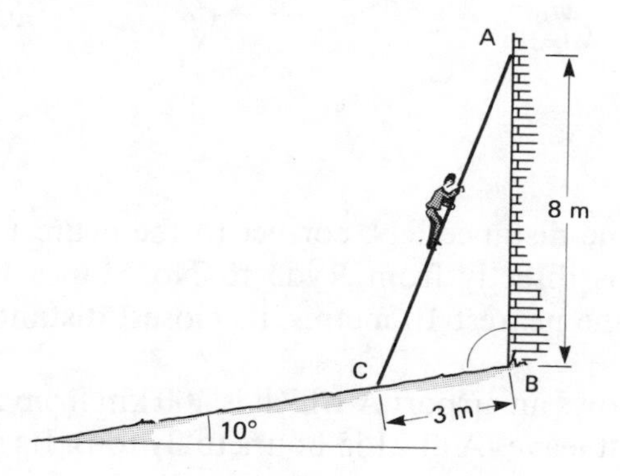

Figure 14.29

(i) Find ∠ ABC.

(ii) Use the cosine rule to find the length of the ladder, AC, correct to 1 decimal place.

(iii) In order that the ladder should not slip, when a person climbs the ladder, the minimum size of ∠ ACB is 70°. Using this angle and the length of the ladder found in part (ii), find the minimum height of the ladder up the wall to 1 decimal place.

(12) A wooden block in the shape of a wedge (Figure 14.30) is used to put behind a wheel of a car to stop the car moving when parked on a hill. If the length of two sides of the triangular cross-section and the included angle are as shown, find the area of the cross-section. If the block is 20 cm wide, what is the volume of wood used?

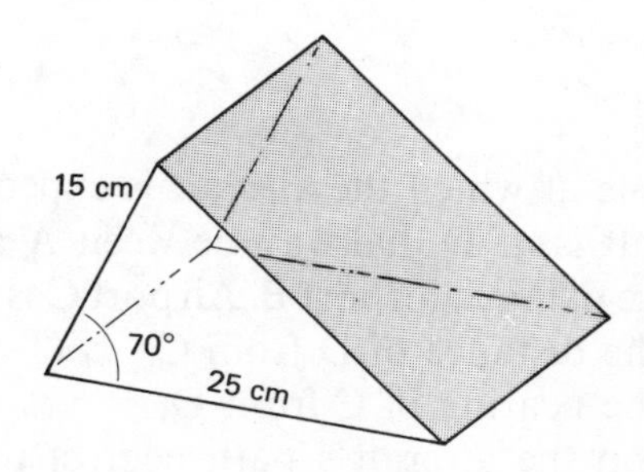

Figure 14.30

(13) Figure 14.31 shows part of the coast of the Isle of Wight and The Solent. The point R represents the position of Ryde, point P the position of Puckpool Point, and point N the position of No Man's Land Fort. RP = 2200 m, $\angle NRP = 29°$ and $\angle RPN = 131°$.

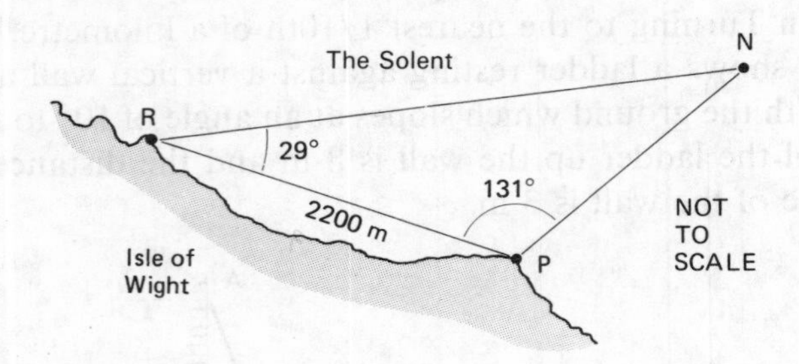

Figure 14.31

(a) Calculate the distance RN, correct to the nearest 10 metres.
(b) A boat sails directly from Ryde to No Man's Land Fort. Calculate, correct to the nearest 10 metres, its closest distance to Puckpool Point.
(MEG)

(14) Figure 14.32 shows an airport A which is 400 km from airport B on a bearing 120°. An aircraft leaves A at 2155 hours to fly to B. Its speed over the ground is 320 km/h.

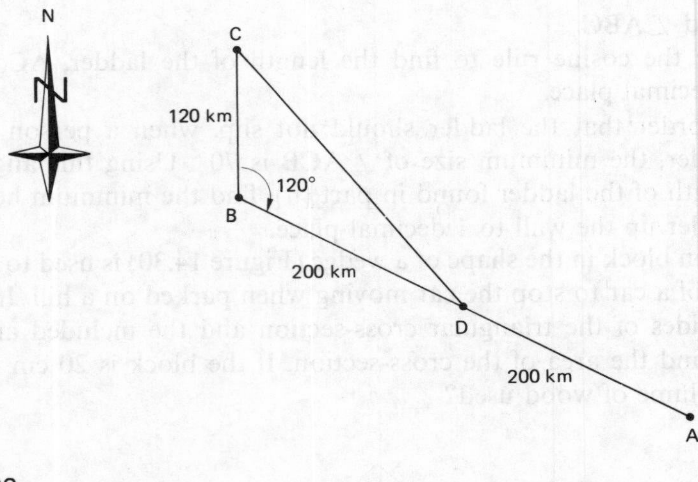

Figure 14.32

(a) Calculate the time at which the aircraft is expected to arrive at B.
(b) When the aircraft is at D, halfway between A and B, it is diverted to airport C because of fog at airport B. Airport C is 120 km due North of B.
(i) Calculate the distance of D from C.
(ii) Calculate the bearing of C from D.
(iii) The point on the aircraft's path nearest to B is X. Calculate the distance of X from B.
(MEG)

14.6 Scale drawings

Some of the questions in the previous exercise could have been tackled by using ruler, protractor and compasses to make a *scale drawing*. Although this technique would not give quite the accuracy that you would obtain by using one of the formulae, it is very useful as a quick and easy-to-understand method for determining estimates of distances and angles.

The following points should be borne in mind when drawing a scale diagram:

(i) If a diagram is not given, a rough sketch allows you to decide where on your drawing paper, the scale diagram is to be placed.
(ii) A scale, if not given, should be chosen which allows full use of the available space.
(iii) Convert all distances using the scale factor.
(iv) State the scale used by the side of the diagram.
(v) North lines should be drawn on all diagrams involving bearings.

Worked example 14.6

Using worked example 14.4, draw scale diagrams to determine (a) the distance of Eve's shot from the hole and (b) the angle of slice of Sandy's shot. Use a scale of 1 cm:25 m.

Using the scale factor of a 1 cm length is equivalent to a distance of 25 m, the measurements on the scale diagrams will be:

Distance	Conversion	Scale diagram
350 m	350/25	14.0 cm
180 m	180/25	7.2 cm
220 m	220/25	8.8 cm
175 m	175/25	7.0 cm

(a) To determine the distance of Eve's ball from the hole we need to:
  (i) Draw a line 14 cm long.
  (ii) Use a protractor at T to measure an angle of 20°.

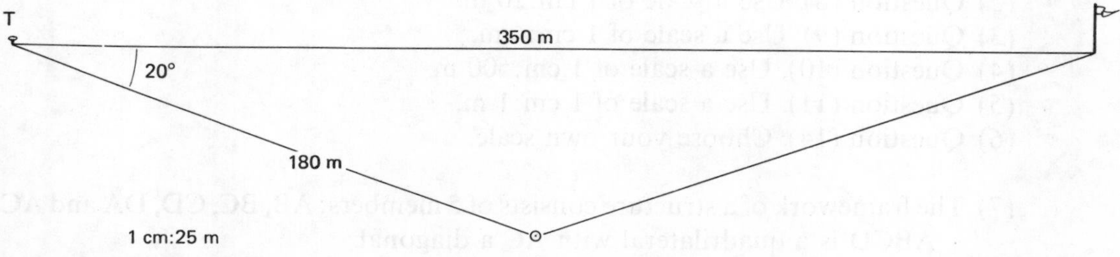

Figure 14.33

(iii) Draw a line 7.2 cm long.
(iv) Measure the distance between the two end points of the lines drawn.

The scale diagram is shown in Figure 14.33.

The measured length is 7.6 cm which converting back to a distance

$$= 7.6 \times 25 \text{ m}$$
$$= 190 \text{ m.}$$

(b) To determine the angle of slice we need to:
(i) Draw a line 14 cm long.
(ii) Use a pair of compasses and placing the point at T, make an arc of radius 8.8 cm.
(iii) Again using the compasses, place the point at the flag and make an arc of radius 7.0 cm.
(iv) Where the two arcs cross, complete the triangle.
(v) The angle at T can now be measured.

The scale diagram is shown in Figure 14.34.

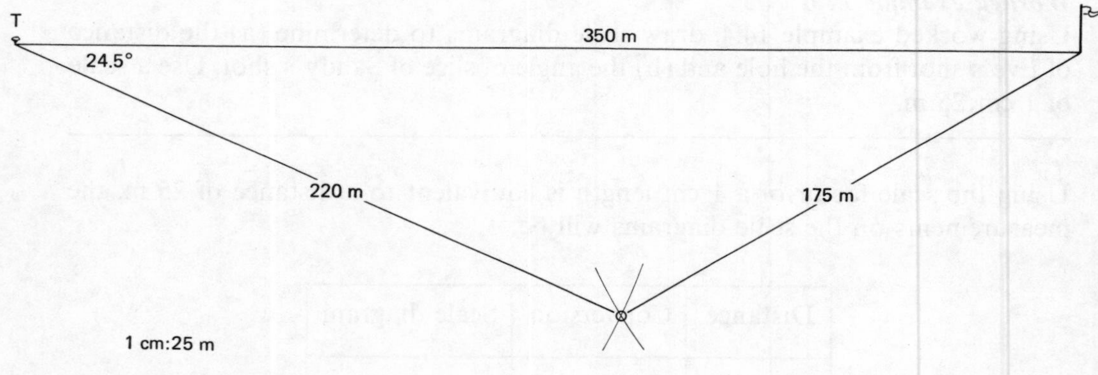

Figure 14.34

The measured angle is 24.5°.

Exercise 14d

Questions (1)–(6) ask for scale drawings for questions referred to in Exercise 14c. Answer the questions from that exercise using your scale drawings.
(1) Question (7). Use a scale of 1 cm:50 m.
(2) Question (8). Use a scale of 1 cm:20 m.
(3) Question (9). Use a scale of 1 cm:4 m.
(4) Question (10). Use a scale of 1 cm:500 m.
(5) Question (11). Use a scale of 1 cm:1 m.
(6) Question (14). Choose your own scale.

(7) The framework of a structure consists of 5 members: AB, BC, CD, DA and AC.
ABCD is a quadrilateral with AC a diagonal.
AB = 900 mm, AC = 850 mm, DC = 860 mm, $\angle$ CAB = 65°, $\angle$ DAC = 55°.

(a) Construct a scale diagram of the framework using a scale of 10 mm to 100 mm.
(b) From your scale drawing determine the actual length of the member BC.

(LEAG)

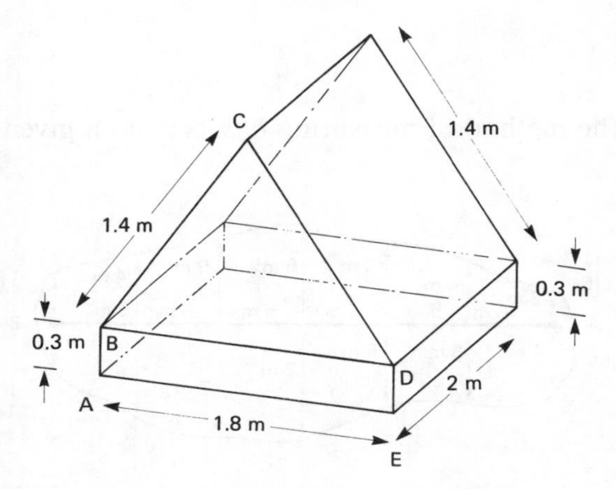

Figure 14.35

(8) Figure 14.35 shows a tent. Using a scale of 5 cm to 1 m, construct accurately a scale drawing of the end ABCDE. From your diagram find:
(a) the height of the tent in metres.
(b) the angle which the sloping edge CB makes with the horizontal.

(NEA)

14.7 Triangulation

The basis of all surveying techniques is the division of the area being surveyed into suitably sized triangles. Consider the field shown in Figure 14.36. How could we set about finding the area of this field?

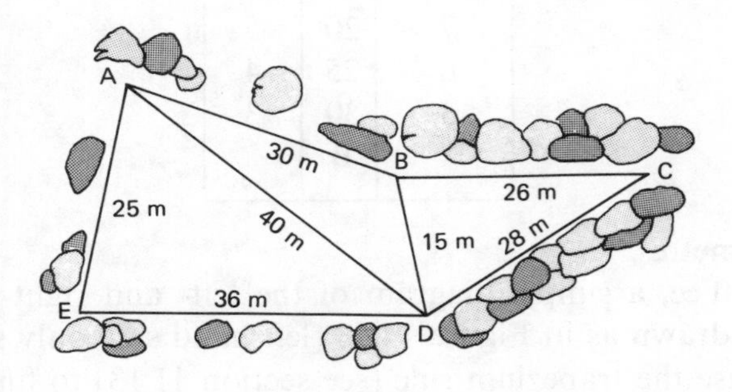

Figure 14.36

Firstly, a number of poles about 2 m long called *ranging rods* are placed in the ground at points A, B, C, D and E. The distances between the poles are then measured using a suitable tape or chain measure. On returning from the field, a scale drawing can be made from which the area of the field can be calculated.

(Obviously, a small area around the outside will not be included.)

Try and make a scale drawing of the field shown in Figure 14.36, and show that the area is approximately 830 m^2.

14.8 Offsets

More frequently, the method of measuring *offsets* from a given base line is used, see Figure 14.37.

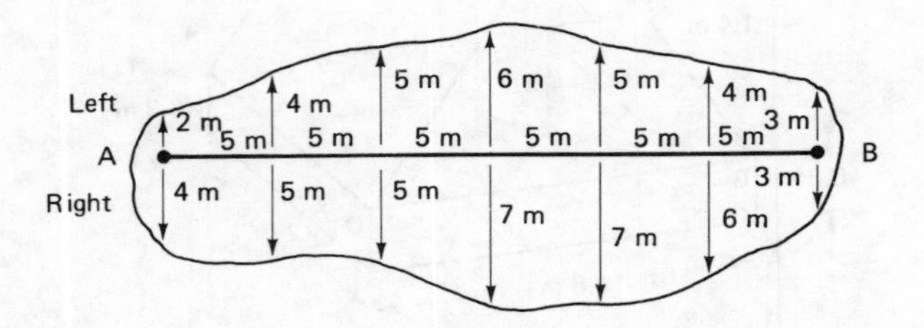

Figure 14.37

The base line AB is laid out using a tape or chain measure, and the distances left and right of this line are then measured, usually at equal intervals—in this case, every 5 m. The results are then written into the surveyor's field book as shown in Table 14.1.

Table 14.1

Right	A	Left
4	0	2
5	5	4
5	10	5
7	15	6
7	20	5
6	25	4
3	30	3
	B	

All distances in metres.

Back in the office, a simple diagram of the left- and right-hand side of the base line can be drawn as in Figure 14.38 (left-hand side only shown).

We can now use the trapezium rule (see section 11.13) to find the area of the

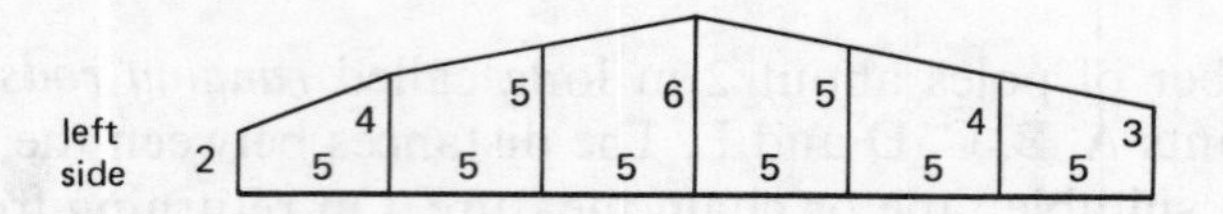

 Figure 14.38

left side. (Note that, if the intervals are not equal, or the same on each side of the base line, the work becomes more complicated.)

$$\text{Area} = \tfrac{5}{2}[2 + 2 \times 4 + 2 \times 5 + 2 \times 6 + 2 \times 5 + 2 \times 4 + 3]\text{m}^2$$

$$= \frac{5}{2}[2 + 8 + 10 + 12 + 10 + 8 + 3]\text{m}^2 = 132.5 \text{ m}^2$$

The right-hand side becomes

$$\text{Area} = \tfrac{5}{2}[4 + 2 \times 5 + 2 \times 5 + 2 \times 7 + 2 \times 7 + 2 \times 6 + 3]\text{m}^2$$

$$= \frac{5}{2}[4 + 10 + 10 + 14 + 14 + 12 + 3]\text{m}^2 = 167.5 \text{ m}^2$$

The area of the field = $167.5 \text{ m}^2 + 132.5 \text{ m}^2 = 300 \text{ m}^2$.

Exercise 14e

Use the following extracts from a surveyor's field book to find the area of each region.

All distances are given in metres.

(1)

Right	A	Left
0	0	0
0	10	5
10	20	18
4	30	12
6	40	3
6	50	2
2	60	0
	B	

(2)

Right	A	Left
2	0	6
2	8	5
4	16	8
8	24	12
3	32	23
1	40	10
0	48	2
	B	

(3)

Right	A	Left
6.4	0	3.9
2.8	5	4.7
5.9	10	5.1
12.3	15	3.4
4.8	20	6.6
2.7	25	6.8
6.3	30	4.2
4.1	35	3.5
2.2	40	4.9
	B	

14.9 Simpson's rule

In the section on offsets, we used the trapezium rule to find an approximation for area. A more accurate answer can be found by using *Simpson's rule.* This states that the area can be found as follows:

$$\text{Area} = \tfrac{1}{3}\ \text{width}(\text{first offset} + \text{last offset} + 2 \times \text{even offsets} + 4 \times \text{odd offsets})$$

or by formula:

$$\text{Area} = \tfrac{1}{3}d[y_0 + y_n + 2(y_2 + y_4 + \cdots) + 4(y_1 + y_3 + \cdots)]$$

The formula works only if you have an even number of strips (greater than 2).

Returning to the area surveyed in Figure 14.37, for the left side of AB:

$$y_0 = 2,\ y_1 = 4,\ y_2 = 5,\ y_3 = 6,\ y_4 = 5,\ y_5 = 4,\ y_6 = 3,\ d = 5$$

Hence the area $= \tfrac{1}{3} \times 5[2 + 3 + 2(5 + 5) + 4(4 + 6 + 4)]$

$= \tfrac{5}{3}[5 + 20 + 56] = 135\ \text{m}^2$

Exercise 14f

Repeat the calculations in Exercise 14e using Simpson's rule, and compare your answers.

14.10 Using contours

Contour lines are used on a map to show points that are at the same height. These lines can be used to construct a cross-section of the region as shown in the following example.

Worked example 14.7

The fifteenth hole at Tariq's local golf course is shown in Figure 14.39. Assuming that the ground is horizontal between X and Q:

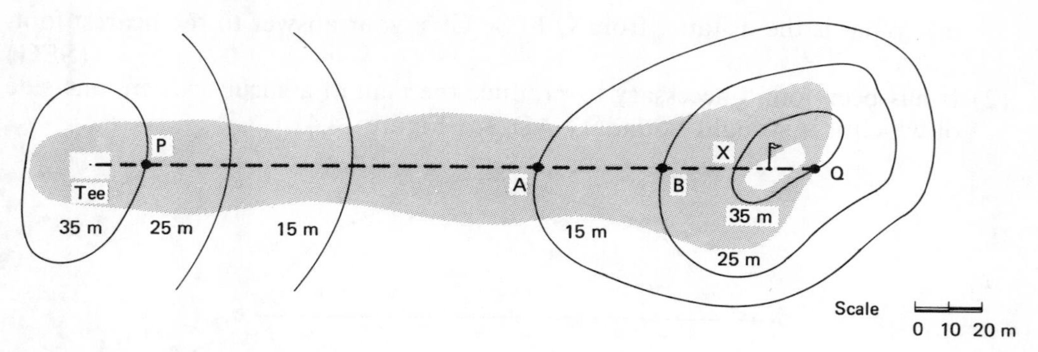

Figure 14.39

(i) Construct a cross-section of this through P and Q.
(ii) Use your cross-section to find the average gradient of the ground between A and B.

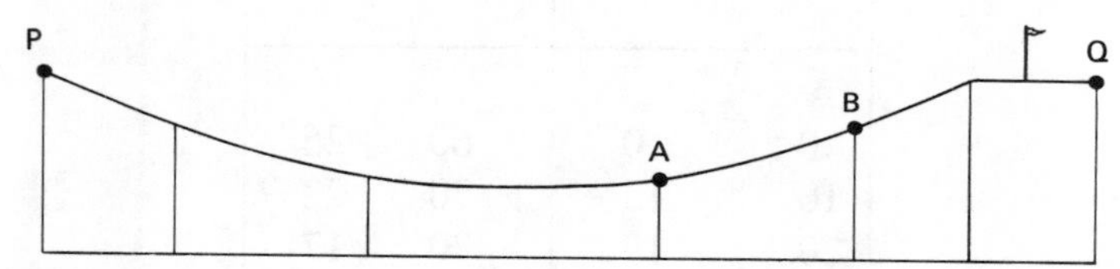

Figure 14.40

The horizontal distances can be measured from the scale drawing, and the vertical distances are read from the contour lines. The cross-section is then drawn as Figure 14.40.

The gradient of AB read from the diagram $= \frac{10}{40} \times 100 = 25\%$.

Exercise 14g

(1) (a) A region PQRS is being measured to contain swings in a park. The area was surveyed by triangulation. The results obtained, together with their conversion to a scale of 1 cm:2 feet, are given in Table 14.2.

Table 14.2

	Ground	Scale 1 cm:2 ft
PQ	19 ft	9.5 cm
QR	14 ft	
RS	27 ft	
SP	20 ft	
PR	22 ft	

Copy and complete the table to show the scaled distances.

(b) Construct an accurate 1 cm:2 ft scale drawing of the quadrilateral PQRS.

(c) What is the distance from Q to S? Give your answer to the nearest foot. (SEG)

(2) It has been found necessary to produce the plan of a shrub garden, one side of which is a straight boundary AD, see Figure 14.41.

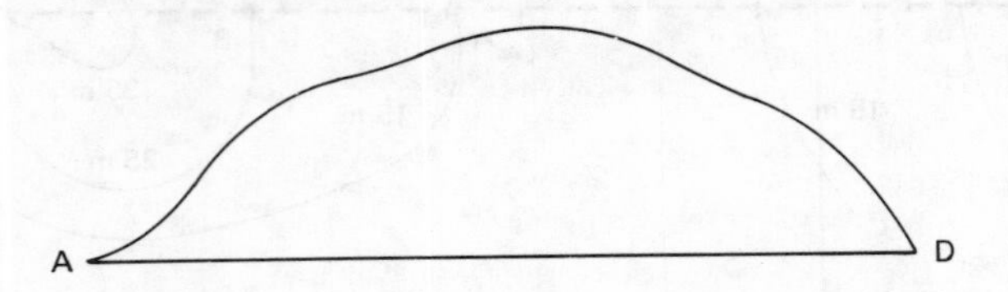

Figure 14.41

Offsets to the boundary of this garden were measured from the line AD, and the results shown in Table 14.3 (in feet) were noted.

Table 14.3

Offset		Offset	
A			
0	0	60	25
10	13	70	22
20	19	80	17
30	24	90	6
40	28	100	0
50	28	D	

(a) Using the line DA as a base, mark these offsets using a scale of 1 cm:10 ft. Draw a smooth curve to show the boundary of the shrub garden.

(b) Use Simpson's rule to estimate the area of the shrub garden. Give your answer in ft^2, correct to the nearest whole number.

(3) Figure 14.42 shows the cross-section of a river, with depth soundings between A and B taken every 2 m. Use either the trapezium rule or Simpson's rule to find the cross-section area of the river at this point. The speed of the river is roughly 4 m/s. Estimate the volume of water that flows every second, giving your answer in litres/s.

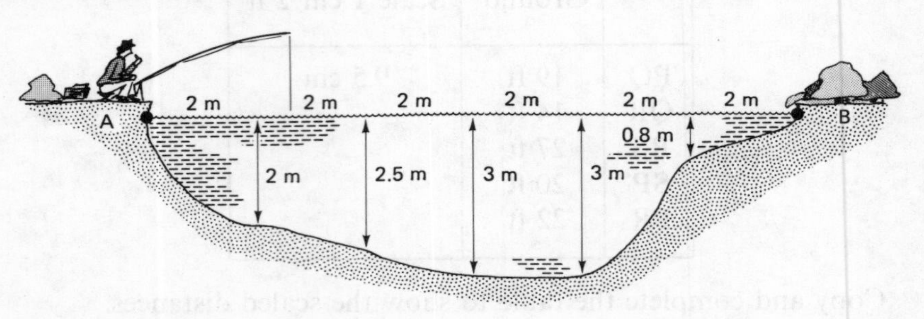

 Figure 14.42

(4) The contour lines near the hill in the South West corner of a playground are shown in Figure 14.43, which is drawn on a scale of 2 cm to 10 ft. Contours are at 1 ft intervals.

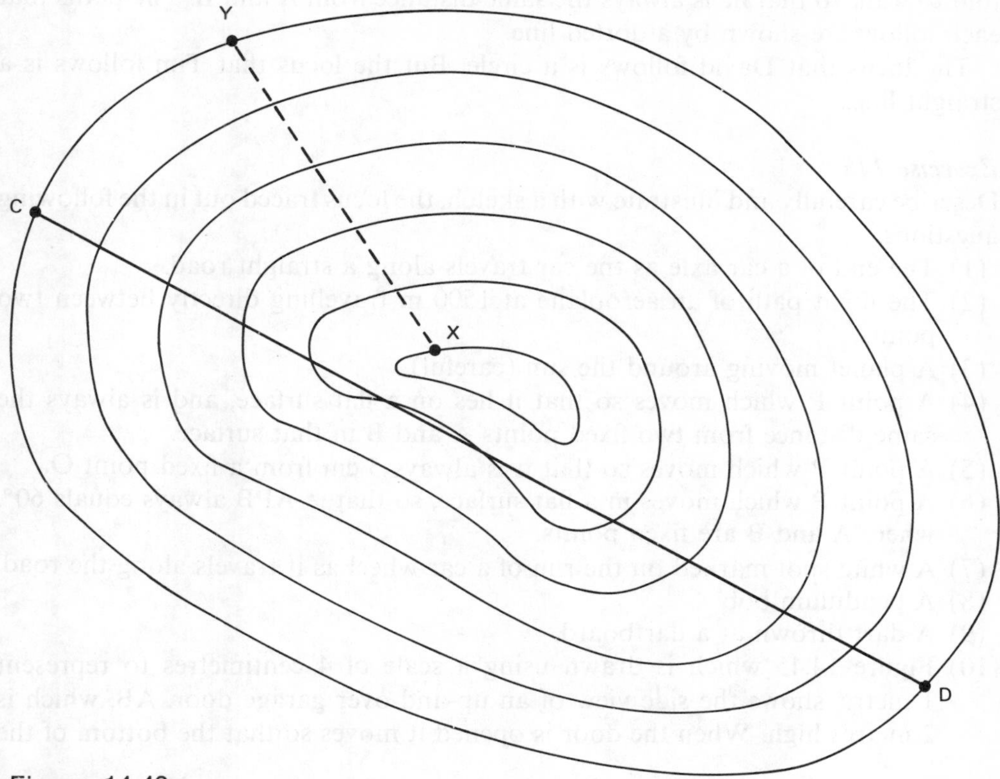

Figure 14.43

The ground slopes uniformly between the points X and Y shown. Calculate the gradient of XY. (SEG)

(5) Draw a cross-section of the playground shown in Figure 14.43 along the line CD. Find the approximate area of this cross-section. (SEG)

14.11 Locus

The word '*locus*' is used to describe a path, region or set of points which is drawn satisfying a fixed set of conditions. The problem could be in 1, 2 or 3 dimensions.

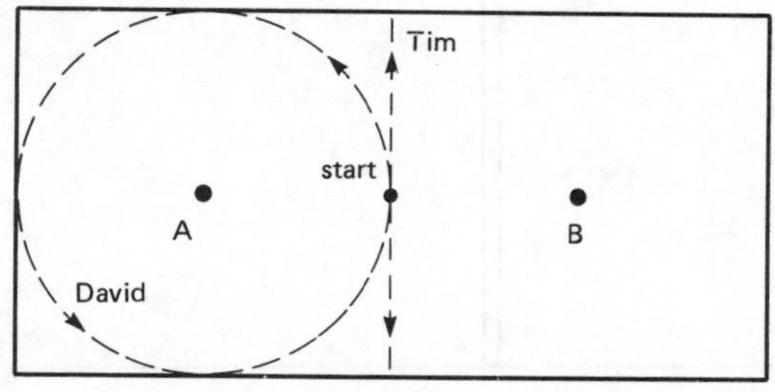

Figure 14.44

Figure 14.44 shows the position of two pegs A and B pushed into the ground 100 m apart. David is told to start half-way between A and B and walk so that he is always 50 m from peg A. Tim also starts half-way between A and B but is told to walk so that he is always the same distance from A and B. The paths that each follow are shown by a dotted line.

The locus that David follows is a circle. But the locus that Tim follows is a straight line.

Exercise 14h

Describe carefully and illustrate with a sketch, the locus traced out in the following questions.

(1) The end of a car axle as the car travels along a straight road.

(2) The flight path of an aeroplane at 1500 m travelling directly between two points.

(3) A planet moving around the sun (careful).

(4) A point P which moves so that it lies on a flat surface, and is always the same distance from two fixed points A and B in that surface.

(5) A point P which moves so that it is always 5 cm from a fixed point O.

(6) A point P which moves in a flat surface, so that $\angle APB$ always equals $60°$, where A and B are fixed points.

(7) A white spot marked on the rim of a car wheel as it travels along the road.

(8) A pendulum bob.

(9) A dart thrown at a dartboard.

(10) Figure 14.45 which is drawn using a scale of 4 centimetres to represent 1 metre, shows the side view of an up-and-over garage door, AB, which is 2 metres high. When the door is opened it moves so that the bottom of the

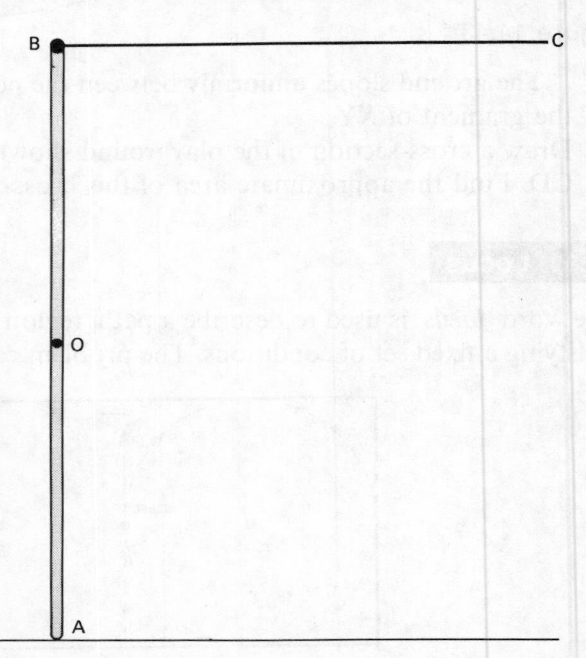

 Figure 14.45

door, A, is always 1 metre from the point O, which is a fixed point on the garage wall midway between the original positions of A and B. In addition, the top of the door, B, is constrained to move along the horizontal line BC.

Copy Figure 14.45 and draw accurately:

(a) the locus of the bottom of the door, A, as the door is fully opened,

(b) the position of the door when the bottom of the door, A, is 1 metre above the ground.

(MEG)

ANSWERS

Exercise 1a

(1) (i) 30 (ii) 26 (iii) 10 (iv) 68 (v) 72 (vi) 64 (vii) $5\frac{4}{7}$ (viii) 240 (ix) 198 (x) 41.8

(2) (i) $(7+4)\times 3$ (ii) $6+(3\times 5)$ (iii) $(3+5)\times(5-3)$ (iv) $6+(8\div 2)$ (v) $(12-2)\div 5$ (vi) $4\times 12\times(3+2)$ (vii) $12-(3\times 2)+4$ (viii) $16+7\times(3+2)$ (ix) $(6\div 8)\times 12$ (x) $((7+8)\div(5\times 5))\times 10$

(3) (i) 27 (ii) 330 (iii) 11 (iv) 24 (v) 144 (vi) 52 (vii) 28 (viii) 8 (ix) 192 (x) 32

Exercise 1b

(1) 10, 60 (2) 2, 240 (3) 5, 200 (4) 12, 72 (5) 10, 300
(6) 4, 84 (7) 20, 240 (8) 2, 800 (9) 20, 600 (10) 14, 56
(11) 9, 540 (12) 12, 120 (13) 3, 36 (14) 6, 72 (15) 2, 950

Exercise 1c

(1) 10, 60 (2) 6, 240 (3) 2, 1600 (4) 22, 132 (5) 5, 1000
(6) 9, 252 (7) 5, 600 (8) 6, 120 (9) 40, 240 (10) 4, 280
(11) 12, 144 (12) 12, 420 (13) 10, 1050 (14) 4, 3400 (15) 5, 1560

Exercise 1d

(1) $\frac{4}{10}, \frac{6}{15}, \frac{8}{20}, \frac{10}{25}, \frac{12}{30}$ (3) $\frac{4}{14}, \frac{6}{21}, \frac{8}{28}, \frac{10}{35}, \frac{12}{42}$ etc.

Exercise 1e

(1) (i) $\frac{2}{3}$ (ii) $\frac{3}{8}$ (iii) $\frac{2}{5}$ (iv) $\frac{1}{4}$ (v) $\frac{1}{5}$ (vi) $\frac{3}{8}$

(2) (i) $\frac{2}{3}$ (ii) $\frac{3}{5}$ (iii) $\frac{4}{5}$ (iv) $\frac{2}{9}$ (v) $\frac{3}{8}$ (vi) $\frac{1}{4}$ (vii) $\frac{1}{7}$ (viii) $\frac{1}{5}$ (ix) $\frac{2}{3}$ (x) $\frac{4}{9}$ (xi) $\frac{6}{7}$ (xii) $\frac{7}{8}$ (xiii) $\frac{5}{6}$ (xiv) $\frac{1}{3}$ (xv) $\frac{2}{7}$ (xvi) $\frac{1}{5}$ (xvii) $\frac{1}{9}$ (xviii) $\frac{1}{8}$ (xix) $\frac{5}{11}$ (xx) $\frac{3}{10}$

(3) (i) $\frac{5}{6}$ (ii) $\frac{11}{15}$ (iii) $1\frac{1}{12}$ (iv) $\frac{1}{15}$ (v) $1\frac{1}{6}$ (vi) $3\frac{3}{8}$ (vii) $1\frac{1}{3}$ (viii) 2 (ix) 12 (x) $\frac{5}{24}$ (xi) $1\frac{1}{10}$ (xii) $\frac{8}{35}$ (xiii) $\frac{8}{9}$ (xiv) $\frac{5}{12}$ (xv) $\frac{5}{6}$ (xvi) $\frac{1}{8}$ (xvii) $\frac{1}{9}$ (xviii) $1\frac{29}{35}$ (xix) $4\frac{9}{10}$ (xx) $10\frac{7}{20}$

Exercise 1f

(1) £1.70 (2) 4.5 litres (3) $\frac{12}{13}$ (4) 28 litres (5) $\frac{3}{4}$ (6) $\frac{7}{10}$
(7) (i) $\frac{9}{25}$ (ii) $\frac{2}{5}$ (iii) $\frac{19}{25}$ (8) (i) $\frac{3}{4}$ (ii) $\frac{13}{16}$ (9) (i) $\frac{1}{3}$ (ii) $\frac{2}{5}$ (iii) $\frac{2}{5}$ (iv) $\frac{3}{20}$ (10) £10.20

Exercise 1g

(1) (i) 60% (ii) 25% (iii) 75% (iv) $12\frac{1}{2}$% (v) 12% (vi) 15% (vii) 62.5% (viii) 30% (ix) 250% (x) 46.5% (xi) $66\frac{2}{3}$% (xii) 55.6% (xiii) $6\frac{2}{3}$% (xiv) 9.1% (xv) $333\frac{1}{3}$%

(2) (i) $\frac{6}{25}$ (ii) $\frac{3}{20}$ (iii) $\frac{1}{3}$ (iv) $\frac{3}{25}$ (v) $\frac{3}{5}$ (vi) $1\frac{1}{2}$ (vii) $\frac{9}{20}$ (viii) $\frac{27}{40}$
(ix) $\frac{7}{25}$ (x) $\frac{33}{100}$ (xi) $1\frac{1}{5}$ (xii) $\frac{2}{25}$ (xiii) $\frac{1}{200}$ (xiv) $\frac{9}{400}$

Exercise 1h

(1) $600 + 80 + 4$ (2) $900 + 6$ (3) $40 + \frac{2}{10}$ (4) $3 + \frac{6}{10} + \frac{8}{100}$
(5) $10\,000 + 100 + 9$ (6) $20 + 8 + \frac{9}{100} + \frac{4}{1000}$ (7) $60 + 3 + \frac{7}{10} + \frac{5}{1000}$
(8) $1 + \frac{1}{1000}$ (9) $6\,000\,000 + 400 + 2$ (10) $6 + \frac{1}{100} + \frac{1}{10000}$
(11) eight hundred (12) eighty
(13) eight (14) eight hundredths
(15) eight tenths
(16) eight thousand
(17) eight hundred
(18) eight (19) eight thousandths
(20) eight hundredths

Exercise 1i

(1) 6.87 (2) 3.10 (3) 28.18 (4) 63.10 (5) 20.09 (6) 31.98 (7) 5.09
(8) 0.01 (9) 0.00 (10) 6.90 (11) 81.6 (12) 2500 (13) 9000 (14) 10.1
(15) 0.0483 (16) 63.8 (17) 185 000 (18) 88.9 (19) 90 900 (20) 12.0
(21) 6890 (22) 64.3 (23) 0.00489 (24) 0.124 (25) 668

Exercise 1j

(1) 0.4 (2) 0.625 (3) 0.25 (4) 0.375 (5) 0.7 (6) 0.3125 (7) 0.12
(8) 0.45 (9) 0.4375 (10) 0.55 (11) 0.027 (12) 0.128

Exercise 1k

(1) rational (2) rational (3) irrational (4) rational (5) rational
(6) irrational (7) rational (8) rational (9) rational (10) rational

Exercise 1l

(1) $-8, -4, -2.5, 3, 6, 10$ (2) 14°C (3) -10°C (3) £12.20 in credit
(5) 19°C (6) £31.83 (7) 36 feet
(8) (i) 1 (ii) -11 (iii) -3 (iv) 13 (v) 8 (vi) 1 (vii) -5
(viii) $-1\frac{1}{2}$ (ix) $17\frac{1}{2}$ (x) $2\frac{1}{4}$ (xi) $2\frac{1}{4}$ (xii) 6 (xiii) -8 (xiv) -33
(xv) 16 (xvi) -13 (xvii) $-13\frac{1}{2}$ (xviii) -8.6 (xix) -4.8 (xx) -4.4
(xxi) -12 (xxii) -1 (xxiii) -3 (xxiv) $8\frac{3}{4}$

Exercise 1m

(1) 64 (2) 512 (3) 36 (4) 256 (5) 8192 (6) 25 (7) 128 (8) 35
(9) 16 (10) 5000 (11) 54 (12) 1 (13) 671 (14) 36 (15) 4

Exercise 1n

(1) 6.3×10^2 (2) 2.8×10^3 (3) 4.875×10^3 (4) 6.8×10^4 (5) 1.26×10^5
(6) 9.6×10^6 (7) 1.259×10^2 (8) 1.2×10^{-1} (9) 3×10^{-3}
(10) 4.65×10^{-4} (11) 1.11×10^{-1} (12) 8.5×10^7 (13) 6.4×10^{-6}
(14) 8.4×10^2 (15) 7.84×10^{-8} (16) 9.98×10^8 (17) 6200 (18) 380

(19) 20 400 (20) 0.016 (21) 0.0036 (22) 0.00064 (23) 9090 (24) 0.106
(25) 633 (26) 0.0206 (27) 30 100 (28) 0.00111 (29) 0.0000141
(30) 6 860 000 (31) 0.00019 (32) 8.4×10^4 (33) 8.71×10^5
(34) 2.389×10^5 (35) 7.2×10^5 (36) 3.36×10^{11} (37) 3.2×10^{13}
(38) 5.2×10^{-2} (39) 3.02×10^{-3} (40) 6.538×10^{-2} (41) 7.6×10^{-2}
(42) 3.43×10^3 (43) 1.58×10^3 (44) 2.83×10^4 (45) 8×10^2 (46) 1.87
(47) 9×10^4 (48) 5.76×10^6 (49) 1.44×10^{-4} (50) 2.24×10^4
(51) 1.23×10^{-7} (52) 9.43×10^{12} km (53) 3.84×10^4 km

Miscellaneous exercise 1

1. (i) 63, 84 (ii) 71 (iii) 121, 169
2. £8.40 3. (i) $\frac{9}{32}$ (ii) $\frac{2}{3}$ (iii) $15\frac{5}{8}$%
4. 270 5. 138.9% 6. (i) shade 21 squares (ii) shade 10 sectors
(iii) shade 7 triangles
7. (i) 369, 123 (ii) 0.9% 8. 1.05×10^{-1} mm
9. (i) 2^{24} (ii) 2^{18} 10. (i) 250 000 (ii) 53 010 (iii) 205 000
(iv) 25 020 000, 2.5×10^5, 5.3×10^4, 2.05×10^5, 2.5×10^7
11. (a) 22 (b) 60% 12. -7°C 13. (a) 180 (b) 80 (c) 65% (d) £960
(e) £1500 (f) 64%

Exercise 2a

(1) (i) 16 (ii) 10 (iii) 21 (iv) $1\frac{3}{4}$ (2) (i) 18 (ii) $10\frac{1}{2}$ (iii) 0 (iv) 2
(3) (i) 48p (ii) 72p (iii) 57p (iv) 17 (4) (i) 6 (ii) 8 (iii) 2 (iv) $2\frac{1}{2}$
(v) $1\frac{3}{4}$ (vi) 0 (5) (i) 0 (ii) -27 (iii) 96 (iv) -27 (v) 26.4
(6) (i) -24 (ii) 192 (iii) 24

Exercise 2b

(1) $10x$ (3) $4t$ (4) $5t^2$ (5) $5pq$ (6) $15m$ (7) $4n$ (8) $3t^2 + 5t$
(9) $12xy$ (10) $2q^2$

Exercise 2c

(1) $4x + 4y$ (2) $2p - 4q$ (3) $3t + 6x$ (4) $4qx + 4qy$
(5) $15t - 6q$ (6) $5a + 7b$ (7) $6x - 30y$ (8) $3z + 6t$ (9) $6x + 6y$
(10) $10a + 5b$ (11) $6 - 6b$ (12) $9x + 14$ (13) $10t - 5$ (14) $6tx + 6ty$
(15) $5q + 7p$ (16) $18x + 6y$ (17) $xa + xb$ (18) $10x + 28$ (19) $6x + 21$
(20) $6x + 18$

Exercise 2d

(1) 8 (2) 35 (3) 2 (4) 11 (5) -9 (6) 15 (7) 5 (8) 3 (9) 4
(10) 7 (11) -1 (12) 3 (13) -10 (14) 7 (15) 1 (16) 2 (17) -1
(18) 0 (19) $\frac{1}{2}$ (20) 24

Exercise 2e

(1) £$4y$ (2) £$52E$ (3) $N + 1$, $N + 2$ (4) $7s$ (5) £$2t + 3s$
(6) (i) £$52e + E$ (ii) $\frac{£52e + E}{52}$ (7) £$\frac{1}{3}P$ (8) £$80c + 23s$, £15, 120 (9) £225

(10) £$x + 2.5y$, £227.50 (11) $180 - \dfrac{360}{n}$ (12) $6x + 4$ cm (13) $20x + 10y$

(14) $3600x + 60y + t$ (15) $4x + 6y + 15$

Exercise 2f

(1) (i) $x + 4$ (ii) $2x + 4 = 38$ (iii) 17, 21 (2) (i) $2x$ (ii) $3x = 54$
(iii) 18, 36 (3) (i) $x + 1$ (ii) 23, 24 (4) 3 (5) 11, 22, 27

Exercise 2g

(1) 5 (2) 8 (3) (i) $\frac{1}{3}$ (ii) 4 (4) 2 (5) 11 (6) $4\frac{1}{2}$ (7) 15
(8) (i) $3\frac{1}{2}$ (ii) 12 (9) (i) 5 (ii) $1\frac{1}{4}$ (10) (i) 25 (ii) $\frac{1}{3}$

Exercise 2h

(1) (i) 30 (ii) 11 (iii) 13 (iv) 2 (v) no (2) 14

Exercise 2i

(1) 61, $3n + 1$ (2) 96, $5n - 4$ (3) 81, $4n + 1$ (4) 135, $7n - 5$
(5) 2^{-19}, $2^{-(n-1)}$ (6) 210, $\dfrac{n(n+1)}{2}$ (7) 6765 (8) 12 (9) 400, n^2

(10) 71 (prime) (11) (i) 16 should be 9 (ii) 30, 16, 36; 45, 25, 45
(iii) square numbers (12) (i) 72, 88, 104 (ii) 10 (iii) $S = 16n + 8$

Miscellaneous exercise 2

1. (a) 68 (b) -10 (c) $F = 2(15 + c)$ (d) 70
2. (a) (i) $x + 13$ (ii) $x + 13 = 2(x - 7)$ (iii) 27
3. (ii) 35 (iii) 590 (iv) $\dfrac{n(3n-1)}{2}$ (v) 4, 7, 10, 13
4. 2 5. (i) 3.36×10^5 (ii) 2.19×10^2
6. (i) 310 cm^2 (ii) 3.95×10^6 km^2 (iii) 620 cm^2 (iv) 1.61 cm
7. (a) (i) 37.68 mm (ii) 226.08 mm (b) 6 s
8. (a) £$450 + 10x$ (b) £$450 + Nx$ (c) £870
9. $x = 2$ 10. (b) (i) $P = 6H$ (ii) $A = H(H + 1)$ (c) 13 (d) 19
11. (b) 0, 3, 9, 18 (c) 0, 1, 3, 6 (d) 10, 15 (e) 63 (f) row 6 column 9
12. (a) (i) $7x$ (ii) $x + 7p$ (iii) $6x + 42p$ (b) $x = 15$, $22p$
13. (a) (i) £$3E$ (ii) £$3E - 5$ (b) $E = 8.5$, £8.50, £25.50, £20.50
14. 3
15. (a) (i) 4 (ii) 12 (c) (i) squares (ii) multiples of 4 (iii) squares
(d) (i) 121 (ii) 48
16. (a) (i) 217.5 (ii) 1.9 (b) (i) 80 (ii) 2010 (c) R becomes negative

Exercise 3a

(1) (i) 840 (ii) 600 000 (iii) 0.84 (iv) 28 000 (v) 64
(2) (i) 8.684 (ii) 10 (iii) 0.085 (iv) 600
(3) (i) 4600 (ii) 0.84 (iii) 5×10^5 (iv) 620 000
(4) (i) 600 (ii) 5800 (iii) 8×10^5 (iv) 6000
(5) (i) 2×10^4 (ii) 0.84 (iii) 4×10^9 (iv) 8×10^7

(6) (i) 6 (ii) 360 (iii) 380 (iv) 2.8
(7) (i) 16.7 (ii) 9.7 (iii) 0.12 (iv) 0.0056
(8) (i) 0.6 (ii) 2000 (iii) 84 (iv) 4.3

Exercise 3b
Exact values (1) 8.53 (2) 7.28 (3) 4.97 (4) 45.5 (5) 96.6 (6) 2025
(7) 21.1 (8) 7.14 (9) 31.1 (10) 0.44 (11) 54.1 (12) 0.23

Exercise 3c
(1) 84 557 (2) 6 cm (3) −6°C (4) 33 m.p.h. (5) 102 km/h (6) 79 911
(7) 26°C (8) 92°C

Exercise 3d
(1) 9 h 38 min (2) 7.05 p.m. (3) 10 (4) 1.20 p.m. (5) 575 (6) 40
(7) 8784 (8) −9°C

Exercise 3e
(1) (i) £121 (ii) £38 (2) (i) 2 h 55 min (ii) 20.49 (iii) 2h 38 min
(iv) 10.06 (v) 6

Exercise 3g
(1) (i) £134.75 (ii) 43 (2) (i) 12 (ii) £39.16 (3) 8 hours (4) $13\frac{3}{4}$ min
(5) £4.44 (6) £3.60 (7) $1\frac{1}{4}$ min (8) 28 h $7\frac{1}{2}$ min, £112.50

Exercise 3h
(1) 450 g (2) 2000 g (will you use them all?) (3) 650 g (4) 1 litre
(5) 25 fl oz if it keeps in tin (6) 3 kg if you can eat them all quickly enough

Exercise 3i
(1) £4.20 (2) £6.14 (3) 97 or 98p

Exercise 3j
(1) 7:1 (2) 6:1 (3) 2:3 (4) 2:5 (5) 5:2 (6) 2:3 (7) 2:3
(8) 10:7:6 (9) 3:4:5 (10) 1:3

Exercise 3k
(1) £15, £18, £12 (2) 3, 8.4, 2.6 (3) £2.40, £2.95, £3.40
(4) £212.50, £522.50, £1015 (5) £6500, £6125, £4875, probably not

Exercise 3l
(1) 56p (2) £10.08 (3) 1.68 m (4) 5 cm^2 (5) £90 000 (6) £5.47
(7) £8.64 (8) 5.06 cm (9) 12.5% (10) 12.5% (11) 22.5% (12) 5%
(13) 9.68% (14) 8.80% (15) 4% (16) 0.103 (17) £1250 (18) 100 300
(19) 0.04%

Miscellaneous exercise 3

1. 31 2. £3, £2.60, £1.60 3. 3.4, 4.6, 5 4. about 2 weeks 5. 1284 cm
6. 3562 7. (i) 57 (ii) 1.8 m 8. (i) 35p (ii) about 4 weeks
9. points 27, 14, 8, 7 10. 99p 11. (a) 48 (b) 37.5 12. (i) $26\frac{2}{3}$ min
(ii) $42\frac{2}{3}$ min 13. (a) £102.50 (b) £147.50 (c) 2160 (d) £80
14. 630 s 15. 1.5×10^5 16. 0.022% 17. (i) 7.45 (ii) 0.18
(iii) Greater London (iv) 3.3 18. (i) 12.57 (ii) 6 (direct) (iii) 1932
19. 431.8, 7849.065, 16.83, 157.48 20. (a) 9.00 a.m. (b) $2\frac{1}{2}$ h (c) 3.45 p.m.

Exercise 4a

(1) £5.12 (2) £10 320 (3) £1.05 (4) £43.20 (5) 5p (6) £32.10
(7) 2 or 3p (8) £133.76 (9) £7150 (10) £1.90 (11) £139.52
(12) 6.25% (13) 64p (14) £125 (15) £96 320 (16) £5.25

Exercise 4b

(1) £153 (2) £34.13 (3) 2 years 8 months (4) 4.2% (5) £3530, £5040
(6) £2720.98 (7) £3561.92 (8) £1106.92 (9) £84 477 (10) £2.08
(11) £140 600

Exercise 4c

(1) £189.42 (2) £8398 (3) £9250 (4) £118.80 (5) 2400 (6) £144

Exercise 4d

(1) £3625, £906.25 (2) £5225, £1306.25 (3) £7290, £1822.50
(4) £9300, £2325 (5) £32 250, £10 005 (6) £108 750, £40 605

Exercise 4e

(1) 46 000 (2) 4077.9 (3) £26.98 (4) £273.60 (5) 101 116
(6) (i) 114 023 (ii) 1411.5 (iii) 1268.3 (iv) 11 323

Exercise 4f

(1) £65.18, £546.48 (2) £139.94, £558.56 (3) £41.23, £164.51
(4) £25.75, £360.88 (5) £59.65, £65.51 (6) £72.97, £291.23
(7) £17.49, £69.74 (8) £157.82, £3269.20 (9) £8.50 (10) £13

Exercise 4g

(1) £135, £15 (2) £24.60, £6.10 (3) £548.80, £50.80
(4) £944, £94 (5) £383, £84 (6) £280, £0 (7) £540, £60
(8) £169.60, £34.60

Exercise 4h

(1) 28.3% (2) 23.1% (3) 21.6% (4) 36.5% (5) 8.2% (6) 112.9%

Exercise 4i

(1) (a) £517.50 (b) £124 200 (2) £787.20 (3) £84 000, £837.90

Exercise 4j

(1) £15.12 (2) £144.22

Exercise 4k

(1) (i) £97.20 (ii) £182.25 (iii) £4.03 (iv) £154.95
(2) (i) £756 (ii) £172 (iii) £100

Exercise 4l

(1) £81.60

Miscellaneous exercise 4

1. (i) £1.72 (ii) 30% 2. £11.91 3. (a) £31.20 (b) £46.80 4. £520.22
5. (i) £12 (ii) £10 (iii) £7.50 (iv) £17.58 6. (a) 52.1% (b) £675
7. (a) (i) £1.88 (ii) 87.02DR (b) (i) £428 (ii) £1.07 (d) 23%
8. (a) (i) 5060 (ii) 7p (b) (i) 27.7 (ii) 10p (c) £64.43 9. (i) £372
(ii) (a) £428.67 (b) £402.58 10. (a) £208 (b) £72.80 (c) £352
(d) £680 (e) £578 11. (b) 24.5% 12. (i) £3 (ii) £4.20 (iii) £1.20
(iv) £24 13. (a) £1665 (b) £647.50 (c) £2660.04
(d) £239.96 more than from the bank
14. (a) £931.76 (b) £1335.26 (c) £1450 (d) £337.02 (e) £2267.02
15. (a) (i) £100 (ii) £540 (iii) £640 (iv) £537 (b) £630
16. (i) 2660 (ii) 9.5; £596.80, 3.8, £141.80
17. (i) £3370 (ii) £3360 (iii) £3120

Exercise 5a

(1) 135° (2) 570° (3) 2160° (4) 18 m, 144 (5) $x = 20$
(6) $a = 20$, $b = 120$, $c = 40$ (7) 137° (8) 36° (9) $x = 55$
(10) $x = 60$, largest angle = 180° (11) 135° (12) 315°, 270°, 585° (13) no, yes
(14) no (15) $a = 61$, $b = 61$, $c = 105$, $d = 68$, $e = 68$, $f = 68$ (16) $x = 100$
(17) 314°, 242°, 150°

Exercise 5b

(1) (iii) 10, 10, 10 (2) $x = 52$ (3) $x = 74$ (5) △PTS is isosceles
(6) (i) 220° (ii) 310° (7) (i) 110° (ii) 290° (iii) 170°
(8) (i) 54° (ii) 63° (9) $a = 45$, $b = 87$, $c = 45$, $d = 48$, $e = 132$
(10) $a = 50$, $b = 40$, $c = 110$

Exercise 5c

(1) 108°, 135°, 144° (2) 6 (4) 8 (5) 2, 5, 9, 14, 54 (6) 15 (7) 30°, 75°
(8) $x = 132$ (9) $x = 60$ (10) $x = 120$

Exercise 5d

(1) $a = 50$, $b = 80$, $c = 26$ (2) trapezium (3) H, O (4) 60°, 90°, 90°
(5) equilateral triangle, square, hexagon (6) jigsaws (9) kite

Exercise 5e

(2) any two of AFE, DEF, FBD and EDC; similar triangles; any of the above and triangle ABC (3) BFC and DCG (4) ADE and BEF (5) 24.25 m
(6) 6 (7) BX = 6 cm (8) BF = 6 cm, GD = 8 cm, DH = 4 cm
(9) (i) 2.4 m (ii) 5/4 (iii) 3 m (iv) no (13.8 m total length)
(10) (a) (i) CDA (ii) RQA or PDQ (b) (i) parallelogram (ii) trapezium

Exercise 5f
(1) (i) 25° (ii) 65° (2) $x = 62$ (3) (i) 50° (ii) 80° (iii) 75°
(4) (i) 62° (ii) 34°, isosceles triangle (5) $a = 90$, $b = 30$, $c = 90$, $d = 60$
(6) (i), (ii) (7) (i) 75° (ii) 37.5° (8) (i) 122° (ii) 29° (iii) 16°
(9) CDE and CBE, ODE and OBE, CDO and CBO (10) (i) DEA and DFC, BED and BFD, BAD and BCD (ii) 70°, 90°, 55°

Exercise 5g
(1) 25/9 (2) 4 cm (3) (i) 1 m (ii) 1.25 m (iii) 1.25 m^2
(4) 25 ml, 32 m^2 (5) 80 g (6) 3 cm (7) (ii)
(8) 5 cm, 5 cm, 4 cm, 6 cm, AWB and DYC (10) (i), (iii) (11) (i), (iii)
(12) (iii)

Miscellaneous exercise 5
1. 210°, 300° 2. $x = 72$ 3. 72°, 72°, 36° 4. (iv) 5. 90°, 36°, 7 sides
6. 60°, 80°, 100°, 120°, trapezium (or could be a cyclic quadrilateral)
7. 70°, 80°, trapezium 8. $C = (9, 7)$, $D = (2, 8)$
10. $a = 67.5°$, $b = 112.5°$, $c = 90°$, $d = 135°$
11. $a = 5$ cm, $b = 90°$, $c = 90°$, $d = 5$ cm 12. (i) 27° (ii) 36° 13. 100 cm^2
14. (a) 12 in^2 (b) 9 in by 12 in (c) 42 in (d) 108 in^2 (e) 9
15. 34°, 68°, 22° 16. 150°

Exercise 6a
(1) $A = 30$ cm^2, $B = 41$ cm^2, $C = 42$ cm^2 (2) (i) 0.25 cm^2 (ii) 30–31
(iii) 7.5 cm^2 approx.

Exercise 6c
(1) 9.94 cm^2 (2) 8.37 cm^2 (3) 5.58 cm^2 (4) 14.62 cm^2 (5) 7.36 cm^2
(6) 8.5 cm^2 (7) 12.79 cm^2 (8) 28.43 cm^2

Exercise 6d
(1) 25.1 cm (2) 3.18 cm (3) 5.64 cm (4) 31.4 m
(5) (i) 320 cm (ii) 1.44 km/h (6) 21.5% (7) 0.285 mm
(8) (i) 204.2 cm (ii) 245

Exercise 6e
(1) 402 cm^3 (2) 268 cm^3 (3) 50 m^3 (4) 1131 cm^2 (5) 1.19 cm
(6) 942.5 cm^2 (7) 238 (8) 12 cm^3 (9) 194.56 cm^2 (10) 1200 cm^3

Exercise 6f
(1) 110.7 (2) 19.32 (3) 4 (4) 78.9 (5) 2.5 (6) 80 (7) 22.08
(8) 0.00129 (9) 1.26×10^5

Miscellaneous exercise 6
1. (a) 50.3 m^2 (b) 249.7 m^2 2. 31 m 3. $\frac{3}{8}$
4. (a) $\frac{1}{4}$ (b) $\frac{1}{8}$ (c) $\frac{1}{16}$ (d) $\frac{3}{16}$ (e) $\frac{3}{8}$ 5. 301.4 cm^2 6. 180
7. (a) 300 mm^2 (b) 266 mm^2 (c) 11.3%
8. (a) 108 cm^2 (b) 48 cm (c) 48 cm^3 9. (a) 39.3 cm^3 (b) 51.1 cm^2

10. (a) 198 cm^3 (b) 1357 mm^3 (c) 145 11. $12\frac{5}{6}$ ft^3
12. (a) (i) 42 cm (ii) 28 cm (iii) 1176 cm^2 (b) (i) 153.9 cm^2
(ii) 923.4 cm^2 (c) 78.5% 13. (a) 1.508×10^5 (b) 1.508×10^6
14. (i) (a) 54π m^3 (b) 170 m^3 (ii) (a) 0.0072π m^3 (b) 0.023 m^3
(iii) 2 h 3 min 15. (i) 10 cm (ii) 4200 cm^3 (iii) 2368 cm^2
16. (a) 0.012 (b) 0.00407 (c) 32 (d) 5.76
17. (a) (i) 6.16 cm^2 (ii) 1.44 cm^3 (b) 852 cm^2
18. (a) (i) 8.2 cm (ii) 17.8 cm (b) 9.8 cm (c) 45.5 cm (d) 6
(e) (i) 1:1.9 (ii) 2.6 cm 19. 81 m^2

Exercise 7a

(1) (i), (iii), (vi) (2) 6 m (3) 13 cm (4) 1.1 m (5) 6.32 km, A and C
(6) (i) 14 cm^2 (ii) AB = 6.32 cm, BC = 4.47 cm, CA = 8.25 cm
(7) 13 m, 56 m (8) 6.5 km (9) 20 cm (10) 18 cm (11) 2.49 cm
(12) 70 m, 510 m

Exercise 7b

(1) $x = 30.3$, $y = 71.6$, $z = 77.3$ (2) (i) 4.3, 6.6 (ii) 8.7, 14.8 (iii) 1.7, 4.6
(3) 9.01 cm, 8.75 cm (4) (i) 2.89 cm (ii) 10 cm (5) 108.9° (6) 12 m
(7) 51.4 m (8) $x = 11.3$ (9) radius = 4.41 cm

Exercise 7c

(1) $x = 38.7$, $y = 54.3$, $z = 91.2$ (2) AB = 6.4, AC = 7.7, DE = 13.2, EF = 21.2, GH = 4.9, HI = 11.0 (3) 64.6°, 70.5° (4) 11.8 m
(5) sin ∠ ADB = 0.5, ∠ BAC, 10.4 cm, BC = 3 cm (6) 2.2 m
(7) BD = 7.52 m, AB = 2.74 m (8) 3.88 km N, 14.49 km E; 6 km N, 10.39 km E; 9.88 km N, 24.88 km E; 068.3°
(9) (a) 1.414 cm (b) 1.732 cm (c) 35.3° (10) (a) 36.9 m (b) 45°
(c) 26.1 m (d) 681 m^2

Miscellaneous exercise 7

1. 8 m, 13 m 2. yes 3. 28.3 m 4. (i) 9 m (ii) 34.8° (iii) 14.6 m
5. (a) 150 m (b) 446 m (c) 19.7°
6. (a) 36 m (b) 27 m (c) 14 m (d) 480 m^2 7. 5 cm
8. (a) 15 cm (b) 90° (c) 3.2 cm 9. (a) 5.2 km (b) 3 km (c) 3.10 p.m.
10. (a) (i) $1\frac{1}{4}$ hrs (ii) 23.10 hrs (b) (i) 301° (ii) 233 km (iii) 103 km
11. (a) 0.8 m (b) 13.2° (c) 2.81 m
12. (i) 3.12 cm (ii) 15.0 cm^2 (iii) 105 cm^3 13. 11 cm 14. 2.06 m^2

Exercise 8a

(1) (2, 1), (3, 1) (2) $A = (10, 8)$, $B = (20, 3)$, $C = (20, 8)$, $D = (22, 8)$
(3) $C = (1, 11)$, $D = (4, 11)$ or $C = (1, 3)$, $D = (4, 3)$ (4) rectangle; 2.2, 4.5; 9.9 u^2
(5) C, (−4, 5), (−1.5, 2.5), (−3.5, 1.5); 4.47 (6) $C = (10, 5)$, AB = 5.8, area = 34 u^2 (7) 180, 240 (9) (a) (i) 1051 (ii) 1254 (iii) 1250
(b) (i) 137 515 (ii) 129 534 (iii) 129 544
(c) (i) Doglane Farm (ii) caravan (site of) (iii) figure 308 (height measurement) (iv) road junction (v) Izaak Walton Hotel

Exercise 8b

(1) (a) (i) 4.5 (ii) 27 (b) 13.5 gallons (2) (a) 50 kg (b) 9 yrs
(c) 11 yrs (d) 46 kg (e) 6 kg (3) (a) £4.50 (b) 100 (d) 175
(4) (i) cyclist = 24 km/h, car = 80 km/h (ii) 18.7 km (iii) 35 min
(5) (i) 20 m/s (ii) 1 m/s^2 (iii) 400 m (6) between 6.30 a.m. and 11.30 a.m.
(7) (b) (i) 615 g (ii) 285 cm^3 (c) 415 g (8) (b) (i) −1.25 m/s^2 (ii) 68 m
(9) (a) 38 m (b) accelerates to maximum speed after 3 seconds
(10) 11.25 a.m.

Exercise 8c

(1) $A = (7, 23)$; (4, 14) (2) $a = -1$, $b = 5$ (3) (a) $A = (12, 0)$ (b) (0, 5)
(c) 13 units (4) (a) $x = 5$, $y = 2$ (b) $x = -1$, $y = 3$ (c) $x = -2$, $y = -5$
(5) (a) 6 (b) $x = 3$, $y = 15$ (c) the point A (d) $y = 3x + 2$
(6) (a) 13 (b) 5.3 (c) 24 (7) (i) 2 (ii) −4
(8) (a) 4, −2, 0, 4 (c) −2.25 (d) 1 (9) (a) 7, 1, 1, 7, 17 (c) 1, 3
(e) (2.85, 5.85); (−0.35, 2.65) (10) (a) 5, 2.67, 2, 2, 2.33, 2.86, 3.5
(c) 1.94 (d) $x = 1.38$, $x = 3.62$

Miscellaneous exercise 8

1. −1 2. (a) (8, 0) (b) 17 (c) 61.9° (d) 14.9° (e) 3.75
3. (i) 16 km/h (ii) 30 min (iii) 75 min (iv) 19.2 km/h
4. (i) 50 s (ii) 40 m/s (iii) 2 m/s^2 (iv) 1400 m
5. (b) 60 → 60.8°F (c) 10.15 a.m. → 7.15 p.m. (d) 12.50 a.m. → 6.40 a.m.
6. (a) 90, 180, 210 (c) 220 km (d) $C = 60 + 0.6x$ 7. (i) (2, 6)
(ii) isosceles (iii) (2, −6) (iv) $y = -3x$ 8. 2, $y = 2x + 1$
9. (a) 9.09, 7.07, 5.31 (c) 6.37 10. (a) 4, 0, 2.25 (b) 2 (c) 2.73, −0.73
11. (a) 0.875, 5.625, 9 (b) (i) −8.75 (ii) $x = 1$, $x = 3.562$
12. (b) 3.5, acceleration of the car after 4 seconds
13. (c) decreases (d) infinity (e) (i) 6.8 (ii) 1.7 (iii) −36

Exercise 9a

(1) (ii) 0 (iii) 1/90 (iv) 1/5 (v) 1/13 (vi) 3/4 (vii) 5/36 (viii) 5/26
(ix) 4/5 (x) 8/11 (xi) 1 (xii) 0 (xiii) 3/20 (xiv) 1/26 (xv) 0
(xvi) 1/1000000 (xvii) 3/4 (xviii) 1/12 (xvix) 2/3 (xx) 4/25
(2) (i) 6/36 = 1/6 (ii) 9/36 = 1/4 (iii) 6/36 = 1/6 (iv) 11/36
(v) 24/36 = 2/3 (3) (i) 15/100 = 3/20 (ii) 51/100 (iii) 93/100

Exercise 9b

(1) (a) 1/6 (b) 1/3 (c) 2/5 (2) there are 2 odd numbers which are also less than 4 (3) 1/2, 1/6 (4) (a) 6 pairs (b) 1/2 (c) 5/6
(5) 1/3, 1/9, 5/9 (6) (i) 1/5 (ii) 1/20 (iii) 7/20 (7) (a) 4/9 (b) 5/9
(8) (a) 1/2 (b) 1/4 (c) 1/4 (d) 1/16 (e) 5/8 (9) (a) (i) 1/4 (ii) 9/16
(b) 49/256 (7/16 × 7/16) (10) (b) 2/3 (c) (i) 1/6 (ii) 1/3

Exercise 9c

(1) (i) 1/3 (ii) 1/10 (iii) 2/5 (iv) 4/9 (v) 2/15 (vi) 3/4 (vii) 3/5

(viii) 5/9 (ix) 6/10 (x) 10/11 (2) (i) 3 to 1 against (ii) 4 to 1 on
(iii) 2 to 1 on (iv) 6 to 1 against (v) 9 to 1 against (vi) 5 to 1 on
(vii) 5 to 2 against (viii) 8 to 3 against (3) 11 to 1 against
(4) 1 to 2 against (5) 1 to 3 against (6) 11 to 1 against
(7) 1/30, 29 to 1 against, 11/30 (8) 39 to 1 against (9) 2 to 1 against
(10) 5 to 1 against

Miscellaneous exercise 9
1. 1/2 2. 20 3. 7/12, 5/12 4. 1/8 5. (a) 1/5 (b) 5/14 6. 1/2, 2/9
7. A→D→C, A→B→C, A→D→E→B→C, A→B→E→D→C, 1/2
8. (a) 3/10 (b) 3/20 (c) 1/2 (d) 3/10 9. (a) 0.72 (b) 0.18
(c) 0.26 (d) 0.98 10. (a) (i) 1/2 (ii) 1/5 (b) (i) 1/20 (ii) 1/10
11. (a)

	First time test		Repeat test		
	Male	Female	Male	Female	Totals
Passed	81	62	60	45	248
Failed	59	48	30	25	162

(b) (i) 98, 66 (ii) the claim is justified (c) 0.42 12. (b) (i) 4/15 (ii) 2/5

Exercise 10a
(1) purchaser may not be the reader; not a random sample; observations of physical facts could be very subjective
(2)

	Drivers	Non-drivers
Male	12	14
Female	16	18

(3) (a) (i) in general would drink more (ii) may drink more expensive drinks
(4) (a) (i) absentees on that day will not form part of the population sampled
(ii) because of the amount of time required, it would be impractical to interview 150 students in this way

Exercise 10b
(2) (a) 60 (b) 130 (c) 160
(3) (a) 46 (b) 538 (c) 60
(d)

0– 99	4
100–199	12
200–299	14
300–399	18
400–499	9
500–599	3

(e) 12 (f) 20%

(4) (i) discrete (ii) continuous (iii) discrete (iv) continuous (v) discrete

(vi) discrete (vii) continuous (5) (b) 15 (c) 55
(6) (a) 34 000 (b) (i) 1983 (ii) 25 000 (8) (a) 110 (b) July
(c) December 82 (d) September (9) (a) 8 (b) 1/8
(10) (i) 250 (ii) 22% (iii) 10 (11) angles: 108°, 97°, 90°, 22°, 43°
(15) no vertical scale (16) (i) joining of plotted points incorrectly
(ii) no heading (iii) vertical scale does not start at zero

Miscellaneous exercise 10

2. a leading question 3. stratified sampling 5. (a) 281 (b) £1352.50
(c) £4.81 (d) closed ticket office (e) Thursday
6. (a) 635 (b) Monday, Wednesday (c) Tuesday 7. (a) continuous
(b) discrete (c) discrete (d) continuous 9. 110°, 90°, 20°, 40°, 100°
10. (a) £8.80 (b) £23.55 (c) £11.10 11. (a) 43 seconds
12. more than half of the 200 cars (152) exceeded 80 km/h therefore a majority broke the law 13. (b) (i) 32 (ii) 84 (iii) 32
14. no quantities and bars are not of uniform width; vertical scale does not start from the origin; no units on the vertical scale; no interpretation of the word 'goodness' 15. I: no vertical scale II: represents information most fairly III: vertical scale does not start from the origin
16. 1: bars not of uniform width, no vertical scale 2: acceptable
3: vertical scale does not start from the origin and is non-uniform

Exercise 11a

(1) 4 (2) −5.76 (3) 12.3125 (4) 1.13 (5) −0.64 (6) 3.25 (7) 1.39
(8) 4.32 (9) 35.95 (10) 10.6 (11) 1.39 (12) −81 (13) 0.04 (14) 5.28
(15) −2.88

Exercise 11b

(1) 6 (2) 12 (3) 9 (4) 4 (5) 30 (6) $6\frac{3}{4}$ (7) $\frac{1}{2}$ (8) 3 (9) 8 (10) 6
(11) 29 (12) $-\frac{8}{11}$

Exercise 11c

(1) $\frac{ty}{4}$ (2) $\frac{2yN}{t}$ (3) $\frac{3}{p}-1$ (4) $y\frac{(r-z)}{3}$ (5) $\frac{y-t}{a-q}$ (6) $\frac{by-a}{1-y}$
(7) $\frac{4N-y}{2}$ (8) $\frac{5z}{t^2}$ (9) $\frac{2}{T-y}$ (10) $\frac{b(q-c)}{a}$

Exercise 11d

(1) 4, 11 (2) 2, 4 (3) 4, 5 (4) 3.4, 2.2 (5) $-\frac{3}{7}$, $1\frac{1}{7}$ (6) 1, 6 (7) $\frac{2}{3}$, 0
(8) −13, −18

Exercise 11e

(1) $8x^2$ (2) $15t^2$ (3) $12q^2$ (4) $12tq$ (5) $2t^3$ (6) $4at$ (7) $15p^2$
(8) $4p^2$ (9) $32x^3$ (10) $12uv$ (11) 6 (12) $3x$ (13) $\frac{2t}{3}$ (14) $2q$
(15) $\frac{2t^2}{p}$ (16) $9x$ (17) $\frac{5}{p}$ (18) $\frac{2}{9x}$ (19) $\frac{1}{2}$ (20) 6

Exercise 11f

(1) $xp + xq + yp + yq$ (2) $ac + ad + 2bc + 2bd$
(3) $3x^2 + 11xy + 6y^2$ (4) $2x^2 + 5x + 2$ (5) $x^2 + 2x - 3$
(6) $2x^2 + 7x + 3$ (7) $x^2 + 8x + 15$ (8) $2t^2 + 7t + 6$ (9) $t^2 - 5t + 6$
(10) $x^2 + x - 30$ (11) $2x^2 - 5x + 2$ (12) $6x^2 + 13x + 5$ (13) $u^2 + 5uv + 6v^2$
(14) $6u^2 + 5uv - 4v^2$ (15) $3x^2 + 16x - 35$ (16) $t^2 - 4$ (17) $9x^2 - 25$
(18) $49x^2 - 28x + 4$ (19) $x^2 + 8x + 16$ (20) $9x^2 - 6x + 1$

Exercise 11g

(1) $4(x+4)$ (2) $3(x-3)$ (3) $t(t+1)$ (4) $a(b+2a)$ (5) $2p(3q+1)$
(6) $ac(1+4ac)$ (7) $3t(3t-1)$ (9) $b(6+5b)$ (10) $p^2(2p+1)$
(11) $(x+1)(x+2)$ (12) $(x+6)(x+4)$ (13) $(x+4)(x+1)$
(14) $(x+6)(x+8)$ (15) $(x+6)(x+3)$ (16) $x(x+3)$ (17) $(x-5)(x-1)$
(18) $(x-4)(x-2)$ (19) $(x+5)(x-3)$ (20) $(2x+1)(x+2)$
(21) $(2x-1)(x+2)$ (22) $(2x-3)(x+1)$ (23) $(x+3)^2$ (24) $(2x+1)^2$
(25) $(x-4)(x+4)$ (26) $(2x-3)(2x+3)$ (27) $2(x-5)(x+5)$
(29) $x(x-1)(x+1)$

Exercise 11h

(1) $-1, 3$ (2) $\frac{1}{2}, -1$ (3) 1 (4) $1.85, -4.85$ (5) $0.36, -8.36$
(6) $1.82, 0.18$ (8) 1 (9) 2, 8 (10) $3, \frac{1}{3}$ (11) 0, 9 (12) $0.41, -2.41$
(13) $2.36, -0.11$ (14) $-1.64, -10.36$ (15) ± 1.15

Exercise 11i

(1) $\dfrac{7x}{12}$ (2) $\dfrac{x}{6}$ (3) $\dfrac{11x}{12}$ (4) $\dfrac{9y}{10}$ (5) $\dfrac{x+4}{4}$ (6) $\dfrac{3}{2y}$ (7) $\dfrac{2n+3m}{6mn}$ (8) $\dfrac{1}{10q}$
(9) $\dfrac{11}{12t}$ (10) $\dfrac{x+2y}{y}$ (11) $\dfrac{2a}{3b}$ (12) $\dfrac{1}{2}$ (13) 1 (14) $\dfrac{a}{d}$ (15) $2x$ (16) 1
(17) $\dfrac{4x}{y}$ (18) 8 (19) $\dfrac{2a}{t}$ (20) $\dfrac{3}{t}$ (21) $1\frac{1}{2}$ (22) $\dfrac{1}{2}$ (23) $\dfrac{1}{x}$ (24) $2x$
(25) q^3

Exercise 11j

(1)
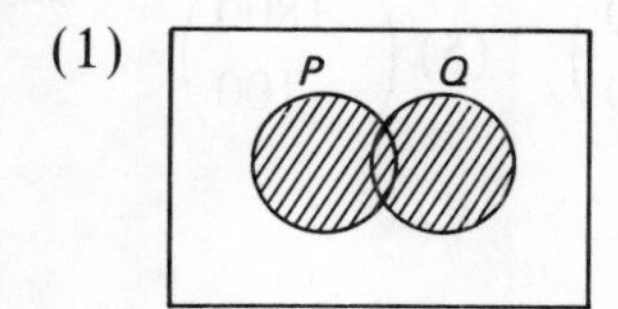

(2)
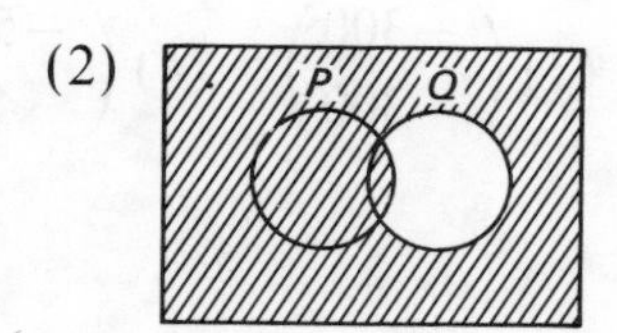

(3)
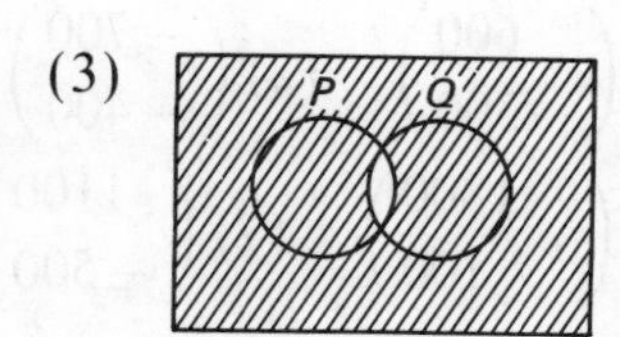

(4)
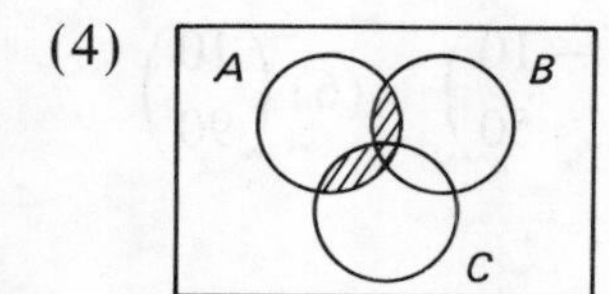

(5)
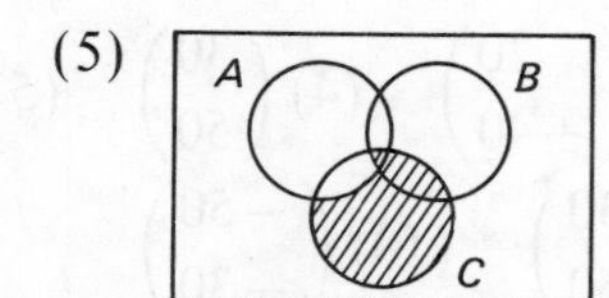

(6)
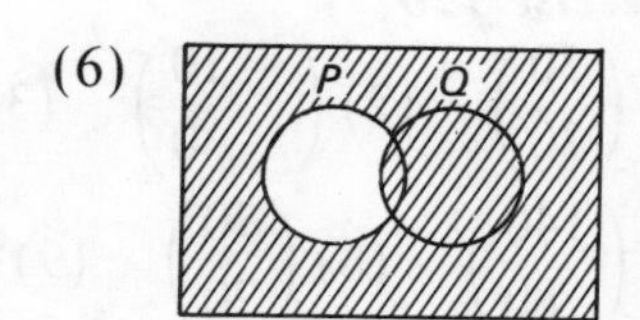

(7) X Y

(8) 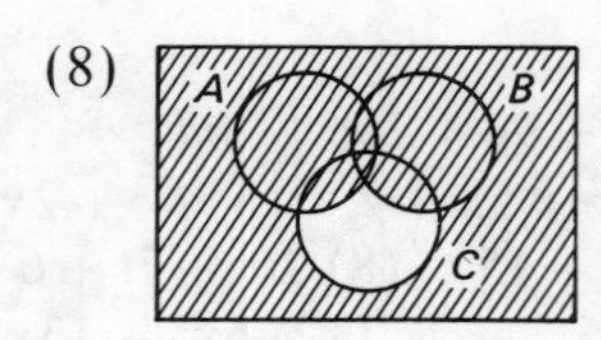

Exercise 11k

(1) (i) -0.531 (ii) 7.531 (iii) 6.000 (iv) -1.000 (v) -0.887 (vi) 7.887 (vii) 0.437 (viii) 3.873

(2) (a) 2.236 (b) 5 (c) 2.2361 (e) $x_{n+1} = \frac{1}{2}\left(x_n + \frac{11}{x_n}\right)$

(3) (a) $x - 2$ (*b*) $\frac{1}{x}, \frac{x-2}{1}$ (d) 2.3333333, 2.4285714, 2.4117647 (e) 2.4142

(4) 72 (5) (ii) 2.15 m/s (iii) 425 m

Miscellaneous exercise 11

1. (a) $9x^2 - 4$ (b) $9x^2 - 4$ 2. (a) 64 m (b) $1 - \frac{2h}{d}$ (c) $\frac{1}{3}$

3. 40, 100 (a) 44p (b) 50, $\frac{40}{(C-40)}$ 4. (a) (i) $\frac{48}{x}$ (ii) $x + 4$ (iii) $\frac{48}{x+4}$ (c) 12 (d) 3 hours

5. (i) £$x + 5$ (ii) $\frac{300}{x+5}$ (iii) $\frac{120}{x-1}$ (iv) $\frac{300}{x+5} + \frac{120}{x-1}$ (v) 25

6. (a) $\frac{12x+1}{(x+3)(2x-1)}$ (b) $\frac{3}{4}$ 7. (a) $9 - x$ (b) $9x - x^2$ (c) 4, 5 8. (c) 4

9. (i) 160 hours (ii) 1060 km 10. (b) $\frac{n(n+1)}{2}$ (c) 1.5, 4.5 11. 2.7

12. 9.28, 3.72 13. (c) 189 14. (b) 1.5, 1.66, 1.7 (c) 1.7071 15. (b) 3 (d) $x_{n+1} = \sqrt{\frac{11}{x_n} - 7x_n}$ 16. (d) 3.66667, 3.36364 (e) 3.45

17. (a) 16.3 (b) 0.308% 18. 840 m 19. (c) (i) triangular (ii) 55 (d) $\frac{n(n+1)}{2}$ 20. (b) 1.3 21. 10 km

Exercise 12a

(1) $\begin{pmatrix} 600 \\ -300 \end{pmatrix}$ (2) $\begin{pmatrix} -700 \\ -400 \end{pmatrix}$ (3) $\begin{pmatrix} -300 \\ -300 \end{pmatrix}$ (4) $\begin{pmatrix} -500 \\ 400 \end{pmatrix}$ (5) $\begin{pmatrix} 1800 \\ -100 \end{pmatrix}$ (6) $\begin{pmatrix} -1800 \\ 100 \end{pmatrix}$ (7) $\begin{pmatrix} 1100 \\ -500 \end{pmatrix}$

Exercise 12b

(1) $\begin{pmatrix} 50 \\ 30 \end{pmatrix}$ (2) $\begin{pmatrix} -10 \\ 30 \end{pmatrix}$ (3) $\begin{pmatrix} -20 \\ -40 \end{pmatrix}$ (4) $\begin{pmatrix} 30 \\ 50 \end{pmatrix}$ (5) $\begin{pmatrix} -10 \\ 50 \end{pmatrix}$ (6) $\begin{pmatrix} 10 \\ 90 \end{pmatrix}$ (7) $\begin{pmatrix} 140 \\ 80 \end{pmatrix}$ (8) $\begin{pmatrix} 70 \\ 70 \end{pmatrix}$ (9) $\begin{pmatrix} 30 \\ 50 \end{pmatrix}$ (10) $\begin{pmatrix} -50 \\ -30 \end{pmatrix}$

Exercise 12c

(1) (i) $11\boldsymbol{i}+5\boldsymbol{j}$ (ii) $-5\boldsymbol{i}-8\boldsymbol{j}$ (iii) $-2\boldsymbol{i}-6\boldsymbol{j}$ (2) (a) $\boldsymbol{b}+\boldsymbol{c}$ (b) $\boldsymbol{a}+\boldsymbol{b}+\boldsymbol{c}$ (c) $\boldsymbol{a}+\boldsymbol{b}$ (3) (a) $\boldsymbol{b}-\boldsymbol{a}$ (b) $-\boldsymbol{b}-\boldsymbol{a}$ (c) $2\boldsymbol{b}$ (4) (a) $\begin{pmatrix}-6\\-2\end{pmatrix}$ (b) $\begin{pmatrix}7\\-2\end{pmatrix}$ (c) $\begin{pmatrix}1\\-4\end{pmatrix}$ (5) $(5, -3)$, $\begin{pmatrix}2\\-4\end{pmatrix}$ (6) $(6, 4)$, $(18, 8)$

Exercise 12d

(1) $\frac{1}{2}\begin{pmatrix}3 & -2\\-5 & 4\end{pmatrix}$ (2) $\begin{pmatrix}6 & 7\\5 & 6\end{pmatrix}$ (4) $\frac{1}{36}\begin{pmatrix}9 & -3\\6 & 2\end{pmatrix}$ (5) $-1\begin{pmatrix}-1 & 0\\0 & 1\end{pmatrix}$ (6) $\frac{1}{78}\begin{pmatrix}11 & 4\\-3 & 6\end{pmatrix}$

Miscellaneous exercise 12

1. (a) $4\boldsymbol{a}$ (b) $3\boldsymbol{a}-2\boldsymbol{b}$ 2. (a) $\boldsymbol{a}+\boldsymbol{c}$, $\frac{2}{3}(\boldsymbol{a}+\boldsymbol{c})$, $\frac{1}{3}(2\boldsymbol{a}-\boldsymbol{c})$ (b) (i) $\frac{1}{2}(2\boldsymbol{a}+\boldsymbol{c})$ (ii) $\boldsymbol{a}+\frac{\boldsymbol{c}}{2}$ (iii) $\boldsymbol{a}+\frac{3}{2}\boldsymbol{c}$ (iv) 1 3. (i) $\frac{1}{2}\boldsymbol{b}$ (ii) $-\boldsymbol{a}$ (iii) $\frac{1}{2}\boldsymbol{b}-\boldsymbol{a}$ (iv) $\boldsymbol{a}+\boldsymbol{b}$ 4. (i) $\boldsymbol{b}-\boldsymbol{a}$ (ii) $\frac{1}{2}(\boldsymbol{b}-\boldsymbol{a})$ (iii) $\frac{1}{2}(\boldsymbol{b}-3\boldsymbol{a})$ 5. (i) $-\boldsymbol{y}$ (ii) $2\boldsymbol{y}$ (iii) $\boldsymbol{y}-\boldsymbol{x}$ 6. (a) (i) $\boldsymbol{x}+\boldsymbol{y}$ (ii) $2(\boldsymbol{x}+\boldsymbol{y})$ (iii) $2\boldsymbol{x}+\boldsymbol{y}$ (iv) $\boldsymbol{x}+2\boldsymbol{y}$ (b) (i) $\frac{3}{3}\boldsymbol{x}+2\boldsymbol{y}$ (ii) $\frac{3}{4}\boldsymbol{x}+\boldsymbol{y}$, N is the mid point of FM 7. (a) (i) $6\boldsymbol{i}+3\boldsymbol{j}$ (ii) $-2\boldsymbol{i}+\boldsymbol{j}$ (iii) $4\boldsymbol{i}+4\boldsymbol{j}$ (b) 5.66 km 8. (a) (i) $\frac{1}{3}\boldsymbol{a}-\boldsymbol{b}$ (ii) $\frac{1}{5}\boldsymbol{a}-\frac{3}{5}\boldsymbol{b}$ (b) $\frac{3}{5}$ (c) (i) 20 cm^2 (ii) 12 cm^2 (iii) 12 cm^2 9. (a) (i) $\frac{3}{4}\boldsymbol{a}$ (ii) $\boldsymbol{b}-\boldsymbol{a}$ (iii) $\frac{1}{4}(\boldsymbol{b}-\boldsymbol{a})$ (iv) $\frac{1}{4}(3\boldsymbol{a}+\boldsymbol{b})$ (v) $\frac{1}{4}\boldsymbol{b}$ 10. (c) 90° clockwise rotation about O (d) $\begin{pmatrix}0 & -1\\1 & 0\end{pmatrix}$ 11. (d) (i) reflection in $y=-x$ (ii) $\begin{pmatrix}0 & -1\\-1 & 0\end{pmatrix}$ 12. (a) 4:1 (b) $\frac{1}{4}\begin{pmatrix}1 & -2\\1 & 2\end{pmatrix}$, $(3, -\frac{1}{2})$ (c) $\begin{pmatrix}0 & 4\\2 & 10\end{pmatrix}$ 13. (b) (ii) enlargement $\times 2$ (d) reflection in $y=2x$ 14. (d) rotation 90° clockwise, $R=PQ$ (e) 4 15. (c) enlargement $\times 2$ (d) $\begin{pmatrix}\frac{1}{2} & 0\\0 & \frac{1}{2}\end{pmatrix}$ 16. (a) $\begin{pmatrix}-2\\2\end{pmatrix}$ (b) $(5, 2)$ 17. $(2, -9)$ 18. (a) $\begin{pmatrix}7 & 0\\0 & 7\end{pmatrix}$ (b) $\frac{1}{7}\begin{pmatrix}2 & 3\\1 & -2\end{pmatrix}$ 19. (d) (i) 4 (ii) 25 (e) (i) $2\frac{1}{2}$

Exercise 13a

(1) (a) (i) 4 (ii) 5 (iii) 6 (b) (i) 19 (ii) 16.5 (iii) 15.5 (2) 18.6, 518.6 (3) $-7/6$, $218\frac{5}{6}$ (4) 26 kg (5) (i) 41 (ii) 64 (iii) 15 (6) 6 (7) 2, 2.5 (8) mode (9) (b) 39, 40 (c) 11/52 (d) $11/52 \times 11/52 = 0.0447$ (10) (i) 0 (ii) 1 (iii) 1.8 (11) (i) 3, 2 (ii) 108 (iii) 2.16 (iv) 0.9 (12) £14.00, £16.00, £14.50, £16.00, £20.00 (a) £16.00, £15.25, £15.50 (b) mean (c) £850, median = £16.00, mean = £15.71 (13) (a) (i) 3, 2 (b) (i) £114 (ii) £3.80 (14) 101–150 miles, 150 miles (15) (b) 28.5 seconds (16) 2.525 kg (18) (i) 72 (ii) 18 (19) 16.7

Exercise 13b

(1) (a) 200 (b) 62 m.p.h. (c) 23 (2) 36.7 min, 42 days (3) 2.9 (4) £3.40 (5) 7.0, 2.3 (a) 10 (b) 3 or 6 (c) 3 (d) 6

(6) (a) 58, 91, 100 (c) 28 (7) (a) 11, 23, 45, 72, 90, 99, 100
(c) (i) 3.1 kg (ii) 0.37 (8) (a) 84 (b) 54 (c) 8 (9) (a) 51 (b) 20 (c) 32
(10) (a) 16 (b) $7\frac{2}{3}$ (c) 35.75 min (e) (i) 9 (ii) 49

Exercise 13c
(1) (a) 38.9, 38.4, 38, 37.3, 36.6, 35.9, 34.7, 33.1
(c) to smooth out daily fluctuations (2) 43.25, 43.50, 44.25
(3) (a) 673, 680, 707 (b) to smooth out termly fluctuations
(4) (i) −ve correlation (ii) +ve correlation (iii) little or no correlation
(5) (iii) 52 (6) (b) as age increases, price decreases (7) (a) (i) 43 seconds
(ii) 62 seconds (b) (ii) negative correlation – the more sunshine there is, the less energy used

Miscellaneous exercise 13
2. 27, 39 3. 5 4. 2.83 ($2\frac{5}{6}$) 5. 183, 186.5, 183.5, 180.25, 178.25, 179.75, 183.25, 187.5, 188.5 6. mode = 6, median = 5, mean = 4.82
7. (a) (i) 25 (ii) 6 (iii) 5 (iv) 5.28 (b) mode = 7, median = 5, mean = 5.36 8. (b) (i) 1 (ii) 0 (c) 0.48 (d) 0.384 9. (a) Tynemouth
(b) Keswick (c) 67 mm (d) 100 mm (e) 19/31 10. 3.2 kg
11. (a) 16.8 cm (c) 19.2 12. (b) (i) 842 (ii) 862 13. (a) (i) 6 (ii) 2.3
(b) (i) 31.6 14. (i) 4550 (ii) 54 (iii) 26 (iv) 20% (v) 41
15. (c) 4, 6, 9, 14, 16, 15 16. (a) 12 (b) 151 (c) 3/11 (e) positive correlation – as the height of a plant increases, so does the number of tomatoes
17. (b) yes (c) 74.1 (d) 76

Exercise 14a
(1) (i) 42° (ii) 138° (2) opposite angles do not add up to 180°
(3) (i) 70° (ii) 110° (4) (i) 15° (ii) ∠OAC (iii) isosceles triangle
(iv) 75° (v) 60° (5) (i) 43° (ii) 100° (6) ∠BAO, ∠BCO, 28°
(7) ∠DAB (i) 25° (ii) 67° (8) 30° (9) (i) 110° (ii) 140° (iii) no
(10) ∠EBC = 75°, ∠ECB = 70° (11) (i) 124° (ii) 98°
(12) (a) BDE or FOD (b) ODC, OBA, OAB, FAE (c) (i) 72° (ii) 18°
(iii) 54°

Exercise 14b
(1) (i) −0.866 (ii) 0.342 (iii) −0.174 (iv) 0.5 (v) 0.985 (vi) −0.940
(vii) 0.940 (viii) −0.342 (2) (i) 0 (ii) −1 (3) 48.6°, 131.4° (4) 143.1°

Exercise 14c
(1) (i) 8.49 (ii) 7.08 (iii) 18.4 (iv) 8.58 (2) (i) 34.2° (ii) 43.0°
(iii) 37.6° (3) (i) 33 (ii) 28.9 (iii) 10.4 (iv) 16.6 (4) (i) 53.1°
(ii) 101.0° (iii) 94.1° (5) (i) 19.8 cm^2 (ii) 167 cm^2 (iii) 68.3 cm^2
(6) 5/12, 24.6° (7) 440 m, 540 m, 410 m (8) (i) 155°, 11° (ii) 127 m
(iii) 53.6 m (9) 42.3 m (10) (i) 6 km, 4 km (ii) 4.6 km
(11) (i) 100° (ii) 9.0 m (iii) 8.6 m (12) 176 cm^2, 3520 cm^3
(13) (a) 4850 m (b) 1070 m (14) (a) 2310 h (b) (i) 280 km (ii) 322°
(iii) 74.2 km

Exercise 14d

(7) (b) 940 mm (8) (a) 1.37 m (b) 50°

Exercise 14e

(1) 670 m^2 (2) 648 m^2 (3) 409.5 m^2

Exercise 14f

(1) 640 m^2 (2) 634.7 m^2 (3) 407.3 m^2

Exercise 14g

(1) 7, 13.5, 10, 11 (c) 33
(2) (b) 1833 ft^2
(3) 22.8 m^2

Exercise 14h

(1) straight line (2) straight line (3) ellipse
(4) line perpendicular to AB through the mid point
(5) sphere or circle (6) arc of a circle
(7) (8) arc of a circle (9) almost a parabola

INDEX